THE ALGEBRAIC THEORY OF SEMIGROUPS

VOLUME II

MATHEMATICAL SURVEYS · *Number 7*

THE ALGEBRAIC THEORY OF SEMIGROUPS

VOLUME II

BY

A. H. CLIFFORD AND G. B. PRESTON

1967

AMERICAN MATHEMATICAL SOCIETY

Providence, Rhode Island

Reprinted with corrections 1971

Library of Congress Catalog Card Number: 61–15686
International Standard Book Number 0-8218-0272-0

Printed in the United States of America

TABLE OF CONTENTS

PREFACE TO VOLUME II

In broad outline, Volume 2 follows the plan predicted in Volume 1. Cross-references to Chapters 6, 7, and 8, from Volume 1, remain correct. The original Chapter 9 has expanded into the present Chapters 9, 10, and 11, and references from Volume 1 should be interpreted accordingly. Chapter 12 is the new name of the chapter referred to as Chapter 10 in Volume 1.

Volumes 1 and 2 should be thought of as a single work presenting a survey of the theory of semigroups. The greater part of Volume 2 deals not with the deeper development of the topics initiated in Volume 1, but with additional branches of the theory to which there was at most passing reference in Volume 1. Most of the subject matter of Volume 2 is taken from papers published prior to the drawing up of the original plan for both volumes. Nevertheless, the chance has been taken, on the topics with which Volume 2 deals, to include what we judge to be the more important developments to date. On the other hand, the theory of matrix representations of semigroups (Chapter 5) has seen important extensions since Volume 1 appeared (see, in particular, Munn [1964b] and the references therein); but we have not presented these in the present volume.

Among the more important recent developments of which we present an extended treatment are B. M. Šaĭn's theory of the representations of an arbitrary semigroup by partial one-to-one transformations of a set (§§7.2, 7.3, 11.4), L. Rédei's theory of finitely generated commutative semigroups, to which we give an introduction (§9.2) (our account is based on a lecture delivered by Professor Rédei at Oxford in 1960; unfortunately we were unable to obtain a copy of Rédei's book, *Theorie der endlich erzeugbaren kommutativen Halbgruppen* [1963], before our manuscript went to the printer), J. M. Howie's theory of amalgamated free products of semigroups (§9.4), and E. J. Tully's theory of representations of a semigroup by transformations of a set (Chapter 11).

In §10.8 we present Malcev's [1952] theory of the congruences on a full transformation semigroup and in §§12.6 and 12.8 his [1937, 1939, 1940] discussion of necessary and sufficient conditions for the embeddability of a semigroup in a group. In both cases our account follows in general plan the argument of Malcev's brilliant papers; but we feel that the considerable amplification our account contains is necessary for complete proofs (which we hope do not now require further amplification).

The material in § 7.4, Chapter 9, §§10.7 and 10.8 was presented by one of the authors in a course of lectures, under the title *Congruences on Semigroups*, to the National Science Foundation Summer Institute in Algebra, June 24–August 16, 1963, at Pennsylvania State University. An account of these lectures, written by Dr. J. M. Howie, was produced by the Department of Mathematics, Pennsylvania State University. Discussions with the students attending this course and with Dr. Howie, who assisted in its presentation, greatly benefited us.

As in Volume 1, various open questions are informally mentioned in the text. Two of those in Volume 1 have been answered: that on page 12 by T. Tamura and N. Graham [1964], and that on page 70 (in the negative) by W. E. Clark [1965].

We gratefully acknowledge the use we have made of comments upon an early draft on Chapters 7 and 8 made variously by Professor W. D. Munn, Dr. J. M. Howie, and Dr. N. R. Reilly. Into Chapter 6 we have been pleased to incorporate improvements suggested by Professor Š. Schwarz.

There have been several editions of the manuscript of Volume 2, and a large number of typists have been involved in its production. The principal among them were Mrs. N. Seddon, Miss Celia Twiddle, and Mrs. J. N. Malcolm; to these and to the others who have helped go our thanks.

For help in correcting the galley proof, the authors are grateful to Mr. R. P. Sullivan who, in addition to having a fine eye for detail, detected some errors in argument that have now been corrected. Finally, we thank the editorial staff of the American Mathematical Society for their unfailing patience and helpfulness.

A. H. C.
G. B. P.

November, 1966
THE TULANE UNIVERSITY OF LOUISIANA
MONASH UNIVERSITY

NOTATION USED IN VOLUME II

Square brackets [ ] are used for alternative readings and for reference to the bibliography.

Let A and B be sets.—

$A \subset B$ (or $B \supset A$) means A is properly contained in B;

$A \subseteq B$ (or $B \supseteq A$) means $A \subset B$ or $A = B$;

$A \backslash B$ denotes the set of elements of A which are not in B;

$A \times B$ means the set of all ordered pairs (a, b) with a in A, b in B.

The signs $\cup$ and $\cap$ are reserved for union and intersection, respectively, of sets and relations.

The signs $\vee$ and $\wedge$ are used for join and meet, respectively, in [semi]lattices.

$|A|$ denotes the cardinal number of the set A.

$\aleph_0$ denotes the smallest infinite cardinal.

The sign $\circ$ is used for composition of relations (§1.4), but is usually omitted for composition of mappings; it is also omitted for composition of relations in §§10.5 and 10.6.

$\square$ denotes the empty set, mapping, or relation.

ι [ι_A] denotes the identity mapping or relation [on the set A]; see below for convention used in §10.8.

$\phi \colon A \rightarrow B$ means that ϕ is a mapping of A into B;

$\phi | C$ means ϕ restricted to C ($C \subseteq A$).

$\langle A \rangle$ denotes the subsemigroup of a semigroup S generated by a subset A of S.

$[A]$ denotes the subgroup of a group G generated by a subset A of G; clearly $[A] = \langle A \cup A^{-1} \rangle$, where $A^{-1} = \{a^{-1} \colon a \in A\}$.

AB means $\{ab \colon a \in A,\ b \in B\}$, when A and B are subsets of a semigroup S.

S^1 [S^0] means the semigroup $S \cup 1$ [$S \cup 0$] arising from a semigroup S by the adjunction of an identity element 1 [a zero element 0], unless S already has an identity [has a zero, and $|S| > 1$], in which case $S^1 = S$ [$S^0 = S$]. (§1.1; §6.1, p. 1.)

$a \rho b$ means $(a, b) \in \rho$, where ρ is a relation on a set X, and a and b are elements of X;

$a\rho$ denotes the set $\{x \in X \colon a \rho x\}$.

S/ρ denotes the factor semigroup of a semigroup S modulo the congruence ρ on S;

$\rho^\natural$ denotes the natural mapping $a \rightarrow a\rho$ of S onto S/ρ. (§1.5.)

ρ^* denotes the congruence on S generated by a relation ρ on S. (§9.2, p. 122.)

Let I be an ideal of a semigroup S.—

I^* denotes the Rees congruence $\iota_S \cup (I \times I)$;

S/I denotes the Rees factor semigroup S/I^*. (§1.5; §10.8, p. 227.)

Let S be a semigroup, and let $a, b \in S$.—

$L(a)$ denotes the principal left ideal S^1a;
$R(a)$ denotes the principal right ideal aS^1;
$J(a)$ denotes the principal two-sided ideal S^1aS^1;
$\mathscr{L}$ means $\{(a, b) \in S \times S\colon L(a) = L(b)\}$;
$\mathscr{R}$ means $\{(a, b) \in S \times S\colon R(a) = R(b)\}$;
$\mathscr{J}$ means $\{(a, b) \in S \times S\colon J(a) = J(b)\}$;
$\mathscr{H}$ means $\mathscr{L} \cap \mathscr{R}$;
$\mathscr{D}$ means $\mathscr{L} \circ \mathscr{R} = \mathscr{R} \circ \mathscr{L}$;
L_a, R_a, J_a, H_a, D_a mean respectively the $\mathscr{L}$, $\mathscr{R}$, $\mathscr{J}$, $\mathscr{H}$, $\mathscr{D}$-class containing a. (§2.1.)
$I(a)$ means $J(a) \backslash J_a$ (which is empty or an ideal of S);
$J(a)/I(a)$ is the principal factor of S corresponding to a. (§2.6.)
$\mathscr{T}_X$ means the semigroup of all transformations of the set X. (§1.1.)
$\mathscr{P}\mathscr{T}_X$ means the semigroup of all partial transformations of a set X. (§11.1, p. 254.)
${}^0\mathscr{T}_X$ means the semigroup of all transformations of the set $X^0 = X \cup 0_X$ leaving 0_X fixed. (§11.1, p. 254.)
$\mathscr{G}_X$ means the group of all permutations of the set X. (§1.1.)
$\mathscr{I}_X$ means the symmetric inverse semigroup on the set X, i.e. the semigroup of all one-to-one partial transformations of X. (§1.9.)
${}^0\mathscr{I}_X$ means the semigroup of all partially one-to-one transformations of $X^0 = X \cup 0_X$ leaving 0_X fixed. (§11.4, p. 263.)
$\mathscr{B}_X$ means the semigroup of all binary relations on X. (§1.12.)
$\mathscr{F}_X$ means the free semigroup on X. (§1.12.)
$\mathscr{F}\mathscr{G}_X$ means the free group on X. (§1.12.)
$\mathscr{C}$ means the bicyclic semigroup. (§1.12.)
$\mathscr{M}^0(G; I, \Lambda; P)$ means the Rees $I \times \Lambda$ matrix semigroup over the group with zero G^0, with $\Lambda \times I$ sandwich matrix P;
$\mathscr{M}(G; I, \Lambda; P)$ means the Rees $I \times \Lambda$ matrix semigroup without zero over the group G, with $\Lambda \times I$ sandwich matrix P. (§3.1.)
M_L, M_R, M_J denote the minimal conditions on the set of principal left, right, two-sided ideals, respectively, of a semigroup. (§§5.3, 5.4.)
M_L^* [M_R^*] denotes the condition on a semigroup S that, for every $\mathscr{J}$-class J of S, the minimal condition holds for the set of principal right [left] ideals generated by the elements of J. (§6.6, p. 30.)
$A_C = A_C(S)$ [${}_CA = {}_CA(S)$] means the right [left] annihilator in a semigroup $S = S^0$ of a subset C of S. (§6.1, p. 5.)
It is also denoted by AC [CA] in Exercises 7–12 for §6.5 (there only).
${}_CA_C = {}_CA_C(S)$ means $A_C \cap {}_CA$. (§6.1, p. 5.)
$\Sigma_r = \Sigma_r(S)$ denotes the right socle of $S = S^0$, the union of 0 and all the 0-minimal right ideals of S;
$\Sigma_l = \Sigma_l(S)$ denotes the dually defined left socle of S. (§6.3, p. 12.)
$\leqq$ and ω both are used to denote the natural partial order on an inverse semigroup. (§7.1, p. 40; §7.2, p. 43.)

$K\omega$ means the ω-closure $\{k\omega: k \in K\}$ of the inverse subsemigroup K of an inverse semigroup S;
π_K denotes the partial right congruence $\{(s, t) \in S \times S: st^{-1} \in K\omega\}$ on S determined by K. (§7.2, p. 44.)
$a^{[-1]}H$ denotes $\{x \in S: ax \in H\}$;
$Ha^{[-1]}$ denotes $\{x \in S: xa \in H\}$. (§7.2, p. 46.)
$\mathscr{R}_H$ denotes the principal right congruence
$\{(a, b) \in S \times S: a^{[-1]}H = b^{[-1]}H\}$ determined by the subset H of S;
${}_H\mathscr{R}$ is its left-right dual. (§7.2, p. 46; §10.2, p. 183.)
$\mathscr{R}_H^*$ denotes the principal partial right congruence
$\{(a, b) \in S \times S: a^{[-1]}H = b^{[-1]}H \neq \square\}$ determined by H;
${}_H\mathscr{R}^*$ is its left-right dual. (§7.2, p. 46; §10.2, p. 183.)
W_H denotes $\{x \in S: x^{[-1]}H = \square\}$;
${}_HW$ denotes $\{x \in S: Hx^{[-1]} = \square\}$. (§7.2, p. 46; §10.2, p. 183.)
τ_H denotes the transitivity relation on the set X determined by an inverse subsemigroup H of $\mathscr{I}_X$. (§7.3, p. 48.)
τ denotes the transitivity relation on $M^- = M \backslash 0_M$, where M is a centered operand over a semigroup. (§11.2, p. 256.)
$\rho_{\mathscr{A}}$ denotes the congruence on an inverse semigroup determined by a kernel normal system $\mathscr{A}$. (§7.4, p. 60.)
$C(X)$ denotes the centralizer of a subset X of S. (§7.6, p. 65.)
$CT\,(A, \mathscr{E}, p, q)$ means the Croisot-Teissier semigroup of transformations of the set A determined by the set $\mathscr{E}$ of equivalences on A and the cardinals p and q satisfying $|A| \geqq p \geqq q$. (§8.2, p. 86.)
$\mathscr{C}\,(S)$ denotes Bruck's simple semigroup containing S. (§8.5, p. 109.)
Let F be the free commutative semigroup on a finite number n of generators, and let G be the free abelian group, containing F, on these n generators.—
$\vee$ and $\wedge$ denote join and meet in G, regarded as a lattice-ordered group;
$\mu^+\,[\mu^-]$ denotes $\mu \vee 0\,[(-\mu) \vee 0]$, $(\mu \in G)$;
M_ρ means the subgroup $\{\alpha - \beta: (\alpha, \beta) \in \rho\}$ of G determined by the congruence ρ on F;
f_ρ means the mapping of M_ρ into the set of ideals of F defined for μ in M_ρ by $\mu f_\rho = \{\xi \in F: (\xi + \mu^+, \xi + \mu^-) \in \rho\}$;
$k(\rho)$ denotes the core $\cap\{\mu f_\rho: \mu \in M_\rho\}$ of ρ;
$\|A\|$ means the norm (=height) of an ideal A of F;
$\|\rho\|$ means $\|k(\rho)\|$. (§9.3, pp. 126, 127, 132, 133.)
$[S_i;\, U]$ and $[S_i;\, U;\, \phi_i]$ are shorthand notations for the semigroup amalgam $[\{S_i;\, i \in I\};\, U;\, \{\phi_i;\, i \in I\}]$. (§9.4, p. 138.)
$\mathscr{G} = \mathscr{G}\,[S_i;\, U;\, \phi_i]$ means the partial groupoid of the semigroup amalgam $[S_i;\, U;\, \phi_i]$. (§9.4, p. 138.)
$\prod^* S_i$ and $\prod^*\{S_i\}$ are shorthand notations for the free product $\prod^*\{S_i: i \in I\}$. (§9.4, p. 141.)
κ_i denotes the canonical embedding of S_i into $\prod^* S_i$. (§9.4, p. 141.)

υ denotes the relation $\{(u\phi_i, u\phi_j): u \in U; i, j \in I\}$ on $\prod^* S_i$, and is defined when $[S_i; U; \phi_i]$ is a semigroup amalgam. (§9.4, p. 141.)

$\prod^*{}_U S_i = \prod^*{}_U\{S_i: i \in I\}$, the free product of the amalgam $[S_i; U; \phi_i]$, is defined to be $\prod^* S_i/\upsilon^*$. (§9.4, p. 142.)

μ_i denotes the canonical homomorphism of S_i into $\prod^*{}_U S_i$. (§9.4, p. 142.)

$\sum S_i$ denotes the direct sum of the semigroups S_i $(i \in I)$. (§9.4, p. 156.)

$\sum_U S_i$ denotes the direct sum of the amalgam $[S_i; U; \phi_i]$. (§9.4, p. 157.)

Let ρ be a relation on a semigroup S.—

ρD means $\{(x, y) \in S \times S: (xs, ys) \in \rho$ for some $s \in S^1\}$;

ρG is the left-right dual of ρD. (§9.5, p. 164.)

ρR means $\{(x, y) \in S \times S: (xs, ys) \in \rho$ for all $s \in S^1\}$;

ρL is the left-right dual of ρR. (§10.1, p. 176.)

ρR^* means $\{(us, vs): (u, v) \in \rho, s \in S^1\}$;

ρL^* is the left-right dual of ρR^*. (§10.1, p. 176.)

ρC means $\{(x, y) \in S \times S: (sxt, syt) \in \rho$ for all $s, t \in S^1\}$. (§9.5, p. 164.)

ρC^* means $\{(sut, svt): (u, v) \in \rho; s, t \in S^1\}$. (§9.5, p. 164.)

ρT means the transitive closure of ρ. (§10.1, p. 176.)

ρ^c means the cancellative congruence generated by ρ. (§9.5, p. 164.)

ρ^c also means the translate of a right congruence ρ on S by an element c of S. (§11.3, p. 261.)

$\alpha(\mathscr{A})$ $[\beta(\mathscr{A})]$ is the minimal [maximal] equivalence on S which admits the family $\mathscr{A}$ of disjoint subsets of S. (§10.1, p. 178.)

P_H denotes the right congruence $\{(a, b) \in S \times S: ua = vb$ for some u, v in $H\}$ determined by the right reversible subsemigroup H of S;

${}_H\mathrm{P}$ is the left-right dual of P_H. (§10.3, pp. 194, 195.)

$H..a$ means the relation $\{(x, y) \in S \times S: xay \in H\}$, with $H \subseteq S$ and $a \in S$;

$\mathscr{P}_H$ means the principal congruence $\{(a, b) \in S \times S: H..a = H..b\}$;

W denotes $\{a \in S: H..a = \square\}$. (§10.4, p. 198.)

$\tau_1 * \tau_2$ denotes the congruence on $P\tau_2$ (where $P = T_1 \cap T_2$) generated by the restrictions to $P\tau_2$ of the congruences τ_1 and τ_2 on the respective subsemigroups T_1 and T_2 of S. (§10.5, p. 204.)

$[\mathscr{P}, N, \{e_i\}, \{f_\lambda\}]$ denotes the congruence on a completely 0-simple semigroup S determined by (i) a permissible partition $\mathscr{P}$ of the rectangle of nonzero $\mathscr{H}$-classes $H_{i\lambda}$ of S, (ii) a normal subgroup N of H_{11}, and (iii) elements $e_i \in H_{i1}, f_\lambda \in H_{1\lambda}$ such that $Nf_\lambda e_i = Nf_\mu e_j$ whenever $H_{i\lambda}$ and $H_{j\mu}$ belong to the same $\mathscr{P}$-class. (§10.7, p. 218.)

Let ξ be a cardinal number.—

ξ' denotes the successor of ξ, i.e., the first cardinal greater than ξ;

I_ξ denotes the ideal of $\mathscr{T}_X$ consisting of all transformations of X of rank less than ξ;

D_ξ denotes the $\mathscr{D}$-class of $\mathscr{T}_X$ consisting of all transformations of rank equal to ξ. (§10.8, p. 227.)

ι denotes the identity congruence on $\mathscr{T}_X$;

ι_X denotes the identity element of $\mathscr{T}_X$. (§10.8, p. 228.)

$\sigma^{\dagger}$ denotes the congruence $\iota \cup [\sigma \cap (D_n \times D_n)] \cup (I_n \times I_n)$ on $\mathscr{T}_X$ determined by the congruence σ on I_{n+1}/I_n, with n a positive integer. (§10.8, p. 228.)

Let α, β be elements of $\mathscr{T}_X$, and ξ an infinite cardinal.—

$X_0 = X_0(\alpha, \beta)$ denotes $\{x \in X : x\alpha \neq x\beta\}$;

dr(α, β), the difference rank of the pair (α, β), is zero if $\alpha = \beta$ and $\max\{|X_0\alpha|, |X_0\beta|\}$ otherwise;

Δ_ξ denotes the congruence $\{(\alpha, \beta) \in \mathscr{T}_X \times \mathscr{T}_X : \mathrm{dr}(\alpha, \beta) < \xi\}$ on $\mathscr{T}_X$ (ξ infinite or $\xi = 1$). (§10.8, p. 228.)

$\eta(\rho)$ means the primary cardinal of the congruence ρ on $\mathscr{T}_X$; $\eta(\iota) = 1$, and, if $\rho \neq \iota$, $I_\eta(\rho)$ is the ρ-class containing transformations of rank 1. (§10.8, p. 231.)

Let λ be a cardinal satisfying $\eta(\rho) \leqslant |X|$.—

λ^* means the smallest cardinal exceeding every cardinal in the set $\{\mathrm{dr}(\alpha, \beta) : (\alpha, \beta) \in \rho, \text{rank } \alpha = \text{rank } \beta = \lambda\}$. (§10.8, p. 234.)

M_S [$_SM$] denotes a right [left] operand over a semigroup S. (§11.1, p. 250.)

$_SM_T$ denotes a bioperand over the semigroups S and T. (§11.1, p. 252.)

$\mathscr{E}(M_S)$ denotes the semigroup of operator endomorphisms of M_S. (§11.1, p. 251; §11.7, p. 279.)

$\mathscr{A}(M_S)$ denotes the group of operator automorphisms of M_S. (§11.1, p. 251; §11.8, p. 281.)

FM denotes the set of fixed elements of M_S. (§11.5, p. 269.)

$\mathscr{C}$-rad S denotes the $\mathscr{C}$-radical of S, which is the intersection of all congruences σ on S such that S/σ is a semigroup of type $\mathscr{C}$;

$\mathscr{C}'$ means the derived type of $\mathscr{C}$; S has type $\mathscr{C}'$ if and only if $\mathscr{C}$-rad $S = \iota_S$. (§11.6, p. 275.)

rad S means $\mathscr{I}$-rad S, where $\mathscr{I}$ is the type of right irreducible semigroups;

rad$^0 S$ is the (rad S)-class of S containing 0, and coincides with—

$N(S)$, the nilradical of S. (§6.6, p. 38; §11.6, pp. 277, 278.)

$\mathscr{N}(\rho)$ means the normalizer in S of the right congruence ρ on S; it is the subsemigroup $\{a \in S : (s, t) \in \rho \text{ implies } (as, at) \in \rho\}$ of S. (§11.6, p. 279.)

Let S be a cancellative semigroup satisfying the quotient condition Z (§12.4, p. 297), and let $a, b \in S$.—

a/b denotes the right quotient $\{(x, y) \in S \times S : xa = yb\}$ of a by b;

$a \backslash b$ denotes the dually defined left quotient of a by b. (§12.4, p. 298.)

$\sigma(I)$ denotes the Malcev system of equations corresponding to the Malcev sequence I. (§12.6, p. 310.)

CHAPTER 6

MINIMAL IDEALS AND MINIMAL CONDITIONS

The first five sections of this chapter deal with semigroups containing 0-minimal ideals. The semigroups we treat will have a (two-sided) zero element, usually denoted by 0, and at least one further element. Recalling our conventions of Volume 1, a semigroup S has this property if $S = S^0$. We will use this as a convenient shorthand and use phrases such as: "let $S = S^0$ be a semigroup" to convey that S is a semigroup, that $|S| > 1$, and that S has a zero element. Again, as in Volume 1, the corresponding results for semigroups with minimal ideals that follow from the results for semigroups with 0-minimal ideals (cf. §2.5) will only rarely be stated.

§6.1 looks at the various ways in which a zero semigroup can be embedded as a 0-minimal ideal in a semigroup. §§6.2–6.4 culminate in structure theorems for the left and right socles of a semigroup. The original ideas and many of the results here are due to Š. Schwarz [1951], but we believe that our main structure theorem (Theorem 6.29) for the union of the left and right socles of a semigroup is essentially new. There is a strong analogy with Dieudonné's theory [1942] of the socle of a ring. In §6.5 are obtained characterizations of the various 0-direct unions of 0-simple semigroups which arise in the treatment of the socles. Here again the first discussion of such decompositions is due to Schwarz [1951]. These characterizations are analogous to those of Croisot [1953], for semigroups without a zero element, that we presented in §4.1.

The final section, §6.6, discusses various minimal conditions, such as M_L, M_R, and M_J, which we met in Chapter 5. The exercises for §6.6 also contain a discussion of some of the radicals that have been defined for semigroups. The topic of radicals is taken up again in §11.6.

6.1 0-MINIMAL ZERO IDEALS

A preliminary objective of this section, Theorem 6.4, gives necessary and sufficient conditions on a semigroup R with zero that R can be embedded as a non-degenerate 0-minimal right ideal of some semigroup S. This is then applied (Theorem 6.7) to the case R a zero semigroup, and an analogous two-sided result (Theorem 6.9) is given. These results are new, and the theory is far from being worked out; note the discussion at the end of the exercises.

Let A be a 0-minimal [left, right] ideal of the semigroup $S = S^0$. Then either $A^2 = A$ or $A^2 = 0$ (§2.5). Clearly, if A is nilpotent, then we cannot have $A^2 = A$ and consequently, a nilpotent 0-minimal [left, right] ideal is a zero (null) semigroup (§1.1). A [left, right] ideal will be called a *zero [left, right] ideal* if it is a zero semigroup.

This section will be concerned primarily with 0-minimal zero ideals. In an obvious sense zero semigroups are trivial semigroups. However semigroups which contain 0-minimal non-nilpotent one-sided ideals in general contain also zero ideals, as we shall see, for example, in §§6.2–6.4. We look here at some of the ways in which a 0-minimal zero ideal of a semigroup S can be embedded in S.

We consider first one-sided ideals and our results here apply also to non-nilpotent ideals. The following lemma extends Lemma 2.31 to include zero ideals.

LEMMA 6.1. *Let $S = S^0$ and let R be a 0-minimal right ideal of S. Then either $rS = R$ for every r in $R\backslash 0$ or else $R = \{0, r\}$ with $rS = 0$.*

PROOF. Since rS is, for r in R, a right ideal of S contained in R, therefore the 0-minimality of R implies that either $rS = R$ or $rS = 0$. Were $r \neq 0$ and $rS = 0$, then $\{0, r\}$ would be a right ideal of S contained in R, so that, again since R is 0-minimal, $R = \{0, r\}$.

A right ideal $R = \{0, r\}$, of S, with $rS = 0S = 0 = S0$, will be said to be *degenerate*.

We need a generalization of the right regular representation (§1.3) of a semigroup. Let R be any right ideal of a semigroup S. For each x in S define the mapping ρ_x of R into R thus:

$$\rho_x\colon r \rightarrow rx \quad (r \in R).$$

We easily verify that $\rho_x \rho_y = \rho_{xy}$ and so

$$\rho_R\colon x \rightarrow \rho_x \quad (x \in S)$$

is a representation of S in $\mathscr{T}_R$. $S\rho_R$ is a semigroup of right translations of R containing the semigroup of inner right translations of R (§1.3).

Let ϕ be any representation of the semigroup S in $\mathscr{T}_X$. Let Y be a nonempty subset of X. Then ϕ (or S, if ϕ is the identity mapping) is said to be *transitive on* Y if, for any y_1, y_2 in Y, there exists s in S such that $y_1(s\phi) = y_2$.

For a 0-minimal right ideal R, ρ_R has the properties in the following lemma.

LEMMA 6.2. *Let R be a non-degenerate 0-minimal right ideal of the semigroup $S = S^0$. Then ρ_R is transitive on $R\backslash 0$ and $S\rho_R$ is a semigroup of right translations of R containing the semigroup of inner right translations of R.*

PROOF. By Lemma 6.1, for r_1 in $R\backslash 0$, $r_1 S = R$. Hence, for r_2 in $R\backslash 0$, there exists s in S such that $r_1 s = r_2$, i.e., $r_1(s\rho_R) = r_2$. In other words, ρ_R is transitive on $R\backslash 0$.

LEMMA 6.3. *Let $R = R^0$ be a semigroup. Let T be a subsemigroup of the semigroup of right translations of R which contains the semigroup of inner right translations of R and which is transitive on $R\backslash 0$. Then R can be embedded as a non-degenerate 0-minimal right ideal in a semigroup S such that $S\rho_R = T$.*

PROOF. Let $\Sigma = T \cup R$ and define an operation $(\circ)$ on Σ thus:

$$\rho \circ \sigma = \rho\sigma,$$

$$\rho \circ r = \rho\rho_r,$$

$$r \circ \rho = r\rho,$$

$$r \circ s = rs,$$

where ρ and σ denote elements of T, and r and s elements of R. By assumption ρ_r, the inner right translation of R by r, belongs to T and so $\rho\sigma$ and $\rho\rho_r$ both make sense as products evaluated in T. The product rs is to be evaluated in R and $r\rho$ denotes the element of R which is the image of r under the right translation ρ of R.

There are eight cases to verify to check that $(\circ)$ is associative on Σ. Consider the case TRT. Here if ρ, $\sigma \in T$ and $r \in R$ then

$$(\rho \circ r) \circ \sigma = (\rho\rho_r) \circ \sigma = (\rho\rho_r)\sigma = \rho(\rho_r\sigma),$$

and

$$\rho \circ (r \circ \sigma) = \rho \circ (r\sigma) = \rho\rho_{r\sigma},$$

by the above definitions, and these are equal by Lemma 1.1.

The other cases may be verified similarly to show that $\Sigma(\circ)$ is a semigroup.

Now any right translation of a semigroup with zero leaves the zero invariant: $0\rho = 0^2\rho = 0(0\rho) = 0$. Further here we have $\rho \circ 0 = \rho\rho_0 = \rho_0$, $\rho \circ \rho_0 = \rho\rho_0 = \rho_0$, $r \circ \rho_0 = r\rho_0 = r \cdot 0 = 0$, $\rho_0 \circ r = \rho_0\rho_r = \rho_{0r} = \rho_0$, $\rho_0 \circ \rho = \rho_0\rho = \rho_0$. Thus $\{0, \rho_0\} = Z$, say, is a two-sided ideal of Σ. Define S to be the quotient semigroup Σ/Z. S is effectively obtained from Σ merely by identifying ρ_0 with 0, and in this sense S is a semigroup containing R and with the zero of R as its zero. Since R was clearly a right ideal of Σ, it remains a right ideal of S. Further the assumed transitivity of T on $R\backslash 0$ clearly implies that if $r \in R\backslash 0$ then $rS = R$. Consequently R is not degenerate and R is a 0-minimal right ideal of S.

Finally, consider $S\rho_R$, i.e. the set of right translations of R determined by multiplying on the right by elements of S. We have to show that $S\rho_R = T$. For $r \in R$, $r\rho_R = \rho_r$, which by assumption belongs to T. For $s \in S\backslash R$, s is effectively an element σ, say, of $\Sigma\backslash R$, and for x in R,

$$x(s\rho_R) = xs = x \circ s = x\sigma;$$

so that $s\rho_R = \sigma \in T$. Similarly, we easily have that $T \subseteq S\rho_R$. Thus $S\rho_R = T$; and this completes the proof of the lemma.

From the previous two lemmas we have immediately the following theorem.

Theorem 6.4. *Let $R = R^0$ be a semigroup. Let T be a subsemigroup of the semigroup of right translations of R. Then R can be embedded as a non-degenerate 0-minimal right ideal of some semigroup S with $S\rho_R = T$ if and only if* (i) *T contains the semigroup of inner right translations of R and* (ii) *T is transitive on $R \backslash 0$.*

Corollary 6.5. *The semigroup $R = R^0$ is a non-degenerate 0-minimal right ideal of some semigroup if and only if its semigroup of right translations is transitive on $R \backslash 0$.*

It will be shown in the next section (Theorem 6.19(5)) that a 0-minimal non-nilpotent right ideal of a semigroup S is contained as a 0-minimal right ideal in a 0-simple subsemigroup of S. Semigroups which are 0-simple and contain both 0-minimal left and 0-minimal right ideals are completely 0-simple (Theorem 2.48). Semigroups which are 0-simple and contain 0-minimal right ideals but which contain no 0-minimal left ideals will be considered in §8.3.

Consider now the case of a zero semigroup $R = R^0$. Here we can improve the above results for we know what the semigroup of right translations of R is: an element of $\mathscr{T}_R$ is a right translation of the zero semigroup R if and only if it maps the zero of R onto itself. (By a remark due to V. V. Vagner [1956], it follows that the semigroup P of right translations of R is isomorphic to the semigroup formed by the single-valued mappings, combined under composition, of subsets of $R \backslash 0$ into $R \backslash 0$, including the "empty mapping". See §11.1.) In particular the semigroup of right translations of R is certainly transitive on $R \backslash 0$ and so we have, as a corollary to Corollary 6.5,

Corollary 6.6. *If $R = R^0$ is a zero semigroup then there exist semigroups which contain R as a 0-minimal right ideal.*

We can also improve the statement of Theorem 6.4 for the case of zero semigroups. Let ϕ be any representation of the semigroup S in $\mathscr{T}_X$. Let Y be a non-empty subset of X. Then Y is said to be *invariant under* ϕ (or, if ϕ is the identity mapping, *invariant under* S) if $Y(s\phi) \subseteq Y$ for all s in S. An element z is said to be *fixed* if $\{z\}$ is invariant. An invariant set Y containing a fixed element z is called *z-transitive* if z is the only fixed element of Y, and if, for every y and y' in Y with $y \neq z$, there exists s in S such that $y(s\phi) = y'$. In this case if S contains a zero element 0, then $y(0\phi)$ is a fixed element of Y, and hence equal to z, for any y in Y.

Consider now the special case in which we take $Y = X = R$ where $R = R^0$ is a zero semigroup. It follows that if R is 0-transitive under a subsemigroup $T = T^0$ of $\mathscr{T}_R$, then the zero of T is the mapping of R onto its zero, i.e. T contains the semigroup of inner right translations of R. We therefore have the following special case of Theorem 6.4.

Theorem 6.7. *Let $R = R^0$ be a zero semigroup. Let $T = T^0$ be a subsemigroup of $\mathscr{T}_R$. Then R can be embedded as a non-denegerate 0-minimal* *right ideal of a semigroup S such that $S\rho_R = T$ if and only if R is* 0*-transitive under T.*

A similar treatment can be given for 0-minimal zero two-sided ideals. The situation is slightly more complicated than in the one-sided case. The next lemma is the analogue, for two-sided ideals, of Lemma 6.1 and is fundamental.

We will use here (and later) the concept of an annihilator of a set. Let C be any non-empty subset of the semigroup $S = S^0$. Define

$$A_C = \{x \in S : Cx = 0\};$$

$${}_CA = \{x \in S : xC = 0\};$$

$${}_CA_C = \{x \in S : xC = 0 \quad \text{and} \quad Cx = 0\},$$

so that ${}_CA_C = {}_CA \cap A_C$. The set $A_C\,[{}_CA, {}_CA_C]$ is called the *right* [*left, two-sided*] *annihilator of C in S*. If we wish to make explicit the dependence on S we will write, for example, $A_C(S)$.

As for one-sided ideals, we will say that the two-sided ideal $M = M^0$ is a *degenerate* ideal of $S = S^0$, if $M = \{0, m\}$ with $0 = 0S = mS = S0 = Sm$.

Lemma 6.8. *Let $S = S^0$ and let M be a non-degenerate* 0*-minimal two-sided ideal of S. Then* (a) *$SmS = M$ and hence $Sm \neq 0$ and $mS \neq 0$ for all m in $M \backslash 0$ or* (b) *$Sm = 0$ and $mS = M$ for all m in $M \backslash 0$ so that M is a* 0*-minimal zero right ideal of S and ${}_MA = S$ or* (c) *$mS = 0$ and $Sm = M$ for all m in $M \backslash 0$ so that M is a* 0*-minimal zero left ideal of S and $A_M = S$.*

Proof. Let $C = \{m \in M : SmS = 0\}$. Clearly C is an ideal of S. Hence, since M is 0-minimal, either (i) $C = 0$ or (ii) $C = M$. In case (i), for m in $M \backslash 0$, SmS is a non-zero ideal of S, and so $SmS = M$; and this is case (a) of the lemma.

In case (ii) $SMS = 0$. We cannot have both $SM = 0$ and $MS = 0$ for in this case M would be degenerate. Suppose that $MS \neq 0$. Then, as a two-sided ideal contained in the 0-minimal ideal M, we have $MS = M$. Consequently, $SM = S(MS) = 0$. Thus either (iib) $MS \neq 0$ and $SM = 0$ or, similarly, (iic) $SM \neq 0$ and $MS = 0$. Suppose that (iib) holds. Then, for m in $M \backslash 0$, $Sm = 0$ and hence, since M is non-degenerate, $mS \neq 0$. mS is thus a non-zero two-sided ideal of S contained in M and hence $mS = M$. Thus case (iib) coincides with case (b) of the lemma. That (iic) coincides with (c) follows dually.

It follows from the previous lemma, that, having already dealt with the immersion of one-sided 0-minimal zero ideals, we can, for two-sided ideals, restrict the discussion to case (a) of this lemma.

If L is a left ideal of the semigroup S define λ_L by the left-right dual of the definition of ρ_R for a right ideal R. Thus $S\lambda_L$ consists of all the left translations of L induced by multiplying L on the left by elements of S.

Theorem 6.9. *Let $M = M^0$ be a zero semigroup. Let $T = T^0$ and $U = U^0$ be subsemigroups of $\mathcal{T}_M$ which have as common zero element the constant mapping of M onto 0. Then M can be embedded as a non-degenerate 0-minimal two-sided ideal of a semigroup S, satisfying $SmS = M$ for m in $M\backslash 0$ and with $S\rho_M = T$ and $S\lambda_M = U$ if and only if* (i) *each element of T commutes with each element of U and* (ii) *UT is transitive on $M\backslash 0$.*

Proof. Assume that M is a 0-minimal zero two-sided ideal of $S = S^0$. Then clearly $S\rho_M = T$ and $S\lambda_M = U$ both contain the constant mapping of M onto 0. Further if $u \in U$, $u = a\lambda_M$, $a \in S$ and if $t \in T$, $t = b\rho_M$, $b \in S$, then $m(ut) = (mu)t = (am)t = (am)b = a(mb) = a(mt) = (mt)u = m(tu)$ for m in M. Thus $ut = tu$, i.e. each element of T commutes with each element of U. Further, if $m \in M\backslash 0$, then $M = SmS = (Sm)S = (mU)S = (mU)T = m(UT)$. Hence, in particular, UT is transitive on $M\backslash 0$.

Conversely, assume that T and U are subsemigroups of $\mathcal{T}_M$ with the constant mapping of M onto its zero as common zero element. Assume further that conditions (i) and (ii) are satisfied by U and T. Let $\overline{T}$ and $\overline{U}$ be two semigroups which are isomorphic to T and U, respectively, and which have the property that they intersect solely in their zero element: $\overline{T} \cap \overline{U} = \{z\}$, say. Let $\overline{T}^* = \overline{T}\backslash z$, $\overline{U}^* = \overline{U}\backslash z$ and put $S = M \cup \overline{T}^* \cup \overline{U}^*$. Let α and $\bar{\alpha}$ denote corresponding elements under the isomorphism between T and $\overline{T}$, respectively, or between U and $\overline{U}$, respectively, as the context demands. Define an operation $(\circ)$ in S thus:

$$\bar{\alpha} \circ \bar{\beta} = \begin{Bmatrix} \bar{\alpha}\bar{\beta}, & \text{if } \bar{\alpha}\bar{\beta} \neq z \\ 0, & \text{if } \bar{\alpha}\bar{\beta} = z \end{Bmatrix}, \quad \text{for } \alpha, \beta \text{ in } T;$$

$$\bar{\mu} \circ \bar{\nu} = \begin{Bmatrix} \bar{\nu}\bar{\mu}, & \text{if } \bar{\nu}\bar{\mu} \neq z \\ 0, & \text{if } \bar{\nu}\bar{\mu} = z \end{Bmatrix}, \quad \text{for } \mu, \nu \text{ in } U;$$

$$\bar{\alpha} \circ \bar{\mu} = 0 = \bar{\mu} \circ \bar{\alpha}, \quad \text{for } \alpha \text{ in } T,\ \mu \text{ in } U;$$

$$\bar{\alpha} \circ m = 0; \qquad m \circ \bar{\alpha} = m\alpha, \quad \text{for } m \text{ in } M,\ \alpha \text{ in } T;$$

$$\bar{\mu} \circ m = m\mu; \qquad m \circ \bar{\mu} = 0, \quad \text{for } m \text{ in } M,\ \mu \text{ in } U;$$

$$m \circ n = mn = 0, \quad \text{for } m, n \text{ in } M.$$

To verify that $S(\circ)$ is associative, there are twenty seven cases to consider according to which of $\overline{T}^*$, $\overline{U}^*$ and M the factors are selected from. We omit the details of this verification. In all except five of the cases, namely, $\overline{T}^*\overline{T}^*\overline{T}^*$, $M\overline{T}^*\overline{T}^*$, $\overline{U}^*\overline{U}^*\overline{U}^*$, $\overline{U}^*\overline{U}^*M$ and $\overline{U}^*M\overline{T}^*$, all products are zero. The case $\overline{U}^*M\overline{T}^*$ rests on the assumption that each element of U commutes with each element of T.

Clearly the definition of $(\circ)$ makes M into a two-sided ideal of S. Further, for m in $M\backslash 0$, $(S \circ m) \circ S = (mU) \circ S = (mU)T = mUT = M$, by the assumed transitivity of UT on $M\backslash 0$. Thus M is a 0-minimal two-sided ideal of S.

The further verification of the fact that $S\rho_M = T$ and $S\lambda_M = U$ is straightforward.

EXERCISES FOR §6.1

1. Let C be a non-empty subset [right ideal] of the semigroup $S = S^0$. Then A_C is a right [two-sided] ideal of S.

2. Let R be a zero right ideal of the semigroup $S = S^0$. Then $R \subseteq A_R$ and the (two-sided) ideal A_R is the $(\rho_R \circ \rho_R^{-1})$-class which is the zero element of $S/(\rho_R \circ \rho_R^{-1})$.

3. Let U be the union of all the 0-minimal zero right ideals of the semigroup $S = S^0$. Suppose that $U \neq \square$. Then U is a zero two-sided ideal of S. Let R be any 0-minimal zero right ideal of S. Then $U \subseteq {}_RA_R$.

4. The extension S of R constructed in the proof of Lemma 6.3 does not come under the types considered in §4.4, when $R^2 = 0$, since then R is not weakly reductive. Nevertheless, it might be conjectured that we could modify the methods of §4.4 by replacing the (now no longer one-to-one) natural mapping of R into its translational hull $\bar{R}$ by some isomorphism ϕ. It is possible to find such isomorphisms in the present case when $|R| > 2$.

Let $T = T^0$ be a semigroup, and let $T^* = T\backslash 0$. Denote the elements of $R\phi$ by $r\phi$ $(r \in R)$. Let $S = R\phi \cup T^*$ and $\bar{S} = \bar{R} \cup T^*$ (where $\bar{R}$ is the translational hull of R). Suppose that an operation $(\circ)$ is defined in S so that $S(\circ)$ is an extension of $R\phi$ by T. Then, for each A in T^*, there exist linked left and right translations λ_A and ρ_A, respectively, of R such that

$$A \circ r\phi = (r\lambda_A)\phi \quad \text{and} \quad (r\phi) \circ A = (r\rho_A)\phi.$$

Then $\theta\colon A \to (\lambda_A, \rho_A)$ is a partial homomorphism of T^* into $\bar{R}$ and, by Theorem 4.19, determines an extension $\bar{S}(*)$ of $\bar{R}$ by T given by

$$A * B = \begin{cases} AB, & \text{if } AB \neq 0 \text{ in } T, \\ (A\theta)(B\theta), & \text{if } AB = 0 \text{ in } T; \end{cases}$$

$$A * \bar{r} = (A\theta)\bar{r}; \qquad \bar{r} * A = \bar{r}(A\theta); \qquad \bar{r} * \bar{s} = \bar{r}\bar{s}$$

where $A, B \in T^*$ and $\bar{r}, \bar{s} \in \bar{R}$.

$S(\circ)$ will be a subsemigroup of $\bar{S}(*)$ if and only if

(1) $A \circ B = (A\theta)(B\theta)$ whenever $AB = 0$ in T;

(2) $(r\lambda_A)\phi = (A\theta)r\phi$ and $(r\rho_A)\phi = (r\phi)(A\theta)$ for all r in R and A in T^*.

In the extension given in the construction in the proof of Lemma 6.3, $\lambda_A = \zeta$ for every A in T^*, where $r\zeta = 0$ for every r in R, while $\{\rho_A\colon A \in T^*\}$ is transitive on $R\backslash 0$. For no isomorphism ϕ of R into $\bar{R}$ will (2) be satisfied, and consequently the extension used in the proof of Lemma 6.3 is not obtainable by the proposed modification of the method of §4.4.

5. Let $R = R^0$ be a semigroup with $R^2 \neq 0$. If also the semigroup of right translations of R is transitive on $R\backslash 0$, then $R^2 = R$.

6. Let $R = R^0$ be a semigroup for which ${}_RA = 0$, and for which the semigroup of right translations of R is transitive on $R\backslash 0$. Then R is right 0-simple (§2.5).

7. (a) Let R be a non-nilpotent 0-minimal right ideal of a semigroup $S = S^0$. Let $A = {}_RA(R)$ and $B = R\backslash A$. Then A is a two-sided ideal of the semigroup R such that $A^2 = 0$, and B is a right simple subsemigroup of R.

(b) If, moreover, R contains a non-zero idempotent, then B is a union of isomorphic groups; cf. Theorem 1.27. (Schwarz [1951], §6.)

8. Let S, R, A, and B be as in Exercise 7(a). For each b in B, define the transformation λ_b of A by $a\lambda_b = ba\ (a \in A)$.

(a) $\lambda_b \lambda_{b'} = \lambda_{b'b}$ for all b, b' in B.

(b) For each b in B, λ_b maps $A\backslash 0$ onto $A\backslash 0$.

(c) If $e^2 = e \in B$, then λ_e is the identity transformation of A.

If $b^2 \neq b \in B$, then 0 is the only fixed point of A under λ_b.

How does one describe the structure of a non-nilpotent 0-minimal right ideal R of some (unspecified) semigroup $S = S^0$? Corollary 6.5 gives a necessary and sufficient condition on R in order that some such S exist, but it does not give the structure of R. For this, in the notation of Exercise 7(a), we must know how B acts on A from the left. Knowing this, Exercise 5 for §8.3 gives a construction which embeds R as a 0-minimal right ideal in a 0-simple semigroup S. Exercise 8 above shows the necessity of the conditions imposed in (2) of Exercise 5 for §8.3.

6.2 The two-sided ideal generated by a 0-minimal right ideal

Our present objective is to investigate the structure of SR, where $R = R^2$ is a 0-minimal right ideal of the semigroup S. Observe firstly that for any x in S, either $xR = 0$ or xR is a 0-minimal right ideal of S (Lemma 2.32) and hence SR is a union of 0-minimal right ideals of S. In fact every 0-minimal right ideal of S contained in SR is of the form xR.

Lemma 6.10. *Let R' be any 0-minimal right ideal of S contained in SR. Then there exists x in S such that $R' = xR$.*

Proof. This result follows immediately from the above remarks and from the fact that any two distinct 0-minimal right ideals of S are 0-disjoint (§2.5).

Lemma 6.11. *Let $x \in SR\backslash 0$. Then $x \in xS$ and xS is a 0-minimal right ideal of S contained in SR. Hence if R' is any 0-minimal right ideal of S contained in SR, then for any $r' \in R'\backslash 0$, $r'S = R'$.*

Proof. The element x must, by the previous lemma, belong to yR for some y in S. Then $x = yr$, for some r in $R\backslash 0$, and $xS = yrS = yR$ (Lemma 2.31), a 0-minimal right ideal of S. Clearly, $x \in xS$.

Equivalently, the result of this lemma could be stated as: *SR does not contain a degenerate 0-minimal right ideal of S.*

LEMMA 6.12. *Let R_1 and R_2 be two 0-minimal right ideals of S contained in SR. Then, $R_1R_2 = R_1$ if and only if $R_2^2 = R_2$; or, equivalently, $R_1R_2 = 0$ if and only if $R_2^2 = 0$.*

PROOF. Since $R_1R_2 \subseteq R_1$ which is a 0-minimal right ideal of S, therefore either $R_1R_2 = R_1$ or $R_1R_2 = 0$. Thus, as asserted, the two conclusions of the lemma are equivalent.

Let $R_2^2 = 0$. Then were $R_1R_2 = R_1$, multiplying on the right by R_2, we would have

$$R_1 = R_1R_2 = (R_1R_2)R_2 = R_1R_2^2 = 0,$$

which is contrary to hypothesis. Hence, in this case, $R_1R_2 = 0$. Conversely, suppose that $R_2^2 = R_2$. There exist, by Lemma 6.10, x_1, x_2 in S such that $R_1 = x_1R$, $R_2 = x_2R$. From $R_2^2 = R_2$ it follows that $Rx_2R = R$, and hence that $R_1R_2 = x_1(Rx_2R) = x_1R = R_1$. This completes the proof of the lemma.

COROLLARY 6.13. *Let R_1 be a 0-minimal right ideal of S contained in SR. Then $SR_1 = SR$ if and only if $R_1^2 = R_1$.*

PROOF. If $R_1^2 = 0$, then $(SR_1)^2 = S(R_1S)R_1 \subseteq SR_1^2 = 0$. But $(SR)^2 \supseteq (RR)^2 = R^2 = R \neq 0$. Hence, when $R_1^2 = 0$, $SR_1 \neq SR$. If $R_1^2 = R_1$, then, by the lemma, $R_1R = R_1$ and $RR_1 = R$. Hence

$$SR = SRR_1 \subseteq SR_1 = SR_1R \subseteq SR;$$

from which it follows that $SR = SR_1$.

The next corollary shows that, within SR, right ideals of S coincide with right ideals of SR. That the analogous assertion does not hold for left ideals is shown by Exercise 1, below.

COROLLARY 6.14. *A subset of SR is a [0-minimal] right ideal of SR if and only if it is a [0-minimal] right ideal of S.*

PROOF. Let C be a right ideal of SR. If $C \neq 0$, let $x \in C\backslash 0$. Then, by Lemma 6.11, xS is a 0-minimal right ideal R', say, of S containing x. Hence, by the lemma, $xSR = R'R = R'$. Consequently, if $C \neq 0$, C is a union of 0-minimal right ideals of S. From this the corollary immediately follows.

COROLLARY 6.15. *$A_R(SR) = A_{SR}(SR)$ is the union of $\{0\}$ and all 0-minimal zero right ideals of S [of SR] in SR; ${}_{SR}A(SR) = 0$.*

PROOF. The first assertion follows immediately from the lemma when we observe also that every non-zero element of SR belongs to a 0-minimal right ideal of S, which, by the previous corollary, is also a 0-minimal right ideal of SR.

As for the left annihilator, we have, for x in $SR\backslash 0$, that $xSR = (xS)R = xS$, by the lemma and by Lemma 6.11. Consequently, $xSR = 0$, for x in SR, implies $x = 0$, because, again by Lemma 6.11, $x \in xS$.

In the previous paragraph we have also proved the following result, which strengthens a previous remark.

COROLLARY 6.16. *Each* 0-*minimal right ideal of* SR *is nondegenerate in* SR.

Denote by $B(= B(R))$ the union of all the 0-minimal right ideals R' of S which are *non-nilpotent* (i.e., here, $(R')^2 = R'$) and are contained in SR. Thus

$$B\backslash 0 = SR\backslash (A_{SR}(SR)).$$

From Corollary 6.13 it follows that $B(R) = B(R_1)$ for any non-nilpotent 0-minimal right ideal R_1 of S contained in SR. B is clearly a subsemigroup of SR.

In analogy with Corollary 6.14 we have

LEMMA 6.17. *A subset of* B *is a* [0-*minimal*] *right ideal of* B *if and only if it is a* [0-*minimal*] *right ideal of* S.

PROOF. Let $x \in B\backslash 0$. Since $B\backslash 0 = SR\backslash (A_{SR}(SR))$, we have that $x(SR) = xB$. The rest of the proof now follows exactly as for Corollary 6.14.

Corollary 6.14 and Lemma 6.17 both provide generalizations of Theorem 2.35.

From this lemma it follows, in particular, that B is a semigroup containing a 0-minimal right ideal (of B). In fact it follows that B is a union of its 0-disjoint 0-minimal right ideals, each of which, by the definition of B, is non-nilpotent. Any semigroup with this property is a union of 0-disjoint, 0-simple (§2.5) subsemigroups (Corollary 6.34). As a special case of this result we have

LEMMA 6.18. B *is a* 0-*simple semigroup containing a* 0-*minimal right ideal.*

PROOF. Let $x \in B\backslash 0$. Then $xS = R'$, say, is, by Lemma 6.11, a 0-minimal right ideal of S. Further, by Lemma 6.12, $R'R = R'$, and so $xB = xSR = R'R = R'$. Hence $BxB = BR'$. From the definition of B and Lemma 6.12 we therefore have $BxB = B$. Thus (Lemma 2.28), B is 0-simple.

The results we have proved about SR combine to give the following theorem due to Schwarz [1951]. We shall contribute a converse in the discussion following the statement of the theorem.

THEOREM 6.19. *Let* R *be a non-nilpotent* 0-*minimal right ideal of the semigroup* $S = S^0$. *Then* SR *is the smallest two-sided ideal of* S *containing* R, *and is a union of* 0-*disjoint nondegenerate* 0-*minimal right ideals of* S. *Let*

$A[B]$ be the union of $\{0\}$ and of all the nilpotent [non-nilpotent] 0-minimal right ideals of S contained in SR. Then

(1) $SR = A \cup B$ and $A \cap B = 0$;

(2) $A = A_R(SR) = A_{SR}(SR)$, and, consequently, $A^2 = 0$ and A is a two-sided ideal of S;

(3) ${}_{SR}A(SR) = 0$;

(4) a subset of $SR[B]$ is a right ideal of $SR[B]$ if and only if it is a right ideal of S; in fact, each non-trivial right ideal of $SR[B]$ is a union of 0-minimal right ideals of S; each 0-minimal right ideal of SR is non-degenerate in SR;

(5) B is a right ideal of S which is a 0-simple semigroup containing a 0-minimal right ideal (of B).

Let R_1 and R_2 be any 0-minimal right ideals of S contained in SR. Then $R_1R_2 = R_1$ if and only if $R_2^2 = R_2$; or, equivalently, $R_1R_2 = 0$ if and only if $R_2^2 = 0$. Further $SR_1 = SR$ if and only if $R_1^2 = R_1$; thus SR is determined by any of the 0-minimal non-nilpotent right ideals of S which it contains.

From the above theorem it follows that, if $A \neq 0$, then A is a union of non-degenerate 0-minimal zero right ideals of SR. 0-minimal zero right ideals were discussed in §6.1 and we proceed now to apply the results of §6.1 to the present case. Let R' be any 0-minimal right ideal of SR contained in A. Then, by Theorem 6.7, $\rho_{R'}$ is a representation of SR, or, since $A^2 = 0$, effectively of B, in $\mathscr{T}_{R'}$ such that R' is 0-transitive under $B\rho_{R'}$. Consequently, ρ_A is a representation of B in $\mathscr{T}_A$, such that A splits up into a set of 0-disjoint sets R' (the 0-minimal right ideals of SR contained in A) on each of which the restriction of ρ_A coincides with the representation $\rho_{R'}$.

As in Theorem 6.7, a converse holds. Let ϕ be any representation of the semigroup S in $\mathscr{T}_X$. As in §11.2 below, we shall say that ϕ is *fully reducible (relative to the element z of X)* if X is a union of subsets of X each of which is z-transitive under ϕ. Such a decomposition of X is necessarily into z-disjoint subsets. In this terminology, the representation ρ_A of B in $\mathscr{T}_A$ is fully reducible relative to 0.

Now let B be any 0-simple semigroup which is the union of its 0-minimal right ideals (each of which, by Lemma 2.34, is therefore non-nilpotent). Let A be any set containing the zero element 0 of B, but otherwise disjoint from B. Let ϕ be any representation of B in $\mathscr{T}_A$ which is fully reducible relative to 0. Let $S = A \cup B$, and define a binary operation $\circ$ in S by:

$$b \circ b' = bb', \quad \text{as defined in } B;$$
$$a \circ b = a(b\phi);$$
$$b \circ a = 0;$$
$$a \circ a' = 0;$$

for any b, b' in B and a, a' in A. As in the proof of Theorem 6.7 it follows that $\circ$ is associative, so that S is a semigroup. Again, from Theorem 6.7, it follows that each 0-transitive subset of A is a non-degenerate 0-minimal zero right ideal of S. Further, let R be any 0-minimal right ideal of B. Then, because B is 0-simple, $B \circ R = BR = B$. And, if R' is any 0-transitive subset of A, then, since $R'(B\phi) = R'$, $R' \circ R = (R'B\phi) \circ R = (R' \circ B) \circ R = R' \circ (B \circ R) = R' \circ B = R'(B\phi) = R'$. Consequently, $A \circ R = A$. Combining this with $B \circ R = B$ gives $S \circ R = S$. Since $B \circ A = 0$ it is clear that the 0-minimality of right ideals in B is preserved in S. Thus $S \circ R$ satisfies all the hypotheses of Theorem 6.19 with the same interpretations of A and of B.

Exercises for §6.2

1. Let S be a semigroup with the following properties:

(1) S has an identity element 1 and its group of units H_1 contains more than one element;

(2) S is right cancellative;

(3) $R = S \backslash H_1 \neq \square$.

Then R is an ideal of S and not every left ideal of R $(= SR)$ is a left ideal of S. (Cf. Exercise 10 for §8.1, where R is also a minimal right ideal of S.)

Then not every left ideal of $SR(= R)$ is a left ideal of S. (A construction of such a semigroup S is given in Exercise 10 for §8.1.)

2. If S is a semigroup containing a minimal right ideal R, then it has a kernel K (§2.5) and $K = SR$. K is the union of all the minimal right ideals of S. Every right ideal of K is also a right ideal of S. (Cf. Exercise 13(b) for §2.7 and Theorem 2.35.)

3. In the notation of Theorem 6.19, any ideal of S properly contained in SR is contained in A.

6.3 The right socle of a semigroup

In this section we consider the union of $\{0\}$ and of all the 0-minimal right ideals of any semigroup $S = S^0$. Following Dieudonné [1942] we will call this union the *right soçle* $\Sigma_r = \Sigma_r(S)$ of S. It was first defined and investigated for semigroups by Schwarz [1951]. As noted by Schwarz (loc. cit., §1), Σ_r is a two-sided ideal of S. For, as a union of right ideals, it is clearly a right ideal. On the other hand, if R is any 0-minimal right ideal of S, then, for any x in S, either $xR = 0$ or xR is also a 0-minimal right ideal of S (Lemma 2.32); in either event, $xR \subseteq \Sigma_r$. Hence $x\Sigma_r \subseteq \Sigma_r$, so that Σ_r is also a left ideal of S. Our results amplify and extend a number of those of Schwarz (loc. cit., §§9, 10); other results of his will be found in the exercises.

The *left socle* $\Sigma_l = \Sigma_l(S)$ of S is the union of $\{0\}$ and of all the 0-minimal left ideals of S. The obvious left-right duals of the results of this section hold for Σ_l. In the next section we consider the structure of the two-sided ideal $\Sigma_r \cup \Sigma_l$ of S.

Lemma 6.20. *If R_1 and R_2 are non-nilpotent 0-minimal right ideals of S, then either $SR_1 = SR_2$ or $SR_1 \cap SR_2 = 0$.*

PROOF. If $R_2 \subseteq SR_1$, then $SR_1 = SR_2$ by Theorem 6.19. Hence we may assume that $R_2 \nsubseteq SR_1$. It follows that $R_2R_1 = 0$; for otherwise $R_2 = R_2R_1 \subseteq SR_1$. Now let $C = SR_1 \cap SR_2$. Then $C(SR_1) \subseteq SR_2SR_1 \subseteq SR_2R_1 = 0$. Thus, since, by Theorem 6.19, ${}_{SR_1}A(SR_1) = 0$, $C = 0$.

Let N be the two-sided annihilator of Σ_r in Σ_r.

LEMMA 6.21. *N is a zero two-sided ideal of S and is the union of $\{0\}$ and of all the 0-minimal right ideals of S which are not contained in any SR with R a non-nilpotent 0-minimal right ideal of S.*

PROOF. N is clearly a two-sided ideal of S; for the two-sided annihilator of any two-sided ideal of S is a two-sided ideal of S. Further, it is clear that if R is a non-nilpotent 0-minimal right ideal of S then $N \cap SR = 0$, for $N \cap SR \subseteq {}_{SR}A(SR) = 0$, by Theorem 6.19(3).

Now let R be any 0-minimal right ideal of S not contained in any SR' with $(R')^2 = R'$. Let R_1 be any 0-minimal right ideal of S. If $RR_1 = R$, then $RR_1^2 = RR_1 = R$ and so $R \subseteq SR_1$ with $R_1^2 = R_1$, contrary to assumption. Hence $RR_1 = 0$. Since this is true for all such R_1, therefore $R\Sigma_r = 0$. Similarly $R_1R = R_1$ implies $R^2 = R \subseteq SR$, again contrary to assumption on R and so also $\Sigma_rR = 0$. Thus $R \subseteq N$. This completes the proof of the lemma.

Let $\mathscr{M}_r$ be the set of all non-nilpotent 0-minimal right ideals of S. If $\mathscr{M}_r$ is non-empty, then by a *basis* of $\mathscr{M}_r$ we mean a subset $\{R_i: i \in I\}$ of $\mathscr{M}_r$ such that (i) if $R \in \mathscr{M}_r$, then $R \subseteq SR_i$ for some $i \in I$, and (ii) if $i \neq j$ in I, then $SR_i \neq SR_j$ (and so $SR_i \cap SR_j = 0$, by Lemma 6.20).

Immediately following from the previous two lemmas we have the following theorem. Observe firstly that the 0-disjoint union of two (or more) two-sided ideals A, B, say, of a semigroup S, forms a semigroup completely determined by A and B. For $AB = BA = 0$, and other products in $A \cup B$ are products either in A or in B. Conversely, if $S = A \cup B$ is the 0-disjoint union of the subsemigroups A and B and also $AB = BA = 0$, then A and B are necessarily two-sided ideals of S. We shall say that S is the *0-direct union* of the subsemigroups $\{S_i: i \in I\}$ if S is their 0-disjoint union and if $S_iS_j = S_jS_i = 0$ for $i \neq j$. The S_i will be said to be *summands* of S in this 0-direct union.

THEOREM 6.22. *Let $S = S^0$ be a semigroup. Let Σ_r be the right socle of S. Then Σ_r is a two-sided ideal of S. Let $\mathscr{M}_r$ be the set of all non-nilpotent 0-minimal right ideals of S. If $\mathscr{M}_r$ is empty, then $\Sigma_r^2 = 0$. Otherwise, let $\{R_i: i \in I\}$ be a basis of $\mathscr{M}_r$. Let N be the two-sided annihilator of Σ_r in Σ_r. Then N is a two-sided ideal of S, and Σ_r is the 0-direct union of N and the set $\{SR_i: i \in I\}$ of two-sided ideals of S.*

As we saw in Theorem 6.19(5) each 0-minimal right ideal of S which is contained in one of the SR_i is a non-degenerate 0-minimal right ideal of that SR_i and so also non-degenerate in Σ_r. This does not apply to the

0-minimal ideals of S contained in N, because Σ_r is the right annihilator of any such ideal. In fact each non-zero element of N determines a degenerate 0-minimal right ideal of Σ_r. Thus Σ_r is a union of 0-minimal right ideals of itself, its two-sided annihilator consisting of the union of $\{0\}$ and its degenerate 0-minimal right ideals (which are also, in fact, two-sided ideals), its other 0-minimal right ideals being (non-degenerate) 0-minimal right ideals of S. Using Exercise 5, below, we can construct a semigroup S in which the cardinal $|N|$ of N has an arbitrary preassigned value, and in which $\Sigma_r \neq N$.

A further decomposition of Σ_r follows from Theorem 6.19. Let A_i be the right annihilator of SR_i in SR_i ($i \in I$). Each A_i is then a two-sided ideal of S. Hence the same is true of $A = \bigcup\{A_i: i \in I\} \cup N$. It easily follows that A is just the right annihilator of Σ_r in Σ_r and is the union of $\{0\}$ and of all the 0-minimal zero right ideals of S. Let B_i be the 0-simple part of SR_i, as in Theorem 6.19, and let $B = \{0\} \cup \{B_i: i \in I\}$. As a union of right ideals B_i of S, B is also a right ideal of S. Clearly each B_i is a two-sided ideal of B and distinct B_i are 0-disjoint. If not equal to $\{0\}$, B is also the union of all the non-nilpotent 0-minimal right ideals of S. We have consequently proved the following theorem.

THEOREM 6.23. *Let $S = S^0$ be a semigroup. Let Σ_r be the right socle of S. Let $A[B]$ be the union of $\{0\}$ and all the nilpotent [non-nilpotent] 0-minimal right ideals of S. Then*

(1) *Σ_r is the 0-disjoint union of A and B;*

(2) *$A = A_{\Sigma_r}(\Sigma_r)$, and, consequently, $A^2 = 0$ and A is a two-sided ideal of S;*

(3) *B is a right ideal of S;*

(4) *B, if $\neq 0$, is the 0-direct union of a set $\{B_i: i \in I\}$ of ideals of B, where each B_i is a right ideal of S and is a 0-simple semigroup containing a 0-minimal right ideal (of B_i).*

If R is a nilpotent right ideal of a semigroup $S = S^0$, say $R^k = 0$, then $S^1R = R \cup SR$ is a nilpotent two-sided ideal of S containing R. For

$$(S^1R)^k = S^1(RS^1)^{k-1}R \subseteq S^1R^k = 0.$$

The 0-*radical* N of S is defined to be the union of all the nilpotent two-sided ideals of S; it contains all the nilpotent one-sided ideals of S. As for rings, N need not be itself nilpotent. For other kinds of radicals, see §11.6 and the exercises for §6.6. We shall say that S is *without nilpotent ideals* if $N = 0$. For such a semigroup, we have the following result due to Schwarz ([1951], §9).

COROLLARY 6.24. *Let $S = S^0$ be a semigroup without nilpotent ideals. Let Σ_r be the right socle of S. Then Σ_r is the 0-direct union of $\{0\}$ and of a set $\{B_i: i \in I\}$ of ideals of S, where each B_i is a 0-simple semigroup containing a 0-minimal right ideal (of B_i).*

Furthermore, each B_i is a 0-simple ideal of S, and if C is any 0-simple ideal of S containing a 0-minimal right ideal (of C), then C is one of the B_i.

PROOF. In the notation of the theorem, the hypothesis that S has no nilpotent ideals implies that $A = 0$. Hence $\Sigma_r = B$ is a 0-direct union of ideals B_i of B, i.e., of ideals of Σ_r.

To see that each B_i is here an ideal of S, we return to the proof of the theorem and observe that each B_i is a 0-simple summand of an SR_i where $R_i = R_i^2$ is a 0-minimal right ideal of S, and that $SR_i = A_i \cup B_i$, where $A_i^2 = 0$ and A_i is an ideal of S. Since S is without nilpotent ideals, $A_i = 0$, and hence $B_i = SR_i$ is a two-sided ideal of S.

Finally, let C be a 0-simple ideal of S containing a 0-minimal right ideal R, say, of C. Since C is 0-simple, $R^2 = R$, and hence $RS = R^2S = R(RS) \subseteq RC \subseteq R$; whence it follows that R is a 0-minimal right ideal of S. Consequently, C is one of the summands B_i of Σ_r.

EXERCISES FOR §6.3

1. In the notation of Theorem 6.23, A is the 0-radical of Σ_r. (Schwarz, [1951], §10.)

2. In the notation of Theorem 6.23, the following statements are equivalent.
   (a) B is a two-sided ideal of S.
   (b) B is the 0-direct union of $\{0\}$ and 0-minimal two-sided ideals of S.
   (c) A is the two-sided annihilator of Σ_r in Σ_r.
   (d) Every non-nilpotent 0-minimal right ideal of S is contained in a 0-minimal two-sided ideal of S. (Schwarz, [1951], §10.)

3. Let $S = S^0$ be a semigroup with non-trivial right socle Σ_r, and such that $ab = 0$ (a, b in S) implies $ba = 0$. Then Σ_r is the 0-direct union of a zero semigroup A and a set of 0-simple semigroups B_i which are unions of their 0-minimal right ideals. Moreover, the B_i (as well as A) are two-sided ideals of S. (Lefebvre [1962], Theorem 3.16 with $W = 0$.)

4. Let P and Q be disjoint semigroups. Let S consist of P, Q, $P \times Q$, and a symbol 0. Define a product in S, preserving that in P and Q, by

$$pq = (p, q)$$
$$p'(p, q) = (p'p, q)$$
$$(p, q)q' = (p, qq')$$

for p, p' in P and q, q' in Q, with all other products defined to be 0. Then S is a semigroup. If Q is a right simple semigroup, then the right socle of S is $0 \cup Q \cup (P \times Q)$. In the notation of Theorem 6.23, $B = 0 \cup Q$ and $A = 0 \cup (P \times Q)$, and we have $SB = A \cup B$.

5. (a) Let F be a finite non-empty set, let $N = F \cup 0$, and define a multiplication $N^2 = 0$ on N so that $N = N^0$ is a zero semigroup. Let T be an infinite cyclic semigroup generated by α. Suppose that the product $N\alpha$

is defined by regarding α as a mapping of N into N which (i) maps 0 into 0, and (ii) permutes the elements of F cyclically. Define $N\alpha^r$ by regarding α^r as the rth iterate of the mapping determined by α. Define also $TN = 0$. Then with these definitions, $S = T \cup N$ is a semigroup with $\Sigma_r(S) = N$.

(b) Let $N = N^0$ be a zero semigroup of arbitrary infinite cardinal. Let T be a semigroup of right translations of $N \backslash 0$ which is transitive on $N \backslash 0$, and suppose that T contains no minimal right ideal. (The existence of such a semigroup T will be established in §8.2.) Then there exists (cf. Theorem 6.7) a semigroup $S = N \cup T$ for which $\Sigma_r(S) = N$.

(c) Let Σ_i, with two-sided annihilator N_i, be the right socle of $S_i = S_i^0$, for $i \in I$. Let S be the 0-direct union of the S_i. Then the right socle Σ of S is the 0-direct union of the Σ_i, and its two-sided annihilator N is the 0-direct union of the N_i for $i \in I$.

6. Let Σ_r be the right socle of a semigroup $S = S^0$. A subset H of $\Sigma_r \backslash 0$ will be called a *cross-section* of Σ_r if it meets each 0-minimal right ideal of S in precisely one element. Following Dubreil [1941], a subset H of S will be called *right* 0-*neat* if, for every a in $S \backslash 0$, there exists x in S such that $ax \in H$. A non-empty subset H of S will be called a 0-*minimal complex* of S if (i) $0 \notin H$, (ii) H is right 0-neat, and (iii) no proper subset of H is right 0-neat.

(a) If H is a 0-minimal complex of S, and $h \in H$, then hS^1 is a non-degenerate 0-minimal right ideal of S, and every non-zero right ideal of S contains some hS^1 with h in H. H is a cross-section of Σ_r.

(b) Let S be a semigroup with zero such that every non-zero right ideal of S contains some non-degenerate 0-minimal right ideal of S. Then any cross-section of Σ_r is a 0-minimal complex of S. (Lefebvre [1962], Theorems 3.4 and 3.7, with $W = 0$.)

6.4 Combined theory of the left and right socles of a semigroup

The last section was devoted to the structure of the right socle Σ_r of a semigroup $S = S^0$, and the left-right duals of the results of that section naturally apply to the left socle Σ_l of S. In the present section we consider both socles together. Theorem 6.29 gives the structure of the two-sided ideal $\Sigma_r \cup \Sigma_l$ of S. We note that the only hypothesis we impose on S is that it contain a zero element. Of course if S contains no 0-minimal right [left] ideal, then $\Sigma_r = 0$ [$\Sigma_l = 0$], and Theorem 6.29 degenerates accordingly, but remains true.

In §§9 and 10 of [1951], Schwarz considers the combined theory, not only of Σ_r and Σ_l, but also of the union Σ_t of all the 0-minimal two-sided ideals of S. We give a couple of Schwarz's results concerning Σ_t in Exercise 7 below. His results concerning Σ_r and Σ_l can all be derived from our general Theorem 6.29.

We will say that an ideal M of a semigroup $S = S^0$ is a *completely* 0-*simple ideal* if M is a completely 0-simple semigroup (§2.7). The following theorem is due to Clifford [1948] and to Rich [1949]. (See also Exercise 2(b) below, and Exercise 6 for §2.7.)

THEOREM 6.25. *Let $L = L^2[R = R^2]$ be a 0-minimal left [right] ideal of the semigroup $S = S^0$. Then $SR = LS$ if and only if $R \cap L \neq 0$ and, in this case, SR is a completely 0-simple ideal of S.*

Conversely, if M is a completely 0-simple ideal of S, then there exists $L = L^2$, a 0-minimal left ideal of S, and $R = R^2$, a 0-minimal right ideal of S, such that $M = SR = LS$, when, necessarily, $R \cap L \neq 0$. $L[R]$ may be taken to be any 0-minimal left [right] ideal of M.

PROOF. By Theorem 6.19, SR is a 0-disjoint union $A \cup B$ where $A = A_{SR}(SR)$ and B is a 0-simple semigroup containing R as a 0-minimal right ideal of B. By the left-right dual of Theorem 6.19, LS is a 0-disjoint union $A' \cup B'$, say, where $A' = {}_{LS}A(LS)$ and B' is a 0-simple semigroup containing L as a 0-minimal left ideal of B'.

Suppose now that $SR = LS = M$, say. Then the right annihilator of SR is also the right annihilator of LS, and by the dual of Theorem 6.19(3), this is trivial. Thus $A = 0$. Similarly $A' = 0$. Hence $M = B = B'$ is a 0-simple semigroup containing both a 0-minimal left ideal and a 0-minimal right ideal. By Theorem 2.48, M is therefore completely 0-simple. It follows, in particular, that $R \cap L \neq 0$ (Corollary 2.52b).

Conversely, let $R \cap L \neq 0$ and let $x \in (R \cap L)\backslash 0$. Then $xS = R$ and $Sx = L$ (Lemma 6.11). Hence $SR = SxS = LS$.

Now let M be any completely 0-simple ideal of S. Let $R[L]$ be a 0-minimal right [left] ideal of M. Let $r \in R\backslash 0$. Then $rS \subseteq rMS \subseteq rM = R = rM \subseteq rS$. Thus $rS = R$ and hence R is a 0-minimal right ideal of S. Similarly L is a 0-minimal left ideal of S. As 0-minimal ideals of M, L and R have the properties $L = L^2$, $R = R^2$ and $L \cap R \neq 0$. Hence, by the first part of the theorem already proved, $SR = LS$ is a completely 0-simple ideal of S having a non-trivial intersection (containing $L \cup R$) with M. Consequently $SR = LS = M$; and this completes the proof of the theorem.

We need three further lemmas to deal with the cases when $R \cap L = 0$.

LEMMA 6.26. *Let $L = L^2[R = R^2]$ be a 0-minimal left [right] ideal of the semigroup $S = S^0$. Let $R \cap L = 0$. Then $SR \cap LS \subseteq A \cap A'$ and $RA' = AL = 0$, where $A = A_{SR}(SR)$ and $A' = {}_{LS}A(LS)$.*

PROOF. *Let* $V = SR \cap LS$. Then $SRV \subseteq SRLS \subseteq S(R \cap L)S = 0$ and hence $V \subseteq A$; similarly $V \subseteq A'$. Since $A' \subseteq LS$ and $RL(\subseteq R \cap L) = 0$ we have $RA' = 0$, and similarly $AL = 0$.

LEMMA 6.27. *Let $R = R^2$ be a 0-minimal right ideal and let L be a 0-minimal zero left ideal of the semigroup $S = S^0$. Then $R \cap L = 0$.*

PROOF. Suppose that $R \cap L \neq 0$ and let $x \in (R \cap L)\backslash 0$. Then $R = xS \subseteq LS$ and so $R^2 \subseteq (LS)^2 = LSLS \subseteq L^2S = 0$, a contradiction. Hence $R \cap L = 0$.

LEMMA 6.28. *Let R be a zero 0-minimal right ideal and L a 0-minimal left ideal of the semigroup $S = S^0$. Then $RL = 0$.*

PROOF. Assume that $RL \neq 0$, and let $x \in RL\backslash 0$, so that, since $RL \subseteq R \cap L$, $x \in R\backslash 0$ and $x \in L\backslash 0$. Now $RL \neq 0$ implies that $RS \neq 0$ and $SL \neq 0$ so that neither R nor L is degenerate. Thus, by Lemma 6.1, $R = xS$ and $L = Sx$. Hence $x \in xS$, and so

$$RL = xS \cdot Sx \subseteq xSx \subseteq xSxS = R^2 = 0;$$

which contradicts the assumption that $RL \neq 0$.

THEOREM 6.29. *Let Σ_l and Σ_r be the left and right socles, respectively, of a semigroup $S = S^0$.*

Let $A[A']$ be the union of $\{0\}$ and all the nilpotent 0-minimal right [left] ideals of S.

Let $C[C']$ be the union of $\{0\}$ and all those non-nilpotent 0-minimal right [left] ideals of S which have a non-zero intersection with some 0-minimal left [right] ideal of S.

Let $D[D']$ be the union of $\{0\}$ and all those non-nilpotent 0-minimal right [left] ideals of S which are 0-disjoint from every 0-minimal left [right] ideal of S.

Let $E = A \cup A'$ and $F = A \cap A'$.

Then the following assertions are true.

(1) *$C = C'$, and C is a two-sided ideal of S which is the 0-direct union of $\{0\}$ and all the completely 0-simple ideals of S.*

(2) *D is a right ideal of S which is the 0-direct union of $\{0\}$ and all the 0-minimal two-sided ideals B_i of D. Each B_i is a right ideal of S containing a 0-minimal right ideal of S, but no 0-minimal left ideal of S. Further, each B_i is a 0-simple semigroup containing a 0-minimal right ideal of itself. D' is a left ideal of S having the left-right duals of the properties stated for D.*

(3) *A, A', E, and F are two-sided ideals of S satisfying*

$$A^2 = A'^2 = AA' = 0,$$
$$CE = EC = DE = ED' = E^3 = 0,$$
$$CF = FC = F^2 = 0.$$

(4)
$$\Sigma_r = C \cup D \cup A,$$
$$\Sigma_l = C \cup D' \cup A',$$
$$\Sigma_r \cup \Sigma_l = C \cup D \cup D' \cup E,$$
$$\Sigma_r \cap \Sigma_l = C \cup F,$$

where the unions are 0-disjoint in all four decompositions.

The 0*-radicals of the semigroups* Σ_r, Σ_l, $\Sigma_r \cup \Sigma_l$, *and* $\Sigma_r \cap \Sigma_l$ *are* A, A', E, *and* F, *respectively.*

REMARKS. Although each B_i described in (2) contains no 0-minimal left ideal of S, it may contain a 0-minimal left ideal of itself. (See Exercise 4 below, and Corollary 6.30.)

The 0-disjoint decompositions in (4) for Σ_r, Σ_l, and $\Sigma_r \cap \Sigma_l$ are in fact 0-direct (see Exercise 8 below), but that for $\Sigma_r \cup \Sigma_l$ fails to be so because we may have $D'D \neq 0$ (Exercise 4).

PROOF. Let $\{R_i: i \in I\}$ be a basis (§6.3) for the set $\mathscr{M}_r$ of all non-nilpotent 0-minimal right ideals of S. Let I' be the set of all i in I such that there exists some 0-minimal left ideal L of S such that $R_i \cap L \neq 0$. By Lemma 6.27, L is non-nilpotent. By Theorem 6.25, for i in I', SR_i is a completely 0-simple ideal of S. Since $R_i \subseteq SR_i$, C is contained in the union C_0 of $\{0\}$ and all the completely 0-simple ideals of S. Conversely, let M be a completely 0-simple ideal of S. Let $R[L]$ be any 0-minimal right [left] ideal of M. Then, by Theorem 6.25, $R[L]$ is a 0-minimal right [left] ideal of S such that $R \cap L \neq 0$, so that $R \subseteq C$. Since M is the union of its 0-minimal right ideals, $M \subseteq C$, whence $C_0 \subseteq C$. Thus $C = C_0$, and clearly the latter is a two-sided ideal of S which is either $\{0\}$ or the 0-direct union of its completely 0-simple summands SR_i $(i \in I')$.

Let $\{L_\lambda: \lambda \in \Lambda\}$ be a basis for the set $\mathscr{M}_l$ of all non-nilpotent 0-minimal left ideals of S. Let Λ' be the set of all λ in Λ such that there exists a (non-nilpotent) 0-minimal right ideal R of S such that $R \cap L_\lambda \neq 0$. By the dual of the foregoing argument, $C' = C_0 = \{0\} \cup \{L_\lambda S: \lambda \in \Lambda'\}$. This completes the proof of part (1) of the theorem.

Let N $[N']$ be the two-sided annihilator of Σ_r in Σ_r [of Σ_l in Σ_l]. Let $I'' = I \backslash I'$ and $\Lambda'' = \Lambda \backslash \Lambda'$. By Theorem 6.22 and its dual, we now have

$$\Sigma_r = \bigcup \{SR_i: i \in I''\} \cup C \cup N,$$

$$\Sigma_l = \bigcup \{L_\lambda S: \lambda \in \Lambda''\} \cup C \cup N'.$$

Let A_i be the right annihilator of SR_i in SR_i; and let A'_λ be the left annihilator of $L_\lambda S$ in $L_\lambda S$. In the notation of Theorem 6.23, $A(= A_{\Sigma_r}(\Sigma_r)) = \bigcup \{A_i: i \in I''\} \cup N$, since $A_i = 0$ for each i in I'. Dually, $A' = \bigcup \{A'_\lambda: \lambda \in \Lambda''\} \cup N'$. By Theorem 6.19, we have $SR_i = A_i \cup B_i$, where B_i is the union of all the non-nilpotent 0-minimal right ideals of S contained in SR_i. For each i in I'', B_i is a 0-simple subsemigroup of SR_i containing 0-minimal right ideals of both S and itself, but containing no 0-minimal left ideal of S, since otherwise SR_i would be a completely 0-simple ideal of S, and we would have $i \in I'$. Dual statements hold for $L_\lambda S = A'_\lambda \cup B'_\lambda (\lambda \in \Lambda'')$.

By their definition, and the foregoing remarks, we have

$$D = \bigcup \{B_i: i \in I''\} \quad \text{and} \quad D' = \bigcup \{B'_\lambda: \lambda \in \Lambda''\}.$$

By Theorem 6.23, D and D' have the properties asserted in part (2) of the theorem.

Clearly

$$\Sigma_r = D \cup C \cup A, \quad \text{and} \quad \Sigma_l = D' \cup C \cup A';$$

and, since $E = A \cup A'$,

$$\Sigma_r \cup \Sigma_l = C \cup D \cup D' \cup E.$$

Clearly C is 0-disjoint from D, D', and E. Suppose that $a \in (D \cap D')\backslash 0$. Then $a \in SR_i \cap L_\lambda S$ for some i in I'', λ in Λ''; and, by the definition of I'' (or Λ''), $R_i \cap L_\lambda = 0$. Hence, by Lemma 6.26, $a \in A_i \cap A'_\lambda$. But $A_i \cap D = 0$. This is a contradiction; and, consequently, $D \cap D' = 0$. From Lemma 6.27 it follows easily that $D \cap A' = 0$. Hence $D \cap E = 0$. Dually, we have $D' \cap E = 0$. Thus C, D, D', and E are 0-disjoint sets, and clearly the same is true of C, D, A and C, D', A'. That

$$\Sigma_r \cap \Sigma_l = C \cup F$$

with $C \cap F = C \cap A \cap A' = 0$, is now immediate. Hence we have established the four 0-disjoint decompositions asserted in part (4) of the theorem.

Since, by Theorem 6.23 and its dual, A and A' are two-sided ideals of S, so also is $E = A \cup A'$. Since C and E are 0-disjoint two-sided ideals of S, we have $CE = EC = 0$. Since D is a right ideal, and E a two-sided ideal, of S, it follows from what we have already proved that $DE \subseteq D \cap E = 0$. Dually, $ED' = 0$.

Now it is immediate from Lemma 6.28 that $AA' = 0$. (The dual $A'A = 0$ does not necessarily hold: see Exercise 5 below.) Hence, since $A^2 = 0$ and $A'^2 = 0$,

$$E^2 = (A \cup A')^2 = A^2 \cup AA' \cup A'A \cup A'^2 = A'A,$$

$$E^3 = (A \cup A')A'A = AA'A \cup A'^2A = 0.$$

This concludes the proof of part (3) of the theorem.

Since E is a nilpotent ideal, it is contained in the 0-radical N of $\Sigma_r \cup \Sigma_l$. But $N \cap R = 0$ for every non-nilpotent 0-minimal right ideal R of S, and hence $N \cap C = N \cap D = 0$. Similarly, $N \cap D' = 0$, and hence $N \subseteq E$. Thus E is the 0-radical of $\Sigma_r \cup \Sigma_l$. The three other cases of part (4) are proved by a similar argument, and this completes the proof of the theorem.

For semigroups without nilpotent ideals, the above theorem can be strengthened.

COROLLARY 6.30. *Let $S = S^0$ be a semigroup without nilpotent ideals which contains both 0-minimal left ideals and 0-minimal right ideals. Let Σ_l and Σ_r be the left and right socles, respectively, of S. Define C, D, D' and E as in the theorem. Then $E = 0$ and*

$$\Sigma = \Sigma_l \cup \Sigma_r = C \cup D \cup D'$$

is a 0-direct union of C, D and D'.

Further, let Σ_λ *and* Σ_ρ *denote the left socle and the right socle, respectively, of* Σ. *Define* C_1, D_1 *and* D_1' *for* Σ *just as* C, D *and* D', *respectively, were defined for* S. *Put* $C^* = C_1 \cap D$, $C^{*\prime} = C_1 \cap D'$. *Then*

$$\Sigma = \Sigma_\lambda \cup \Sigma_\rho = C_1 \cup D_1 \cup D_1'$$

is a 0*-direct union of* C_1, D_1 *and* D_1'; *and*

$$C_1 = C \cup C^* \cup C^{*\prime},$$

$$D = D_1 \cup C^*,$$

$$D' = D_1' \cup C^{*\prime},$$

each union being 0*-direct.*

Furthermore, D_1 $[D_1']$ *if not equal to* $\{0\}$ *is the* 0*-airect union of* 0*-simple semigroups each of which contains a* 0*-minimal right [left] ideal of itself but contains no* 0*-minimal left [right] ideal of itself.*

PROOF. From the theorem, part (3), $E^3 = 0$; hence, since, by assumption, S is without nilpotent ideals, $E = 0$. Again, by the theorem, D is a right ideal and C is a two-sided ideal of S. Hence $DC \subseteq D \cap C = 0$. Consequently, $(CD)^2 = CDCD = 0$, whence the two-sided ideal CD of S is 0, by hypothesis. Similarly, $D'C = 0 = C'D$. Again, $DD' \subseteq D \cap D' = 0$ and that $D'D = 0$ follows exactly as for CD. Hence we have shown that the union of C, D and D' is 0-direct. That the decompositions given for C_1, D and D' are also 0-direct unions now follows easily.

It remains to show that the 0-simple semigroups, if any, whose union forms D_1, cannot contain 0-minimal left ideals. Suppose, on the contrary, that L is a 0-minimal left ideal of B_1, where B_1 is a 0-simple two-sided ideal of D_1, so that $D_1 = B_1 \cup D^*$, say, is a 0-direct union. Let $l \in L\backslash 0$. Then, since $\Sigma = C_1 \cup B_1 \cup D^* \cup D_1'$ is a 0-direct union, $\Sigma l = B_1 l = L$. Thus L is a 0-minimal left ideal of Σ. Further $L^2 = B_1 l B_1 l = B_1 l = L$; and hence, by Theorem 6.25, B_1 is a completely 0-simple ideal of Σ. Thus $B_1 \subseteq C_1$, which is contrary to the assumption that $B_1 \subseteq D_1$. Dual remarks apply to D_1'; and this completes the proof of the corollary.

EXERCISES FOR §6.4

1. Let S be a semigroup without zero. Let S contain a minimal left ideal and a minimal right ideal. As a corollary to Theorem 6.25 (applied to S^0) it follows that S has a kernel K and that K is completely simple. From Theorem 6.19(4) and its dual it follows that every right [left] ideal of K is a right [left] ideal of S. (Clifford [1948].)

2. (a) Let $R[L]$ be a 0-minimal right [left] ideal of S such that $RL \neq 0$. Then neither R nor L is degenerate, $R \cap L \neq 0$, $R^2 = R$ and $L^2 = L$.

 (b) Let $R[L]$ be a 0-minimal right [left] ideal of S such that $RL \neq 0$ and $LR \neq 0$. Then LR is a completely 0-simple ideal of S. (Rich [1949].)

3. Let S be a regular Rees $I \times \Lambda$ matrix semigroup $\mathscr{M}^0(G; I, \Lambda; P)$ (§3.1, p. 88) over G^0, where G is a group, with sandwich matrix $P = (p_{\lambda i})$. Let $p_{\kappa j} = 0$ for a particular $\kappa \in \Lambda$, $j \in I$. Let

$$L_\kappa = \{(a)_{i\kappa}: a \in G^0, i \in I\},$$
$$R_j = \{(a)_{j\lambda}: a \in G^0, \lambda \in \Lambda\}.$$

Then $R_j[L_\kappa]$ is a non-nilpotent 0-minimal right [left] ideal of S such that

$$R_j L_\kappa = R_j \cap L_\kappa = H_{j\kappa} \cup \{0\} \neq 0;$$

but $L_\kappa R_j = 0$. Thus R_j and L_κ satisfy the hypotheses of the first part of Theorem 6.25 but not those of Rich's Theorem (Exercise 2(b)).

4. In Exercise 4 for §6.3, let Q be right simple and P left simple. In the notation of Theorem 6.29, $C = \{0\}$, $D = Q \cup \{0\}$, $D' = P \cup \{0\}$, and $A = A' = E = F = D'D = (P \times Q) \cup \{0\}$. If Q is also left simple, i.e. if Q is a group, then D is a completely 0-simple semigroup, but is not a completely 0-simple ideal of S. This example also shows that it is possible for a semigroup to contain non-nilpotent 0-minimal right and left ideals R and L, respectively, such that $RL = 0$ and $LR \neq 0$.

5. Let P and Q be disjoint semigroups. Let e, 0 be symbols denoting elements not in P or Q. Let T consist of P, Q, $P \times Q$, e, $P \times e$, $e \times Q$ and 0. Define a product in T according to the following table, where p and p' are arbitrary elements of P, and q and q' of Q.

	p'	q'	(p', q')	e	(p', e)	(e, q')
p	pp'	(p, q')	(pp', q')	(p, e)	(pp', e)	(p, q')
q	0	qq'	0	0	0	0
(p, q)	0	(p, qq')	0	0	0	0
e	0	(e, q')	0	e	0	(e, q')
(p, e)	0	(p, q')	0	(p, e)	0	(p, q')
(e, q)	0	(e, qq')	0	0	0	0

Then T is a semigroup. (The semigroup S of Exercise 4 is a subsemigroup of T. Since $T = \langle S, e\rangle$ we need only apply Light's associativity test (§1.2) to the single element e.) Now assume that P is left simple and Q right simple. In the notation of Theorem 6.29, $A'A = (P \times Q) \cup \{0\}$ and $E^2 \neq 0$.

6. (a) If the right socle Σ_r and the left socle Σ_l of a semigroup S coincide, then $\Sigma_r = \Sigma_l = C \cup E$, where C is $\{0\}$ or a 0-direct union of a set of completely 0-simple semigroups, E is a zero semigroup, and C and E are 0-disjoint ideals of S.

(b) If S is without nilpotent ideals, then $\Sigma_r = \Sigma_l$ if and only if every 0-minimal two-sided ideal of S that contains a 0-minimal left ideal of S also contains a 0-minimal right ideal of S, and vice versa. $\Sigma_r(=\Sigma_l)$ is then $\{0\}$ or a 0-direct union of completely 0-simple semigroups. (Schwarz [1951], §9.)

7. Let Σ_t be the union of $\{0\}$ and all the 0-minimal two-sided ideals of a semigroup $S = S^0$. If S is without nilpotent ideals, then $\Sigma_r \subseteq \Sigma_t$; and $\Sigma_r = \Sigma_t$ if and only if every 0-minimal two-sided ideal of S contains a 0-minimal right ideal of S. (Schwarz [1951], §9.)

8. In the notation of Theorem 6.29, $CD = DC = CD' = D'C = DD' = 0$, while $A'A(=E^2)$ and $D'D$ are contained in F.

9. Let $S = \mathscr{M}^0(G; I, \Lambda; P)$ be the Rees $I \times \Lambda$ matrix semigroup over the group with zero G^0 with non-zero sandwich matrix P. Denote by M and N the left and right annihilators, respectively, of S. Set $M \cap N = K$, $(S \backslash (M \cup N)) \cup \{0\} = C$, $(M \backslash K) \cup \{0\} = A$, $(N \backslash K) \cup \{0\} = B$, $C \cup B = U$ and $C \cup A = V$. Then

(i) K is the two-sided annihilator of S;

(ii) C is a completely 0-simple semigroup;

(iii) M, N and K are each zero semigroups and two sided ideals of S;

(iv) V is its own left socle, C is the union of the non-nilpotent 0-minimal left ideals of V and A is the union of $\{0\}$ and of the nilpotent 0-minimal left ideals of V;

(v) C is a left ideal of V and $CA = A$;

(vi) the left-right duals of (iv) and (v) hold for U, C and B;

(vii) $BA = K$;

(viii) $C = S$ if and only if the matrix P is regular.

6.5 0-DIRECT UNIONS OF 0-SIMPLE SEMIGROUPS

In this section we consider various ways of characterizing semigroups, such as those we met as subsemigroups of the socles of a semigroup, which are 0-direct unions of 0-simple semigroups. The main theorems, Theorems 6.32 and 6.33, are essentially due to Schwarz ([1951], §2), and likewise Corollaries 6.34 and 6.37 (loc. cit., §§9, 10). These results may be compared with those of Croisot [1953] on decompositions of a semigroup into simple semigroups discussed in §4.1.

We will say that a semigroup $S = S^0$ is *bi-0-stratified* if $x \in SxyS$ and $y \in SxyS$ for any x, y in S such that $xy \neq 0$. Analogously we say that S is [*left*] *right* 0-*stratified* if [$y \in Sxy$] $x \in xyS$ for any x, y in S such that $xy \neq 0$. Concepts equivalent to these concepts were introduced by Schwarz (loc. cit., §2); see Exercise 1 below.

In terms of Green's equivalence relations (§2.1), S is left [right, bi-] 0-stratified if and only if $y \mathscr{L} xy$ [$x \mathscr{R} xy$, $x \mathscr{J} xy \mathscr{J} y$] for any x, y in S such that $xy \neq 0$. For example, to verify the last of these, observe firstly that $J_{xy} \leqq J_x$ for any x, y. Further, if $xy \neq 0$ and $x \in SxyS$, then clearly $x \cup xS \cup Sx \cup SxS \subseteq SxyS$, so that $J_x \leqq J_{xy}$. Thus $J_x = J_{xy}$; and similarly $J_{xy} = J_y$. Conversely, suppose that $J_x = J_{xy} = J_y$. Then y, $x \in xy \cup xyS \cup Sxy \cup SxyS$

and hence there exist a, b, c, d in S^1 such that $x = axyb$ and $y = cxyd$. Hence

$$\begin{aligned} x &= axyb = ax \cdot cxyd \cdot b, \\ &= axc \cdot axyb \cdot ydb = (axca)xy(bydb), \\ &\in SxyS; \end{aligned}$$

and similarly $y \in SxyS$. And this proves our assertion.

Extending a terminology of Dubreil [1941] (see also §9.4) we say that a subset X of the semigroup $S = S^0$ is [*left, right*] *0-consistent* if $xy \in X \backslash 0$ implies [$x \in X$, $y \in X$] both $x \in X$ and $y \in X$.

LEMMA 6.31. *A semigroup $S = S^0$ is left [right, bi-] 0-stratified if and only if every left [right, two-sided] ideal of S is right [left, —] 0-consistent.*

PROOF. Let S be bi-0-stratified and let I be a two-sided ideal of S. Let $xy \in I \backslash 0$. Since S is bi-0-stratified $x, y \in SxyS \subseteq I$. Thus I is 0-consistent. Similarly if S is left [right] 0-stratified then each left [right] ideal is right [left] 0-consistent.

Conversely, if every two-sided ideal is 0-consistent, then x, y belong to $J(xy) = xy \cup xyS \cup Sxy \cup SxyS$ if $xy \neq 0$. Thus $x = axyb$ and $y = cxyd$ for some a, b, c, d belonging to S^1. Hence, repeating an earlier argument, $x, y \in SxyS$ when $xy \neq 0$, i.e. S is bi-0-stratified. The one-sided cases are dealt with similarly.

We are now ready for our first main theorem. The notion of 0-direct union was defined just before Theorem 6.22. We remark that if S is a 0-direct union of 0-minimal ideals M_i of S, then every 0-minimal ideal of S is one of the M_i, and so the M_i in any such 0-direct decomposition of S are unique. It follows from this, and the proof of Theorem 6.32, that if (B) holds, then the 0-simple components S_j of S and the zero semigroup Z are uniquely determined by S. The S_j are the non-nilpotent M_i, and Z is the union of the nilpotent M_i.

THEOREM 6.32. *The following are equivalent conditions on a semigroup $S = S^0$.*

(A) *S is a 0-direct union of 0-minimal ideals of S.*
(B) *S is a 0-direct union of 0-simple semigroups and of a zero semigroup.*
(C) *S is bi-0-stratified.*
(D) *Every two-sided ideal of S is 0-consistent.*

PROOF. The equivalence of (C) and (D) is part of Lemma 6.31. Since a 0-minimal ideal of S is either a 0-simple semigroup or a zero semigroup (Theorem 2.29) it follows that (A) implies (B). Assume that (B) holds, so that S is the 0-direct union of the 0-simple semigroups $S_i (i \in I)$ and of the zero semigroup Z, say. Let $xy \in S \backslash 0$. Then, clearly, neither x nor y can belong to Z. Further, since $S_iS_j = 0$ if $i \neq j$, x and y must belong to the same S_i, to S_k, say. Then $SxyS = S_kxyS_k = S_k$, since S_k is 0-simple. Thus $x, y \in SxyS$, i.e. S is bi-0-stratified. Thus (B) implies (C).

We now show that (B) implies (A). Firstly, let M be an ideal of S contained in S_i. Then M is an ideal of S_i, and since S_i is 0-simple either $M = 0$ or $M = S_i$. Thus S_i is a 0-minimal ideal of S. Let M_z denote $\{0, z\}$ for $z \in Z\backslash 0$. Then Z is the 0-direct union of the M_z ($z \in Z\backslash 0$) and we easily verify that each M_z is a 0-minimal (degenerate) ideal of S. Thus S is a 0-direct union of 0-minimal ideals; which proves that (B) implies (A).

Assume now that (C) holds. Denote by $\{S_i : i \in I\}$ the set of distinct sets $J_x \cup 0$ ($x \neq 0$), where, as usual, J_x denotes the $\mathscr{J}$-class containing x. Then $S_i \cap S_j = 0$ if $i \neq j$. Further if $J_x \neq J_y$ then $J_xJ_y = 0$; for if $u \in J_x$, $v \in J_y$ and $uv \neq 0$, then $J_x = J_u = J_{uv} = J_v = J_y$, as was remarked earlier. Hence $S_iS_j = 0$ if $i \neq j$. Let $y, z \in J_x$; if $yz \neq 0$, then again $J_y = J_{yz} = J_z = J_x$ so that $yz \in J_x$. Thus each S_i is a subsemigroup of S and S is thus the 0-direct union of the subsemigroups S_i. We have already shown that $I(x) = J(x)\backslash J_x$ is equal to $\{0\}$ for any $x \neq 0$. Consequently the S_i are isomorphic to the principal factors of S which we know to be either 0-simple or zero semigroups (Lemma 2.39). The 0-direct union of zero semigroups is clearly a zero semigroup. Thus S is the 0-direct union of 0-simple semigroups and of a zero semigroup. Thus (C) implies (B).

This completes the proof of the theorem.

We now obtain similar results dealing with 0-direct unions of 0-simple semigroups each of which contains a 0-minimal one-sided ideal.

Theorem 6.33. *The following are equivalent conditions on a semigroup* $S = S^0$.

(A) *S is the union of 0-minimal right ideals (in other words, S is its own right socle Σ_r).*

(B) *S is right 0-stratified.*

(C) *Every right ideal of S is left 0-consistent.*

Proof. That (B) and (C) are equivalent is part of Lemma 6.31. Assume (A), and let x and y be elements of S such that $xy \neq 0$. By (A), $x \in R$ for some 0-minimal right ideal R of S. Since $xy \neq 0$, R is not degenerate. By Lemma 6.1, $R = xyS$, and hence $x \in xyS$. Thus (A) implies (B).

Assume conversely that (B) holds, and let $x \in S\backslash 0$. If $xS = 0$, then x belongs to the degenerate 0-minimal right ideal $\{0, x\}$. Assume $xS \neq 0$. Then $xy \neq 0$ for some y in S, and $x \in xyS$ by hypothesis, whence $x \in xS$. But xS is a 0-minimal right ideal of S. For if $r \in xS\backslash 0$, then $r = xy \neq 0$ for some y in S, and $x \in xyS = rS$, whence $xS \subseteq rS$. Consequently every element $x \neq 0$ of S belongs to some 0-minimal right ideal of S, that is, (A) holds.

For semigroups without nilpotent ideals we have a stronger result.

Corollary 6.34. *The following are equivalent conditions on a semigroup $S = S^0$ without nilpotent ideals.*

(A) *S is the union of its 0-minimal right ideals.*

(B) *S is the 0-direct union of 0-simple semigroups each containing a 0-minimal right ideal.*

(C) *S is right 0-stratified.*

(D) *Every right ideal of S is left 0-consistent.*

Proof. We merely have to show that when S is without nilpotent ideals then (B) is equivalent to any one of (A), (C), (D). That (A) implies (B) follows directly from Corollary 6.24.

Now assume that (B) holds so that S is the 0-direct union of the set $\{S_i: i \in I\}$ of 0-simple ideals of S, each S_i containing a 0-minimal right ideal. Let $x, y \in S$ with $xy \neq 0$. Then, for some k in I, $x, y \in S_k$ and $xyS_k = xyS$. But xyS_k is the 0-minimal right ideal of S_k containing xy, and, since $xyS_k \subseteq xS_k$, $xyS_k = xS_k$ the 0-minimal right ideal of S_k containing x. Thus $x \in xyS$; and this shows that S is right 0-stratified. Thus (B) implies (C).

This completes the proof of the corollary.

From this corollary there follows

Corollary 6.35. *Let $S = S^0$ be a semigroup without nilpotent ideals. If S is right 0-stratified, then S is bi-0-stratified.*

We now state explicitly the application of Corollary 6.34 to the case of a semigroup without zero (cf. §8.2). We say that S is *right* [*left*] *stratified* if $x \in xyS$ [$x \in Syx$] for all x, y in S. Thus, if S has no zero element, S is right stratified if and only if S^0 is right 0-stratified. A subset X of S is said to be [*left, right*] *consistent* if $xy \in X$ implies [$x \in X$, $y \in X$] both $x \in X$ and $y \in X$. Thus, if S has no zero element and $X \subseteq S$, then X is a left consistent subset of S if and only if $X \cup \{0\}$ is a left 0-consistent subset of S^0.

We can now infer the following characterizations of simple semigroups containing minimal right ideals (cf. Croisot [1953]).

Theorem 6.36. *The following are equivalent conditions on a semigroup S.*

(A) *S is simple and has a minimal right ideal.*

(B) *S is the union of its minimal right ideals.*

(C) *S is right stratified.*

(D) *Every right ideal of S is left consistent.*

We turn now to semigroups with zero that possess both 0-minimal right and 0-minimal left ideals.

Theorem 6.37. *The following are equivalent conditions on a semigroup $S = S^0$.*

(A) *S is both its own left socle and its own right socle.*

(B) *S is the 0-direct union of a zero semigroup and of completely 0-simple semigroups.*

(C) *S is both left and right 0-stratified.*

(D) *Every right ideal of S is left 0-consistent and every left ideal of S is right 0-consistent.*

PROOF. That (C) and (D) are equivalent follows from Lemma 6.31. The equivalence of (A) and (C) follows directly from Theorem 6.33 and its dual.

Assume that (A) holds. In the notation of Theorem 6.29, we have

$$S = \Sigma_r = C \cup D \cup A,$$

$$S = \Sigma_l = C \cup D' \cup A'$$

where $E = A \cup A'$ is a zero semigroup. Further

$$S = \Sigma_r \cup \Sigma_l = C \cup D \cup D' \cup E;$$

and each of these three expressions of S as a union is 0-disjoint. Hence it follows that $D = D' = \{0\}$ and $S = C \cup E$. By Theorem 6.29, C is the 0-direct union of completely 0-simple semigroups. Thus (A) implies (B). That (B) implies (A) is evident. This completes the proof of the theorem.

When applied to semigroups without nilpotent ideals (cf. Exercise 4, below) Theorem 6.37 gives a characterization of 0-direct unions of completely 0-simple semigroups. We now derive a further characterization of such semigroups.

When without nilpotent ideals, the semigroups of Theorem 6.37 are regular, i.e. $a \in aSa$ for any $a \in S$ (§1.9). In a regular semigroup every principal (hence every 0-minimal) one-sided ideal has an idempotent generator (Lemma 1.13). Part of the next lemma was given as Exercise 10 for §2.7.

LEMMA 6.38. *Let $S = S^0$ be a regular semigroup and let e be a non-zero idempotent of S. Then the following assertions are equivalent.*

(1) *e is primitive.*

(2) *Se is a 0-minimal left ideal of S.*

(3) *eS is a 0-minimal right ideal of S.*

PROOF. Assume that (1) holds and let L be a left ideal of S contained in Se. By Lemma 1.13 there exists an idempotent f such that $Sf \subseteq L$ with $f \neq 0$ if $L \neq 0$. Then $f = f^2 \in Sf \subseteq Se$. Hence $f = xe$, for some x in S. Since $f^2 = xexe$, $ef = exe \neq 0$. Now, clearly, $e(ef) = ef = (ef)e$, i.e. $ef \leqq e$. Since e is primitive therefore $ef = e$. Hence $e \in Sf$ and so $Se \subseteq Sf$. Thus $Se = Sf$ and Se is a 0-minimal left ideal. Consequently we have shown that (1) implies (2). That (1) implies (3) follows by symmetry.

To complete the proof of the lemma, it will suffice, again appealing to symmetry, to show that (2) implies (1). Suppose then that Se is a 0-minimal left ideal and that $0 \neq f^2 = f \leqq e$. Then $fe = f$ and so $0 \neq Sf \subseteq Se$. By the 0-minimality of Se, therefore $Sf = Se$. Hence $e = xf$, for some x in S, and so $ef = xf^2 = xf = e$. But $f \leqq e$ implies $ef = f$. Thus $0 \neq f^2 = f \leqq e$

implies that $e = f$, i.e., e is primitive and we have shown that (2) implies (1).

A regular semigroup is said to be *primitive* if each of its non-zero idempotents is primitive. Theorem 6.39 gives the structure of all primitive regular semigroups, since the structure of completely 0-simple semigroups is known, by the Rees Theorem. This result was essentially conjectured by H. Schneider in 1959 in a letter to one of the authors (see Preston [1969]). It was also found (and proved) independently by P. S. Venkatesan in [1963] and [1966]. The equivalence of (A), (B), and (C) is due to O. Steinfeld [1966], who also gives further equivalent conditions involving 0-minimal quasi-ideals (see Exercise 17 for §2.7).

THEOREM 6.39. *The following conditions on a semigroup $S = S^0$ are equivalent.*

(A) *S is a 0-direct union of completely 0-simple subsemigroups.*
(B) *S is a union of 0-minimal right ideals of the form eS ($e^2 = e$).*
(C) *S is regular and the union of its 0-minimal right ideals.*
(D) *S is primitive regular.*

PROOF. We first show the equivalence of (B), (C), and (D). From the left-right symmetry of (D), it will follow that the duals of (B) and (C) are also equivalent to (D).

Assume (B), and let $a \in S \backslash 0$. Then $a \in eS$ ($e^2 = e$), where eS is 0-minimal. By Lemma 6.1, $aS = eS$, so $e = ax$ for some x in S. From $a \in eS$ we have $ea = a$, hence $axa = ea = a$, showing that S is regular. Thus (B) implies (C).

Assume (C), and let e be a non-zero idempotent of S. Then e belongs to some 0-minimal right ideal R of S, and $R = eS$. By Lemma 6.38, e is primitive, and this establishes (D).

Assume (D), and let $a \in S \backslash 0$. Since S is regular, $axa = a$ for some x in S. Setting $e = ax$, we have $e^2 = e \neq 0$ and $a \in eS$. By Lemma 6.38, eS is 0-minimal, and hence (B) holds.

It is straightforward to verify that (A) implies (D). Conversely, assume (D). Then both (B) and its dual hold. From Corollary 6.34 ((A) implies (B)) and its dual, and the remark about uniqueness prior to Theorem 6.32, it follows that S is the 0-direct union of 0-simple semigroups S_i each containing a 0-minimal right ideal and also a 0-minimal left ideal. By Theorem 2.48, each S_i is completely 0-simple.

We remark that the result given in Exercise 11 for §2.7 is an immediate corollary of Theorem 6.39. In particular, every primitive regular semigroup without zero is completely simple.

We state a theorem to make explicit what the summands in (A) are. Its proof is immediate from the remarks prior to Theorem 6.32.

THEOREM 6.40. *Let $S = S^0$ be a primitive regular semigroup. Then S is the 0-direct union of the set of its completely 0-simple ideals.*

Exercise 6 below is a corollary of Theorem 6.39, except for the condition (C) therein, and serves to characterize primitive inverse semigroups. The study of these is taken up in §7.7.

Exercises for §6.5

1. The following assertions about a semigroup $S = S^0$ are equivalent.

(a) S is bi-0-stratified.

(b) If $x, y \in S \backslash 0$, and $x \in J(y)$, then $y \in J(x)$.

(c) Every non-zero $\mathscr{J}$-class of S is minimal in the partially ordered set of all non-zero $\mathscr{J}$-classes of S (cf. §5.3).

An analogous proposition holds for each of the one-sided cases.

(Condition (b) and its one-sided analogues are those used by Schwarz in §2 of [1951].)

2. A semigroup is bistratified ($x, y \in SxyS$) if and only if it is simple.

3. The following conditions on a semigroup $S = S^0$ are equivalent.

(A) S is a 0-direct union of 0-simple semigroups (each of which is then necessarily a 0-minimal two-sided ideal of S).

(B) S is without nilpotent ideals and is bi-0-stratified.

(C) S is without nilpotent ideals and every two-sided ideal of S is 0-consistent.

4. The semigroup $S = S^0$ is a 0-direct union of completely 0-simple semigroups if and only if it is without nilpotent ideals and is both left and right 0-stratified.

5. Any non-zero idempotent in $\Sigma_r \cup \Sigma_l$ is primitive. (Schwarz [1951], §7.)

6. The following conditions on a semigroup $S = S^0$ are equivalent.

(A) S is an inverse semigroup which is the union of its 0-minimal right ideals.

(B) S is a 0-direct union of completely 0-simple inverse semigroups, i.e. of Brandt semigroups (see Theorem 3.9).

(C) If $a \in S \backslash 0$, then there exists a unique element x of S such that $axa = a$.

(D) S is an inverse semigroup in which every non-zero idempotent is primitive.

(Preston [1954b, 1969] for the equivalence of (A), (B) and (D); Venkatesan [1962] for the equivalence of (B), (C) and (D).)

Exercises 7–12 below are due to Š. Schwarz [1960]. For present purposes, it will be more convenient to denote by AX [XA] the right [left] annihilator of a subset X of a semigroup $S = S^0$, than to use our previous notation A_X [$_XA$]. A semigroup $S = S^0$ is called *dual* if $A(RA) = R$ for every right ideal R of S, and $(AL)A = L$ for every left ideal L of S. A left ideal L of S is called *maximal* if $L \neq S$, and no proper left ideal of S properly contains L.

7. Let S be a dual semigroup. Then the following assertions, together with their left-right duals, hold for S.

(a) $(\bigcap_i R_i)A = \bigcup_i(R_iA)$, for any set of right ideals R_i of S.

(b) $R_1 \cap R_2 = 0$ implies $R_1A \cup R_2A = S$.

(c) $SA = 0$.

(d) If R is a 0-minimal right ideal of S, then RA is a maximal left ideal of S.

(e) The mapping $R \rightarrow RA$ is an anti-isomorphism of the (complete) lattice of all right ideals of S onto that of all left ideals of S, and $L \rightarrow AL$ is inverse thereto. The sublattice of all two-sided ideals of S is anti-isomorphic to itself.

8. A 0-direct union $S = \bigcup_i S_i$ ($S_i \cap S_j = 0$ if $i \neq j$) of a set of semigroups S_i is dual if and only if each component S_i of S is dual.

9. Let N be the 0-radical of a dual semigroup S, and let J be a two-sided ideal of S such that $J \cap N = 0$.

(a) $JA = AJ$, and S is the 0-direct union of J and AJ.

(b) J and AJ are dual semigroups.

10. For a semigroup $S = S^0$, the following are equivalent.

(A) S is a dual semigroup without nilpotent ideals, such that every nonzero two-sided ideal of S contains a 0-minimal two-sided ideal of S.

(B) S is a 0-direct union of 0-simple dual semigroups.

11. Let S be a dual semigroup, and e an idempotent of S such that Se is a 0-minimal left ideal of S. Then, for every x in S, either $ex = x$ or $ex = 0$.

12. Let S be a completely 0-simple semigroup. Then S is dual if and only if it is inverse (and hence a Brandt semigroup, by Theorem 3.9).

6.6 M_R, M_L AND SIMILAR MINIMAL CONDITIONS

The minimal conditions M_R, M_L and M_J, first discussed by Green [1951], were mentioned and used in Chapter 5, (p. 148). Recall that M_R [M_L, M_J] is the condition that every non-empty set of principal right [left, two-sided] ideals of a semigroup S contains a minimal element (relative to ordering by inclusion). Equivalently, S satisfies M_R if and only if every properly descending chain of principal right ideals of S is finite. Recall also that the ordering of principal right [left, two-sided] ideals can be interpreted as an ordering of $\mathscr{R}$- [$\mathscr{L}$-, $\mathscr{J}$-] classes. We write $R_a \leqq R_b$ [$L_a \leqq L_b$, $J_a \leqq J_b$] if and only if $R(a) \subseteq R(b)$ [$L(a) \subseteq L(b)$, $J(a) \subseteq J(b)$]. Thus S satisfies M_R if and only if every non-empty set of $\mathscr{R}$-classes contains a minimal element relative to $\leqq$.

We shall be more concerned with two conditions M_L^* and M_R^*, weaker than M_L and M_R, respectively. A semigroup S satisfies M_L^* [M_R^*] if and only if, for each $\mathscr{J}$-class the set of all $\mathscr{L}$- [$\mathscr{R}$-] classes contained in the $\mathscr{J}$-class has a minimal element. These conditions were first discussed by Munn [1957].

Very similar conditions, the conditions of left and right stability, were also introduced by Wallace and Koch [1957]. For the (topological) semigroups actually treated by Wallace and Koch the condition of left [right] stability coincides with the condition M_L^* [M_R^*]. We shall modify the Wallace-Koch definition of stability to make this identification always possible (see Exercise 1 below for the original definition of Wallace and Koch). The further conditions that a semigroup be left [right] elementary are due to Schützenberger [1957]. The section ends with some examples which illustrate some of the matters discussed.

Let us define the quasi-orderings (i.e. reflexive, transitive, binary relations, sometimes termed pre-orderings) λ, ρ on a semigroup S, thus:

$$\lambda = \{(x, y): L(x) \subseteq L(y)\},$$
$$\rho = \{(x, y): R(x) \subseteq R(y)\}.$$

Then, clearly, $\mathscr{L} = \lambda \cap \lambda^{-1}$, $\mathscr{R} = \rho \cap \rho^{-1}$. The next lemma is due to Munn [1957].

Lemma 6.41. *Condition M_L^* holds in a semigroup S if and only if $\lambda \cap \mathscr{J} = \mathscr{L}$, i.e., if and only if each $\mathscr{L}$-class in any $\mathscr{J}$-class is minimal in the set of $\mathscr{L}$-classes in that $\mathscr{J}$-class.*

Proof. Clearly, if $\lambda \cap \mathscr{J} = \mathscr{L}$, then M_L^* is satisfied. Conversely, let us assume that M_L^* is satisfied. Let $(a, b) \in \lambda \cap \mathscr{J}$, so that $L_a \leqq L_b \subseteq J_a = J_b$. By assumption there exists an $\mathscr{L}$-class, L_c, say, minimal in the set of $\mathscr{L}$-classes of S contained in J_a. Now $L_a \leqq L_b$ implies there exists x in S^1 such that $a = xb$. Since $J_b = J_c$, there exist m, n in S^1 such that $b = mcn$. Then $a = xb = xmcn$. Put $xmc = d$. Now $L_d \leqq L_c$. If $L_d < L_c$, then since L_c is minimal in $J_a = J_c$, it follows that $J_d < J_c$, when also $J_a = J_{dn} \leqq J_d < J_c$, a contradiction. Hence $L_d = L_c$. Thus there exists l in S^1 such that $lxmc = c$. Then

$$b = mcn = m(lxmc)n = ml(xb) = (ml)a.$$

The equations $a = xb$ and $b = (ml)a$ together imply that $L_a = L_b$. Thus $\lambda \cap \mathscr{J} \subseteq \mathscr{L}$. Since the reverse inequality always holds, this proves the lemma.

For Munn's elegant proof of this lemma, see the Remark following Lemma 6.44.

A semigroup is said to be *left [right] stable* if $a \in Sab$ [$a \in baS$], for a, b in S, implies that $L(a) = L(ab)$ [$R(a) = R(ba)$]. A semigroup is *stable* if it is both left and right stable. The next lemma is due to Wallace and Koch (loc. cit.).

Lemma 6.42. *The semigroup S is left stable if and only if it satisfies the condition M_L^*.*

Proof. Let S be left stable and suppose that $L_a \leqq L_b \subseteq J_a = J_b$. Since $J_a = J_b$, there exist x, y in S^1 such that $b = xay$. If $y = 1$, then $b = xa$ and

$L_b \leqq L_a$, so that $L_a = L_b$. If $y \in S$, then since $L_a \leqq L_b$ implies that $a = sb$ with $s \in S^1$, we have $a = sxay \in Say$. Since S is left stable therefore $L(a) = L(ay)$. Hence $L(a) \subseteq L(b) = L(xay) \subseteq L(ay)$ implies that $L(a) = L(b)$. Thus L_b is minimal in the set of $\mathscr{L}$-classes in J_a. This shows that left stability implies M_L^*.

Conversely, let S satisfy condition M_L^* and suppose that $a \in Sab$. Then $a \cup Sa \cup aS \cup SaS \subseteq Sab \cup SabS$, so that $J(a) \subseteq J(ab)$. Since always $J(ab) \subseteq J(a)$, we have $J(a) = J(ab)$. Now $a \in Sab$ immediately implies $L_a \leqq L_{ab}$, and hence, from Lemma 6.41, since $J_a = J_{ab}$, $L_a = L_{ab}$, i.e., $L(a) = L(ab)$. Thus S is left stable; and this completes the proof of the lemma.

A semigroup S is said to be *left* [*right*] *elementary* if the equation $\lambda \cap \mathscr{R} = \mathscr{L} \cap \mathscr{R}$ [$\mathscr{L} \cap \rho = \mathscr{L} \cap \mathscr{R}$] holds in S. It is said to be *elementary* if it is both left and right elementary.

LEMMA 6.43. *The semigroup S is left elementary if and only if $\lambda \cap \mathscr{D} = \mathscr{L}$, i.e., if and only if each $\mathscr{L}$-class in any $\mathscr{D}$-class is minimal in the set of $\mathscr{L}$-classes in that $\mathscr{D}$-class.*

PROOF. Let S be left elementary and let $(a, b) \in \lambda \cap \mathscr{D}$. Since $(a, b) \in \mathscr{D} = \mathscr{L} \circ \mathscr{R}$, there exists c in S such that $L_a = L_c$ and $R_c = R_b$. Thus $(c, b) \in \lambda \cap \mathscr{R} = \mathscr{L} \cap \mathscr{R}$, by hypothesis. Hence $L_b = L_c = L_a$. This shows that $\lambda \cap \mathscr{D} \subseteq \mathscr{L}$. Since the reverse inclusion always holds, we have $\lambda \cap \mathscr{D} = \mathscr{L}$, as required.

The converse immediately follows from the fact that $\mathscr{R} \subseteq \mathscr{D}$.

A semigroup is semisimple if each of its principal factors is either 0-simple or simple (§2.6, p. 74). We shall say that a semigroup is *completely semisimple* if each of its principal factors is either completely 0-simple or completely simple [Munn, 1957]. If $a \in S$ and $I(a) \neq \square$, then L_a^0 will denote the subset $L_a \cup \{I(a)\}$ of $S/I(a)$.

LEMMA 6.44. *If $I(a) \neq \square$, then the $\mathscr{L}$-class L_a of the semigroup S is minimal in the set of $\mathscr{L}$-classes in J_a if and only if L_a^0 is a* (0-*minimal*) *left ideal of $S/I(a)$.*

If $I(a) = \square$, then L_a is minimal in the set of $\mathscr{L}$-classes in J_a if and only if L_a is a (*minimal*) *left ideal of S.*

PROOF. L_a is clearly minimal in the set of $\mathscr{L}$-classes in J_a if and only if there exists no left ideal of S properly contained between $L(a) \cup I(a)$ and $I(a)$. From this the assertions in the lemma immediately follow.

REMARK. Lemma 6.41 can be deduced as a corollary to Lemma 6.44. For, if $I(a) \neq \square$, $J(a)/I(a)$ is a 0-minimal two-sided ideal of $S/I(a)$ (cf. Lemma 2.39). Hence, by Theorem 2.33, $J(a)/I(a)$ contains a 0-minimal left ideal of $S/I(a)$ if and only if it is a union of 0-minimal left ideals of $S/I(a)$. Thus, by the preceding lemma, J_a contains a minimal $\mathscr{L}$-class if and only if it is a union of minimal $\mathscr{L}$-classes. Similar remarks apply to the case $I(a) = \square$.

The first part of the next theorem rounds out the theorem of Wallace and Koch [1957] that $\mathscr{J} = \mathscr{D}$ in a stable semigroup. The equivalence of (A) and (D) when S is semisimple is due to Munn [1957].

THEOREM 6.45. *The following conditions on a semigroup S are equivalent.*

(A) *S satisfies both M_L^* and M_R^*.*

(B) *S is elementary and $\mathscr{J} = \mathscr{D}$.*

(C) *S is stable.*

If S is semisimple, then any one of these conditions is equivalent to

(D) *S is completely semisimple.*

PROOF. The equivalence of (A) and (C) follows from Lemma 6.42 and its left-right dual. That (B) implies (A) follows from Lemmas 6.41 and 6.43. By the same lemmas, (A) implies that S is elementary. All that remains for the first part of the theorem is to show that (A) implies $\mathscr{J} = \mathscr{D}$.

Assume (A), and let $a\mathscr{J}b$ $(a, b \in S)$. Then $b = xay$ for some x, y in S^1. Then $J_a = J_{xa} = J_b$. From $L_{xa} \leqq L_a$ and Lemma 6.41, we conclude that $L_{xa} = L_a$. From $R_b \leqq R_{xa}$ and the dual of Lemma 6.41, we conclude that $R_b = R_{xa}$. Hence $a\mathscr{L}xa$ and $xa\mathscr{R}b$, whence $a\mathscr{D}b$.

Now assume that S is semisimple and satisfies (A). The principal factors of S are the semigroups $J(a)/I(a)$ [$J(a)$, if $I(a) = \square$]. By Lemma 6.44 and its dual, $S/I(a)$ contains 0-minimal left and right ideals which are contained in $J(a)/I(a)$ (and a similar assertion holds if $I(a) = \square$). Consequently, by Corollary 2.50, since $J(a)/I(a)$ is a 0-minimal two-sided ideal of $S/I(a)$, and, by hypothesis is 0-simple, $J(a)/I(a)$ is completely 0-simple. Similarly, when $I(a) = \square$, we conclude that $J(a)$ is completely simple. Thus (D) holds.

Conversely, assume (D). It suffices to prove any one of the three equivalent conditions (A), (B), (C), and we shall prove (A). By hypothesis, each principal factor $J(a)/I(a)$ of S is completely [0-] simple. It follows from Theorem 2.35 [and its dual] that the $\mathscr{L}$- [$\mathscr{R}$-] classes in J_a with respect to $J(a)/I(a)$ are also $\mathscr{L}$- [$\mathscr{R}$-] classes with respect to S. Since every non-zero $\mathscr{L}$-class and every non-zero $\mathscr{R}$-class in a completely [0-] simple semigroup is minimal in its $\mathscr{J}$-class, (A) follows by definition of M_L^* and M_R^*.

As a corollary we have the following result of Green [1951].

COROLLARY 6.46. *Let S be a semisimple semigroup satisfying both M_L and M_R. Then S is completely semisimple and hence $\mathscr{J} = \mathscr{D}$ in S.*

A semigroup which is regular is necessarily semisimple for then each principal factor contains non-zero idempotents. On the other hand for any set X, $\mathscr{T}_X$ is regular (Exercise 1 for §2.2), but if X is infinite then $\mathscr{T}_X$ is not completely semisimple (see Example 3 below). A further result of Munn [1957] is

LEMMA 6.47. *Let S be a regular semigroup. Then S satisfies M_L^* if and only if it satisfies M_R^*.*

PROOF. This follows directly from Lemma 6.38. For, by Lemma 6.44, M_L^* holds if and only if the principal factors of S all contain 0-minimal (or minimal) left ideals, and, by Lemma 6.38 (and its analogue for semigroups without zero) and Lemma 1.13, a regular semigroup contains a 0-minimal [minimal] left ideal if and only if it contains a 0-minimal [minimal] right ideal.

Consequently we have shown that the following theorem holds.

THEOREM 6.48. *The following assertions about a semigroup S are equivalent.*

(A) *S is completely semisimple.*
(B) *S is regular and satisfies one of M_L^* and M_R^*.*
(C) *S is regular, either right or left elementary and $\mathcal{J} = \mathcal{D}$.*

We conclude with a discussion of the interdependence of the conditions M_L, M_R and M_J. Firstly, we give a result of Green [1951].

THEOREM 6.49. *Let S satisfy M_L and M_R. Then S satisfies M_J.*

PROOF. Suppose that S does not satisfy M_J. Then there exists a properly descending infinite chain of $\mathcal{J}$-classes, $J_1 > J_2 > J_3 > \cdots > J_n > \cdots$. Let $a_i \in J_i$ $(i = 1, 2, \ldots)$. Then since $J_i > J_{i+1}$, there exist x_i, y_i in S^1 such that $a_{i+1} = x_i a_i y_i$ $(i = 1, 2, \ldots)$. Thus

$$a_{i+1} = x_i x_{i-1} \cdots x_1 a_1 y_1 y_2 \cdots y_i.$$

Now

$$R(a_1) \supseteq R(a_1 y_1) \supseteq \cdots \supseteq R(a_1 y_1 y_2 \cdots y_n) \supseteq \cdots,$$

and

$$L(x_1) \supseteq L(x_2 x_1) \supseteq \cdots \supseteq L(x_n x_{n-1} \cdots x_1) \supseteq \cdots.$$

Hence if M_L and M_R hold in S (and so also in S^1) there must exist an integer m such that

$$R(a_1 y_1 y_2 \cdots y_m) = R(a_1 y_1 y_2 \cdots y_m y_{m+1})$$

and

$$L(x_m x_{m-1} \cdots x_1) = L(x_{m+1} x_m \cdots x_1).$$

Thus there exist u, v in S^1 such that

$$a_1 y_1 y_2 \cdots y_{m+1} v = a_1 y_1 y_2 \cdots y_m,$$

and

$$u x_{m+1} x_m \cdots x_1 = x_m x_{m-1} \cdots x_1.$$

Hence

$$u x_{m+1} \cdots x_1 a_1 y_1 \cdots y_{m+1} v = x_m \cdots x_1 a_1 y_1 \cdots y_m,$$

i.e. $u a_{m+2} v = a_{m+1}$. But this implies that $J_{m+1} \leqq J_{m+2}$. This is a contradiction. Hence, when M_L and M_R both hold in S, so also does M_J.

Apart from the dependence exhibited in the above theorem, the conditions

M_L, M_R and M_J are independent. The existence of simple semigroups without either minimal left or minimal right ideals (for example, see Exercises 8–10 for §2.1 (Andersen [1952])) shows that M_J does not imply either M_L or M_R.

The example of Munn (Example 1, below) shows that M_L [M_R] may hold in a semigroup without either M_R [M_L] or M_J holding. Further M_L [M_R] and M_J may both hold without M_R [M_L] holding: for example the Baer-Levi semigroups discussed in §8.1 are right simple, and so satisfy M_R and M_J, but do not satisfy M_L.

Example 1. Let S be a set consisting of an element 0 and of all ordered pairs (i, j) where i and j are positive integers such that $i < j$. Define a product in S by the rules

$$(i, j)(r, s) = \begin{cases} (i, s) & \text{if } j = r, \\ 0 & \text{if } j \neq r; \end{cases}$$

$$0x = 0 = x0, \quad \text{for all } x \text{ in } S.$$

Then S is a semigroup with this product (in fact S is a subsemigroup of a semigroup of matrix units (§2.7, Exercise 7). The left ideal generated by (i, j) is

$$\{0\} \cup \{(r, j) \colon 1 \leqq r \leqq i\};$$

since this is a finite set, S satisfies M_L. Similarly, the right ideal R_{ij} generated by (i, j) is

$$R_{ij} = \{0\} \cup \{(i, s) \colon s \geqq j\},$$

and the two-sided ideal I_{ij} generated by (i, j) is

$$I_{ij} = \{0\} \cup \{(r, s) \colon 1 \leqq r \leqq i, s \geqq j\}.$$

The strictly descending sequences

$$R_{ij} \supset R_{i,j+1} \supset R_{i,j+2} \supset \cdots,$$

$$I_{ij} \supset I_{i,j+1} \supset I_{i,j+2} \supset \cdots,$$

can both be continued indefinitely. Thus S satisfies M_L but neither M_R nor M_J.

Note further that I_{ij}, $I_{i,j+1}$, $I_{i,j+2}, \cdots$, is an infinite strictly descending sequence of left ideals of S. Thus the condition M_L is weaker than the minimal condition on the set of all the left ideals of a semigroup. (Munn [1957].)

Example 2. Let S be the 0-direct union of the infinite set of 0-simple semigroups $\{S_i \colon i \in I\}$. Then S satisfies M_J but does not satisfy the minimal condition on the set of all two-sided ideals of S.

EXAMPLE 3. For any set X the semigroup $\mathcal{T}_X$ is regular (§2.2, Exercise 1(c)). Further, in $\mathcal{T}_X$, $\mathcal{D} = \mathcal{J}$ (Theorem 2.9(i)). Let $|X|$ be infinite and let D be the $\mathcal{D}$-class determined by $|X|$, i.e. let D consist of all elements α of $\mathcal{T}_X$ such that $|X\alpha| = |X|$ (Theorem 2.9). The $\mathcal{L}$-classes of $\mathcal{T}_X$ in D are in one-to-one correspondence with the subsets Y of X such that $|Y| = |X|$, the $\mathcal{L}$-class $L(Y)$ corresponding to Y consisting of all α such that $X\alpha = Y$ (Theorem 2.9). We easily have that $L(Y_1) \leqq L(Y_2)$ if and only if $Y_1 \subseteq Y_2$. It follows that D contains no minimal $\mathcal{L}$-class. Hence $\mathcal{T}_X$ is regular but not completely semisimple. (Preston [1958].)

EXAMPLE 4. Let S be a semigroup in which every element is of finite order. Then $\mathcal{D} = \mathcal{J}$ in S. For let $(a, b) \in \mathcal{J}$ so that there exist x, y, z, u in S^1 such that

$$xay = b, \qquad zbu = a.$$

Let r be an integer chosen so that $(uy)^r$ is idempotent. Then, since $(xz)^r b(uy)^r = b$, we have $b(uy)^r = b$. Hence, putting $c = bu$, $(b, c) \in \mathcal{R}$. Similarly, putting $d = zb$, $(b, d) \in \mathcal{L}$. Hence, also, $(bu, du) \in \mathcal{L}$, i.e. $(c, a) \in \mathcal{L}$. Thus, $(a, b) \in \mathcal{D}$. (Green [1951].)

EXAMPLE 5. (a) Let $J(a) \supseteq J(b)$, where a, b are elements of the semigroup S. Then, if the principal factor $J(b)/I(b)$ is 0-simple (or, simple, when $I(b) = \square$), $b \in J_b a J_b$.

(b) Let S be a semigroup which is semisimple and which satisfies M_L. Let $J(a_1) \supseteq J(a_2) \supseteq \cdots \supseteq J(a_n) \supseteq \cdots$ be a descending sequence of principal ideals. Then there exist x_i, y_i in J_{a_i} such that $a_i = x_i a_{i-1} y_i$, $i = 2, 3, \cdots$. Condition M_L implies that, for some integer n, $L(a_n) = L(x_{n+1}a_n) = L(x_{n+2}x_{n+1}a_n) = \cdots$. It follows that $J(a_n) = J(a_{n+1}) = J(a_{n+2}) = \cdots$. Thus a semisimple semigroup satisfying M_L also satisfies M_J. (Munn [1957].)

EXERCISES FOR §6.6

1. According to the original definition of Wallace and Koch [1957] a semigroup S is *stable* if

(1) $a, b \in S$ and $Sa \subseteq Sab$ imply $Sa = Sab$ and
(2) $a, b \in S$ and $aS \subseteq baS$ imply $aS = baS$.

A semigroup stable in this sense is always stable in the sense of the text, but not conversely. For semigroups with an identity element and for regular semigroups the two definitions of stability coincide.

2. A semigroup S is stable (in fact in the stronger sense of Wallace and Koch given in Exercise 1) if it has any one of the following properties:

(a) S is commutative;
(b) S is a union of groups;
(c) S is periodic.

(Wallace and Koch [1957].)

3. Let A be an ideal of the semigroup S. Then S satisfies M_L [M_R] if and only if both A and S/A satisfy M_L [M_R]. (Munn [1957]; T. Saitô, Math. Japon. **13**(1968), 95–104.) (See Appendix.)

4. Let $\mathscr{C}$ be the set of cyclic subsemigroups of S. $\mathscr{C}$ can be partially ordered by inclusion. Let M_C denote the corresponding minimal condition, i.e., M_C holds in S if every non-empty subset of $\mathscr{C}$ contains a minimal element. Then M_C holds in S if and only if every element of S is of finite order. (Green [1951].)

5. Let S be a semigroup satisfying M_J. Then, for any a in S, $J(a^k) = J(a^{2k})$ for some integer k (depending on a). Consequently, putting $b = a^k$ and $c = a^{2k}$, $c = b^2 \in J_b$. Hence, in a semigroup satisfying M_J, some power of each element belongs to a simple or 0-simple principal factor. (Green [1951].)

6. The following conditions on a semigroup S are equivalent.

(A) S is completely semisimple.
(B) S satisfies M_L^* and each $\mathscr{J}$-class of S contains an idempotent.
(C) S is regular and satisfies M_L^*.

(Munn [1957].)

7. A semigroup is a union of groups if and only if it is left regular (cf. Theorem 4.2) and it satisfies M_R^*. (Munn [1957].)

8. If a completely semisimple semigroup satisfies any one of the conditions M_L, M_R, M_J, then it also satisfies the others. (Munn [1957].)

In the exercises which follow we consider some of the ideal radicals that have been defined for semigroups in analogy with the radicals of a ring. In §11.6 we shall examine radicals of a semigroup S which are congruences on S. The radicals we consider here involve an ideal I (the radical) of the semigroup S such that, under the right conditions, S/I has some property and such that I is the least ideal giving S/I this property. There is an analogy with maximal homomorphic images of a semigroup with some given property (cf. §1.5, §4.3, and §11.6). For further details we refer to the authors cited.

9. Let (α) be some property which holds in a semigroup if and only if each principal factor of the semigroup has property (β) where (β) is some property which holds for the one-element semigroup. Let I be the intersection of all ideals A of S such that S/A has property (α). Then, if I is nonempty, S/I has property (α), and we may call I the (α)-*radical*. For example, let (α) be the property of semisimplicity and let (β) be the property of being simple or 0-simple. Then the corresponding (α)-radical is what Munn has termed the *upper radical*. (Munn [1957].)

10. Let M be a two-sided ideal of the semigroup S. An ideal A of S is said to be M-*potent* if $A^n \subseteq M$ for some integer n. Any one-sided M-potent ideal is contained in a two-sided M-potent ideal. Let N_M be the union of

all M-potent ideals of S. N_M is called the *M-radical* of S. If N_M is M-potent, i.e. if S contains a maximal M-potent ideal, then S/N_M contains no nilpotent ideals. (Schwarz [1943] and [1951].)

11. Suppose that the semigroup S has an upper radical N' and an M-radical N_M (cf. Exercises 9 and 10). Then $N_M \subseteq N' \cup M$. In particular, if S has a kernel K, then $N_K \subseteq N'$. (Munn [1957].)

12. Let S be the semigroup of three elements z, a, b with multiplication table

	z	a	b
z	z	z	z
a	z	a	a
b	z	a	$a.$

Then $K = \{z\}$ is the kernel of S, $N_K = K$ and $N' = S$ (cf. Exercise 11). (Munn [1955].)

13. An ideal A of a semigroup $S = S^0$ is said to be a *nilideal* if some power of each element in A is zero. Let $N(S)$ denote the union of all the nilideals of S. Then $N_K \subseteq N(S)$. (Clifford [1949].) Further $N_K = N(S)$ if and only if $N(S/N_K) = 0$, the zero of S/N_K. (Luh [1960].)

14. Let $S = S^0$ be a semigroup which is its own right socle. Let N_K be the K-radical, where $K = \{0\}$ is the kernel of S (Exercise 10). In the notation of Theorem 6.23, $N_K = A$. Further S/N_K has no nilpotent ideals (other than $\{0\}$) and Corollary 6.34 applies. (Schwarz [1951].)

15. Let S be a semigroup satisfying both M_L and M_R and also satisfying the ascending chain condition for two-sided ideals, i.e. such that any set of two-sided ideals contains a maximal element. Then S satisfies M_J and so contains a kernel K, which, since S satisfies M_L^* and M_R^*, is completely simple. The maximal condition on two-sided ideals implies that the K-radical N_K is K-potent. Theorem 6.29 then applies to S/N_K giving $\Sigma_l \cup \Sigma_r = C$, since D, D' and E are here each $\{0\}$, i.e., $\Sigma_l \cup \Sigma_r$ (in S/N_K) is the 0-direct union of completely 0-simple semigroups. (Cf. Schwarz [1943] and [1951], and Munn [1957].)

CHAPTER 7

INVERSE SEMIGROUPS

Inverse semigroups are regular semigroups in which each element has a unique inverse. An introduction to their properties was given in §1.9. In particular, the fundamental embedding theorem (Theorem 1.20) was there proved: any inverse semigroup can be embedded in a symmetric inverse semigroup $\mathscr{I}_X$.

After a section treating the natural partial order on an inverse semigroup, we take up again the theme of representations of inverse semigroups by one-to-one partial transformations of a set. The results we present are due to B. M. Šaĭn [1962] and §§7.2 and 7.3 are devoted to their presentation. A key part in the argument is played by the principal (partial) right congruences of Dubreil, a detailed discussion of which is given in §10.2. The properties of these partial right congruences relevant to our immediate purpose are given in §7.2. They will be recognised as the generalization to inverse semigroups, appropriate to representation theory, of the right congruences on a group which correspond to the decomposition of a group into right cosets modulo a subgroup.

In §7.3, it is shown that the transitive representations of an inverse semigroup, by one-to-one partial transformations of a set, are those determined by the partial right congruences introduced in §7.2. Any representation is shown to split uniquely into transitive representations. The section ends with a characterization of equivalent representations. The general problem of representing an arbitrary semigroup by one-to-one partial transformations is taken up again in §11.4, where once more the results are due to Šaĭn [1961].

The basic properties of homomorphisms and congruences on an inverse semigroup are discussed in §7.4. Fundamental is the fact that any homomorphic image of an inverse semigroup is also an inverse semigroup (Theorem 7.36). A congruence on an inverse semigroup is determined by the congruence classes containing idempotents (Theorem 7.38). The precise nature of this determination is the subject of the latter half of this section.

§7.5 discusses semilattices of inverse semigroups, while §7.6 is devoted to homomorphisms which separate idempotents. The final section, §7.7, treats homomorphisms of inverse semigroups onto primitive inverse semigroups, i.e. inverse semigroups in which each non-zero idempotent is primitive. Such homomorphisms have proved important in recent work on matrix representations of inverse semigroups.

7.1 The natural partial order on an inverse semigroup

One-to-one partial transformations (§1.9, p. 29) of the set X, regarded as subsets of $X \times X$, are partially ordered by inclusion. Any inverse semigroup has a faithful representation as a semigroup of one-to-one partial transformations (Theorem 1.20). Consequently any inverse semigroup has a partial ordering induced by such a representation. If α and β are one-to-one partial transformations of X then we easily see that $\alpha \subseteq \beta$ if and only if $\alpha\beta^{-1} = \alpha\alpha^{-1}$. Since the partial ordering can be thus equationally defined it follows that the same partial ordering is induced on an inverse semigroup by any faithful representation as a semigroup of one-to-one partial transformations. We now proceed to discuss this partial order abstractly following V. V. Vagner [1952].

The relation $\leqq$ defined on the inverse semigroup S by $a \leqq b$ if and only if $ab^{-1} = aa^{-1}$ is called the *natural partial order* on S. From the above remarks we see that $\leqq$ is in fact a partial order on S. Below we will re-derive this directly from the definition, without appealing to the representation as a semigroup of one-to-one partial transformations. Throughout the remainder of this section $\leqq$ will denote the natural partial order on an inverse semigroup S. It will also be denoted by ω in §7.2.

Lemma 7.1. *If a, b belong to the inverse semigroup S then $a \leqq b$ if and only if any one of the following equations hold*: (1) $ab^{-1} = aa^{-1}$; (1′) $ba^{-1} = aa^{-1}$; (2) $a^{-1}b = a^{-1}a$; (2′) $b^{-1}a = a^{-1}a$; (3) $ab^{-1}a = a$; (3′) $a^{-1}ba^{-1} = a^{-1}$.

Proof. For each k, $k = 1, 2, 3$, (k') is obtained from (k) by taking inverses, and so (k) is equivalent to (k').

Equation (1) immediately gives $(ab^{-1})a = (aa^{-1})a$. Thus, since $aa^{-1}a = a$, (1) implies (3). Further (3′) implies that $(a^{-1}b)(a^{-1}b) = a^{-1}b$, i.e. $a^{-1}b$ is an idempotent. Hence $(a^{-1}b)(a^{-1}a) = a^{-1}aa^{-1}b = a^{-1}b$, since idempotents commute. Thus (3′) implies that $a^{-1}b = a^{-1}ba^{-1}a = a^{-1}a$, i.e. (3′) implies (2).

To complete the proof it will suffice to show that (2′) implies (1). Observe firstly that (2′) implies that ab^{-1} is idempotent. For we have $a(b^{-1}a)b^{-1} = a(a^{-1}a)b^{-1} = ab^{-1}$. It then follows that $ab^{-1} = (aa^{-1}a)b^{-1} = (aa^{-1})(ab^{-1}) = (ab^{-1})(aa^{-1}) = a(b^{-1}a)a^{-1} = a(a^{-1}a)a^{-1} = aa^{-1}$, i.e. that (2′) implies (1).

Lemma 7.2. *The binary relation $\leqq$ is a compatible partial ordering on the inverse semigroup S.*

Proof. It is clear that the relation $\leqq$ is reflexive. Suppose that $a \leqq b$ and $b \leqq a$. Then $a = ab^{-1}a$, $ab^{-1} = bb^{-1}$ and $b^{-1}a = b^{-1}b$, using Lemma 7.1. Hence

$$a = (ab^{-1})a = b(b^{-1}a) = bb^{-1}b = b.$$

Thus $\leqq$ is anti-symmetric.

Suppose that $a \leqq b$ and $b \leqq c$. Then using Lemma 7.1, as appropriate, we have

$$ac^{-1}a = a(a^{-1}a)c^{-1}a = aa^{-1}(bc^{-1})a$$
$$= a(a^{-1}b)(b^{-1}a) = aa^{-1}aa^{-1}a = a,$$

so that $a \leqq c$. Thus $\leqq$ is transitive.

It remains to show that the partial ordering $\leqq$ is compatible with the product on S. Let $a \leqq b$ and let x be any element of S. Then

$$(xa)(xb)^{-1} = x(ab^{-1})x^{-1} = x(aa^{-1})x^{-1} = (xa)(xa)^{-1},$$

and

$$(ax)^{-1}(bx) = x^{-1}(a^{-1}b)x = x^{-1}(a^{-1}a)x = (ax)^{-1}(ax);$$

so that $xa \leqq xb$ and $ax \leqq bx$.

This completes the proof of the lemma.

Lemma 7.3. *The binary relation $\leqq$ on the inverse semigroup S is compatible with inversion in S i.e. $a \leqq b$, for a, b in S, implies $a^{-1} \leqq b^{-1}$.*

Proof. This follows directly from Lemma 7.1: in detail, from the fact that (1) implies (2).

We note here that the natural partial ordering of an inverse semigroup was used in Volume 1 in Rees' proof of Theorem 1.23 on embedding a right reversible cancellative semigroup in a group.

We will say that a is *less than* or *under* b and that b is *greater than* or *over* a if $a \leqq b$. If e and f are two idempotents then $e \leqq f$ if and only if $ef\ (= fe) = e$. Thus the natural partial order on an inverse semigroup S coincides, when restricted to the semilattice of idempotents of S, with the natural partial ordering $\leqq$ of this semilattice (§1.8, p. 24).

We complete this section by proving a generalization of a result by A. E. Liber [1954]. We need a preliminary lemma.

Lemma 7.4. *An inverse semigroup S is a union of groups if and only if the left and right units* of each element of S are equal.*

Proof. If S is an inverse semigroup which is a union of groups then the inverse of any element lies in the group to which the element belongs. It follows that the left and right units of the element then coincide with the identity element of this group, and so are equal.

Conversely, if $aa^{-1} = e = a^{-1}a$, then a belongs to the maximal subgroup H_e of S (Exercise 3 for §1.7).

For the case of a chain of length 2 or of length 3 the following theorem is due to A. E. Liber [loc. cit.]. (See Appendix.) Exercise 3 below shows that it is not in general true for infinite E.

Theorem 7.5. *Let S be an inverse semigroup with a finite set E of idempotents. If E forms a chain (under its natural ordering), then S is a union of groups.*

---

* Recall (Vol. 1, p. 30) that $aa^{-1}[a^{-1}a]$ is called the left [right] unit of a.

PROOF. Let $a \in S$. Let $e = aa^{-1}$, $f = a^{-1}a$. Then $ea = af = a$, $a^{-1}e = fa^{-1} = a^{-1}$. By hypothesis, either $e \leqq f$ or $f \leqq e$, say the latter. Then $ef = fe = f$, and so

$$ae = afe = af = a, \qquad ea^{-1} = (ae)^{-1} = a^{-1}.$$

Thus e is the identity of the inverse subsemigroup S_a of S generated by a. From Lemma 1.31 it then follows that if $f \neq e$, then S_a is a bicyclic semigroup. But a bicyclic semigroup has an infinite number of distinct idempotents. This is contrary to the hypotheses on S. Consequently we must have $f = e$; and the theorem follows from the preceding lemma.

EXERCISES FOR §7.1

1. If the set of idempotents of a semigroup S forms a semilattice then the set T of regular elements of S forms a subsemigroup and T is an inverse semigroup. [Vagner [1953].)
2. Let a, b belong to the inverse semigroup S. Let e be the right unit of ab and f the right unit of b. Then $e \leqq f$.
3. In a bicyclic semigroup the set of idempotents is a chain under its natural ordering, and the bicyclic semigroup is not a union of groups. (Cf. Theorem 7.5.)
4. Let S be an inverse semigroup. Let $x \leqq b$ and $y \leqq b$; then there exists z such that $z \leqq x$ and $z \leqq y$ (for example, $z = yy^{-1}x$). Thus the partially ordered set $\{x \in S \colon x \leqq b\}$ is directed (downwards).

7.2 PARTIAL RIGHT CONGRUENCES ON AN INVERSE SEMIGROUP

In §7.3 we give an account of B. M. Šaĭn's [1962] general theory of representations of an inverse semigroup by one-to-one partial transformations of a set. For this theory Šaĭn introduced a generalization to inverse semigroups of the coset decompositions of a group. We discuss these equivalences in this section.

It will be convenient to change our notation for the natural partial order on an inverse semigroup. Throughout this section ω will denote the natural partial order on the inverse semigroup in question. Let S be an inverse semigroup. Then, if $s \in S$, $s\omega$ denotes the set $\{x \in S \colon s\,\omega\,x\}$ (see §1.4, p. 14). Extending this notation, if H is any subset of S, we write

$$H\omega = \bigcup\{h\omega \colon h \in H\}.$$

We will call $H\omega$ the *closure* of H (under ω). A subset H which is its own closure will be said to be *closed*.

LEMMA 7.6. *Let H be a subset of the inverse semigroup S. Let $s \in S$. Then*

$$(H\omega)s \subseteq (Hs)\omega.$$

PROOF. Let $z \in (H\omega)s$. Then $z = xs$ where, for some h in H, $h\,\omega\,x$. Since ω is compatible (Lemma 7.2), therefore $hs\,\omega\,xs = z$, i.e. $z \in (Hs)\omega$.

LEMMA 7.7. *Let H be a subsemigroup of the inverse semigroup S. Let $h \in H$. Then*

$$(H\omega)h \subseteq H\omega.$$

PROOF. This follows immediately from the preceding lemma, since $Hh \subseteq H$.

LEMMA 7.8. *Let H be a subset of the inverse semigroup S. Let $s \in S$. Then*

$$(Hs)\omega = ((H\omega)s)\omega.$$

PROOF. From Lemma 7.6 we have that

$$((H\omega)s)\omega \subseteq (Hs)\omega^2 = (Hs)\omega.$$

Conversely, since $H \subseteq H\omega$, the opposite inclusion is immediate.

Recall that an inverse subsemigroup of the inverse semigroup S is a subsemigroup H which admits inversion, i.e. which is such that $H^{-1} \subseteq H$ (§1.9, p. 30).

LEMMA 7.9. *Let H be an inverse subsemigroup of the inverse semigroup S. Then $H\omega$ is a closed inverse subsemigroup of S.*

PROOF. It is clear that $H\omega$ is closed, because ω is transitive. Let $x, y \in H\omega$ so that there exist h_1, h_2 in H such that $h_1\,\omega\,x$ and $h_2\,\omega\,y$. From the compatibility of ω we then have that $h_1h_2\,\omega\,xy$. Since H is a subsemigroup therefore $xy \in H\omega$. This shows that $H\omega$ is a subsemigroup.

Further, consider $x \in H\omega$, so that $h\,\omega\,x$ for some h in H. By Lemma 7.3, $h^{-1}\,\omega\,x^{-1}$, whence, since $h^{-1} \in H$, we conclude $x^{-1} \in H\omega$. This completes the proof of the lemma.

An extension of our terminology and notation about equivalences will be useful. Let S be a set and T a subset of S. Let ρ be an equivalence on T. Then ρ will also be called a *partial equivalence on* S, T will be called the *domain* of ρ, $S \backslash T$ its *contra-domain* and, where convenient, the set T/ρ of ρ-classes will be denoted by S/ρ. It is easily verified that a binary relation ρ on S is a partial equivalence on S if and only if it is symmetric and transitive. The domain of the partial equivalence ρ on S is the set $S\rho$ ($= \{x \in S: s\,\rho\,x$ for some s in $S\}$).

Now let S be a semigroup and let ρ be a partial equivalence on S. Then ρ is said to be *right* [*left*] *compatible* on S if, for each s in S, $a\,\rho\,b$ implies either that $as\,\rho\,bs$ [$sa\,\rho\,sb$] or that neither as nor bs [sa nor sb] belongs to the domain of ρ. A partial equivalence on S which is right [left] compatible on S is called a *partial right* [*left*] *congruence on* S. Let τ be any right congruence on S and let T be the union of a set of τ-classes. Let ρ denote τ restricted to T. Then ρ is a partial right congruence on S with contra-domain $S\backslash T$. Conversely, if ρ is a partial right congruence on S with domain T, then there exist right congruences τ on S whose restriction to T is ρ. This result is given as Corollary 10.5.

Note the following detail of technique which will be needed several times. If e is any idempotent of the inverse semigroup S and $x, y \in S$, then $xe\,\omega\,x$, $ex\,\omega\,x$ and $xey\,\omega\,xy$.

THEOREM 7.10. *Let K be an inverse subsemigroup of the inverse semigroup S. Let ω denote the natural partial order on S. Put $K\omega = H$ and define π_K thus:*

$$\pi_K = \{(s, t) \in S \times S \colon st^{-1} \in H\}.$$

Then π_K is a partial right congruence on S and the domain of π_K is the set $D_K = \{s \in S \colon ss^{-1} \in H\}$.

The equivalence classes modulo π_K are the sets $(Ks)\omega$ $(= (Hs)\omega)$ for s in D_K. $(Hs)\omega$ is the equivalence class which contains s and, in particular, H is one of the π_K-classes.

PROOF. It follows from Lemma 7.9 that H is a closed inverse subsemigroup of S. In particular, since $H^{-1} \subseteq H$, $st^{-1} \in H$ implies $ts^{-1} \in H$. Thus π_K is symmetric. Let (s, t) and (t, u) belong to π_K. Then $st^{-1} \in H$ and $tu^{-1} \in H$ and so, since H is a subsemigroup, $st^{-1}tu^{-1} \in H$. Since $t^{-1}t$ is an idempotent, $st^{-1}tu^{-1}\,\omega\,su^{-1}$. Hence $su^{-1} \in H\omega = H$, since H is closed. Thus $(s, u) \in \pi_K$; and this shows that π_K is transitive. Hence π_K is a partial equivalence on S.

It is clear that D_K is the domain of π_K. Let $(a, b) \in \pi_K$, so that $ab^{-1} \in H$, and let $s \in S$. If $as \in D_K$, then $as(as)^{-1} \in H$. Hence $as(as)^{-1} \cdot ab^{-1} = ass^{-1}a^{-1}ab^{-1} = aa^{-1}ass^{-1}b^{-1} = as(bs)^{-1} \in H$. Thus $(as, bs) \in \pi_K$ and $bs \in D_K$. This shows that π_K is right compatible.

It remains to show that the sets $(Ks)\omega$, $s \in D_K$, are the π_K-classes. Observe firstly, that from Lemma 7.8 we have, for any s in S, $(Ks)\omega = ((K\omega)s)\omega$, i.e. $(Ks)\omega = (Hs)\omega$. Let $x \in (Hs)\omega$, $s \in D_K$. Then there exists h in H such that $hs\,\omega\,x$. Hence $hss^{-1}\,\omega\,xs^{-1}$. Now $ss^{-1} \in H$, since $s \in D_K$. Thus $xs^{-1} \in H\omega = H$, i.e. $x\,\pi_K\,s$. Consequently $(Hs)\omega$ is contained in the π_K-class which contains s. Suppose, conversely, that $x\,\pi_K\,s$, i.e. that $xs^{-1} \in H$. Then $xs^{-1}s \in Hs$, and, since $s^{-1}s$ is an idempotent, $x \in (Hs)\omega$. This completes the proof of the theorem.

In the case when the inverse semigroup S of Theorem 7.10 is a group, then K is a subgroup, the natural partial order reduces to equality and π_K is the right congruence on S whose equivalence classes are the right cosets of K. In general we will say that the sets $(Ks)\omega$, for s in D_K, are the *right ω-cosets* of K.

We now amplify Šaĭn's account and obtain a characterization of the partial right congruences of Theorem 7.10.

Lemma 7.11. *Under the hypotheses of Theorem 7.10, the partial right congruence* π_K *satisfies the three conditions*:

(i) *precisely one* π_K*-class, namely* $K\omega$, *contains idempotents*;

(ii) *each* π_K*-class is closed* (*under* ω);

(iii) π_K *is right cancellative, in the sense that* $(ax, bx) \in \pi_K$ *implies that* $(a, b) \in \pi_K$.

Proof. (i) Let e be an idempotent belonging to some π_K-class. Then $(e, e) \in \pi_K$ and so $e = ee^{-1} \in H$ $(= K\omega)$. Hence H is the only π_K-class containing any idempotents. That H does in fact contain idempotents follows from the fact that H is an inverse semigroup.

(ii) This is clear from the theorem because $\omega^2 = \omega$.

(iii) Let $(ax, bx) \in \pi_K$, i.e. suppose that $ax(bx)^{-1} \in H$. Then $axx^{-1}b^{-1} \in H$, and since xx^{-1} is idempotent and H is closed, therefore $ab^{-1} \in H$, i.e. $(a, b) \in \pi_K$. Thus π_K is right cancellative.

A partial right congruence ρ on an inverse semigroup which has properties (i), (ii) and (iii) of the preceding lemma, i.e. which is such that (i) precisely one ρ-class contains idempotents, (ii) each ρ-class is closed and (iii) is right cancellative will be called a *principal* partial right congruence on S.

The converse of Lemma 7.11 holds and we have

Theorem 7.12. (i) *Let* K *be an inverse subsemigroup of the inverse semigroup* S. *Let* ω *denote the natural partial order on* S. *Define* π_K *thus*:

$$\pi_K = \{(s, t) \in S \times S: st^{-1} \in K\omega\}.$$

Then π_K *is a principal partial right congruence on* S.

(ii) *Let* ρ *be a principal partial right congruence on* S. *Let* K *be the* ρ*-class which contains idempotents. Then* K *is a closed inverse subsemigroup of* S *and* $\rho = \pi_K$.

Proof. Part (i) of the theorem is Lemma 7.11.

To prove part (ii) consider an element (s, t) of the principal partial right congruence ρ. Since $t = tt^{-1}t$ and ρ is a partial right congruence, $(st^{-1}t, tt^{-1}t) \in \rho$. Since ρ is right cancellative, therefore $(st^{-1}, tt^{-1}) \in \rho$. But tt^{-1} is an idempotent and so belongs to the ρ-class K which contains idempotents. Hence also $st^{-1} \in K$. Thus $(s, t) \in \rho$ implies $st^{-1} \in K$.

To show the converse we first prove that K is a closed inverse subsemigroup of S. That K is closed is one of the hypotheses on ρ. If $k \in K$, then $(k, k) \in \rho$, so that, by the result just proved, $kk^{-1} \in K$. Let $h, k \in K$. Then, since $kk^{-1} \in K$ and K is a ρ-class, $(h, kk^{-1}) \in \rho$. Hence $(hk, k) \in \rho$, since $k = kk^{-1}k \in K$. Thus $hk \in K$. This shows that K is a subsemigroup. To see that K admits inverses, consider $k \in K$, so that $(k, kk^{-1}) \in \rho$, whence $(kk^{-1}, kk^{-1}k^{-1}) \in \rho$. From $(kk^{-1})k^{-1}\ \omega\ k^{-1}$ and $kk^{-1}k^{-1} \in K$, it follows, since K is closed, that $k^{-1} \in K$.

Thus K is a closed inverse subsemigroup of S and $st^{-1} \in K$ if and only if

$(s,t)\in\pi_K$. Hence we have shown that $\rho\subseteq\pi_K$. Conversely, let $(s,t)\in\pi_K$. Then $st^{-1}\in K$. By Theorem 7.10, $(s,t)\in\pi_K$ implies that $tt^{-1}\in K$. Thus $(st^{-1},tt^{-1})\in\rho$. Hence $(s,t)\in\rho$, by right cancellation. Thus $\pi_K\subseteq\rho$; and we deduce that $\rho=\pi_K$.

This completes the proof of the theorem.

We now give a characterization of principal partial right congruences, alternative to that of Šaĭn, which identifies them with principal partial right congruences in the sense of Dubreil [1941].

Let S be any semigroup. Let H be a subset of S. For any element a in S define $a^{[-1]}H$ and $Ha^{[-1]}$ thus:

$$a^{[-1]}H=\{x\in S\colon ax\in H\}$$
$$Ha^{[-1]}=\{x\in S\colon xa\in H\}.$$

Other notations in use are $H.\cdot a$ for $a^{[-1]}H$, and $H\cdot.a$ for $Ha^{[-1]}$. Define the binary relations $\mathscr{R}_H$ and $\mathscr{R}_H^*$ thus:

$$\mathscr{R}_H=\{(a,b)\in S\times S\colon a^{[-1]}H=b^{[-1]}H\},$$
$$\mathscr{R}_H^*=\{(a,b)\in S\times S\colon a^{[-1]}H=b^{[-1]}H\neq\square\}.$$

We then have the following result (Dubreil [1941]).

LEMMA 7.13. *Let H be a subset of the semigroup S. Then $\mathscr{R}_H$ is a right congruence on S and $\mathscr{R}_H^*$ is a partial right congruence on S. The contra-domain W_H of $\mathscr{R}_H^*$ is, if non-empty, an $\mathscr{R}_H$-class and $\mathscr{R}_H^*$ coincides with the restriction of $\mathscr{R}_H$ to the domain of $\mathscr{R}_H^*$.*

PROOF. It is clear that $\mathscr{R}_H$ is an equivalence on S and that $\mathscr{R}_H^*$ is a partial equivalence on S. Let $(a,b)\in\mathscr{R}_H$. Then $a^{[-1]}H=b^{[-1]}H$, and so, for any fixed element c in S, $acx\in H$ if and only if $bcx\in H$, in other words, $(ac,bc)\in\mathscr{R}_H$. Thus $\mathscr{R}_H$ is a right congruence on S.

Let $W_H=\{x\in S\colon x^{[-1]}H=\square\}$. Then W_H is clearly the contra-domain of $\mathscr{R}_H^*$ and, if non-empty, is an $\mathscr{R}_H$-class. It is further clear that $\mathscr{R}_H^*$ is $\mathscr{R}_H$ restricted to $S\backslash W_H$.

The relation $\mathscr{R}_H$ is called the *principal right congruence* (determined by H) and $\mathscr{R}_H^*$ is called the *principal partial right congruence* (determined by H). Their general discussion will be taken up again in §10.2. Meanwhile the following lemma justifies our terminology.

LEMMA 7.14. *Let H be a closed inverse subsemigroup of the inverse semigroup S. Then $\pi_H=\mathscr{R}_H^*$, and the π_H-classes are the non-empty sets $Ha^{[-1]}$, $a\in S$. When $Ha^{[-1]}$ is non-empty it is the π_H-class containing a^{-1}.*

PROOF. Let $(a,b)\in\pi_H$, i.e. suppose that $ab^{-1}\in H$. Then $b^{-1}\in a^{[-1]}H$, so that $a^{[-1]}H\neq\square$. For any x such that $ax\in H$ we then have $ax(ax)^{-1}\cdot ab^{-1}\in H$. Hence $aa^{-1}axx^{-1}b^{-1}=axx^{-1}b^{-1}\in H$. Thus $(ax,bx)\in\pi_H$. Since H is a π_H-class (Lemma 7.11(i)), therefore $bx\in H$. Similarly it follows that $bx\in H$ implies $ax\in H$. Hence $a^{[-1]}H=b^{[-1]}H\neq\square$, i.e. $(a,b)\in\mathscr{R}_H^*$. Thus $\pi_H\subseteq\mathscr{R}_H^*$.

Conversely, let $(a, b) \in \mathscr{R}_H^*$. Then, for some $x, ax, bx \in H$. Hence $(ax, bx) \in \pi_H$, and since π_H is right cancellative (Lemma 7.11(iii)), $(a, b) \in \pi_H$. Thus $\mathscr{R}_H^* \subseteq \pi_H$ which, combined with the earlier inequality, gives $\mathscr{R}_H^* = \pi_H$.

Let $(a, b) \in \pi_H$, so that $ax, bx \in H$, for some x. Then $a, b \in Hx^{[-1]}$. Let $y \in Hx^{[-1]}$, so that $yx \in H$. Then $ax(yx)^{-1} = axx^{-1}y^{-1} \in H$. Hence, since H is closed, $ay^{-1} \in H$, i.e. $(a, y) \in \pi_H$. Thus the π_H-class containing a is $Hx^{[-1]}$.

Further, let $Ha^{[-1]} \neq \square$. Then, for some z, $za \in H$, i.e. $z(a^{-1})^{-1} \in H$, i.e. $(z, a^{-1}) \in \pi_H$. Hence, as before, $Ha^{[-1]}$ is the π_H-class containing a^{-1}.

Exercises for §7.2

1. Let ρ be both a partial left and a partial right congruence on S. Let A, B be two ρ-classes. Then either AB is contained in a ρ-class or AB is contained in the contra-domain of ρ.

2. Let F be a closed subset of an inverse semigroup S. Suppose that F contains an idempotent and that $FF^{-1}F \subseteq F$. Then F is an inverse subsemigroup of S.

3. Let H be a closed inverse subsemigroup of the inverse semigroup S. Let D be the domain of π_H. Then, either $S \backslash D$ is empty or $S \backslash D$ is a right ideal of S. Hence $\pi_H \cup ((S \backslash D) \times (S \backslash D))$ is a right congruence on S whose restriction to D is π_H.

4. A subset H of a semigroup S is said to be *strong* (§10.2) if, for any a, b in S, $a^{[-1]}H \cap b^{[-1]}H \neq \square$ implies that $a^{[-1]}H = b^{[-1]}H$.

Let H be an inverse subsemigroup of the inverse semigroup S. Then H is strong if and only if it is closed.

7.3 Representations by one-to-one partial transformations

Recall that $\mathscr{I}_X$ denotes the symmetric inverse semigroup on X, i.e. the semigroup of all one-to-one partial transformations of the set X. Theorem 1.20 shows that any inverse semigroup S can be represented faithfully as a semigroup of one-to-one partial transformations of a set, in fact that S can be embedded in $\mathscr{I}_S$. In this section we give a general discussion of representations, not necessarily faithful, of inverse semigroups as semigroups of one-to-one partial transformations.

The results we give are due to B. M. Šaĭn [1962]. A comparison may be made with §11.4, where Šaĭn's [1961] theory of representations of an arbitrary semigroup by one-to-one partial transformations is presented. Another approach, for inverse semigroups, has been given by N. R. Reilly [1965]; we describe this briefly at the end of §7.6. (See Appendix.)

It will simplify matters if, for this section, we agree that by a *representation* of an inverse semigroup S we mean a homomorphism ϕ of S into some symmetric inverse semigroup $\mathscr{I}_X$.

It is generally true, as we shall see below (§7.4), that a homomorphic image of an inverse semigroup is itself an inverse semigroup and that inverses are mapped by a homomorphism onto inverses. For the special case of a representation this result can be easily shown as follows.

LEMMA 7.15. *Let $\phi: S \to \mathscr{I}_X$ be a representation of S. Then for any s in S, $(s\phi)^{-1} = s^{-1}\phi$. Hence $S\phi$ is an inverse subsemigroup of $\mathscr{I}_X$.*

PROOF. Let $s \in S$. Then $s = ss^{-1}s$ and $s^{-1}ss^{-1} = s^{-1}$. Hence

$$s\phi = s\phi \cdot s^{-1}\phi \cdot s\phi \quad \text{and} \quad s^{-1}\phi \cdot s\phi \cdot s^{-1}\phi = s^{-1}\phi.$$

Thus, in the inverse semigroup $\mathscr{I}_X$, $s^{-1}\phi$ is the unique inverse of $s\phi$ (§1.9, p. 28). However $(s\phi)^{-1}$ is the inverse of $s\phi$ in $\mathscr{I}_X$; and thus $(s\phi)^{-1} = s^{-1}\phi$.

Let H be an inverse subsemigroup of $\mathscr{I}_X$. Define

$$\tau_H = \{(a, b) \in X \times X: (a, b) \in \eta \text{ for some } \eta \text{ in } H\}.$$

Then τ_H is called the *transitivity relation* of H.

LEMMA 7.16. *If H is an inverse subsemigroup of $\mathscr{I}_X$, then its transitivity relation τ_H is a partial equivalence on X.*

PROOF. The transitivity of τ_H follows from the fact that $H^2 \subseteq H$, using the definition of multiplication in $\mathscr{I}_X$. Similarly, $H^{-1} \subseteq H$ implies that τ_H is symmetric.

The equivalence classes of τ_H are called the *transitivity classes* of H. When τ_H is the universal equivalence on its domain, i.e. when $\tau_H = (X\tau_H) \times (X\tau_H)$, then H is said to be *transitive* (on $X\tau_H$). Thus H is transitive if and only if, for each pair of elements x_1, x_2 in $X\tau_H$, there is some element h of H which, regarded as a partial transformation of X, maps x_1 onto x_2. H is said to be an *effective* subsemigroup of $\mathscr{I}_X$ if $X\tau_H = X$.

Let ϕ be a representation of the inverse semigroup S in $\mathscr{I}_X$. Then ϕ is said to be *transitive* if $S\phi$ is transitive, to be *effective* if $S\phi$ is effective.

Let $\phi: S \to \mathscr{I}_X$ and $\psi: S \to \mathscr{I}_Y$ be two representations of S. Then ϕ and ψ are said to be *equivalent* if there is a one-to-one mapping θ, say, of X onto Y such that for $x, x' \in X$, and for $s \in S$, $(x, x') \in s\phi$ if and only if $(x\theta, x'\theta) \in s\psi$, i.e., $(x(s\phi))\theta = (x\theta)s\psi$, whenever either side is defined.

Let $\phi_i: S \to \mathscr{I}_{X_i}$, $i \in I$, be a family of representations of the inverse semigroup S, and suppose that the sets X_i are pairwise disjoint. If not satisfied, this latter condition can be ensured by replacing each ϕ_i by an equivalent representation. Then by the *direct sum* or, simply, the *sum* of the representations ϕ_i is meant the representation ϕ defined by

$$s\phi = \bigcup\{s\phi_i: i \in I\}.$$

Then $\phi: S \to \mathscr{I}_X$, where $X = \bigcup\{X_i: i \in I\}$. It is easily seen that $S\phi$ is isomorphic to a subdirect product of all the $S\phi_i$, $i \in I$ (see remarks preceding Exercise 10 for §9.4). Further, $s\phi = t\phi$ if and only if $s\phi_i = t\phi_i$ for all i.

Hence

$$\phi \circ \phi^{-1} = \bigcap \{\phi_i \circ \phi_i^{-1} : i \in I\}.$$

Again, if we put $S\phi = H$ and $S\phi_i = H_i$, $i \in I$, then

$$\tau_H = \bigcup \{\tau_{H_i} : i \in I\}.$$

THEOREM 7.17. *An effective representation of an inverse semigroup S is the sum of a uniquely determined family of transitive effective representations of S.*

PROOF. Let $\phi: S \to \mathscr{I}_X$ be an effective representation of S. Put $S\phi = H$ and let τ_H be the transitivity relation of H. Let X_i, $i \in I$, be the transitivity classes of H. Then, since ϕ is effective,

$$X = \bigcup \{X_i : i \in I\};$$
$$(1) \qquad \tau_H = \bigcup \{X_i \times X_i : i \in I\}.$$

For each i in I define the mapping $\phi_i: S \to \mathscr{I}_{X_i}$, thus:

$$s\phi_i = s\phi \bigcap (X_i \times X_i), \quad s \in S.$$

Finally, put $S\phi_i = H_i$, $i \in I$.

Then each ϕ_i is an effective and transitive representation of S. To show firstly that ϕ_i is a representation consider $(st)\phi_i$, for s, t in S. Then $(a,b) \in (st)\phi_i$ if and only if $(a,b) \in X_i \times X_i$ and $(a,b) \in (st)\phi = (s\phi)(t\phi)$. Now $(a,b) \in (s\phi)(t\phi)$ if and only if $(a,c) \in s\phi$ and $(c,b) \in t\phi$ for some c in X. $(a,c) \in s\phi$ implies that $(a,c) \in \tau_H$, and so, since $a \in X_i$, also $c \in X_i$. Thus $(a,b) \in (st)\phi_i$ if and only if, for some c, $(a,c) \in s\phi \cap (X_i \times X_i) = s\phi_i$ and $(c,b) \in t\phi \cap (X_i \times X_i) = t\phi_i$. In other words, $(a,b) \in (st)\phi_i$ if and only if $(a,b) \in (s\phi_i)(t\phi_i)$. Thus ϕ_i is a representation of S.

Let a, b be any two elements of X_i. Since $(a,b) \in \tau_H$, there exists s in S such that $(a,b) \in s\phi$. Hence $(a,b) \in s\phi_i$. Thus $\tau_{H_i} = X_i \times X_i$, i.e. H_i, and so ϕ_i, is transitive. This also shows simultaneously that ϕ_i is effective.

The definition of the ϕ_i and of the X_i implies that

$$s\phi = \bigcup \{s\phi_i : i \in I\}.$$

Thus ϕ is the sum of the family $\{\phi_i\}$ of effective transitive representations of S.

It remains to prove the uniqueness of the family $\{\phi_i\}$. To this end suppose that ϕ is also the sum of the family $\{\psi_j : j \in J\}$ of effective transitive representations of S, where $\psi_j: S \to \mathscr{I}_{Y_j}$ and the Y_j are pairwise disjoint sets. It follows immediately, from the definition of a sum of representations that

$$X = \bigcup \{Y_j : j \in J\}.$$

Put $S\psi_j = K_j$. Since ψ_j is transitive and effective, for each j,

$$\tau_{K_j} = Y_j \times Y_j.$$

Consequently,

$$\tau_H = \bigcup\{\tau_{K_j}: j \in J\} = \bigcup\{Y_j \times Y_j: j \in J\}. \tag{2}$$

Comparing (1) and (2) it now follows that the set $\{Y_j\}$ coincides with the set $\{X_i\}$. From the further fact that, for each j,

$$s\psi_j = s\phi \bigcap (Y_j \times Y_j), \quad s \in S,$$

it now follows the families $\{\psi_j\}$ and $\{\phi_i\}$ coincide.

Having split up any effective representation into a sum of transitive effective representations we now study further the transitive representations and obtain Šaĭn's characterization of such representations using the principal partial right congruences of the preceding section.

Let $\phi: S \to \mathscr{I}_X$ be a representation of S. Let $x \in X$ and put

$$S_x = \{s \in S: (x, x) \in s\phi\}.$$

S_x is called the set of elements of S which *fix* x (under ϕ).

LEMMA 7.18. *Let $\phi: S \to \mathscr{I}_X$ be a representation of the inverse semigroup S. Let $x \in X$. Then S_x, if non-empty, is a closed inverse subsemigroup of S. Clearly, S_x is non-empty if ϕ is effective.*

PROOF. Clearly if $s, t \in S_x$ then $(x, x) \in (s\phi)(t\phi) = (st)\phi$ and so $st \in S_x$. Further $(x, x) \in s\phi$ implies $(x, x) \in (s\phi)^{-1} = s^{-1}\phi$ (Lemma 7.15). Thus S_x is an inverse subsemigroup of S.

Now suppose that $s\,\omega\,t$, where ω is the natural order on S, and that $s \in S_x$. Then $ts^{-1} = ss^{-1}$ and so $(t\phi)(s^{-1}\phi) = (s\phi)(s^{-1}\phi)$. Since $(x, x) \in s\phi \cap s^{-1}\phi$, it follows that $(x, x) \in t\phi$. Thus S_x is closed.

This completes the proof of the lemma.

Let H be any closed inverse subsemigroup of the inverse semigroup S. Let π_H denote the principal partial right congruence determined by H. Let $\mathscr{X}$ be the set of right ω-cosets of H, i.e. $\mathscr{X}$ is the set of π_H-classes, the set of all $(Hs)\omega$ for s such that $ss^{-1} \in H$ (Theorem 7.10), where ω is the natural partial order on S. Define $\phi_H: S \to \mathscr{B}_{\mathscr{X}}$, where $\mathscr{B}_{\mathscr{X}}$ denotes the semigroup of all binary relations on $\mathscr{X}$ (§1.4), by

$$s\phi_H = \{(X, (Xs)\omega): X \in \mathscr{X} \text{ and } (Xs)\omega \in \mathscr{X}\}, \quad \text{for } s \in S. \tag{3}$$

LEMMA 7.19. *Let H be a closed inverse subsemigroup of the inverse semigroup S. Then ϕ_H, as defined by* (3), *is an effective transitive representation of S in $\mathscr{I}_{\mathscr{X}}$ where $\mathscr{X}$ is the set of right ω-cosets of H. The subset H of S is the set of elements of S which fix the element H of $\mathscr{X}$.*

PROOF. Since π_H is a partial right congruence, for any s in S, either Xs is contained in the contra-domain of π_H or Xs is contained in an element of $\mathscr{X}$. In the latter case, we easily see that this element is $(Xs)\omega$. For let $X = (Ht)\omega$. Then, applying Lemma 7.8 to the set Ht, we obtain $(Xs)\omega =$

$(((Ht)\omega)s)\omega = ((Ht)s)\omega = (Hts)\omega$. Now $(Hts)\omega$ is an element of $\mathscr{X}$ if and only if $(Ht)\omega s = Xs$ is contained in an element of $\mathscr{X}$; which proves our assertion.

From the fact that π_H is right cancellative (Lemma 7.11) it follows that $s\phi_H$ is one-to-one (we say that the empty relation is one-to-one). Hence ϕ_H maps S into $\mathscr{I}_{\mathscr{X}}$. The element $X = (Ht)\omega$ of $\mathscr{X}$ is mapped by $t^{-1}\phi_H$ into $(Htt^{-1})\omega = H$ and, in turn, if $ss^{-1} \in H$ then H is mapped into $(Hs)\omega$ by $s\phi_H$. Once we have shown that ϕ_H is a representation of S, this shows, simultaneously, that ϕ_H is both transitive and effective.

Let $s, t \in S$. Then $(X, Y) \in (st)\phi_H$ if and only if $Y = (Xst)\omega$ and Xst is contained in a π_H-class. Since π_H is right cancellative, Xst is contained in a π_H-class only if Xs is contained in a π_H-class. Hence $(X, Y) \in (st)\phi_H$ if and only if $(X, Z) \in s\phi_H$ and $(Z, Y) \in t\phi_H$, where $Z = (Xs)\omega \in \mathscr{X}$. Consequently, $(st)\phi_H = (s\phi_H)(t\phi_H)$. Thus ϕ_H is a representation.

It remains to determine the subset S_H of S which fixes the element H of $\mathscr{X}$. Clearly, since $Hh \subseteq H$ for h in H, $H \subseteq S_H$. Conversely, let s be any element of S_H. Then $(Hs)\omega = H$. By Theorem 7.10, $(Hs)\omega$ is the π_H-class which contains s. Hence $s \in H$. Thus $S_H \subseteq H$. Consequently, $S_H = H$, as required.

LEMMA 7.20. *Let $\phi: S \to \mathscr{I}_Y$ be a transitive effective representation of S. Let y be an element of Y and put $H = S_y$, the set of elements of S which fix y. Then ϕ is equivalent to ϕ_H.*

PROOF. Lemma 7.18 shows that S_y is a closed inverse subsemigroup of S, so the notation ϕ_H is justified. Let $\mathscr{X}$ denote the set of right ω-cosets of H. Since ϕ is effective and transitive, for any y' in Y there exists an element s of S such that $y(s\phi) = y'$. Define θ, thus:

$$y'\theta = (Hs)\omega, \quad \text{if} \quad y' \in Y \quad \text{and} \quad y' = y(s\phi).$$

Note, firstly, that $(Hs)\omega \in \mathscr{X}$. For $(y, y) \in (s\phi)(s\phi)^{-1} = (ss^{-1})\phi$. Hence $ss^{-1} \in H$, whence (Theorem 7.10) $(Hs)\omega \in \mathscr{X}$. In fact θ is a one-to-one mapping of Y onto $\mathscr{X}$.

To prove this, consider firstly s and t such that $y' = y(s\phi) = y(t\phi)$. Then $(y, y') \in s\phi \cap t\phi$. Thus $st^{-1} \in H$, i.e. $(s, t) \in \pi_H$ and so, by Theorem 7.10, $(Hs)\omega = (Ht)\omega$. This shows that θ is a mapping, and since ϕ is effective and transitive, θ maps Y into $\mathscr{X}$.

Now suppose that $y'\theta = y''\theta$, i.e. that $(Hs)\omega = (Ht)\omega$, say, where $y' = y(s\phi)$ and $y'' = y(t\phi)$. Then, by Theorem 7.10, $(s, t) \in \pi_H$ so that $st^{-1} \in H$. Hence $(y, y) \in (st^{-1})\phi = (s\phi)(t^{-1}\phi) = s\phi(t\phi)^{-1}$. Since $(y, y') \in s\phi$ and $(y, y'') \in t\phi$, it follows that $y' = y''$. This shows that θ is a one-to-one mapping.

Let $(Hs)\omega$ be any element of $\mathscr{X}$. Then $ss^{-1} \in H$ and so $(y, y) \in (ss^{-1})\phi = (s\phi)(s\phi)^{-1}$. Thus, for some y' in Y, $(y, y') \in s\phi$, whence it follows that $y'\theta = (Hs)\omega$. Thus θ is onto $\mathscr{X}$.

To show that ϕ and ϕ_H are equivalent it will suffice to show that for any s in S, $(y', y'') \in s\phi$ if and only if $(y'\theta, y''\theta) \in s\phi_H$. Choose t' and t'' so that

$(y, y') \in t'\phi$, $(y, y'') \in t''\phi$. Then $y'\theta = (Ht')\omega$ and $y''\theta = (Ht'')\omega$. Now $((Ht')\omega, (Ht'')\omega) \in s\phi_H$ if and only if $(Ht'')\omega = (Ht's)\omega$, i.e. if and only if $(t'', t's) \in \pi_H$ i.e. if and only if $t''(t's)^{-1} \in H$, i.e. if and only if $(y, y) \in (t''\phi)((t's)\phi)^{-1}$. But, since $(y, y'') \in t''\phi$, this holds if and only if $(y, y'') \in (t's)\phi = (t'\phi)(s\phi)$. Since $(y, y') \in t'\phi$, this holds if and only if $(y', y'') \in s\phi$. Thus ϕ and ϕ_H are equivalent; and this completes the proof of the lemma.

As a converse to Lemma 7.20 we have

LEMMA 7.21. *Let $\phi: S \to \mathscr{I}_Y$ be a representation of S. Let H be a closed inverse subsemigroup of S. Suppose that ϕ is equivalent to ϕ_H. Then there exists an element y_0, say, of Y, such that $H = S_{y_0}$, the set of elements of S which fix y_0.*

PROOF. By assumption there is a one-to-one mapping θ, say, of the right ω-cosets of H onto Y. H is a right ω-coset of H (Theorem 7.10). Put $H\theta = y_0$. Then, from the definition of equivalence, it follows that

$$(H(s\phi_H))\theta = y_0(s\phi),$$

and, since θ is one-to-one,

$$H(s\phi_H) = H \quad \text{if and only if } y_0(s\phi) = y_0.$$

Thus s fixes the right ω-coset H if and only if s fixes the element y_0. Hence, by Lemma 7.19, $H = S_{y_0}$.

Combining Lemmas 7.19–21 we have the following characterization of effective transitive representations of inverse semigroups. The representation ϕ_H of Lemma 7.19 will be called *the representation of S on the right ω-cosets of H.*

THEOREM 7.22. *Let H be any closed inverse subsemigroup of the inverse semigroup S. Then the representation of S on the right ω-cosets of H is effective and transitive. Furthermore, H is the set of all elements of S which fix the right ω-coset H.*

Let $\phi: S \to \mathscr{I}_Y$ be an effective transitive representation of S. Let $y \in Y$ and let $H = S_y$, the set of all elements of S which fix y under ϕ. Then H is a closed inverse subsemigroup and ϕ is equivalent to ϕ_H. If K is any closed inverse subsemigroup of S such that ϕ is equivalent to ϕ_K, then K is the set of all elements of S which fix some element of Y.

We now derive from this approach another proof of the fact that every inverse semigroup S has a faithful representation (as a semigroup of one-to-one partial transformations of a set). To this end we first find a characterization of the congruence $\phi_H \circ \phi_H^{-1}$. The relation defined in Lemma 7.23 is the principal congruence of Croisot [1957], which will be discussed in §10.4.

LEMMA 7.23. *Let H be a closed inverse subsemigroup of the inverse semi-*

group S, and let ϕ_H denote the representation of S on the right ω-cosets of H. Then

$$\phi_H \circ \phi_H^{-1} = \{(x, y) \in S \times S\colon sxt \in H \text{ if and only if } syt \in H, \text{ for } s, t \text{ in } S\}.$$

PROOF. From the definition (3) of ϕ_H it follows that $((Hs)\omega, (Ht^{-1})\omega) \in x\phi_H$ if and only if $sx \, \pi_H \, t^{-1}$, i.e. if and only if $sxt \in H$. Whence the assertion of the lemma immediately follows.

Let e be an idempotent of the inverse semigroup S. Then $e\omega$ is a closed inverse subsemigroup of S (Lemma 7.9). For $H = e\omega$, write ϕ_e for ϕ_H. If e and f are both idempotents then $e\omega = f\omega$ if and only if $e = f$.

THEOREM 7.24. *Let E denote the set of idempotents of the inverse semigroup S. Let ϕ be the sum of the family $\{\phi_e\colon e \in E\}$ of representations of S. Then ϕ is faithful.*

PROOF. Let $x\phi = y\phi$. Then $x\phi_e = y\phi_e$ for all $e \in E$. Hence, by Lemma 7.23, $sxt \in e\omega$ if and only if $syt \in e\omega$ for each e in E. Hence, in particular, $xx^{-1} \, \omega \, xx^{-1} \cdot x \cdot x^{-1}$ if and only if $xx^{-1} \, \omega \, xx^{-1} \cdot y \cdot x^{-1}$. Since $xx^{-1}yx^{-1} \, \omega \, yx^{-1}$, therefore $xx^{-1} \, \omega \, yx^{-1}$. Thus $x = xx^{-1}x \, \omega \, yx^{-1}x \, \omega \, y$, i.e. $x \, \omega \, y$. Similarly, $y \, \omega \, x$. Hence $x = y$; and this proves the theorem.

Let H and K be two closed inverse subsemigroups of the inverse semigroup S. Then H and K are said to be *conjugate* if ϕ_H is equivalent to ϕ_K.

From Theorem 7.22 we have immediately

LEMMA 7.25. *The closed inverse subsemigroups H and K of the inverse semigroup S are conjugate if and only if H is the set of all elements of S which fix some right ω-coset of K (under the representation on the right ω-cosets of K). In this case K is also the set of all elements of S which fix some right ω-coset of H (under the representation on the right ω-cosets of H).*

LEMMA 7.26. *Let H be a closed inverse subsemigroup of the inverse semigroup S. Let $aa^{-1} \in H$, so that $(Ha)\omega$ is a right ω-coset of H. Then the elements of $a^{-1}Ha$ fix the ω-coset $(Ha)\omega$.*

PROOF. This is clear because $(Ha)(a^{-1}Ha) \subseteq Ha$.

This lemma leads to a further characterization of conjugate subsemigroups.

THEOREM 7.27. *Let H and K be closed inverse subsemigroups of the inverse semigroup S. Then H and K are conjugate if and only if there is an element a in S such that*

$$a^{-1}Ha \subseteq K \quad \text{and} \quad aKa^{-1} \subseteq H.$$

Any such element a necessarily satisfies $aa^{-1} \in H$, $a^{-1}a \in K$.

PROOF. Suppose H and K are conjugate. By Lemma 7.25, there is a right ω-coset of H, $(Ha)\omega$, say, such that K is the set of elements of S which fix $(Ha)\omega$. From Lemma 7.26 we then infer that $a^{-1}Ha \subseteq K$. That $(Ha)\omega$ is a right ω-coset of H further gives that $aa^{-1} \in H$.

Again, from the fact that K fixes $(Ha)\omega$, $(HaK)\omega = (Ha)\omega$, and this means that each element of aK is equivalent to a modulo the principal partial right congruence π_H. Thus $aKa^{-1} \subseteq H$. Further, since $a^{-1}a$ clearly fixes $(Ha)\omega$, $a^{-1}a \in K$.

Conversely, suppose that there exists an element a such that $a^{-1}Ha \subseteq K$ and $aKa^{-1} \subseteq H$. Then $a(a^{-1}Ha)a^{-1} \subseteq aKa^{-1} \subseteq H$. Let e be any idempotent in H. Then $aa^{-1} \cdot e \cdot aa^{-1} = aa^{-1}e \in H$, and from the fact that H is closed, it follows that $aa^{-1} \in H$. (It similarly follows that $a^{-1}a \in K$.) Hence $(Ha)\omega$ is a right ω-coset of H. To complete the proof that H and K are conjugate it suffices to show that K is the set of elements of S which fixes $(Ha)\omega$.

Let x fix $(Ha)\omega$. Then $(Hax)\omega = (Ha)\omega$ and hence $axa^{-1} \in H$. Thus $a^{-1}axa^{-1}a \in a^{-1}Ha \subseteq K$, whence, since K is closed, $x \in K$. Conversely, if $k \in K$, then $aka^{-1} \in H$, i.e. $ak \,\pi_H\, a$, i.e. $(Hak)\omega = (Ha)\omega$, i.e. k fixes $(Ha)\omega$. The proof is complete.

If H is a closed inverse subsemigroup of S with no conjugates other than itself, then H is said to be *self-conjugate*. Denote, as in Theorem 7.10, the domain $\{s: ss^{-1} \in H\}$ of π_H by D_H. When H is self-conjugate, D_H is an inverse subsemigroup of S and ϕ_H differs only trivially from a representation of D_H. Further, $S\phi_H$ is a group. This we proceed to show.

LEMMA 7.28. *Let H be a closed inverse subsemigroup of the inverse semigroup S. Then H is self-conjugate if and only if $s^{-1}Hs \subseteq H$ for all s in D_H.*

PROOF. Let $s \in D_H$. Then $(Hs)\omega$ is a right ω-coset of H. By Lemma 7.26, $s^{-1}Hs$ fixes $(Hs)\omega$ and so by Theorem 7.22 and Lemma 7.25, $s^{-1}Hs$ is contained in a conjugate of H. If H is self-conjugate, it follows that $s^{-1}Hs \subseteq H$.

Conversely, suppose that $s^{-1}Hs \subseteq H$ for all s in D_H. Let K be a conjugate of H. Then, by Lemma 7.25, there exists a right ω-coset $(Ha)\omega$, say, of H, which is fixed by K. Thus if $k \in K$, $(Hak)\omega = (Ha)\omega$ and so $aka^{-1} \in H$. By hypothesis, since $a \in D_H$, $a^{-1}Ha \subseteq H$. Hence $a^{-1}(aka^{-1})a \in a^{-1}Ha \subseteq H$. But $a^{-1}a \cdot k \cdot a^{-1}a \,\omega\, k$, and so because H is closed, $k \in H$. Thus $K \subseteq H$.

Since $a^{-1}Ha \subseteq H$ and $aa^{-1} \in H$, therefore $a^{-1}aa^{-1}a = a^{-1}a \in H$. Hence $a^{-1} \in D_H$ and thus $aHa^{-1} \subseteq H$. Hence for any h in H, $ah \,\pi_H\, a$, i.e. $(Hah)\omega = (Ha)\omega$. Thus every element of H fixes $(Ha)\omega$. Thus $H \subseteq K$. Consequently $H = K$ and this shows that H is self-conjugate.

LEMMA 7.29. *Let H be a self-conjugate closed inverse subsemigroup of the inverse semigroup S. Then D_H is a closed inverse subsemigroup of S.*

PROOF. Let $s \in D_H$. By Lemma 7.28, $s^{-1}Hs \subseteq H$. Hence, since $ss^{-1} \in H$, $s^{-1}s = s^{-1}(ss^{-1})s \in H$. Thus $s^{-1} \in D_H$.

Let $s, t \in D_H$. Then $s^{-1} \in D_H$, as just shown, and so, by Lemma 7.28,

$sHs^{-1} \subseteq H$. Hence, since $tt^{-1} \in H$, $st(st)^{-1} = stt^{-1}s^{-1} \in sHs^{-1} \subseteq H$. Thus $st \in D_H$.

It remains to show that D_H is closed. To this end, consider $s \in D_H$ and t such that $s\,\omega\,t$. By Lemma 7.3, $s^{-1}\,\omega\,t^{-1}$ and so by Lemma 7.2 $ss^{-1}\,\omega\,tt^{-1}$. Hence, since $ss^{-1} \in H$ and H is closed, $tt^{-1} \in H$, i.e. $t \in D_H$. This completes the proof of the lemma.

Let S be any semigroup and U a subset of S. Then U is said to be *left* [*right*] *unitary in* S if $u \in U$ and $ux \in U$ [$xu \in U$], for x in S, together imply that $x \in U$. A subset U which is both left and right unitary in S is said to be *unitary in* S.

LEMMA 7.30. *Let H be a self-conjugate closed inverse subsemigroup of the inverse semigroup S. Then D_H is unitary in S.*

PROOF. Let u, $ux \in D_H$. Then $uxx^{-1}u^{-1} \in H$ and, by Lemma 7.28,

$$u^{-1}u \cdot xx^{-1} = u^{-1}(uxx^{-1}u^{-1})u \in u^{-1}Hu \subseteq H.$$

Since H is closed, therefore $xx^{-1} \in H$, i.e. $x \in D_H$. Thus D_H is left unitary.

That D_H is right unitary follows similarly.

LEMMA 7.31. *Let H be a self-conjugate closed inverse subsemigroup of the inverse semigroup S. Let $s \in S \backslash D_H$. Then $s\phi_H = \square$.*

PROOF. Let X, Y be right ω-cosets of H. Then $(X, Y) \in s\phi_H$ if and only if, for some x in D_H, $X = (Hx)\omega$ and $Y = (Hxs)\omega$. Thus $(X, Y) \in s\phi_H$ if and only if, for some x in D_H, $xs \in D_H$. Since D_H is unitary, by the preceding lemma, $xs \in D_H$ implies $s \in D_H$. This is contrary to assumption, and the lemma follows.

LEMMA 7.32. *Let H be a self-conjugate closed inverse subsemigroup of the inverse semigroup S. Let $s \in D_H$. Then $s\phi_H$ is a permutation of the set of right ω-cosets of H.*

PROOF. We know already that $s\phi_H$ is a one-to-one partial transformation of the set of right ω-cosets of H (Lemma 7.19). Let X be any right ω-coset of H, so that $X = (Hx)\omega$ for some x in D_H. By Lemma 7.29, $xs \in D_H$, and hence $(Hxs)\omega = Y$, say, is a right ω-coset of H. Thus, for any right ω-coset X of H there exists Y such that $(X, Y) \in s\phi_H$.

Similarly we see that $(Hxs^{-1})\omega = Z$, say, is a right ω-coset of H and that $(Z, X) \in s\phi_H$. This completes the proof of the lemma.

The previous four lemmas combine to give the following theorem.

THEOREM 7.33. *Let H be a self-conjugate closed inverse subsemigroup of the inverse semigroup S. Let $D_H = \{s\colon ss^{-1} \in H\}$. Let ϕ_H denote the representation of S on the right ω-cosets of H.*

Then

(i) D_H *is a closed unitary inverse subsemigroup of* S;

(ii) *if* $D_H = S$, *then* $S\phi_H$ *is a group of permutations of the right* ω*-cosets of* H;

(iii) *if* $S \backslash D_H = W_H \neq \square$, *then* $W_H\phi_H$ *is the empty mapping,* $D_H\phi_H$ *is a group of permutations of the right* ω*-cosets of* H *and hence* $S\phi_H$ *is a group with zero.*

The general topic of homomorphisms of a semigroup onto a group is taken up in Chapter 10. (See Exercises for §10.2.) Homomorphisms of inverse semigroups onto a group are considered further in §7.7.

Exercises 3 and 4 below present another characterization of self-conjugate closed inverse subsemigroups of an inverse semigroup.

Exercises for §7.3

1. Let S be a semilattice. Let $\phi: S \to \mathscr{I}_X$ be a transitive effective representation of S. Then $|X| = 1$. Hence S has no faithful transitive representation if $|S| > 2$. (Šaĭn, [1962].)

2. Let S denote a semilattice. Then no two distinct closed inverse subsemigroups of S are conjugate. If H is any closed inverse subsemigroup of S, π_H is the universal congruence on its domain, H.

3. A subsemigroup of an inverse semigroup is a closed inverse subsemigroup if and only if it is unitary.

4. A unitary subsemigroup H of an inverse semigroup S is self-conjugate if and only if, for any $x, y \in S$, $xy \in H$ implies $yx \in H$, i.e. if and only if H is reflexive (see §10.2).

5. Let R be an $\mathscr{R}$-class of an inverse semigroup S. For each a in S define a partial transformation ρ_a of R as follows. The domain of ρ_a is $R \cap Sa^{-1}$ ($= R \cap Saa^{-1}$), and for each x in $R \cap Sa^{-1}$, define $x\rho_a = xa$.

(a) Let e be the idempotent element of R (Corollary 2.19(i)). Then $x \in R \cap Saa^{-1}$ if and only if $ex = x$, $xx^{-1} = e$, and $xaa^{-1} = x$.

(b) ρ_a is a one-to-one mapping of $R \cap Saa^{-1}$ onto $R \cap Sa^{-1}a$, and $\rho_{a^{-1}}$ is its inverse.

(c) $\rho_a\rho_b = \rho_{ab}$ for all a, b in S. (As in the proof of Theorem 1.20, the difficult part of this is the equality of the two domains.)

(d) $\rho_R: a \to \rho_a$ is a transitive representation of S by one-to-one partial transformations. As R ranges over the $\mathscr{R}$-classes of S, ρ_R ranges over the transitive components (Theorem 7.17) of the Vagner-Preston representation of Theorem 1.20.

(e) If S is 0-bisimple, and $R \neq 0$, then ρ_R is faithful. (Reilly [1965], Theorem 1.31.) (See Appendix.)

7.4 Homomorphisms of inverse semigroups

We show firstly in this section that any homomorphic image of an inverse semigroup is also an inverse semigroup. Then we show that a congruence on an inverse semigroup is uniquely determined by the set of congruence classes containing idempotents. These results are due to V. V. Vagner [1953], and, independently, to G. B. Preston [1954a]. There are several ways in which a congruence can be constructed in terms of the congruence classes which determine it. We give here the method due to Preston [1954a].

Lemma 7.34. *Let $\phi: S \to S'$ be a homomorphism of the inverse semigroup S onto S'. Let e' be an idempotent of S'. Then $e'\phi^{-1}$ is an inverse subsemigroup of S.*

Proof. Let $a, b \in e'\phi^{-1}$. Then $(ab)\phi = a\phi \cdot b\phi = (e')^2 = e'$. Thus $ab \in e'\phi^{-1}$. Hence $e'\phi^{-1}$ is a subsemigroup of S. Further, let $a \in e'\phi^{-1}$ and put $x' = a^{-1}\phi$. From $aa^{-1}a = a$ and $a^{-1}aa^{-1} = a^{-1}$ we obtain $e'x'e' = e'$ and $x'e'x' = x'$. From $(aa^{-1})(a^{-1}a) = (a^{-1}a)(aa^{-1})$ we have $(e'x')(x'e') = (x'e')(e'x')$. Hence, since $e'e' = e'$,

$$\begin{aligned} e' &= e'x'e' = e'(x'e'x')e' = e'(x'e'e'x')e' \\ &= e'(e'x'x'e')e' = e'x'x'e' = x'e'x' = x'. \end{aligned}$$

Consequently, $a^{-1}\phi = e'$, i.e. $a^{-1} \in e'\phi^{-1}$. This completes the proof of the lemma.

Lemma 7.35. *Let $\phi: S \to S'$ be a homomorphism of the regular semigroup S onto S'. Then S' is regular.*

Proof. Let $a' = a\phi$ be any element of S'. Let x be an inverse of a in S. Then, clearly, $x\phi$ is an inverse in S' of a'. Since each element of S' has an inverse therefore S' is regular.

Theorem 7.36. *A homomorphic image of an inverse semigroup is an inverse semigroup. Moreover, in any homomorphism, the inverse of an element is mapped onto the inverse of the image of that element.*

Proof. Let $\phi: S \to S'$ be a homomorphic mapping of the inverse semigroup S onto the semigroup S'. By Lemma 7.35, S' is regular. By Lemma 7.34, every idempotent of S' is the image of an idempotent of S. It follows, since the idempotents of S commute one with another, that the idempotents of S' form a semilattice. That S' is an inverse semigroup now follows from Theorem 1.17.

It remains to show that for $a \in S$, $a^{-1}\phi = (a\phi)^{-1}$. From $aa^{-1}a = a$ and $a^{-1}aa^{-1} = a^{-1}$ we have

$$a\phi(a^{-1}\phi)a\phi = a\phi \quad \text{and} \quad a^{-1}\phi(a\phi)a^{-1}\phi = a^{-1}\phi.$$

Thus $a^{-1}\phi$ is the unique inverse of $a\phi$, i.e. $a^{-1}\phi = (a\phi)^{-1}$.

COROLLARY 7.37. *Let A be an ideal of the semigroup S. Then S is an inverse semigroup if and only if A and the Rees quotient S/A are both inverse semigroups.*

PROOF. Let S be an inverse semigroup and let A be an ideal of S. Since S/A is a homomorphic image of S, therefore, by the theorem, S/A is an inverse semigroup. Further, let $a \in A$. Then, since A is an ideal, $a^{-1}aa^{-1} \in A$. But $a^{-1}aa^{-1} = a^{-1}$. Hence $A^{-1} \subseteq A$. Thus A is an inverse semigroup.

Conversely, suppose that A is an ideal of the semigroup S and that A and S/A are both inverse semigroups. Let $a \in S$. If $a \in A$ then there exists a unique x in A such that $axa = a$ and $xax = x$. Further, since A is an ideal, $xax \in A$ for any x in S. Hence there is a unique x in S satisfying these equations. If $a \in S \backslash A$, then since S/A is an inverse semigroup, there exists a unique x in $S \backslash A$ such that $axa = a$ and $xax = x$. Further any such x in S must belong to $S \backslash A$, again since A is an ideal. Hence S is an inverse semigroup.

Let S be any semigroup and let $\mathscr{A} = \{A_i : i \in I\}$ be a set of pairwise disjoint subsets of S. $\mathscr{A}$ is said to be an [*left, right*] *admissible set of subsets of* S or, to be [*left, right*] *admissible in* S, if there exists a [left, right] congruence ρ on S such that each set $A_i, i \in I$, is a ρ-class. Conversely, any such ρ is said to *admit* $\mathscr{A}$. If $\mathscr{A}$ is [left, right] admissible and there is precisely one [left, right] congruence on S which admits $\mathscr{A}$ then $\mathscr{A}$ is said to be a [*left, right*] *normal set of subsets of* S or, to be [*left, right*] *normal in* S. For example, if S is a group and if ρ is a congruence on S, then any non-empty set of ρ-classes is normal in S. The general topic of admissible and normal sets will be taken up again in §10.1. At present we wish to show that the set consisting of all those ρ-classes which contain idempotents, where ρ is a congruence on an inverse semigroup S, is normal in S. This follows from the more general result of the next theorem.

THEOREM 7.38. *Let S be a regular (in particular, an inverse) semigroup and let ρ be a congruence on S. Let $\mathscr{A}$ be the set of ρ-classes which contain idempotents. Then $\mathscr{A}$ is normal in S.*

PROOF. Let σ be any congruence on S which admits $\mathscr{A}$. It will suffice to show that $\sigma = \rho$.

Let $(x, y) \in \sigma$. Since S is regular, there exist a, b in S such that $xax = x$, $axa = a$, $yby = y$, $byb = b$. It follows, in particular, that xa and by are idempotents.

Since σ is right compatible, $(xa, ya) \in \sigma$, and from left compatibility, similarly, $(bx, by) \in \sigma$. Now xa is an idempotent. Hence, by the assumption that σ admits $\mathscr{A}$, the σ-class containing xa coincides with the ρ-class containing xa. Thus $(xa, ya) \in \rho$; and a similar argument gives $(bx, by) \in \rho$.

Consequently, using the left and right compatibility of ρ, as appropriate, we have

$$\begin{aligned} x &= xax, \\ (xax, yax) &\in \rho, \\ yax &= ybyax, \\ (ybyax, ybxax) &\in \rho, \\ ybxax &= ybx, \\ (ybx, yby) &\in \rho, \\ yby &= y. \end{aligned}$$

Whence, from the transitivity of ρ, $(x, y) \in \rho$.

Consequently, $\sigma \subseteq \rho$. The argument can be repeated with σ and ρ interchanged, giving $\rho \subseteq \sigma$. Whence the theorem follows.

For inverse semigroups a stronger result holds.

THEOREM 7.39. *Let S be an inverse semigroup and let ρ be a left congruence on S. Let $\mathscr{A}$ be the set of ρ-classes which contain idempotents. Then $\mathscr{A}$ is left normal in S.*

PROOF. Let σ be any left congruence on S which admits $\mathscr{A}$. It will suffice to show that $\sigma = \rho$.

Let $(x, y) \in \sigma$. Proceeding as in the proof of the previous theorem but using this time only left compatibility we have

$$(x^{-1}x, x^{-1}y) \in \rho \quad \text{and} \quad (y^{-1}x, y^{-1}y) \in \rho.$$

Hence, in turn, we deduce

$$\begin{aligned} x &= xx^{-1}x, \\ (xx^{-1}x, xx^{-1}y) &\in \rho, \\ xx^{-1}y &= yy^{-1}xx^{-1}y, \end{aligned}$$

since idempotents in S commute,

$$\begin{aligned} (yy^{-1}xx^{-1}y, yy^{-1}xx^{-1}x) &\in \rho, \\ yy^{-1}xx^{-1}x &= yy^{-1}x, \\ (yy^{-1}x, yy^{-1}y) &\in \rho, \\ yy^{-1}y &= y. \end{aligned}$$

Hence, from the transitivity of ρ, $(x, y) \in \rho$.

As in the proof of Theorem 7.38, the theorem now follows.

By Theorem 7.38 and the definition of normality, a congruence ρ on an inverse semigroup S is uniquely determined by the normal set $\mathscr{A}$ of ρ-classes containing idempotents. We proceed to obtain a characterization of such sets $\mathscr{A}$ and derive a construction for the associated congruences.

The set $\mathscr{A} = \{A_i: i \in I\}$ is defined to be a *kernel normal system* of the inverse semigroup S if

(K1) each A_i is an inverse subsemigroup of S;

(K2) $A_i \cap A_j = \square$ if $i \neq j$;

(K3) each idempotent in S is contained in some element of $\mathscr{A}$;

(K4) for each $a \in S$ and $i \in I$, $a^{-1}A_i a \subseteq A_j$, for some j; we shall write $j = ia$, so that $a^{-1}A_i a \subseteq A_{ia}$;

(K5) if $a, ab, bb^{-1} \in A_i$, then $b \in A_i$.

Some variation is possible in the choice of these conditions. For some alternatives see Exercises 6 and 8 below.

Let ϕ be a homomorphism of the inverse semigroup S. Then the set of idempotents of $S/(\phi \circ \phi^{-1})$ will be called the *kernel* of ϕ (and of $\phi \circ \phi^{-1}$).

LEMMA 7.40. *Let ρ be a congruence on the inverse semigroup S. Then the kernel of ρ is a kernel normal system of S.*

PROOF. Let $\mathscr{A} = \{A_i: i \in I\}$ be the set of idempotents of S/ρ, i.e. $\mathscr{A}$ is the kernel of ρ. We have to show that $\mathscr{A}$ satisfies conditions K1–K5.

Condition K2 is immediate. Condition K1 follows from Lemma 7.34. Denote by $\rho^\natural$ the natural mapping of S onto S/ρ. Any idempotent in S is mapped by $\rho^\natural$ onto an idempotent of S/ρ; whence it follows that condition K3 holds. That condition K4 holds for $\mathscr{A}$ is almost obvious.

There remains condition K5. Let $a, ab, bb^{-1} \in A_i$ for some i in I, so that

$$a\rho^\natural = (a\rho^\natural)(b\rho^\natural) = (b\rho^\natural)(b\rho^\natural)^{-1} = A_i$$

an idempotent of S/ρ, recalling that, by Theorem 7.36, $b^{-1}\rho^\natural = (b\rho^\natural)^{-1}$. Whence it follows that

$$b\rho^\natural = (b\rho^\natural)(b\rho^\natural)^{-1}(b\rho^\natural) = A_i(b\rho^\natural) = (a\rho^\natural)(b\rho^\natural) = A_i.$$

Thus $b \in A_i$. This shows that condition K5 holds and completes the proof of the lemma.

We now proceed in the converse direction. Any kernel normal system $\mathscr{A} = \{A_i: i \in I\}$ of S determines a congruence $\rho_{\mathscr{A}}$ on S defined thus:

$$(1) \qquad \rho_{\mathscr{A}} = \{(a, b) \in S \times S: aa^{-1}, bb^{-1}, ab^{-1} \in A_i, \quad \text{for some } i \text{ in } I\}.$$

To show this some preliminary lemmas will be helpful. We use the notation of the statement of the conditions K1–K5.

LEMMA 7.41. *If $aa^{-1} \in A_i$, then $a^{-1}a \in A_{ia}$.*

PROOF. Using condition K4,

$$a^{-1}a = a^{-1}(aa^{-1})a \in a^{-1}A_i a \subseteq A_{ia}.$$

LEMMA 7.42. *If $ab^{-1} \in A_i$, then $A_{ia} = A_{ib}$.*

PROOF. Since A_i is an inverse subsemigroup of S (K1), $ab^{-1} \in A_i$ implies $ba^{-1} \in A_i$. Hence $(ab^{-1})(ba^{-1}) \in A_i$ and $(ba^{-1})(ab^{-1}) \in A_i$. Therefore

$$a^{-1}a \cdot b^{-1}b = a^{-1}a \cdot b^{-1}b \cdot a^{-1}a$$
$$= a^{-1}(ab^{-1} \cdot ba^{-1})a \in A_{ia},$$

and

$$b^{-1}b \cdot a^{-1}a = b^{-1}b \cdot a^{-1}a \cdot b^{-1}b$$
$$= b^{-1}(ba^{-1} \cdot ab^{-1})b \in A_{ib}.$$

Hence $A_{ia} \cap A_{ib} \neq \square$; so, by condition K2, $A_{ia} = A_{ib}$.

LEMMA 7.43. *If* $aa^{-1}, bb^{-1}, ab^{-1} \in A_i$, *then* $a^{-1}a, b^{-1}b, a^{-1}b \in A_{ia}$ $(= A_{ib})$.

PROOF. By Lemma 7.41, $a^{-1}a \in A_{ia}$, $b^{-1}b \in A_{ib}$, and by Lemma 7.42, $A_{ia} = A_{ib}$. It remains to show that $a^{-1}b \in A_{ia}$.

Put $x = b^{-1}b$, $y = a^{-1}b$. We have shown that $x \in A_{ia}$. Further

$$xy = b^{-1} \cdot ba^{-1} \cdot b \in b^{-1}A_ib \subseteq A_{ib} = A_{ia};$$

and

$$yy^{-1} = a^{-1} \cdot bb^{-1} \cdot a \in a^{-1}A_ia \subseteq A_{ia}.$$

Hence $x, xy, yy^{-1} \in A_{ia}$. Thus, by condition K5, $y = a^{-1}b \in A_{ia}$.

LEMMA 7.44. (i) $(ia)b = i(ab)$.
(ii) *If* $aa^{-1} \in A_i$, *then* $(ia)a^{-1} = i$.

PROOF. (i) $a^{-1}A_ia \subseteq A_{ia}$ implies that

$$(ab)^{-1}A_i(ab) = b^{-1}(a^{-1}A_ia)b \subseteq b^{-1}A_{ia}b \subseteq A_{(ia)b}.$$

But $(ab)^{-1}A_i(ab) \subseteq A_{i(ab)}$; hence from condition K2 it follows that $(ia)b = i(ab)$.

(ii) If $aa^{-1} \in A_i$ then $(aa^{-1})^{-1}A_i(aa^{-1}) \subseteq A_i$. However $(aa^{-1})^{-1}A_i(aa^{-1}) \subseteq A_{i(aa^{-1})}$; whence, by condition K2, $i = i(aa^{-1})$. From (i) it then follows that $(ia)a^{-1} = i$.

We can now prove our assertion about $\rho_{\mathscr{A}}$.

LEMMA 7.45. *Let* $\mathscr{A} = \{A_i : i \in I\}$ *be a kernel normal system of the inverse semigroup* S *and define the binary relation* $\rho_{\mathscr{A}}$ *as in* (1). *Then* $\rho_{\mathscr{A}}$ *is a congruence on* S.

PROOF. $\rho_{\mathscr{A}}$ is clearly reflexive by K3, and condition K1 implies that $\rho_{\mathscr{A}}$ is symmetric.

To show that $\rho_{\mathscr{A}}$ is transitive consider (a, b) and (b, c) in $\rho_{\mathscr{A}}$. Then there exists i in I such that aa^{-1}, bb^{-1}, cc^{-1}, ab^{-1}, bc^{-1} all belong to A_i. Thus to show that $(a, c) \in \rho_{\mathscr{A}}$ it suffices to show that $ac^{-1} \in A_i$.

Let $x = cb^{-1}$, $y = ac^{-1}$. Then $x = (bc^{-1})^{-1} \in A_i$. Further, $xy = cb^{-1}ac^{-1} \in cA_{ia}c^{-1}$, by Lemma 7.43. But, by Lemma 7.42, $A_{ia} = A_{ib} = A_{ic}$. Hence $cA_{ia}c^{-1} = cA_{ic}c^{-1} \subseteq A_{(ic)c^{-1}} = A_i$, by Lemma 7.44(ii). Thus $xy \in A_i$.

Finally, $yy^{-1} = ac^{-1}ca^{-1} \in aA_{ic}a^{-1}$, by Lemma 7.41. But $A_{ic} = A_{ia}$, and hence $aA_{ic}a^{-1} = aA_{ia}a^{-1} \subseteq A_{(ia)a^{-1}} = A_i$, by Lemma 7.44(ii). Thus $yy^{-1} \in A_i$. Hence we have shown that x, xy, $yy^{-1} \in A_i$, and so, by condition K5, $ac^{-1} = y \in A_i$. Consequently $\rho_{\mathscr{A}}$ is an equivalence on S.

Let $aa^{-1}, bb^{-1}, ab^{-1} \in A_i$. Then $(ca)(ca)^{-1} = c \cdot aa^{-1} \cdot c^{-1} \in cA_ic^{-1} \subseteq A_{ic^{-1}}$, and, similarly, $(cb)(cb)^{-1}$ and $(ca)(cb)^{-1}$ both belong to $A_{ic^{-1}}$. Consequently, $(a, b) \in \rho_{\mathscr{A}}$ implies $(ca, cb) \in \rho_{\mathscr{A}}$. Now Lemma 7.43 states that $(a, b) \in \rho_{\mathscr{A}}$ implies that $(a^{-1}, b^{-1}) \in \rho_{\mathscr{A}}$. Hence $(a, b) \in \rho_{\mathscr{A}}$ implies that $(c^{-1}a^{-1}, c^{-1}b^{-1}) \in \rho_{\mathscr{A}}$ which, in turn, implies that $(ac, bc) \in \rho_{\mathscr{A}}$. Thus $\rho_{\mathscr{A}}$ is both left and right compatible. Hence $\rho_{\mathscr{A}}$ is a congruence on S; and the proof of the lemma is complete.

We now show that $\rho_{\mathscr{A}}$ admits $\mathscr{A}$. In fact we have

LEMMA 7.46. *Let $\mathscr{A} = \{A_i : i \in I\}$ be a kernel normal system of the inverse semigroup S. Then $\mathscr{A}$ is the kernel of $\rho_{\mathscr{A}}$.*

PROOF. We must first show that each A_i is a $\rho_{\mathscr{A}}$-class. Let $a, b \in A_i$. Then, by condition K1, $aa^{-1}, bb^{-1}, ab^{-1} \in A_i$ and hence $(a, b) \in \rho_{\mathscr{A}}$. Suppose now that $a \in A_i$ and $(a, b) \in \rho_{\mathscr{A}}$. Then, for some j in I, $aa^{-1}, bb^{-1}, ab^{-1} \in A_j$. But $a \in A_i$ implies $aa^{-1} \in A_i$. Hence $i = j$ and $a, ab^{-1} \in A_i$. Further, by Lemma 7.43, $a^{-1}a, b^{-1}b \in A_{ia}$. But $a \in A_i$ implies that $a^{-1}a \in A_i$. Hence $i = ia$ and $b^{-1}b \in A_i$. Hence $a, ab^{-1}, b^{-1}(b^{-1})^{-1} \in A_i$ so that, by condition K5, $b^{-1} \in A_i$. Consequently, by condition K1, $b \in A_i$. This shows that A_i is a $\rho_{\mathscr{A}}$-class.

It is clear that each A_i is an idempotent of $S/\rho_{\mathscr{A}}$, for each A_i is an inverse semigroup and so contains an idempotent. Conversely, Lemma 7.34 implies that each idempotent of $S/\rho_{\mathscr{A}}$ is a $\rho_{\mathscr{A}}$-class containing an idempotent of S. From condition K3 it therefore follows that each idempotent of $S/\rho_{\mathscr{A}}$ is an element of $\mathscr{A}$. Consequently $\mathscr{A}$ is the kernel of $\rho_{\mathscr{A}}$.

LEMMA 7.47. *Let $\mathscr{A}$ be the kernel of the congruence ρ on the inverse semigroup S. Then $\rho = \rho_{\mathscr{A}}$.*

PROOF. This follows immediately from Lemma 7.46 and Theorem 7.38.

The previous lemmas combine to provide a proof of the following theorem.

THEOREM 7.48. *Let $\mathscr{A} = \{A_i : i \in I\}$ be a kernel normal system of the inverse semigroup S. Then $\rho_{\mathscr{A}}$, as defined by* (1), *is a congruence on S, and $\mathscr{A}$ is the kernel of $\rho_{\mathscr{A}}$.*

Conversely, let $\phi: S \to S'$ be a homomorphism of the inverse semigroup S onto the semigroup S'. Let $\mathscr{A}$ be the kernel of ϕ. Then $\mathscr{A}$ is a kernel normal system of S and $\rho_{\mathscr{A}} = \phi \circ \phi^{-1}$.

EXERCISES FOR §7.4

1. Let π, ρ, σ be congruences on a semigroup S such that (i) $\pi \subseteq \rho$, (ii) S/π is regular, and (iii) $c^2 \; \pi \; c$ $(c \in S)$ implies $c\rho = c\sigma$. Then $\sigma \subseteq \rho$.

As corollaries to this result we have:

(a) Theorem 7.38.

(b) Let ρ and σ be congruences on a regular semigroup S such that S/ρ and S/σ are regular. Let σ admit the set of idempotents of S/ρ. Then $\sigma = \rho$. (Cf. Preston [1961].)

2. Let $\phi: S \to T$ be a homomorphism of the inverse semigroup S onto T. Let H be an inverse subsemigroup of T. Then $H\phi^{-1}$ is an inverse subsemigroup of S. In particular, let $\mathscr{A} = \{A_i: i \in I\}$ be a kernel normal system of S. Then $A = \bigcup\{A_i: i \in I\}$ is an inverse subsemigroup of S.

3. Let $\phi: S \to T$ be a homomorphism of the inverse semigroup S onto the semigroup T. Let $\mathscr{B} = \{B_i: i \in I\}$ be a kernel normal system of T. Let $A_i = B_i\phi^{-1}$ and $\mathscr{A} = \{A_i: i \in I\}$. Then $\mathscr{A}$ is a kernel normal system of S.

4. Let $\mathscr{A} = \{A_i: i \in I\}$ be a kernel normal system of the inverse semigroup S. Then for any i, j in I, there exists k in I such that $A_iA_j \subseteq A_k$ and $A_jA_i \subseteq A_k$.

5. Let S be the inverse semigroup of one-to-one mappings consisting of the mappings

$$\square, \begin{pmatrix}1\\1\end{pmatrix}, \begin{pmatrix}2\\2\end{pmatrix}, \begin{pmatrix}1\\2\end{pmatrix}, \begin{pmatrix}2\\1\end{pmatrix}, \begin{pmatrix}1 & 2\\1 & 2\end{pmatrix}.$$

Let

$$A_1 = \left\{\square, \begin{pmatrix}1\\1\end{pmatrix}, \begin{pmatrix}2\\2\end{pmatrix}\right\},$$

$$A_2 = \left\{\begin{pmatrix}1 & 2\\1 & 2\end{pmatrix}\right\}.$$

Then $\mathscr{A} = \{A_1, A_2\}$ satisfies conditions K1–K4 and further satisfies the condition that A_iA_j and A_jA_i are both contained in the same A_k, for any i,j. However condition K5 is not satisfied by $\mathscr{A}$.

6. Condition K5 can be replaced by the condition K5′: $a, ab^{-1}, bb^{-1} \in A_i$ implies that $b \in A_i$. In fact K1 and K5′ together imply K5, while K1, K2, K4 and K5 together imply K5′.

7. Let $E = \{e_i: i \in I\}$ be the set of idempotents of the inverse semigroup S. Put $A_i = \{e_i\}$. Then $\mathscr{A} = \{A_i: i \in I\}$ is a kernel normal system of S and $\rho_{\mathscr{A}}$ is the identical relation on S.

8. Let $\mathscr{A} = \{A_i: i \in I\}$ be a set of subsets of the inverse semigroup S which satisfies conditions K1, K2, K3, K5′ (of Exercise 6) and also the condition K4′: If $aa^{-1}, bb^{-1}, ab^{-1} \in A_i$, then, for any j in I, there exists k in I such that

$$aA_ja^{-1} \subseteq A_k \quad \text{and} \quad aA_jb^{-1} \subseteq A_k.$$

Then $\mathscr{A}$ is a kernel normal system of S.

In fact conditions K1, K2, K3, K4′ and K5′ are equivalent to conditions K1–K5. (Preston [1954a] and Lyapin [1960a, Chapter 7].)

7.5 Semilattices of inverse semigroups

Since a kernel normal system $\mathscr{A} = \{A_i: i \in I\}$ of an inverse semigroup S is the set of idempotents of the quotient semigroup $S/\rho_{\mathscr{A}}$, it follows that A, the union of the elements of $\mathscr{A}$ is a semilattice of the inverse semigroups $A_i, i \in I$ (§1.8, p. 26). Denote by E_i the set of idempotents of the semigroup A_i, and let E be the set of idempotents of A (and so also of S). Then, clearly, since A is a semilattice of the A_i, E is a semilattice of the semigroups $E_i, i \in I$. The reverse of this implication holds. As a corollary to Theorems 4.5 and 4.11 (§4.2) it follows that a semigroup which is a union of groups is a semilattice of groups if and only if its idempotents commute, i.e. if and only if it is an inverse semigroup. The following theorem (Preston [1956]), combined with Theorem 7.52, is a generalization of this result.

Theorem 7.49. *Let S be an inverse semigroup which is the union of the disjoint inverse semigroups $A_i, i \in I$. Let E_i denote the set of idempotents of A_i, and put $E = \bigcup\{E_i: i \in I\}$.*

Then S is a semilattice of the semigroups $A_i, i \in I$, if and only if E is a semilattice of the semilattices $E_i, i \in I$.

Proof. Suppose that E is a semilattice of the semilattices $E_i, i \in I$. Then, for any i, j in I, there exists k in I, such that E_iE_j $(= E_jE_i) \subseteq E_k$.

Let $a \in A_i$, $b \in A_j$ and suppose that $abb^{-1} \in A_p$. Then

$$abb^{-1}(abb^{-1})^{-1} = a \cdot bb^{-1} \cdot bb^{-1} \cdot a^{-1} = abb^{-1}a^{-1} \in E_p,$$

and

$$(abb^{-1})^{-1}(abb^{-1}) = bb^{-1} \cdot a^{-1}a \cdot bb^{-1} = bb^{-1}a^{-1}a \in E_p.$$

But

$$bb^{-1} \cdot a^{-1}a \in E_iE_j \subseteq E_k$$

for some k in I; and so, since distinct A_i are disjoint, $k = p$.

Now suppose that $ab \in A_q$. Then

$$ab(ab)^{-1} = abb^{-1}a^{-1} \in E_q.$$

It has already been shown that $abb^{-1}a^{-1} \in E_p$. Hence $q = p = k$. Thus for any a in A_i, b in A_j, $ab \in A_k$ where $E_iE_j \subseteq E_k$. This shows that S is a semilattice of the semigroups $A_i, i \in I$.

As we have already commented, the converse is evident. The proof of the theorem is complete.

Corollary 7.50. *If an inverse semigroup is the union of a band I of inverse semigroups, then I is a semilattice.*

Corollary 7.51. *An inverse semigroup is a band of groups if and only if it is a semilattice of groups.*

In fact, more strongly, a semigroup which is a semilattice of inverse semigroups is itself an inverse semigroup.

THEOREM 7.52. *Let S be a semilattice of inverse semigroups $A_i, i \in I$. Then S is an inverse semigroup.*

PROOF. Since each A_i is regular, it is immediate that S also is regular. It will therefore suffice to prove that each element of S has a unique inverse.

Let $a \in S$ and let x be any inverse of a: $axa = a$, $xax = x$. There exist j, k in I such that $a \in A_j$, $x \in A_k$. Since S is a semilattice of the $A_i, i \in I$, there exists p in I such that

$$A_j A_k \subseteq A_p, \qquad A_k A_j \subseteq A_p.$$

Then

$$a = axa \in A_j A_k A_j \subseteq A_p,$$

$$x = xax \in A_k A_j A_k \subseteq A_p.$$

But $a \in A_j$. Thus $p = j$ and $x \in A_j$. Hence any inverse of a lies in A_j and, since A_j is an inverse semigroup, therefore a has only one inverse in S.

This completes the proof.

COROLLARY 7.53. *Any semigroup which is a semilattice of groups is an inverse semigroup.*

7.6 HOMOMORPHISMS WHICH SEPARATE IDEMPOTENTS

An inverse semigroup with only one idempotent is a group. Consequently if ϕ is a homomorphism of an inverse semigroup S which preserves (i.e. separates) idempotents, i.e. which induces an isomorphism of the semilattice of idempotents of S, then each of the elements A_i of the kernel $\mathscr{A} = \{A_i: i \in I\}$ of ϕ is a group. Conversely if each element of a kernel normal system $\mathscr{A} = \{A_i: i \in I\}$ of an inverse semigroup S is a group, then the corresponding natural homomorphism $\rho_{\mathscr{A}}^{\natural}$ preserves idempotents. For kernel normal systems of this kind we obtain a simple characterization [Preston 1956]. Observe firstly, that the union of the elements of any kernel normal system $\mathscr{A}$ of S is clearly an inverse subsemigroup of S, for it is the inverse image under the homomorphism $\rho_{\mathscr{A}}^{\natural}$ of the set of idempotents of $S/\rho_{\mathscr{A}}$ (cf. Exercise 2 for §7.4).

Let X be a subset of the semigroup S. Then the *centralizer $C(X)$ of X in S* is defined by

$$C(X) = \{s \in S: sx = xs \quad \text{for all } x \text{ in } X\}.$$

THEOREM 7.54. *Let E be the set of idempotents of the inverse semigroup S. Denote by $H_e, e \in E$, the maximal subgroup of S with identity element e. For each $e \in E$ choose a subgroup N_e of H_e and let $\mathscr{N} = \{N_e: e \in E\}$.*

Put

$$N = \bigcup \{N_e: e \in E\}.$$

Then $\mathcal{N}$ *is a kernel normal system of* S *if and only if* (i) N *is a subsemigroup of* S *and* (ii) $a^{-1}Na \subseteq N$ *for all* a *in* S.

If this is the case, then N *is the semilattice* E *of the groups* N_e; *and, for every* a *in* S, $a^{-1}N_e a \subseteq N_f$, *with* $f = a^{-1}ea$. *In particular,* $N \subseteq C(E)$.

PROOF. If $\mathcal{N}$ is a kernel normal system, then, as already observed, (i) is clearly satisfied. Further, defining condition K4 for a kernel normal system (§7.4) ensures that (ii) also holds.

Conversely, suppose that $\mathcal{N}$ satisfies conditions (i) and (ii). Without taking into account (i) and (ii), conditions K1–K3 of §7.4, defining a kernel normal system, are automatically satisfied because of the definition of $\mathcal{N}$. We merely have to show that conditions K4 and K5 follow from the assumption of (i) and (ii).

From Theorem 7.49 and (i) it follows that N is the semilattice E of the semigroups $N_e, e \in E$. To show that K4 holds, it suffices to show that $a^{-1}N_e a \subseteq N_f$, with $f = a^{-1}ea$ $(e \in E, a \in S)$. Let $n \in N_e$, and let $b = a^{-1}na$. Then $bb^{-1} = a^{-1}n(aa^{-1})n^{-1}a = a^{-1}(aa^{-1})nn^{-1}a$, since $n \in C(E)$, by Lemma 4.8. Hence $bb^{-1} = a^{-1}ea = f$. Similarly, $b^{-1}b = f$. From (ii) it follows that $b \in N$; whence $b \in H_f \cap N = N_f$.

Observe that we have incidentally shown that, when (i) and (ii) hold, then the assertion of the final paragraph of the theorem holds.

To verify condition K5, consider a, b in S such that a, ab, bb^{-1} all belong to N_e for some e in E. Then $a^{-1} \in N_e$ and $bb^{-1} = e = a^{-1}a$. Hence $b = (bb^{-1})b = a^{-1}(ab)$, the product of two elements of N_e, also belongs to N_e. Thus condition K5 holds.

The proof of the theorem is complete.

We will call a kernel normal system of which each element is a group a *group kernel normal system.* If N_e, with identity e, is an element of a group kernel normal system, then it immediately follows from Theorem 7.54 that N_e is a normal subgroup of H_e, the maximal subgroup containing N_e. This is also obvious from the fact that the restriction to H_e of the natural homomorphism determined by the kernel normal system is a homomorphism of H_e with kernel N_e.

For group kernel normal systems the strong analogy with the group situation shown in Theorem 7.54 extends further. In fact the equivalence classes of the congruence $\rho_{\mathcal{N}}$ are the 'cosets' of the elements of the kernel normal system $\mathcal{N}$. Precisely, we have (Preston [1954a]):

THEOREM 7.55. *Let* $\mathcal{N} = \{N_e : e \in E\}$ *be a group kernel normal system of the inverse semigroup* S, *where* e *denotes the identity of* N_e. *Let* a *be any element of* S. *Then the* $\rho_{\mathcal{N}}$*-class containing* a *is* $N_f a = aN_g$, *where* $aa^{-1} = f$ *and* $a^{-1}a = g$.

PROOF. Let $b \in N_f a$. Then $b = na$ for some n in N_f. Hence $bb^{-1} = naa^{-1}n^{-1} = nfn^{-1} = nn^{-1} = f$. Further $ab^{-1} = a(na)^{-1} = aa^{-1}n^{-1} = fn^{-1} =$

$n^{-1} \in N_f$. Thus aa^{-1}, bb^{-1} and ab^{-1} all belong to N_f, i.e. $(a, b) \in \rho_{\mathcal{N}}$. Consequently $N_f a$ is contained in a single $\rho_{\mathcal{N}}$-class.

Now let $b \in S$ and $(b, a) \in \rho_{\mathcal{N}}$. Then aa^{-1}, bb^{-1} and ab^{-1} all belong to N_e for some e in E. But $aa^{-1} = f$. Hence $e = f$ and $bb^{-1} = f$ and $ab^{-1} = n \in N_f$. Further $(b, a) \in \rho_{\mathcal{N}}$ implies similarly that $a^{-1}a = b^{-1}b$ (Lemma 7.43). Thus $b = bb^{-1}b = ba^{-1}a = (ab^{-1})^{-1}a = n^{-1}a \in N_f a$. Since $a = fa \in N_f a$, this shows that $N_f a$ is the $\rho_{\mathcal{N}}$-class containing a.

The left-right dual of this argument shows that $N_f a = aN_g$. This completes the proof of the theorem.

It is to be noted that the 'cosets' such as $N_f a$ are not to be confused with ω-cosets in the sense of Šaĭn's theory of §7.2. The right ω-cosets of N_f are the sets $(N_f x)\omega$, for x such that $xx^{-1} \in N_f$, i.e. $xx^{-1} = f$, where ω denotes the natural partial order. Thus they are the closures, under ω, of the 'cosets' being considered here.

Let E be the set of idempotents of the inverse semigroup S and let $H_e, e \in E$, denote the maximal subgroup of S with identity e. Exercise 1 below shows that in general $\{H_e : e \in E\}$ is not a (group) kernel normal system of S. Thus it is not sufficient to take arbitrary normal subgroups N_e of each H_e to form a group kernel normal system (see also Exercise 2, below).

Let $\mathcal{M} = \{M_e : e \in E\}$ and $\mathcal{N} = \{N_e : e \in E\}$, where e is the identity of both N_e and M_e, be two group kernel normal systems of the inverse semigroup S. Define

$$\mathcal{M}\mathcal{N} = \mathcal{M} \vee \mathcal{N} = \{M_e N_e : e \in E\},$$

$$\mathcal{M} \wedge \mathcal{N} = \{M_e \cap N_e : e \in E\}.$$

THEOREM 7.56. *$\mathcal{M}\mathcal{N}$ and $\mathcal{M} \wedge \mathcal{N}$ are group kernel normal systems of S. Further*

$$\rho_{\mathcal{M}} \circ \rho_{\mathcal{N}} = \rho_{\mathcal{M}\mathcal{N}} = \rho_{\mathcal{N}} \circ \rho_{\mathcal{M}},$$

and

$$\rho_{\mathcal{M}} \cap \rho_{\mathcal{N}} = \rho_{\mathcal{M} \wedge \mathcal{N}}.$$

PROOF. We begin by proving that $\mathcal{M}\mathcal{N}$ is a group kernel normal system. Since each $M_e N_e$ is a group we merely have to show that conditions (i) and (ii) of Theorem 7.54 are satisfied.

Let $e, f \in E$ and consider $M_e N_e \cdot M_f N_f$. Let $m \in M_e$, $n \in N_e$, $m' \in M_f$, $n' \in N_f$. Using $M, N \subseteq C(E)$, we have

$$\begin{aligned} mnm'n'(mnm'n')^{-1} &= mnm'n'(m'n')^{-1}(mn)^{-1} \\ &= mnf(mn)^{-1} = mn(mn)^{-1}f \\ &= ef. \end{aligned}$$

Similarly, the right unit of $mnm'n'$ is also ef. Hence $M_e N_e \cdot M_f N_f \subseteq H_{ef}$, the maximal subgroup with identity ef. Since $N \subseteq C(E)$ we have $mnm'n' = mfnm'n'$. Hence

$$mnm'n' = mm'(m')^{-1}nm'n'.$$

But $mm' \in M_eM_f \subseteq M_{ef}$ and $(m')^{-1}nm' \in (m')^{-1}N_em' \subseteq N_{ef}$, since $(m')^{-1}em' = ef$. Further, $N_{ef}n' \subseteq N_{ef}N_f \subseteq N_{ef}$. These facts combine to show that

$$mnm'n' \in M_{ef}N_{ef},$$

i.e. that $M_eN_e \cdot M_fN_f \subseteq M_{ef}N_{ef}$; and this shows that $\mathscr{M}\mathscr{N}$ satisfies condition (i) of Theorem 7.54.

To prove condition (ii), let $a \in S$, $e \in E$ and consider $a^{-1}M_eN_ea$. Since $N \subseteq C(E)$, we have $a^{-1}M_eN_ea = a^{-1}M_ea \cdot a^{-1}N_ea$, so that $a^{-1}M_eN_ea \subseteq M_fN_f$, where $a^{-1}ea = f$; and this proves that condition (ii) holds.

This completes the proof that $\mathscr{M}\mathscr{N}$ is a group kernel normal system.

We now show that $\mathscr{M}\mathscr{N}$ is the kernel of $\rho_{\mathscr{M}} \circ \rho_{\mathscr{N}}$. Let $(a, b) \in \rho_{\mathscr{M}} \circ \rho_{\mathscr{N}}$, so that $(a, x) \in \rho_{\mathscr{M}}$ and $(x, b) \in \rho_{\mathscr{N}}$ for some x in S. Thus $aa^{-1} = xx^{-1} = bb^{-1} = e$, say, and $ax^{-1} \in M_e$, $xb^{-1} \in N_e$, and $a^{-1}a = x^{-1}x$. Hence $ab^{-1} = aa^{-1}ab^{-1} = ax^{-1}xb^{-1} \in M_eN_e$. Thus aa^{-1}, bb^{-1}, $ab^{-1} \in M_eN_e$. Consequently, $(a, b) \in \rho_{\mathscr{M}\mathscr{N}}$, so that $\rho_{\mathscr{M}} \circ \rho_{\mathscr{N}} \subseteq \rho_{\mathscr{M}\mathscr{N}}$.

Conversely, let $(a, b) \in \rho_{\mathscr{M}\mathscr{N}}$, so that, for some e in E, $aa^{-1} = bb^{-1} = e$ and $ab^{-1} \in M_eN_e$. Let $ab^{-1} = mn$ where $m \in M_e$, $n \in N_e$. Put $x = nb$. Then $xx^{-1} = nbb^{-1}n^{-1} \in N_e$; thus $xx^{-1} = e$. Further $ax^{-1} = a(nb)^{-1} = ab^{-1}n^{-1} = mnn^{-1} = me = m$, and $xb^{-1} = nbb^{-1} = ne = n$. Hence $aa^{-1} = xx^{-1} = e$ and $ax^{-1} \in M_e$, i.e. $(a, x) \in \rho_{\mathscr{M}}$; and $xx^{-1} = bb^{-1} = e$ and $xb^{-1} \in N_e$, i.e. $(x, b) \in \rho_{\mathscr{N}}$. Thus $(a, b) \in \rho_{\mathscr{M}} \circ \rho_{\mathscr{N}}$, so that $\rho_{\mathscr{M}\mathscr{N}} \subseteq \rho_{\mathscr{M}} \circ \rho_{\mathscr{N}}$.

Combining this with the earlier inequality we have

$$\rho_{\mathscr{M}\mathscr{N}} = \rho_{\mathscr{M}} \circ \rho_{\mathscr{N}}.$$

Let $e, f \in E$. Then $M_eM_f \subseteq M_{ef}$ and $N_eN_f \subseteq N_{ef}$; hence $(M_e \cap N_e)(M_f \cap N_f) \subseteq M_{ef} \cap N_{ef}$. Consequently, $\mathscr{M} \wedge \mathscr{N}$ satisfies condition (i) of Theorem 7.54. Let $a \in S$ and put $a^{-1}ea = g$. Then $a^{-1}M_ea \subseteq M_g$ and $a^{-1}N_ea \subseteq N_g$; hence $a^{-1}(M_e \cap N_e)a \subseteq M_g \cap N_g$. Consequently, $\mathscr{M} \wedge \mathscr{N}$ satisfies condition (ii) of Theorem 7.54. Hence, by Theorem 7.54, $\mathscr{M} \wedge \mathscr{N}$ is a (group) kernel normal system.

Now $(a, b) \in \rho_{\mathscr{M}} \cap \rho_{\mathscr{N}}$ if and only if, for some e in E, $aa^{-1} = bb^{-1} = e$ and $ab^{-1} \in M_e \cap N_e$, i.e. if and only if $(a, b) \in \rho_{\mathscr{M} \wedge \mathscr{N}}$. Thus

$$\rho_{\mathscr{M}} \cap \rho_{\mathscr{N}} = \rho_{\mathscr{M} \wedge \mathscr{N}}.$$

This completes the proof of the theorem.

If σ is a congruence on an inverse semigroup S such that each congruence class contains at most one idempotent then σ will be called an *idempotent separating* congruence. If σ is idempotent separating then the kernel of σ is a group kernel normal system. The previous theorem shows that the set of idempotent separating congruences on an inverse semigroup forms a sublattice of the lattice of all congruences on the semigroup. For, by Theorem 7.56, when $\mathscr{M}$ and $\mathscr{N}$ are group kernel normal systems, $\rho_{\mathscr{M}} \circ \rho_{\mathscr{N}}$ is the union of $\rho_{\mathscr{M}}$ and $\rho_{\mathscr{N}}$ in this lattice. Further the construction given in Theorem 7.56 for $\rho_{\mathscr{M}} \circ \rho_{\mathscr{N}}$ and for $\rho_{\mathscr{M}} \cap \rho_{\mathscr{N}}$ shows that this sublattice of idempotent separating congruences is a sublattice of the direct product of the lattices of normal subgroups of the maximal subgroups of the semigroup. The lattice of normal subgroups of a group is modular. Modularity is preserved under the formation of direct products and under the operation of taking sublattices. Consequently it follows that the lattice of idempotent separating congruences on an inverse semigroup is modular.

We have already seen that the set of maximal subgroups of an inverse semigroup does not in general form a kernel normal system of the semigroup. However there does exist a unique maximal idempotent separating congruence. The following result is due to J. M. Howie [1964b].

LEMMA 7.57. *Let S be an inverse semigroup with E as its set of idempotents. Define the relation μ, thus:*

$$\mu = \{(x, y) \in S \times S\colon x^{-1}ex = y^{-1}ey \text{ for all } e \text{ in } E\}.$$

Then μ is an idempotent separating congruence on S. Further, if σ is any idempotent separating congruence on S, then $\sigma \subseteq \mu$.

PROOF. Clearly, μ is an equivalence on S. Let $(x, y) \in \mu$. Let z be any element of S. Then, for any e in E, $z^{-1}ez \in E$. Hence $x^{-1}(z^{-1}ez)x = y^{-1}(z^{-1}ez)y$, i.e. $(zx)^{-1}e(zx) = (zy)^{-1}e(zy)$ for all e in E. Thus $(zx, zy) \in \mu$ and so μ is left compatible. Again $x^{-1}ex = y^{-1}ey$ implies that $z^{-1}(x^{-1}ex)z = z^{-1}(y^{-1}ey)z$, i.e. $(xz)^{-1}e(xz) = (yz)^{-1}e(yz)$. Hence μ is also right compatible. Thus μ is a congruence on S.

Let $e, f \in E$ and suppose that also $(e, f) \in \mu$. Then, in particular, $e^{-1}ee = f^{-1}ef$ and $e^{-1}fe = f^{-1}ff$, i.e. $e = ef$ and $ef = f$. Thus $e = f$. This shows that μ is idempotent separating.

Now let σ be any idempotent separating congruence on S. Let $(x, y) \in \sigma$. This implies $(x^{-1}, y^{-1}) \in \sigma$, for this holds for any congruence on S. Hence, in turn, $(ex, ey) \in \sigma$ and $(x^{-1}ex, y^{-1}ey) \in \sigma$, for any element e in E. But if $e \in E$, then $x^{-1}ex$ and $y^{-1}ey$ are each idempotents. Since σ is idempotent separating, therefore $x^{-1}ex = y^{-1}ey$, for all e in E, i.e. $(x, y) \in \mu$. Thus $\sigma \subseteq \mu$, and this completes the proof of the lemma.

We have thus shown the following theorem, which is new except for Howie's contribution of the 1-element (Lemma 7.57). (See Appendix.)

THEOREM 7.58. *The set of all idempotent separating congruences on an inverse semigroup forms a modular sublattice, with 1-element and 0-element, of the lattice of all congruences on the semigroup.*

The validity of analogues of the Jordan-Hölder-Schreier theorems for modular lattices (cf. Birkhoff [1948], Chapter VI) means that it is possible to develop such theorems for the kernels of idempotent separating congruences on an inverse semigroup. An indication of how to do this for a wider class of congruences is contained in the paper by Preston [1954a].

There are other interesting ways, besides the method of Šaĭn described in

§7.3, to construct representations of inverse semigroups. One such is given in Exercise 5 below. Another is given by N. R. Reilly in his dissertation [1965]. We give a brief description of this. (See Appendix.)

Let S be an inverse semigroup, and let G_e be a subgroup of S with identity element e. A subset A of S is called a *coset* of G_e in S if $A = G_e a$ for some element a of S such that $aa^{-1} = e$. Let $\mathscr{B}$ be a set of disjoint subgroups of S, and let $\mathscr{C}$ be the set of all cosets of the members of $\mathscr{B}$. Two members of $\mathscr{C}$ are shown to be either equal or disjoint. For each element a of S, a partial transformation π_a of $\mathscr{C}$ is defined as follows. Let the domain of π_a be $\{G_e b \in \mathscr{C}\colon G_e \in \mathscr{B},\ bb^{-1} = e,\ b \in Sa^{-1}\}$, and for each $G_e b$ therein, define $(G_e b)\pi_a = G_e(ba)$. Then the mapping $\pi\colon a \to \pi_a$ is a representation of S by one-to-one partial transformations of $\mathscr{C}$. It reduces to the Vagner-Preston representation (Theorem 1.20) when $\mathscr{B}$ consists of all the one-element subgroups of S. If every $\mathscr{D}$-class of S contains a member of $\mathscr{B}$, then $\pi \circ \pi^{-1}$ separates idempotents. In fact, if $\mathscr{B} = \{H_e\colon e \in E'\}$, where E' is a set of idempotents which meets every $\mathscr{D}$-class of S, then $\pi \circ \pi^{-1}$ is precisely the maximum idempotent-separating congruence μ on S.

Exercises for §7.6

1. Let S be the inverse semigroup consisting of the matrices

$$I = \begin{pmatrix} 1 & 0 \\ 0 & 1 \end{pmatrix}, \quad A = \begin{pmatrix} 0 & 1 \\ 1 & 0 \end{pmatrix}, \quad E_{11} = \begin{pmatrix} 1 & 0 \\ 0 & 0 \end{pmatrix}, \quad E_{12} = \begin{pmatrix} 0 & 1 \\ 0 & 0 \end{pmatrix},$$
$$E_{21} = \begin{pmatrix} 0 & 0 \\ 1 & 0 \end{pmatrix}, \quad E_{22} = \begin{pmatrix} 0 & 0 \\ 0 & 1 \end{pmatrix}, \quad 0 = \begin{pmatrix} 0 & 0 \\ 0 & 0 \end{pmatrix}.$$

Let E be the set of idempotents of S. For each e in E let N_e be the maximal subgroup of S with identity e. Let $\mathscr{N} = \{N_e\colon e \in E\}$. Then $\mathscr{N}$ satisfies condition (ii) of Theorem 7.54. But $\mathscr{N}$ is not a kernel normal system of S.

2. Let S be the inverse semigroup, a subsemigroup of the symmetric inverse semigroup on three symbols, consisting of all one-to-one mappings into $\{1, 2, 3\}$ of subsets of $\{1, 2, 3\}$ of cardinal less than or equal to two. Put $e = \begin{pmatrix} 1 & 2 \\ 1 & 2 \end{pmatrix}$, $f = \begin{pmatrix} 2 & 3 \\ 2 & 3 \end{pmatrix}$. Then $H_e = \left\{e, \begin{pmatrix} 2 & 1 \\ 1 & 2 \end{pmatrix}\right\}$ and $H_f = \left\{f, \begin{pmatrix} 3 & 2 \\ 2 & 3 \end{pmatrix}\right\}$ are the maximal subgroups of S with identities e and f respectively. Let $N_e = H_e$ and $N_f = \{f\}$. Then N_e and N_f are normal subgroups of H_e and H_f, respectively. Put $a = \begin{pmatrix} 1 & 2 \\ 3 & 2 \end{pmatrix}$. Then $a^{-1}N_e a = H_f$. Consequently no homomorphism of S can, when restricted to H_e and H_f, have kernels N_e and N_f, respectively.

3. Let μ be the maximum idempotent separating congruence on the inverse semigroup S. Let E be the set of idempotents of S. For e in E, let $N_e = C(E) \cap H_e$, where H_e is the maximal subgroup of S with identity e. Then $\{N_e\colon e \in E\}$ is the kernel of μ and $\bigcup \{N_e\colon e \in E\} = C(E)$.

4. Let μ be the maximum idempotent separating congruence on the inverse semigroup S. Let E be the set of idempotents of S. Then (i) $S/\mu \cong E$ if and only if S is a union of groups; (ii) $S/\mu \cong E$ if and only if E is contained in the centre of S. (Howie [1964a].)

5. Let S be an inverse semigroup with E as its set of idempotents. For e in E denote eSe by S_e. Let $a \in S$ and put $aa^{-1} = f$, $a^{-1}a = g$. Define α_a thus:

$$\alpha_a\colon s \to s\alpha_a = a^{-1}sa \qquad (s \in S_f).$$

Then α_a is an isomorphism of S_f onto S_g, and $\alpha\colon a \to \alpha_a$ ($a \in S$) is a representation of S by isomorphisms between subsemigroups of S. Further the kernel of α is $\{B_e\colon e \in E\}$ with $B_e = H_e \cap C(E \cup S_e)$. (Cf. Preston [1957].)

6. Let S be an inverse semigroup with E as its set of idempotents. For $a \in S$ define β_a as the mapping which maps e onto f, $e, f \in E$, if and only if e is the left unit and f is the right unit of eaf. Then β_a is a one-to-one mapping of a subset of E into E, and $\beta\colon a \to \beta_a$ ($a \in S$) is a representation of S by one-to-one mappings. Further $\beta \circ \beta^{-1} = \mu$, the maximum idempotent separating congruence S. (Preston [1954c] and Howie [1964a].)

7.7 Homomorphisms onto primitive inverse semigroups

We present in this section the results of W. D. Munn [1964a] and Preston [1967] on homomorphisms of inverse semigroups onto primitive inverse semigroups. Recall that a regular semigroup $S = S^0$ is said to be *primitive* when each of its non-zero idempotents is primitive (§6.5). By Theorem 6.39, a primitive regular semigroup $S = S^0$ is the 0-direct union of a uniquely determined set of completely 0-simple ideals. If S is also an inverse semigroup then each of these completely 0-simple ideals is a Brandt semigroup by Theorem 3.9. (See also Exercise 6 for §6.5.) The results on homomorphisms of inverse semigroups onto Brandt semigroups are due to Munn. Munn's determination of such homomorphisms is crucial in his solution of the problem of determining all irreducible representations (by finite matrices over a field) of an arbitrary inverse semigroup [1964b]. The extension to homomorphisms onto an arbitrary primitive inverse semigroup is due to Preston and plays a similar part in the complete determination of all representations of an arbitrary inverse semigroup [1967].

We first make some observations on the occurrence of primitive inverse semigroups as ideals.

A non-zero ideal A of a regular semigroup $S = S^0$ will be called a *primitive ideal* if every non-zero idempotent of A is primitive in S. A primitive ideal is necessarily regular, for any ideal of a regular semigroup is regular. It suffices for the ideal A to be primitive that each of its non-zero idempotents is primitive in A. For if $e^2 = e$, $f^2 = f$, $ef = fe = f$ and $e \in A$, then, because A is an ideal, $f \in A$.

THEOREM 7.59. *For a regular semigroup $S = S^0$, the following conditions are equivalent.*

(A) *The right socle Σ_r of S is $\neq 0$.*
(A′) *The left socle Σ_l of S is $\neq 0$.*
(B) *S contains a primitive idempotent.*
(C) *S contains a primitive ideal.*

If any one of these equivalent conditions holds, then $\Sigma_r = \Sigma_l$, and the socle $\Sigma(=\Sigma_r = \Sigma_l)$ of S is the 0-direct union of completely 0-simple ideals of S. The socle Σ of S is the maximum primitive ideal of S; it contains all the primitive idempotents of S, and the primitive ideals of S are just the non-zero ideals of Σ.

PROOF. By definition of right socle (§6.3), condition (A) is equivalent to the existence of a 0-minimal right ideal R of S. Since S is regular, $R = eS$ for some idempotent $e \neq 0$ of S. By Lemma 6.38, eS is 0-minimal if and only if e is primitive. Thus (A) and (B) are equivalent. By the same lemma (or duality), (A′) and (B) are equivalent.

Assuming (A), every non-zero idempotent of Σ_r generates a 0-minimal right ideal of S, and so is primitive. Thus Σ_r is a primitive ideal of S, and (C) holds. Conversely, let A be any primitive ideal of S. If $a \in A \backslash 0$, and $axa = a$, then ax is a non-zero idempotent of A, hence primitive. Thus (C) implies (B).

Assume now that (A) holds. Each 0-minimal right ideal eS of S (e primitive) meets the 0-minimal left ideal Se of S in the element $e \neq 0$, and consequently, in the notation of Theorem 6.29, the sets A, A', D, D' are all 0. We conclude from that theorem that $\Sigma_r = C = \Sigma_l$, and that Σ is the 0-direct union of all the completely 0-simple ideals of S.

Let I be any primitive ideal of S, and let $a \in I \backslash 0$. Then $a = axa$ for some x in S, and $e = ax$ is an idempotent $\neq 0$ of I. Hence e is primitive, so eS is a 0-minimal right ideal of S, whence $eS \subseteq \Sigma$. Hence $e \in \Sigma$; but then $a = ea \in \Sigma$ also. Hence $I \subseteq \Sigma$. This argument also shows that every primitive idempotent is contained in Σ.

Conversely, if I is any non-zero ideal of Σ, then I is the union of some or all of the completely 0-simple summands of Σ. I is therefore an ideal of S, and evidently primitive.

Having given one of the ways that primitive inverse semigroups may be encountered, we now proceed to consider homomorphisms onto primitive inverse semigroups. First we give a self-evident lemma which enables us to simplify our discussion.

LEMMA 7.60. *Let ρ be a congruence on the semigroup S such that $S/\rho = (S/\rho)^0$. Put $V = 0(\rho^\natural)^{-1}$. Then V is an ideal of S and $(S/V)^0 = S/V$.*

Define $\bar{\rho}$ on S/V thus

$$\bar{\rho} = [\rho \cap (S\backslash V \times S\backslash V)] \cup [(V, V)].$$

Then $\bar{\rho}$ is a congruence on S/V and $(S/V)/\bar{\rho}$ is isomorphic to S/ρ under the natural mapping,

$$x\bar{\rho} \to x\rho \qquad (x \in S\backslash V),$$
$$V \to V.$$

In particular, the zero V of S/V is a $\bar{\rho}$-class.

Conversely, if V is an ideal of S, and σ is a congruence on S/V such that $\{V\}$ is a σ-class, then

$$\rho = [\sigma \cap (S\backslash V \times S\backslash V)] \cup (V \times V)$$

is a congruence on S such that $V = (0\rho^{\natural})^{-1}$ and $\bar{\rho} = \sigma$.

The construction of the preceding lemma shows that in considering homomorphisms of a semigroup S onto a semigroup $T = T^0$ there is no loss of generality in supposing, firstly, that $S = S^0$ and, secondly, that the zero of S is the only element mapped onto the zero of T. For the general case is easily reduced to this special case, as will be explained more fully in the remarks following Theorem 7.70. Accordingly, we introduce a definition.

A congruence ρ on an inverse semigroup $S = S^0$ will be called *0-restricted* if $\{0\}$ is a ρ-class. A homomorphism ϕ of $S = S^0$ will be called *0-restricted* if the congruence $\phi \circ \phi^{-1}$ is 0-restricted. A homomorphic image under a 0-restricted homomorphism will be called a *0-restricted homomorphic image.*

We now proceed to show that an inverse semigroup possesses a 0-restricted primitive homomorphic image if and only if it is categorical at zero, in the sense of the following definition. A semigroup $S = S^0$ is *categorical at zero* if, for any a, b, c in S, $ab \neq 0$ and $bc \neq 0$ together imply that $abc \neq 0$.

LEMMA 7.61. *A primitive inverse semigroup is categorical at zero.*

PROOF. As noted above, a primitive inverse semigroup S is the 0-direct union of a set of completely 0-simple inverse semigroups, i.e. of Brandt semigroups, $B_i, i \in I$, say. Let $a, b, c \in S$. Suppose that $ab \neq 0$ and $bc \neq 0$. Then a, b, c must belong to the same summand B_i whence, by definition (§3.3, condition A2, p. 101), $abc \neq 0$. This completes the proof of the lemma.

LEMMA 7.62. *Let ρ be a 0-restricted congruence on the semigroup $S = S^0$ and such that S/ρ is a primitive inverse semigroup. Then S is categorical at zero.*

PROOF. Let $abc = 0$ in S. Put $\phi = \rho^{\natural}$. Then $(abc)\phi = (a\phi)(b\phi)(c\phi) = 0\phi$. Since the homomorphism ϕ is onto S/ρ, 0ϕ is the zero of S/ρ. By the preceding lemma S/ρ is categorical at zero. Hence $(a\phi)(b\phi) = 0$ or $(b\phi)(c\phi) = 0$, i.e. $(ab)\phi = 0$ or $(bc)\phi = 0$. Since ρ is 0-restricted we infer that either $ab = 0$ or $bc = 0$. This shows that S is categorical at zero.

The next lemma is of a technical nature and will facilitate the discussion.

LEMMA 7.63. *Let $S = S^0$ be an inverse semigroup which is categorical at zero. Then, for e, g, x, c in S,*

(i) *if $e^2 = e$, $g^2 = g$, $ex \neq 0$, $gx \neq 0$, then $egx \neq 0$;*
(ii) *if $e^2 = e$, $ex \neq 0$, $cex = 0$, then $cx = 0$; and*
(iii) *if $e^2 = e$, $ex \neq 0$, $exc = 0$, then $xc = 0$.*

PROOF. (i) Suppose $egx = 0$. Then $egxx^{-1} = 0$, i.e., since idempotents commute, $exx^{-1}g = 0$. Since S is categorical at zero, therefore either $exx^{-1} = 0$ or $xx^{-1}g = gxx^{-1} = 0$. Thus either ex $(= exx^{-1}x) = 0$ or gx $(= gxx^{-1}x) = 0$. This is contrary to assumption. Hence $egx \neq 0$.

(ii) If $cex = 0$, then $cexx^{-1} = 0$, i.e. $cxx^{-1}e = 0$. Since S is categorical at zero, therefore either $cx = 0$ or $xx^{-1}e = 0$. But $xx^{-1}e = 0$ implies $ex = (exx^{-1})x = (xx^{-1}e)x = 0$, contrary to assumption. It follows that $cx = 0$.

(iii) This follows directly from the fact that S is categorical at zero.

Now let $S = S^0$ be any inverse semigroup and (following Munn (l.c.)) define on S the binary relation β thus:

(1) $\beta = \beta(S) = \{(x, y) \in S \times S\colon ex = ey \neq 0, \text{ for some } e = e^2 \text{ in } S\} \cup \{(0, 0)\}$.

LEMMA 7.64. *Let $S = S^0$ be an inverse semigroup which is categorical at zero. Then $\beta = \beta(S)$, as defined by* (1), *is a 0-restricted congruence on S and S/β is primitive.*

Furthermore, if σ is any 0-restricted congruence on S such that S/σ is primitive, then $\beta \subseteq \sigma$.

PROOF. Clearly β is always reflexive and symmetric. Suppose that $(x, 0) \in \beta$; then it is immediate that $x = 0$. Consequently, when (x, y) and (y, z) both belong to β either all three of x, y, z are zero or none of x, y, z is zero. If all are zero then clearly $(x, z) \in \beta$. If none is zero there exist idempotents e, g, say, such that $ex = ey \neq 0$ and $gy = gz \neq 0$. From the preceding lemma, part (i), it then follows that $egy \neq 0$. Hence

$$egx = g(ex) = gey = e(gy) = egz \neq 0,$$

which implies that $(x, z) \in \beta$. We have therefore shown that β is an equivalence relation on S.

We now show that β is compatible. Consider $(x, y) \in \beta$. If $x = y = 0$, then $(cx, cy) = (0, 0) \in \beta$ for any c in S. Otherwise there exists an idempotent e such that $ex = ey \neq 0$. Let c be any element of S. If $cex = 0$, then $cey = 0$ and, by part (ii) of the preceding lemma, $cx = 0 = cy$ and so $(cx, cy) \in \beta$. If $cex \neq 0$, then

$$\begin{aligned}(cec^{-1})cx = c(ec^{-1}c)x = (cc^{-1}c)ex = cex \\ = cey = (cec^{-1})cy \neq 0.\end{aligned}$$

Thus $f = cec^{-1}$ is an idempotent such that $fcx = fcy \neq 0$. Hence $(cx, cy) \in \beta$. This completes the proof that β is left compatible.

That β is right compatible follows similarly (and more simply) using part (iii) of the preceding lemma. Thus we have shown that β is a congruence on S. That β is 0-restricted amounts to an observation that has already been made.

Let E, F be non-zero idempotents of S/β. By Lemma 7.34, there exist idempotents e,f of S with $e \in E$ and $f \in F$. If $ef \neq 0$, then $e(ef) = ef \neq 0$ shows that $(ef, f) \in \beta$; and similarly it follows that $(ef, e) \in \beta$. Hence $(e, f) \in \beta$. Accordingly, either $EF = 0$ or $E = F$. In other words, every non-zero idempotent of S/β is primitive, i.e. S/β is a primitive inverse semigroup.

It remains to show that β is the finest 0-restricted congruence on S such that S/β is primitive. To this end let σ be any 0-restricted congruence on S such that S/σ is primitive. Let $(x, y) \in \beta$. If $x = y = 0$, then $(x, y) \in \sigma$. Otherwise there exists an idempotent e in S such that $ex = ey \neq 0$. It follows, since σ is 0-restricted, that

$$(e\sigma^\natural)(x\sigma^\natural) = (e\sigma^\natural)(y\sigma^\natural) \neq 0$$

in one of the Brandt semigroup summands of S/σ. From the definition of a Brandt semigroup (§3.3, p. 101, Condition A1) it then follows that $x\sigma^\natural = y\sigma^\natural$, i.e. that $(x, y) \in \sigma$. Thus we have shown that $\beta \subseteq \sigma$; and this completes the proof of the lemma.

Corollary 7.65. *Let $S = S^0$ be a primitive inverse semigroup. Then $\beta = \beta(S)$, defined by* (1), *is the identical relation on S.*

The preceding lemmas combine to provide a proof of the following theorem.

Theorem 7.66. *Let $S = S^0$ be an inverse semigroup. Then S possesses a 0-restricted primitive homomorphic image if and only if S is categorical at zero.*

When S is categorical at zero the binary relation $\beta = \beta(S)$, defined by (1), *is the unique finest 0-restricted congruence on S with the property that its quotient semigroup is a primitive inverse semigroup.*

Since any homomorphic image of a primitive inverse semigroup is also a primitive inverse semigroup, we may describe S/β in the language of Proposition 1.7 as the *maximal primitive homomorphic image of S.*

In his discussion of homomorphisms onto completely 0-simple semigroups, Munn [1964a] introduces the following condition. We shall say that the zero of the semigroup $S = S^0$ is *indecomposable* if, for any two ideals A and B of S, $A \cap B = 0$ implies either $A = 0$ or $B = 0$. The zero of S will be said to be *decomposable* if it is not indecomposable.

Lemma 7.67. *Let ϕ be a 0-restricted homomorphism of the semigroup $S = S^0$ such that $S\phi$ is 0-simple. Then S has an indecomposable zero.*

Proof. Suppose that A and B are ideals of S with $A \cap B = 0$. Since $S\phi$ is 0-simple and $A\phi$ is an ideal of $S\phi$, therefore either $A\phi = 0$ or $A\phi = S\phi$.

Similarly, either $B\phi = 0$ or $B\phi = S\phi$. When $A\phi = B\phi = S\phi$, then $S\phi = (S\phi)^2 = (A\phi)(B\phi) = (AB)\phi \subseteq (A \cap B)\phi = 0\phi$, which is impossible. Consequently, either $A\phi = 0$ or $B\phi = 0$, whence it follows, since ϕ is 0-restricted, that either $A = 0$ or $B = 0$. Thus the zero of S is indecomposable.

A completely 0-simple homomorphic image of a primitive inverse semigroup S is obtained by projecting S onto one of its summands, i.e. by mapping all summands except one onto zero and by mapping the remaining summand isomorphically. Such a homomorphism is 0-restricted if and only if S has only one summand, i.e. if and only if S is completely 0-simple. More generally, we have the following corollary of Lemma 7.67.

COROLLARY 7.68. *Let $S = S^0$ be a primitive regular semigroup. Then S possesses a 0-restricted completely 0-simple homomorphic image if and only if it has only one completely 0-simple summand, i.e. if and only if S is completely 0-simple.*

We now prove Munn's theorem [1964a].

THEOREM 7.69. *Let $S = S^0$ be an inverse semigroup. Then S possesses a 0-restricted completely 0-simple homomorphic image if and only if* (i) *S is categorical at zero and* (ii) *the zero of S is indecomposable.*

Furthermore, when S satisfies (i) *and* (ii), *then the binary relation $\beta = \beta(S)$, defined by* (1), *is the unique finest 0-restricted congruence on S with the property that its quotient semigroup is completely 0-simple. Thus S/β is the maximal Brandt homomorphic image of S.*

PROOF. Let ϕ be a 0-restricted homomorphism of S such that $S\phi$ is completely 0-simple. In particular $S\phi$ is then primitive and so, by Theorem 7.66, S is categorical at zero. Lemma 7.67 gives that the zero of S is indecomposable. Thus S satisfies conditions (i) and (ii).

Conversely, suppose that S satisfies (i) and (ii). By Theorem 7.66, condition (i) implies that S/β is a primitive inverse semigroup. Let A and B be two completely 0-simple summands of S/β. Then $A_1 = A(\beta^\natural)^{-1}$ and $B_1 = B(\beta^\natural)^{-1}$ are ideals of S. If A and B are distinct then $A \cap B = 0$ and, since β is 0-restricted (Theorem 7.66), $A_1 \cap B_1 = 0$. This is contrary to the assumption (ii). Hence $A_1 = B_1$ and so also $A = B$. Consequently S/β has only one completely 0-simple summand, i.e. S/β is itself completely 0-simple. We have therefore shown that when S satisfies (i) and (ii) then S has a 0-restricted completely 0-simple homomorphic image.

Let S satisfy (i) and (ii) and suppose that σ is a 0-restricted congruence on S such that S/σ is completely 0-simple. In particular S/σ is then primitive whence, from Theorem 7.66, it follows that $\beta \subseteq \sigma$. We have already seen that β is 0-restricted and that S/β is completely 0-simple. Accordingly, the proof of the theorem is complete.

We make use of the foregoing to describe all congruences ρ on an inverse semigroup S such that S/ρ is primitive. Such a congruence ρ will be called *primitive.* (S/ρ is also an inverse semigroup, by Theorem 7.36.)

An ideal V of a semigroup S will be called *categorical* if $V \neq S$ and

$$abc \in V \ (a, b, c \text{ in } S) \text{ implies } ab \in V \quad \text{or} \quad bc \in V.$$

An ideal V of S is categorical if and only if S/V is categorical at zero. For each categorical ideal V of S, we define

$$\beta_V = \{(x, y) \in S \times S: ex = ey \notin V, \quad \text{for some } e = e^2 \text{ in } S\} \cup (V \times V).$$

By Lemma 7.64, and the converse part of Lemma 7.60, β_V is a congruence on S such that $0(\beta_V^{\natural})^{-1} = V$, S/β_V is primitive, and every primitive congruence ρ on S such that $0(\rho^{\natural})^{-1} = V$ contains β_V.

The congruence ρ on S induces a 0-restricted congruence ρ' on S/β_V defined as follows:

$$(x\beta_V, y\beta_V) \in \rho' \text{ if and only if } (x, y) \in \rho. \tag{2}$$

Conversely, given a 0-restricted congruence ρ' on S/β_V, (2) serves to define a congruence ρ on S such that $0(\rho^{\natural})^{-1} = V$. We call ρ the *natural extension* of ρ' to S.

The foregoing implies the following theorem.

THEOREM 7.70. *Let ρ be any primitive congruence on an inverse semigroup S, and let $V = 0(\rho^{\natural})^{-1}$. Then V is a categorical ideal of S, β_V is a primitive congruence on S, and ρ is the natural extension to S of a 0-restricted congruence on the primitive inverse semigroup S/β_V.*

This theorem reduces the problem of determining all primitive congruences on an inverse semigroup S to those of determining (i) all categorical ideals V of S, and (ii), for each such V, all 0-restricted congruences on the primitive inverse semigroup S/β_V.

The second of these is handled as follows. Let $P = \bigcup\{B_i: i \in I\}$ be any primitive inverse semigroup, the B_i being its Brandt components. Let ρ be any 0-restricted congruence on P. Suppose $(x, y) \in \rho$, where $x \in B_i$, $y \in B_j$, and $i \neq j$ in I. Then $(xx^{-1}x, xx^{-1}y) \in \rho$. But $xx^{-1}x = x$ and $xx^{-1}y = 0$. Since ρ is 0-restricted, we conclude that $x = 0$, and hence $y = 0$ also. Thus $\rho = \bigcup\{\rho_i: i \in I\}$ where each $\rho_i = \rho \cap (B_i \times B_i)$ is a congruence on B_i. Conversely, given a congruence ρ_i on B_i, for each i in I, their union is a congruence on P. The problem of finding all 0-restricted congruences on a primitive inverse semigroup is thus reduced to that of finding all such on a Brandt semigroup, and this will be solved in §10.7.

We will now explain the relevance of primitive inverse semigroups to the problem of determining the representations, by finite matrices over a field, of an arbitrary inverse semigroup $S = S^0$. Let ϕ be such a representation of the semigroup S. Considerations of matrix rank show that $S\phi$ contains primitive idempotents and so contains, by Theorem 7.59, a unique maximal primitive ideal, P, say. S then contains $P\phi^{-1}$ as an ideal. Denote by V the ideal $V = 0\phi^{-1}$. Then V is an ideal of $P\phi^{-1}$ and, by Theorem 7.66, $(P\phi^{-1})/V$ is categorical at zero. Let us assume, for the purposes of this summary account, that $V = 0$. Then $P\phi^{-1}$ is categorical at zero and the representation of $P\phi^{-1}$, induced by the restriction of ϕ to $P\phi^{-1}$, is effectively a representation of $(P\phi^{-1})/\beta$, a 0-direct union of Brandt semigroups, where $\beta = \beta(P\phi^{-1})$ is defined by (1). A representation of a 0-direct union of Brandt semigroups decomposes into the representations induced on its summands. We gave an account of the theory of representations of completely 0-simple semigroups in §5.4, and for the special case of a Brandt semigroup the representations are completely determined in terms of the representations of its structure group. (See Exercise 4 for §5.4. This exercise is deficient in one respect: to give the whole story the following conclusion (d) should be appended to the (a), (b) and (c) of the exercise: (d) Every proper extension is basic.) Thus the representations of $P\phi^{-1}$ can be determined. The next step is to show that the representation ϕ decomposes into two components ϕ_P and ϕ^*, say, where ϕ_P is determined completely by the restriction of ϕ to $P\phi^{-1}$. The representation ϕ^* can then be treated similarly. If this process terminates after n steps, so that n components, such as ϕ_P, of ϕ are obtained, then there is an associated ideal series of $2n$ terms

$$0 \subseteq V \subset P\phi^{-1} \subseteq \cdots,$$

the first two terms of which are V (we drop the assumption now that $V = 0$) and $P\phi^{-1}$. Such ideal series can be abstractly characterized. Conversely, it can then be shown that any such ideal series determines a representation of S determined by arbitrarily assigned representations of the quotient semigroups such as $(P\phi^{-1})/V$.

The reader is referred to Preston [1967] for the details.

Exercises for §7.7

1. An element $e \neq 0$ of a semigroup $S = S^0$ is called a *categorical identity* (or *unit*) if, for every x in S, ex is either x or 0, and xe is either x or 0. In the case $ex = x$, we call e a *categorical left unit* of x, and dually for *categorical right unit* of x. A semigroup $C = C^0$ is called a *small category with zero* if it satisfies the following two conditions.

I. Each non-zero element of C has a categorical left unit and a categorical right unit.

II. C is categorical at zero.

In the following, let C be a small category with zero.

(a) Each non-zero element a of C has a unique categorical left unit $e_l(a)$ and a unique categorical right unit $e_r(a)$.

(b) Let $a, b \in C\backslash 0$. Then $ab \neq 0$ if and only if $e_r(a) = e_l(b)$.

($C\backslash 0$ is a small abstract category; see, for example, MacLane [1965], p. 41. By "small" we mean that $C\backslash 0$ is a "set" as opposed to a "class".

Conversely, if K is a small abstract category, we can adjoin a new symbol 0 to K, and define $ab = 0$ if ab is undefined in K, together with $a0 = 0a = 00 = 0$ for all a in K. The resulting system $C = K \cup 0$ is a small category with zero as defined above.)

2. A non-zero element a of a small category C with zero (see Exercise 1) is called *invertible* if there exists an element a' in C such that $aa' = e_l(a)$ and $a'a = e_r(a)$. For a semigroup $S = S^0$, the following conditions are equivalent.

(A) S is a small category with zero, every non-zero element of which is invertible.

(B) S is a primitive inverse semigroup. (Cf. Exercise 6 for §6.5.) (Hoehnke [1962], p. 146.)

3. (a) Let $\{X_i\colon i \in I\}$ be a set of disjoint sets. Let S_{ij} $(i, j \in I\}$ be the set of all mappings of X_i into X_j, and let S be the union of all the S_{ij}, together with the empty mapping 0. Let the binary operation on S be iteration of mappings. Then S is a small category with zero (Exercise 1). The categorical units of S are the identity transformations of X_i $(i \in I)$. If $\alpha \in S_{ij}$ and $\beta \in S_{kl}$, then $\alpha\beta \neq 0$ if and only if $j = k$.

(b) Let P_{ij} be the set of all one-to-one mappings of X_i onto X_j, with $P_{ij} = \square$ if $|X_i| \neq |X_j|$. Let P be the union of all the P_{ij}, together with 0. Then P is a small category with zero, every non-zero element of which is invertible (Exercise 2). The decomposition of P into Brandt semigroups corresponds to the partition of $\{X_i\colon i \in I\}$ into sets of equal cardinal.

4. Let B be a Brandt semigroup with structure group G and set of non-zero idempotents I. Let H be a group which contains G as a subgroup of index $|I|$ in H. Let S consist of the empty mapping together with all mappings of right cosets of G (in H) onto right cosets of G induced by right multiplications by elements of H. Let the operation of S be that of iteration of mappings. Then S is a Brandt semigroup isomorphic to B. Consequently, every small category with zero in which every non-zero element is invertible (Exercise 2 above) can be faithfully represented as a category of one-to-one mappings.

5. A completely 0-simple semigroup is categorical at zero. Hence a primitive regular semigroup is also categorical at zero.

6. Let S be an inverse semigroup, and define the relation $\sim$ on S as follows: $x \sim y$ if and only if there exists an idempotent e in S such that $ex = ey$. Then $\sim$ is a congruence on S, and $S/\sim$ is a group. Furthermore, if σ is any congruence on S such that S/σ is a group, then $\sim\, \subseteq \sigma$. Hence every inverse semigroup possesses a maximal group homomorphic image (Proposition 1.7). (The congruence $\sim$ used in Rees' proof of Theorem 1.23, the Ore embedding theorem for right reversible cancellative semigroups, is the same as the congruence $\sim$ of this exercise.) (Munn [1961].)

7. Let $S = S^0$ be a regular semigroup and, for each λ in Λ, let M_λ be a non-zero ideal of S which is categorical at zero. Let $M = \bigcup\{M_\lambda\colon \lambda \in \Lambda\}$. Then $M = M^0$ is a regular semigroup which is categorical at zero. (Munn [1964a].)

8. Let $S = S^0$ be an inverse semigroup which is categorical at zero and with indecomposable zero. Let M be a non-zero ideal of S. Define $\beta_1 = \beta_1(S)$ and $\beta_2 = \beta_2(M)$ by equation (1) applied to the semigroups S and M, respectively. Then $S/\beta_1 \cong M/\beta_2$. (Munn [1964a].)

9. Let us say that a semigroup $S = S^0$ of $n \times n$-matrices over a field is *homogeneous* if (a) it contains the zero matrix and (b) all its non-zero elements are of the same rank.

(i) Let S be a Brandt semigroup of $n \times n$-matrices over a field whose zero element is the zero matrix. Then S is homogeneous.

(ii) Let S be a homogeneous semigroup of $n \times n$-matrices over a field. Then S is primitive. (Munn [1964b].)

10. (i) Let S be the semigroup of three idempotents, $S = \{e, f, 0\}$, where 0 is a zero element and $ef = fe = 0$. Then S is categorical at zero and the zero of S is decomposable.

(ii) Let X be the set of positive integers and denote, as usual, by $\mathscr{T}_X$ the full transformation semigroup on X. The set F, say, of all elements of $\mathscr{T}_X$ of infinite rank is a $\mathscr{D}$-class of $\mathscr{T}_X$ (Lemma 2.8). The set $I = \mathscr{T}_X \backslash F$, consisting of all elements of $\mathscr{T}_X$ of finite rank, is an ideal of $\mathscr{T}_X$ (Theorem 2.9(ii)). Hence $S = \mathscr{T}_X/I$ is a 0-simple semigroup. Thus the zero of S is indecomposable. Define the elements α, β of $\mathscr{T}_X$, thus:

$$x\alpha = \begin{cases} x, & \text{if } x \text{ is odd,} \\ 1, & \text{if } x \text{ is even;} \end{cases}$$

$$x\beta = \begin{cases} 2, & \text{if } x \text{ is odd,} \\ x, & \text{if } x \text{ is even.} \end{cases}$$

Then, denoting by ϵ the identical mapping of X, in S we have $\alpha\epsilon\beta = \alpha\beta = 0$, whereas $\alpha\epsilon = \alpha \neq 0$ and $\epsilon\beta = \beta \neq 0$. Thus S is not categorical at zero. (Munn [1964a].)

11. Let S be the semigroup of Example 1 of §6.6. Then $S = S^0$ and it is easily verified that S is categorical at zero and that the zero of S is indecomposable. For any a in S, $a^2 = 0$ and, consequently, S has no completely 0-simple homomorphic image.

CHAPTER 8

SIMPLE SEMIGROUPS

Semigroups without proper ideals (of various kinds) have frequently turned up in our discussion. In Chapter 2 we met the concepts of simple and 0-simple semigroups. In Chapters 2 and 3 we developed in detail the properties of completely simple [0-simple] semigroups. Such semigroups were characterized as simple [0-simple] semigroups containing both a minimal [0-minimal] left ideal and a minimal [0-minimal] right ideal (Theorem 2.48). Chapter 6 contains a discussion of the properties of semigroups which can be decomposed into unions of simple semigroups of various kinds. In §2.7 we saw that any semigroup determines its set of principal factors, each principal factor being simple, 0-simple, or a zero semigroup. A semigroup with a principal ideal series can be built up by successive extensions from its (unique) sequence of principal factors. §§4.4 and 4.5 discuss such extensions. Semisimple semigroups play an essential rôle in the representation theory of Chapter 5. Completely semisimple semigroups are discussed in §6.6.

Simple semigroups are thus of fundamental importance. Little is known, however, about simple [0-simple] semigroups not satisfying other explicitly given conditions. Similarly, we know only little about the general bisimple semigroup.

In this chapter we give a survey of certain classes of simple semigroups (excluding, except incidentally, completely simple semigroups). We first develop the properties of right simple semigroups without idempotents following the papers of M. Teissier [1953a, b], modifying the treatment in the light of the later work of R. Croisot [1954a]. Croisot deals with the more general situation of simple semigroups without idempotents and containing minimal left (or right) ideals. We give his main theorem (Theorem 8.18), which shows that any such semigroup can be embedded in another of the same kind, which we might term a master semigroup, which is constructed by a definite method. This theorem does at any rate give us some idea of the sort of objects these semigroups are. We do not know in detail how these semigroups are embedded in the master semigroups.

§8.3 discusses 0-simple semigroups containing 0-minimal one-sided ideals but containing no non-zero idempotents. The incompleteness of the results reflects the present state of knowledge here.

Bisimple inverse semigroups are the subject of §8.4. Theorem 8.44, due to Clifford [1953], shows that a bisimple inverse semigroup with an identity is completely determined by a subsemigroup of generators that is of somewhat simpler structure.

The final two sections, §§8.5 and 8.6, deal with two embedding theorems. One is a construction due to R. H. Bruck [1958] embedding any semigroup in a simple semigroup. Another is a theorem of Preston [1959] embedding any semigroup in a regular bisimple semigroup.

8.1 Baer-Levi semigroups

In their paper [1932], R. Baer and F. Levi construct a right cancellative right simple semigroup which is not a group. The semigroup they construct is the semigroup of all one-to-one mappings α of a countable set, I, say, into itself with the property that $I\backslash I\alpha$ is not finite. More generally, we shall say that S is a *Baer-Levi semigroup of type* (p, q) *on the set* A if $|A| = p$ and if S is the semigroup of all one-to-one mappings η (combined under composition) of A into A, having the property that $A\backslash A\eta$ is of infinite cardinal q.

The first objective of this section is to show (Theorem 8.5) that any right cancellative, right simple semigroup which is not a group can be embedded in a Baer-Levi semigroup. This result, due to M. Teissier [1953a], indicates the scope of the enigmatic "Class IV" of semigroups which we encountered in §1.11 (p. 39). The proof we give follows that of Croisot [1954a]. We then proceed to show (Theorem 8.8) that a semigroup can be embedded in a right simple, right cancellative semigroup without idempotents if and only if it is right cancellative and without idempotents. This result is due to P. M. Cohn [1956a]. The proof given here is new. Actually we show directly that the embedding is into a Baer-Levi semigroup, so Theorem 8.5 is a corollary of Theorem 8.8. It seemed worth while, however, to give also Croisot's simpler proof of the former.

Lemma 8.1. *For any two infinite cardinals* p, q *such that* $p \geqq q$, *there exists a Baer-Levi semigroup of type* (p, q).

Proof. Let A be of cardinal p and let ξ, η be one-to-one mappings of the set A into A such that $A\backslash A\eta$ and $A\backslash A\xi$ are each of cardinal q. It suffices to show that $A\backslash A\xi\eta$ is of cardinal q.

Since η is a one-to-one mapping η maps $A\backslash A\xi$ in a one-to-one manner onto $A\eta\backslash A\xi\eta$, and so $|A\backslash A\xi| = |A\eta\backslash A\xi\eta|$. Now $A\backslash A\xi\eta = (A\backslash A\eta) \cup (A\eta\backslash A\xi\eta)$, the union of two disjoint sets each of cardinal q. Since q is an infinite cardinal, therefore $|A\backslash A\xi\eta| = q$.

Theorem 8.2. *Let* S *be a Baer-Levi semigroup. Then* S *is a right cancellative, right simple semigroup without idempotents.*

Proof. Let S be the semigroup of one-to-one mappings η of the set A, of cardinal p, into A, such that $A\backslash A\eta$ is of cardinal q.

For any η in S, $|A\backslash A\eta| = |A\eta\backslash A\eta^2|$. Hence, since q is an infinite cardinal, $\eta^2 \neq \eta$. Thus S contains no idempotents.

Let $\alpha, \beta \in S$. To show that S is right simple we must show that there exists γ in S such that $\alpha\gamma = \beta$. We construct γ as follows. For x in A let γ map $x\alpha$ upon $x\beta$. This defines γ for elements of $A\alpha$. Since $|A \backslash A\alpha| = |A \backslash A\beta| = q$ there exist one-to-one mappings δ of $A \backslash A\alpha$ into $A \backslash A\beta$ such that $(A \backslash A\beta) \backslash ((A \backslash A\alpha)\delta)$ is of cardinal q. For elements of $A \backslash A\alpha$ let γ coincide with any such δ. Then γ so constructed belongs to S and $\alpha\gamma = \beta$.

It remains to prove that S is right cancellative. Suppose that $\alpha, \beta, \gamma \in S$ and that $\alpha\beta = \gamma\beta$. Let $x \in A$. Then $x\alpha\beta = x\gamma\beta$, and so, since β is a one-to-one mapping, $x\alpha = x\gamma$. Since this holds for all x in A, therefore $\alpha = \gamma$.

We need two lemmas preliminary to our first embedding theorem.

LEMMA 8.3. *Let S be a right simple semigroup without idempotents. Then the equation $xy = y$ cannot hold between any two elements x, y in S. (Cf. §1.3, proof of Lemma 1.0.)*

PROOF. Suppose that $xy = y$ where $x, y \in S$. Since S is right simple, $yS = S$. Hence there exists z in S such that $yz = x$. This implies that $x^2 = x(yz) = (xy)z = yz = x$, so that x is an idempotent, contrary to the assumption that S contains no idempotents.

LEMMA 8.4. *Let S be a right simple, right cancellative semigroup without idempotents. Then, for any s in S, $|S \backslash Ss| = |S|$.*

PROOF. Since every finite semigroup contains an idempotent (§1.6), S must be infinite. Let s be any element of S. Define a transformation ϕ of S as follows. For each x in S, let $x\phi$ be an element x' of S such that $xx' = s$. At least one such x' exists, since S is right simple, and we choose one of these to be $x\phi$. If $x\phi = y\phi$, then $x(x\phi) = s = y(y\phi) = y(x\phi)$, and so $x = y$ since S is right cancellative. Hence ϕ is one-to-one. Now $S\phi \cap Ss = \square$. For if $z \in S\phi \cap Ss$, then $z = x\phi = ys$ for some x, y in S. Then $s = x(x\phi) = xys$. But this is impossible by Lemma 8.3. Hence $S\phi \subseteq S \backslash Ss$ and

$$|S| = |S\phi| \leqq |S \backslash Ss| \leqq |S|,$$

so that

$$|S \backslash Ss| = |S|.$$

THEOREM 8.5. *Let S be a right cancellative, right simple semigroup without idempotents. Then S can be embedded in a Baer-Levi semigroup of type (p, p), where $p = |S|$.*

PROOF. By Lemma 1.0 the extended (right) regular representation of S is a faithful representation of S as a semigroup of one-to-one mappings of S^1 into itself. The element s of S is represented by the mapping

$$\rho_s : \begin{cases} 1 \rightarrow s \\ x \rightarrow xs, \qquad x \in S. \end{cases}$$

Furthermore,

$$|S^1 \backslash S^1\rho_s| = |S^1 \backslash (s \cup Ss)| = |S \backslash Ss|,$$

since $|S|$ is infinite. Hence, by Lemma 8.4,

$$|S^1| = |S| = |S \backslash Ss| = |S^1 \backslash S^1 \rho_s|.$$

Thus ρ_s is an element of the Baer-Levi semigroup of type (p, p) consisting of mappings of S^1 into itself. The faithfulness of the extended regular representation implies that S is embedded as required.

The next two lemmas are preliminary to the proof of Cohn's theorem. Recall that if I is an inverse semigroup and $A \subseteq I$, then the inverse semigroup generated by A is the intersection of all inverse subsemigroups of I that contain A.

LEMMA 8.6. *Let I be an infinite inverse semigroup with identity element* 1. *Let S be a subsemigroup of I such that S generates the inverse semigroup I, and such that if $a \in S$ then $aa^{-1} = 1$ and $a^{-1}a \neq 1$. Set $p = |I|$, and denote by $\mathscr{Q}_I$ the Baer-Levi semigroup of type (p, p) on the set I. Let ϕ denote the (right) regular representation of I.*

Then $\phi \mid S$, the restriction of ϕ to S, is an isomorphism of S into $\mathscr{Q}_I$.

PROOF. Since S is a set of generators of I, I consists of the finite products of the elements of S and their inverses. Consequently, $|S| = |I| = p$. Further, $S^{-1} \cap S = \square$, where $S^{-1} = \{b : b^{-1} \in S\}$; for $b^{-1} \in S$ implies that $bb^{-1} \neq 1$. Hence also $|S^{-1}| = p$.

Let $a \in S$ and consider Ia. Then $Ia \cap S^{-1} = \square$. For let $b \in S$ and suppose by way of contradiction, that $b^{-1} = xa$, where $x \in I$. Then $bb^{-1} = (xa)^{-1}xa = a^{-1}x^{-1}xa$. But $a^{-1}x^{-1}xa \leqq a^{-1}a$, where $\leqq$ denotes the natural partial order on I (§7.1). Hence $a^{-1}x^{-1}xa \neq 1$, i.e. $bb^{-1} \neq 1$, contrary to the supposition that $b \in S$. Thus $S^{-1} \subseteq I \backslash Ia$; whence we have that $|I \backslash Ia| = p$.

Now $Ia = I(a\phi)$ and so the lemma will follow once we show that, for $a \in S$, $a\phi$ is a one-to-one mapping of I into I. But this is true because if $x(a\phi) = y(a\phi)$ for x, y in I, then $x = xaa^{-1} = (x(a\phi))a^{-1} = (y(a\phi))a^{-1} = yaa^{-1} = y$.

Let S be a right cancellative semigroup without idempotents. We recall the definition (§1.9, p. 32) of the *inverse hull of* S. Let $\rho: s \rightarrow \rho_s$ $(s \in S)$ be the extended regular representation of S. By Lemma 1.0, each ρ_s is a one-to-one mapping of S^1 into S^1, and ρ is faithful. Then the inverse hull of S is the inverse subsemigroup $\Sigma = \Sigma(S)$ of $\mathscr{I}_{S^1}$ generated by $S\rho$. Here $\mathscr{I}_{S^1}$ denotes the symmetric inverse semigroup consisting of all one-to-one partial transformations of S^1.

LEMMA 8.7. *Let S be a right cancellative semigroup without idempotents and let ρ denote its extended regular representation.*

Then the inverse hull Σ of S contains as identity element 1 *the identical mapping on S^1. Further, if $a \in S\rho$, then $aa^{-1} = 1$ and $a^{-1}a \neq 1$.*

PROOF. Each element of $S\rho$ is a one-to-one mapping of S^1 into S^1. Consequently, if $a \in S\rho$, then $aa^{-1} = 1$ which is clearly the identity element of Σ. Further $1 \notin \Sigma a$, therefore $a^{-1}a \neq 1$.

THEOREM 8.8. *A semigroup can be embedded in a right simple, right cancellative semigroup without idempotents if and only if it is right cancellative and without idempotents.*

PROOF. The necessity of the condition is clear To prove sufficiency, consider a right cancellative semigroup S without idempotents. Let ρ denote the extended regular representation of S and Σ the inverse hull of S. Since a semigroup without idempotents is infinite, Lemma 8.7 shows that the hypotheses of Lemma 8.6 are satisfied taking Σ and $S\rho$ as the semigroups I and S, respectively, of Lemma 8.6. Consequently, by Lemma 8.6, $S\rho$ and hence S, since ρ is faithful, can be mapped isomorphically into $\mathcal{Q}_{\Sigma}$, the Baer-Levi semigroup on Σ of type $(|\Sigma|, |\Sigma|)$. The theorem now follows from Theorem 8.2.

Since, in the above proof, we clearly have $|\Sigma| = |S|$, Theorem 8.5 follows as a corollary to Theorem 8.8.

A right simple semigroup whose principal left ideals are totally ordered by inclusion is termed by Cohn [1956b] a *sesquilateral left division semigroup.* Such a semigroup is necessarily right cancellative. For these semigroups Cohn proves the theorem: *a semigroup whose principal left ideals are totally ordered by inclusion can be embedded in a sesquilateral left division semigroup without idempotents if and only if it is right cancellative and without idempotents.*

EXERCISES FOR §8.1

1. If S is a right simple semigroup without idempotents then each $\mathscr{L}$-class in S contains only one element.
2. If S is a right simple semigroup without idempotents then every equation $ax = b$ has an infinity of solutions x in S.
3. If S is a right simple semigroup without idempotents then every element of S is contained in an infinity of distinct principal left ideals.
4. Let S be a right simple semigroup. If $Sx = Syx$ for any x, y in S then S is a right group.
5. A right simple right cancellative semigroup with an idempotent is a group.
6. If S is a right simple semigroup without idempotents then so also is Sx for any x in S.
7. Let S_1, S_2 be two right simple semigroups one of which contains no idempotents. Then $S_1 \times S_2$ is right simple without idempotents.
8. Let S be a right simple semigroup without idempotents. Let M be the set $S \times \Lambda$ where Λ is any set. Let $\phi\colon \Lambda \to S$ be any mapping of Λ into S. Define a product in M by the rule: $(s, \lambda)(t, \mu) = (s(\lambda\phi)t, \mu)$. Then M becomes, with this product, a right simple semigroup without idempotents. (M is effectively a matrix semigroup over S, cf. §3.1.) If S is right cancellative then in general M is not right cancellative.

9. With the notation of Exercise 8 suppose there is also a mapping

$$\chi\colon \Lambda \to \Lambda.$$

Define a new product in M thus:

$$(s, \lambda)(t, \mu) = (s(\lambda\phi)t, \mu\chi).$$

Show that if

(i) $\chi^2 = \chi$ and

(ii) $x(\mu\chi)\phi y = x(\mu\phi)y$ for all x, y in S, and all μ in Λ

then M becomes a semigroup relative to the new multiplication. Further M is necessarily idempotent-free and is right simple if and only if χ is a mapping onto Λ and thus, by (i), is the identity mapping.

10. Let A be an infinite set and R the Baer-Levi semigroup of type $(|A|, |A|)$ on A. Let H_1 be a non-trivial group of permutations of A with identity element 1. Set $S = H_1 \cup R$. Then S forms a semigroup under the operation of iteration of mappings.

Then S is right cancellative, R is a two-sided ideal of S and a minimal right ideal of S. Let $r \in R$, $g \in H_1$, $g \neq 1$; then $gr \notin Rr$. Thus $r \cup Rr$ is a left ideal of R but not a left ideal of S. (Cf. Exercise 1 for §6.2.) (See Appendix.)

8.2 Croisot-Teissier semigroups

In this section we consider a generalization of the Baer-Levi semigroups of §8.1 which will play the same rôle for idempotent-free simple semigroups which are the unions of their minimal right ideals as the Baer-Levi semigroups played for right simple right cancellative semigroups without idempotents in Theorem 8.5. This type of semigroup was first introduced by M. Teissier in considering right simple semigroups without idempotents which were not necessarily right cancellative [1953b]. The generalization to the situation dealt with here is due to R. Croisot [1954a].

Let p and q be infinite cardinals with $p \geqq q$, and let A be a set with $|A| \geqq p$. Suppose $\mathscr{E} = \{\mathscr{E}_i\colon i \in I\}$ is a set of distinct equivalences defined on A such that each quotient set $A/\mathscr{E}_i$, $i \in I$, is of cardinal p. A subset $B \subseteq A$ will be said to be *well separated by* $\mathscr{E}$ if (i) $|B| = p$ and (ii) $\mathscr{E}_i \cap (B \times B) = \iota_B$, the identical relation on B, for all i in I. For each i in I let T_i denote the set of all mappings η_i of A into A such that (i) $\eta_i \circ \eta_i^{-1} = \mathscr{E}_i$ and (ii) there exists a set B $(= B(\eta_i)$, in general depending on $\eta_i)$ well separated by $\mathscr{E}$ such that $A\eta_i \subseteq B$ and $|B \backslash A\eta_i| = q$. If any well separated subset of A exists then it is clear that each of the sets T_i is then non-empty. When A contains a subset well separated by $\mathscr{E}$ we will denote the set of mappings $\bigcup\{T_i\colon i \in I\}$ by $\mathrm{CT}(A, \mathscr{E}, p, q)$.

Lemma 8.9. *Any set of mappings* $\mathrm{CT}(A, \mathscr{E}, p, q)$ *forms, under composition, an idempotent-free semigroup in which each subset* T_i *is a right ideal.*

PROOF. Let $\xi \in T_i$, $\eta \in T_j$ so that there exist sets X and Y, well separated by $\mathscr{E}$, such that $A\xi \subseteq X$, $A\eta \subseteq Y$ and $|X \backslash A\xi| = |Y \backslash A\eta| = q$. Then $\xi\eta \in T_i$. For, firstly, $\xi\eta \circ (\xi\eta)^{-1} = \xi \circ (\eta \circ \eta^{-1}) \circ \xi^{-1} = \xi \circ \mathscr{E}_j \circ \xi^{-1} = \xi \circ \iota_X \circ \xi^{-1} = \xi \circ \xi^{-1} = \mathscr{E}_i$. Secondly, putting $Z = X\eta$, since X is well separated, η is a one-to-one mapping when restricted to X and so $|Z| = |X| = p$. Further $Z \subseteq A\eta \subseteq Y$, so that Z is contained in a well separated set. Hence, since $|Z| = p$, Z is also well separated.

Now $A\xi \subseteq X$ and so $A\xi\eta \subseteq X\eta = Z$. Further, since η is a one-to-one mapping when restricted to X, therefore $|X \backslash A\xi| = |(X \backslash A\xi)\eta| = |X\eta \backslash A\xi\eta| = |Z \backslash A\xi\eta|$. Thus $|Z \backslash A\xi\eta| = q$; and this completes the proof that $\xi\eta \in T_i$.

Hence we have shown that $\mathrm{CT}(A, \mathscr{E}, p, q)$ forms a semigroup under composition and that each subset T_i is a right ideal thereof. That $\mathrm{CT}(A, \mathscr{E}, p, q)$ contains no idempotents follows easily. For suppose $\xi^2 = \xi$ with $A\xi$ contained in the well separated set X such that $|X \backslash A\xi| = q$. Since ξ is a one-to-one mapping when restricted to X, therefore $|X\xi \backslash A\xi^2| = q$. But $X\xi \subseteq A\xi$ and so if $\xi^2 = \xi$ then $X\xi \backslash A\xi^2 = \square$. This is impossible because, by hypothesis, q is an infinite cardinal.

Semigroups of the type $\mathrm{CT}(A, \mathscr{E}, p, q)$ will be called *Croisot-Teissier semigroups*. Before proceeding we give an existence theorem. For this purpose we have the following example due to R. Sikorski (verbal communication). Let A be the Cartesian product of $|I|$ copies of the set E of cardinal p, so that we can write $A = \{(a_i): i \in I,\ a_i \in E\}$. For each j in I define $\mathscr{E}_j = \{(a, b): a = (a_i),\ b = (b_i),\ a_j = b_j\}$. Then $|A/\mathscr{E}_j| = |E| = p$. Further $B = \{(b_i): b_i = b_j \text{ for all } i, j \text{ in } I\}$ is clearly a well separated subset of A. It is now clear that for any infinite $q \leqq p$ a semigroup $\mathrm{CT}(A, \mathscr{E}, p, q)$ exists. Moreover, the cardinal $|I|$ of the set of right ideals T_i (cf. preceding lemma) can be chosen arbitrarily. In Sikorski's example, it follows by symmetry that the T_i form a set of isomorphic semigroups. Exercise 7 below shows that this is not so in general.

The Croisot-Teissier semigroups we are principally interested in are those for which $p = q$, i.e. the semigroups $\mathrm{CT}(A, \mathscr{E}, p, p)$. These semigroups turn out to be simple and their right ideals T_i turn out to be minimal right ideals. The following lemma, which could be inferred from the results of §6.3, facilitates the proof of these assertions.

LEMMA 8.10. *A semigroup which is the union of its minimal right ideals is simple.*

PROOF. Let $\{R_i: i \in I\}$ be the set of minimal right ideals of S, so that $S = \bigcup\{R_i: i \in I\}$. Let $x \in S$ and let $y \in R_i$. Then yxS is a right ideal of S contained in R_i. Hence $yxS = R_i$. Consequently $S = \bigcup\{R_i: i \in I\} = \bigcup\{yxS: y \in S\} = SxS$. Thus S is simple (Lemma 2.28).

We now have

THEOREM 8.11. *Each semigroup* $\mathrm{CT}(A, \mathscr{E}, p, p)$ *is a simple idempotent-free semigroup which is the union of its minimal right ideals* T_i, $i \in I$.

PROOF. Because of Lemmas 8.9 and 8.10, it suffices to prove that each right ideal T_i is a minimal right ideal. Thus we have to show that given ξ, η in T_i there exists ζ in S such that $\eta = \xi\zeta$.

Corresponding to ξ and η there exist well separated subsets X and Y, respectively, of A such that $A\xi \subseteq X$, $A\eta \subseteq Y$, and $|X\backslash A\xi| = |Y\backslash A\eta| = p$. When restricted to $A\xi$ define ζ by $\zeta\colon x\xi \to x\eta$ $(x \in A)$. Since $\xi \circ \xi^{-1} = \mathscr{E}_i = \eta \circ \eta^{-1}$, ζ, so defined, is single-valued (in fact ζ restricted to $A\xi$ is a one-to-one mapping of $A\xi$ onto $A\eta$). Now extend ζ to the rest of A as follows. Since $A\xi \subseteq X$ and X is well-separated there is at most one element of each $\mathscr{E}_i$-class in $A\xi$. For each x in A let ζ map the whole $\mathscr{E}_i$-class containing $x\xi$ onto $x\eta$. Let C be the set of those $\mathscr{E}_i$-classes which have no elements in common with $A\xi$. Then $|C| \leqq p$, since $|A/\mathscr{E}_i| = p$. Since p is infinite and $|Y\backslash A\eta| = p$ there exists a one-to-one mapping δ, say, of the set C into $Y\backslash A\eta$ such that $(Y\backslash A\eta)\backslash C\delta$ is also of cardinal p. Choose any such mapping δ and then define the mapping ζ on the elements of A in the $\mathscr{E}_i$-classes belonging to C by agreeing that ζ will map all the elements in any such $\mathscr{E}_i$-class onto the element that the class is mapped onto by δ. Then we have $\zeta \circ \zeta^{-1} = \mathscr{E}_i$, $A\zeta \subseteq Y$ and $|Y\backslash A\zeta| = |(Y\backslash A\eta)\backslash C\delta| = p$. Thus $\zeta \in S$. Further it is clear that $\xi\zeta = \eta$. This completes the proof of the theorem.

We now prove a sequence of results which culminate in the result that every simple idempotent-free semigroup with a minimal right ideal can be embedded in a semigroup $\mathrm{CT}(A, \mathscr{E}, p, p)$. Simple semigroups with minimal right ideals, but which are not idempotent-free, are in fact completely simple. This is merely the application to semigroups without zero of the result (Schwarz [1951], Theorem 7.2) of Exercise 12 for §2.7. To make our discussion self-contained we digress to prove this result. The next lemma is essentially Exercise 14 for §2.7.

LEMMA 8.12. *Let e be an idempotent in the semigroup S. The following assertions are then mutually equivalent.*

(A) *eSe is a group.*
(B) *Se is a minimal left ideal.*
(B′) *eS is a minimal right ideal.*
(C) *e is primitive and SeS is simple.*

PROOF. To show that (A) implies (B), suppose that $eSe = G$ is a group and that L is a left ideal of S contained in Se. Let $x \in L$. Then $ex \in L$ since L is a left ideal and $xe = x$, since $L \subseteq Se$. Hence $ex = y \in G$, and so, if y^{-1} denotes the inverse of y in the group G, $Se = Sy^{-1}y \subseteq L$, again because L is a left ideal. Thus Se contains properly no left ideal of S and we have shown that (A) implies (B).

Conversely, to show that (B) implies (A), assume that Se is a minimal left ideal of S. Then since Se is minimal $Se \cdot exe = Se$ for any x in S. Thus

$eSe \cdot exe = eSe$ for any exe in eSe. Hence eSe is a left simple semigroup with a left identity, and so is a group.

That (A) is equivalent to (B′) now follows by symmetry. It remains to prove the equivalence of (C) to each of (A) – (B′).

Suppose that (A) holds. Then clearly SeS is simple. Suppose that $f = f^2$ and that $f \leqq e$ so that $fe = ef = f$. Then $f = efe \in eSe$; since a group has only one idempotent therefore $f = e$. Thus e is primitive and we have shown that (A) implies (C).

That (C) implies (A) follows from Lemma 2.47.

The next lemma may be inferred from the results of §6.2 and §6.3 (cf. Lemma 6.18). It includes a converse of Lemma 8.10.

LEMMA 8.13. *Let S be a simple semigroup containing a minimal right ideal. Then S is the disjoint union of its minimal right ideals; xS is the minimal right ideal containing x; every minimal right ideal is a right simple semigroup.*

PROOF. Let R be a minimal right ideal of S. For any x in S, xR is a minimal right ideal of S (Lemma 2.32). For any r in R, $S = SrS \subseteq SR$. Thus $S = SR = \bigcup \{xR : x \in S\}$ is a union of minimal right ideals. Since the intersection of two right ideals is, if non-empty, a right ideal, two distinct minimal right ideals must be disjoint. Further the minimal right ideal which contains x must contain the right ideal xS and so must coincide with xS.

Now let T be a right ideal of the minimal right ideal R of S. Then $TR \subseteq T$. But $(TR)S = T(RS) \subseteq TR$ so that TR is a right ideal of S. Thus $TR = R \subseteq T$. Hence $T = R$ and R is right simple.

REMARK. If R is a minimal right ideal of an arbitrary semigroup S then it is clear that the above proof shows that R is right simple.

We now have as a corollary the result mentioned above (Schwarz [1951], Koch [1953], Wallace [1955], Munn [1957], Saitô and Hori [1958]).

THEOREM 8.14. *Let S be a simple semigroup containing a minimal right ideal. Then S is completely simple if and only if S contains an idempotent.*

PROOF. The necessity of the condition follows from the definition of a completely simple semigroup (§2.7). The sufficiency follows because if e is an idempotent of S, then, by Lemma 8.13, eS is the minimal right ideal of S containing e. Hence, by Lemma 8.12, e is primitive and so S is completely simple.

After this diversion on completely simple semigroups we now return to the main subject of this section, simple semigroups with minimal one-sided ideals which are not completely simple. And in view of Theorem 8.14 we are therefore considering simple semigroups without idempotents.

The following generalization of Lemma 8.3 plays an important part in the sequel.

LEMMA 8.15. *Let S be an idempotent-free simple semigroup containing a minimal right ideal. Then the equation $xy = y$ cannot hold for x, y in S.*

PROOF. Suppose that x belongs to the minimal right ideal R of S (Lemma 8.13). If $xy = y$ then $y \in R$. Hence $xy = y$ for x, y in the right simple semigroup R (Lemma 8.13). This is impossible by Lemma 8.3.

The next two lemmas are part of the proof of our embedding theorem. The first lemma is a direct generalization of Lemma 8.4.

LEMMA 8.16. *Let S be a simple idempotent-free semigroup containing a minimal right ideal. Then*

$$|Ss| = |Ss \backslash Sts| = |Su \backslash Svu| = |Su|$$

for any s, t, u, v in S.

PROOF. Let $R_i, i \in I$, be the minimal right ideals of S, so that $S = \bigcup\{R_i: i \in I\}$, (Lemma 8.13). Since for any $s \in S$, $R_i s \subseteq R_i$, and the R_i are disjoint, the result in the lemma will follow if we show that $R_i s$ and $R_i s \backslash R_i ts$ are sets of equal cardinal independent of the choice of s and t in S. This we proceed to show.

Let $s, u \in S$. Then, by Theorem 6.36, there exists x in S such that $s = sux$. Hence $R_i s = R_i sux$ and so $|R_i s| \geqq |R_i su| \geqq |R_i sux| = |R_i s|$. Hence $|R_i s| = |R_i su|$. But $R_i s \subseteq R_i$, and so $|R_i su| \leqq |R_i u|$. Thus $|R_i s| \leqq |R_i u|$. Similarly we have $|R_i u| \leqq |R_i s|$. Whence it follows that $|R_i s| = |R_i u|$ for any s, u in S.

It remains to prove, for any s, t in S, that $|R_i s| = |R_i s \backslash R_i ts|$. Suppose that $t \in R_j$. For each element u in $R_j s$ there exist elements x' in R_i such that $ux' = t$. This is so because $u = xs$ for some x in R_j and hence $uR_i = xsR_i = R_j$, since $x \in R_j$. For each u in $R_j s$ select a definite element u', say, in R_i such that $uu' = t$. Then the mapping μ: $u \rightarrow u's$ is a one-to-one mapping of $R_j s$ into $R_i s$. For suppose that $u\mu = v\mu$, i.e. $u's = v's$. Now, by Lemma 8.13, there exists z in S such that $s = su'sz$. Let x, y be such that $u = xs$, $v = ys$ $(u, v \in R_j s)$. Then $u = xs = (xs)u'sz = (uu')sz = tsz = v(v's)z = vu'sz = y(su'sz) = ys = v$. Thus $u\mu = v\mu$ implies that $u = v$; and so μ is one-to-one.

In fact μ maps $R_j s$ into $R_i s \backslash R_i ts$. For suppose $u \in R_j s$ and $u\mu \in R_i ts$. Then $u\mu = u's = zts$ where $z \in R_i$. Hence $ts = uu's = (uz)(ts)$ and such an equation is impossible by Lemma 8.15.

Hence $|R_j s| \leqq |R_i s \backslash R_i ts| \leqq |R_i s|$. By symmetry, $|R_j s| = |R_i s|$, for all $i, j \in I$. But then the foregoing inequality becomes an equality; which concludes the proof of the lemma.

LEMMA 8.17. *Let S be a simple idempotent-free semigroup containing a minimal right ideal R, say. Define the relation $\mathscr{E}$ on S thus:*

$$\mathscr{E} = \{(x, y): xr = yr \text{ for some } r \text{ in } R\}.$$

Then $\mathscr{E}$ is an equivalence on S and, for any t in R, $\mathscr{E} \cap (St \times St) = \iota_{St}$ and $|S/\mathscr{E}| = |St|$.

PROOF. The fact that $xr = yr$ for some r in R if and only if $xr = yr$ for all r in R immediately implies that $\mathscr{E}$ is an equivalence on S.

Now let $t \in R$ and let $x, y \in St$. If $(x, y) \in \mathscr{E}$, then $xr = yr$ for all r in R. Hence, putting $x = ut$, $y = vt$, we have $utr = vtr$ for $tr \in R$. Hence $ur' = vr'$ for all r' in R and in particular $ut = vt$, i.e. $x = y$. Thus $\mathscr{E} \cap (St \times St) = \iota_{St}$.

Consider now the mapping $\mu: s \to st$ of S onto St (for some t in R). Then clearly $\mu \circ \mu^{-1} = \mathscr{E}$ so that $|S/\mathscr{E}| = |St|$.

We can now prove Croisot's embedding theorem.

THEOREM 8.18. *Let S be an idempotent-free simple semigroup containing a minimal right ideal. Let $p = |Ss|$ for some s in S. Then S can be embedded in a semigroup* CT$(A, \mathscr{E}, p, p)$ *for some set A.*

Furthermore, the semigroup CT$(A, \mathscr{E}, p, p)$ *can be chosen and the embedding of S can be carried out in such a manner that each minimal right ideal of* CT$(A, \mathscr{E}, p, p)$ *contains precisely one minimal right ideal of S.*

PROOF. We know that $S = \bigcup \{R_i: i \in I\}$ where each R_i is a minimal right ideal of S (Lemma 8.13). Let A be the set $S \cup \Theta$, where $\Theta = \{\theta_0\} \cup \{\theta_i: i \in I\}$ and θ_0, θ_i do not belong to S. Let $\mathscr{E} = \{\mathscr{E}_i: i \in I\}$ be a set of equivalences on A defined thus:

$$\mathscr{E}_i = \{(x, y) \in S \times S: xr_i = yr_i \ (r_i \in R_i)\} \cup ((\Theta \backslash \theta_i) \times (\Theta \backslash \theta_i)) \cup \{(\theta_i, \theta_i)\}.$$

By Lemma 8.17, $|A/\mathscr{E}_i| = |Sr_i \cup \{\theta_i, \theta_0\}| = |Sr_i| = |Ss|$ for any s in S, the latter equality holding by Lemma 8.16. Thus $|A/\mathscr{E}_i| = p$ for each i in I. Further, if $i \neq j$ then $(\theta_0, \theta_j) \in \mathscr{E}_i \backslash \mathscr{E}_j$ and hence $\mathscr{E}_j \neq \mathscr{E}_i$; thus $\mathscr{E}$ is a set of distinct equivalences on A.

Let $X = Sr_i \cup \{r_i, \theta_0\}$, for some $r_i \in R_i$. Then X is well separated by $\mathscr{E}$. For, firstly, $|X| = p$. Further $\mathscr{E}_i$ reduces to the identical relation on Sr_i, by Lemma 8.17, and since $r_i \notin Sr_i$ (Lemma 8.15) $\mathscr{E}_i \cap (X \times X) = \iota_X$. Suppose now that $(x, y) \in \mathscr{E}_j$, $j \neq i$, and that $x, y \in X$. There are five possibilities: (i) $x = ur_i$, $y = vr_i$; (ii) $x = ur_i$, $y = r_i$ $[x = r_i$, $y = vr_i]$; (iii) $x = ur_i$, $y = \theta_0$ $[x = \theta_0$, $y = vr_i]$; (iv) $x = r_i$, $y = \theta_0$ $[x = \theta_0$, $y = r_i]$; (v) $x = y$. Since also $(x, y) \in \mathscr{E}_j$ possibilities (iii) and (iv) can be immediately excluded. In case (i) $(x, y) \in \mathscr{E}_j$ implies either that $x = y$ or that $xr_j = yr_j$ for r_j in R_j. But then since $r_i r_j = t_i \in R_i$, $ut_i = vt_i$ and so $us = vs$ for all s in R_i. In particular $x = ur_i = vr_i = y$, i.e. $x = y$. In case (ii) $(x, y) \in \mathscr{E}_j$ similarly implies that $r_i r_j = t_i \in St_i$ for t_i in R_i which is impossible by Lemma 8.15. Thus $(x, y) \in \mathscr{E}_j \cap (X \times X)$ implies $x = y$. Hence X is, as asserted, well separated by $\mathscr{E}$.

Denote by $\mathscr{S}$ the Croisot-Teissier semigroup $\mathscr{S} = \text{CT}(A, \mathscr{E}, p, p)$ determined by A and $\mathscr{E}$ and, as before, denote by T_i, $i \in I$, the minimal right ideals of $\mathscr{S}$ (Lemma 8.11). Then each element s in S determines an element $s\alpha$ in

$\mathscr{S}$ defined thus:

$$x(s\alpha) = \begin{cases} xs & (x \in S), \\ s & (x = \theta_i \text{ and } s \in R_i), \\ \theta_0 & (x = \theta_j, j \neq i). \end{cases}$$

Then the mapping

$$\alpha: S \to \mathscr{S}$$

is an isomorphic embedding of S in $\mathscr{S}$.

We must first show that $s\alpha \in \mathscr{S}$. Let $s \in R_i$. Then in fact $s\alpha \in T_i$. For it is clear from the definition of $\mathscr{E}_i$, since $xs \neq s$ for any x in $\mathscr{S}$, that $x(s\alpha) = y(s\alpha)$ if and only if $(x, y) \in \mathscr{E}_i$. Further, since S is simple, $S^2 = S$ and so there exist v, u in S such that $s = vu$. Hence, $A(s\alpha) = Ss \cup \{s, \theta_0) = Svu \cup \{vu, \theta_0\} \subseteq Su \cup \{\theta_0\} \subseteq Su \cup \{u, \theta_0\} = A'$, say, which we have shown to be a well separated set. Since $|A' \backslash A(s\alpha)| = |Su \backslash Svu| = p$ (Lemma 8.16), therefore $s\alpha \in \mathscr{S}$.

We now show that α is an isomorphism. Suppose $s\alpha = t\alpha$ for s, t in S. Let $s \in R_i$, $t \in R_j$. Then $\theta_i(s\alpha) = s$, and if $i \neq j$, $\theta_i(t\alpha) = \theta_0$. Thus $s\alpha = t\alpha$ implies $i = j$. It then follows further that $s = \theta_i(s\alpha) = \theta_i(t\alpha) = t$. Hence α is one-to-one.

To show that α is a homomorphism consider $(s\alpha)(t\alpha)$ where $s \in R_i$, $t \in R_j$. Then for x in S, $x(s\alpha)(t\alpha) = (xs)(t\alpha) = xst = x((st)\alpha)$. Further, since s and st both belong to R_i, $\theta_i(s\alpha)(t\alpha) = s(t\alpha) = st = \theta_i((st)\alpha)$; and for $k \neq i$, $\theta_k(s\alpha)(t\alpha) = \theta_0(t\alpha) = \theta_0 = \theta_k((st)\alpha)$. Thus $(s\alpha)(t\alpha) = (st)\alpha$.

Finally, we have already observed that if $s \in R_i$ then $s\alpha \in T_i$, i.e. that $R_i\alpha \subseteq T_i$. From the fact that $\mathscr{E}_i \neq \mathscr{E}_j$, if $i \neq j$, it follows that $T_i \cap T_j = \square$, if $i \neq j$. Thus each minimal right ideal of $\mathscr{S}$ contains precisely one minimal right ideal of $S\alpha$.

This completes the proof of the theorem.

We can now extend Theorem 8.8 to semigroups which are not right cancellative. The extension is again due to Cohn [1956a] and the proof we give is new.

THEOREM 8.19. *Let S be a semigroup. Then S can be embedded in a right simple semigroup without idempotents if and only if* (a) *S is without idempotents and* (b) *$xu = yu$ for x, y, u in S implies $xv = yv$ for all v in S.*

PROOF. The necessity of the conditions is clear.

To prove sufficiency consider a semigroup S satisfying conditions (a) and (b). On S define the equivalence ϵ thus:

$$\epsilon = \{(x, y): xu = yu \text{ for some } u \text{ in } S\}.$$

Condition (b) ensures that ϵ is an equivalence on S.

Let I be any set such that $|I| = p$ where $p = |S/\epsilon|$. Set $A = S^1 \times I$ and define the relation $\mathscr{E}$ on A thus:

$$\mathscr{E} = \{((x, i), (y, i)): i \in I,\ x, y \in S^1 \text{ and } xu = yu \text{ for some } u \text{ in } S\}.$$

Then $|A/\mathscr{E}| = p$. For, by condition (b), $\phi_u: x \to xu$ $(x \in S)$ satisfies $\phi_u \circ \phi_u^{-1} = \epsilon$. Hence, for any u in S, $p = |S/\epsilon| = |Su|$. Since Su is a semigroup without idempotents, therefore p is infinite. Whence it follows that

$$|A/\mathscr{E}| = |Su \times I| = p^2 = p,$$

for any u in S.

Further, for any u in S, set

$$A_u = S^1u \times I \quad \text{and} \quad B_u = A_u \cup (\{1\} \times I).$$

Then B_u is a subset of A well-separated by $\mathscr{E}$. For simplicity of typography, we write $\mathscr{E}$ instead of $\{\mathscr{E}\}$. For let $((x, i), (y, j)) \in \mathscr{E} \cap (B_u \times B_u)$. Then $i = j$, $xv = yv$ for some v in S and $x, y \in S^1u \cup \{1\}$. Suppose that $x = au$, $y = bu$, $a, b \in S^1$. Then $auv = buv$, where $uv \in S$. Hence, if $a, b \in S$, then, by condition (b), $au = bu$, i.e. $x = y$. If $a = b = 1$, then again $x = y$. If $a = 1$ and $b \in S$ then $u = bu$, for u, b in S. This is impossible, for it implies that $bu = b^2u$, so that, by condition (b), $b^2 = b^3$ which gives $b^4 = b^2$. Thus b^2 is an idempotent, contrary to condition (a). The only remaining possibilities are $x = y = 1$ or $x = 1$, $y \in S^1u$. The latter possibility leads to an equation $v = yv$ in S and we have just shown that this is incompatible with conditions (a) and (b). Hence, in all cases, $x = y$. This shows that B_u is well separated by $\mathscr{E}$.

Now define, for each u in S, the mapping $\rho_u: A \to A$ thus:

$$(x, i)\rho_u = (xu, i), \quad \text{where } x \in S^1, i \in I.$$

Then $A\rho_u = A_u$. Further $B_u \backslash A_u = \{1\} \times I$ and so $|B_u \backslash A_u| = p$. Since B_u is well separated, and since, clearly, $\rho_u \circ \rho_u^{-1} = \mathscr{E}$, this shows that each ρ_u belongs to the right simple semigroup without idempotents $\mathrm{CT}(A, \mathscr{E}, p, p)$.

The proof of the theorem will be complete if we show that $\rho: u \to \rho_u$ is an isomorphism of S into $\mathrm{CT}(A, \mathscr{E}, p, p)$. But this is evident.

It is to be remarked that, for the special case of right cancellative semigroups, the above proof provides an alternative proof of Theorem 8.8. For, when S is right cancellative, $\mathscr{E}$ is the identity relation on A, and $\mathrm{CT}(A, \mathscr{E}, p, p)$ reduces to the Baer-Levi semigroup of type (p, p) on A, which is right cancellative.

In addition to the embedding theorems we have mentioned, Cohn [1956a] also obtains a characterization of the subsemigroups of a right group.

Semigroups similar to the Croisot-Teissier semigroups are constructed by Saitô and Hori [1958]; and they prove a similar embedding theorem. Their results are stated in Exercises 8 and 9, below.

Exercises for §8.2

1. Let S be a simple idempotent-free semigroup containing a minimal right ideal. Then $\mathscr{L} = \mathscr{H} = \iota_S$ and hence $\mathscr{R} = \mathscr{D}$.

Further, $\mathscr{R}$ is a congruence on S and $S/\mathscr{R}$ is a left zero semigroup of cardinal equal to the cardinal of the set of minimal right ideals of S.

2. If S is a simple, idempotent-free semigroup and contains a minimal right ideal, then S satisfies the modified cancellation law: $axy = bxy$ implies $ax = bx$.

3. Let S be an idempotent-free simple semigroup containing a minimal right ideal. Then, for any s in S, Ss is an idempotent-free simple right cancellative semigroup containing a minimal right ideal.

4. With the assumptions and notation of Lemma 8.17, $\mathscr{E}$ is a left congruence on S. If S is also right simple show that $\mathscr{E}$ is then a congruence on S and that $S/\mathscr{E}$ is a right simple, right cancellative, idempotent-free semigroup. (M. Teissier [1953a].)

5. With the assumptions and the notation of the proof of Theorem 8.18 $R_i\alpha = T_i \cap S\alpha$ for each i in I.

6. Let S be a right simple semigroup without idempotents. Let M be the set $\Lambda \times S$ where Λ is any set. Let $\phi\colon \Lambda \rightarrow S$ be any mapping of Λ into S. Define a product in M by the rule $(\lambda, s)(\mu, t) = (\lambda, s(\mu\phi)t)$. Then M becomes, with this product, a simple semigroup without idempotents containing a minimal right ideal. In fact the minimal right ideals of S are the subsemigroups $\{\lambda\} \times S$ of M (cf. Exercise 8 for §8.1).

7. Let A be an infinite set. Let $\mathscr{E}_1$ be the identical relation on A. Let $\mathscr{E}_2$ be an equivalence on A such that one $\mathscr{E}_2$-class contains two elements whereas each of the other $\mathscr{E}_2$-classes contain only one element. Let $\mathscr{E} = \{\mathscr{E}_1, \mathscr{E}_2\}$, $|A| = p$ and let $\mathscr{S} = \mathrm{CT}(A, \mathscr{E}, p, p)$. Let T_i $(i = 1, 2)$ be the set of all β in $\mathscr{S}$ such that $\beta \circ \beta^{-1} = \mathscr{E}_i$. Then T_1 is right cancellative and T_2 is not right cancellative. $\mathscr{S}$ is a simple semigroup which is the union of the two non-isomorphic right ideals T_1 and T_2. (Teissier [1953b].)

8. Let $\{K_{i\lambda}\colon i \in I, \lambda \in \Lambda\}$ be a set of disjoint sets, where I and Λ are arbitrary index sets, such that for each i in I, $|K_{i\lambda}| = m_\lambda$, an infinite cardinal. Set $K_i = \bigcup\{K_{i\lambda}\colon \lambda \in \Lambda\}$ and $K_i^* = K_i \cup \{\theta_i\}$, where the θ_i, $i \in I$, are all distinct and $\theta_i \notin K = \bigcup\{K_i\colon i \in I\}$. For each λ in Λ, let m'_λ be an infinite cardinal with $m'_\lambda \leqq m_\lambda$.

Denote by R_i^j the set of one-to-one mappings $\alpha_i^j\colon K_i^* \rightarrow K_j$ satisfying $K_{i\lambda}\alpha_i^j \subseteq K_{j\lambda}$ and $|K_{j\lambda} \backslash K_{i\lambda}\alpha_j^i| = m'_\lambda$, for all λ in Λ. Let $i \rightarrow \lambda_i$ be a fixed mapping of I into Λ. Now define R_i to be the set of all elements $\alpha_i = (\alpha_i^j)$ of the cartesian product $\Pi\{R_i^j\colon j \in I\}$ such that $\theta_i\alpha_i^j \in K_{j\lambda_i}$. Let $S = \bigcup\{R_i\colon i \in I\}$. Define the product $\alpha_i\beta_k$ of two elements $\alpha_i = (\alpha_i^j)$ and $\beta_k = (\beta_k^j)$ of S (α_i in R_i, β_k in R_k) to be the element $\gamma_i = (\gamma_i^j)$ of R_i given by $\gamma_i^j = \alpha_i^k\beta_k^j$ (iteration of mappings).

Then S becomes thereby a simple semigroup which is the union of its minimal right ideals R_i, $i \in I$. Further, S satisfies the condition that $xa = ya$ for all a in S implies that $x = y$. (Saitô and Hori [1958].)

9. Let S be an idempotent-free simple semigroup containing a minimal right ideal. Suppose further that S satisfies the condition that $xa = ya$ for

all a in S implies that $x = y$. Let R_i, $i \in I$, be the distinct minimal right ideals of S. For each i in I define the relation μ_i thus:

$$\mu_i = \{(x, y): xr_i = yr_i \text{ for some } r_i \in R_i\}.$$

Then μ_i is an equivalence relation on S.

Let $K_i = S/\mu_i$ be the set of μ_i-classes; we may and shall regard K_i and K_j as disjoint if $i \neq j$, say by attaching the label i to each element of K_i. Denote by K_{ij} the set of all elements of K_i contained in R_j. Then $K_i = \bigcup\{K_{ij}: j \in I\}$. Set $K_i^* = K_i \cup \{\theta_i\}$, where the $\theta_i, i \in I$, are all distinct and $\theta_i \notin \bigcup\{K_i: i \in I\}$.

Let $a \in S$. Then, for some i in I, $a \in R_i$ and a determines the mappings $a_i^j: K_i^* \to K_j$ defined thus:

$$xa_i^j = \begin{cases} a\mu_j^\natural, & \text{if } x = \theta_i, \\ (sa)\mu_j^\natural, & \text{if } x = s\mu_i^\natural. \end{cases}$$

Denote by a_i the vector $(a_i^j: j \in I)$.

Then $a \to a_i$, where $a \in R_i$, is an isomorphic mapping of S into the semigroup constructed, as in Exercise 8, from the K_i, where the set Λ is now to be identified with I and $i \to \lambda_i$ is the identity mapping. The cardinals m_i and m_j', $i \in I$ and $j \in I$, involved in this construction, are to be taken as all equal (to $|Rs| = |Rs \backslash Rts|$, where R is a minimal right ideal of S and $s, t \in S$: see Lemma 8.16 and its proof). (Saitô and Hori [1958].)

8.3 0-SIMPLE SEMIGROUPS CONTAINING 0-MINIMAL ONE-SIDED IDEALS: GLUSKIN'S EQUIVALENCE

We recall that a 0-simple semigroup S is a semigroup with zero, 0, such that $SsS = S$ for each non-zero element s of S. Equivalently, S is 0-simple if it is not the zero semigroup containing only two elements and if it contains no proper two-sided ideal, i.e. it properly contains no ideal other than the ideal consisting of the zero alone (§2.5).

It is possible to prove the analogues of many of the results of the preceding section. The methods of proof of these results are similar to those of their analogues. We begin with an analogue of Lemma 8.12. Observe that the result here does not hold for an arbitrary semigroup $S = S^0$ (see Exercise 1, below, and compare Lemma 6.38).

LEMMA 8.20. *Let S be a 0-simple semigroup containing a non-zero idempotent e. Then the following assertions about S are equivalent.*

(A) *eSe is a group with zero.*
(B) *Se is a 0-minimal left ideal of S.*
(B′) *eS is a 0-minimal right ideal of S.*
(C) *e is primitive.*

PROOF. To show that (A) implies (B), suppose that $eSe = G^0$ is a group with zero and that L is a non-zero left ideal of S contained in Se. Since L is non-zero and S is 0-simple therefore $S = SLS \subseteq LS$, and so $S = LS$. Because $eS = eLS \neq 0$, therefore $eL \neq 0$. Let $y \in eL \backslash 0$. Since $L \subseteq Se$, therefore $y \in G$. Let y^{-1} be the inverse of y in G. Then $e = y^{-1}y \in y^{-1}eL \subseteq L$. Hence $Se \subseteq L$. Consequently, $L = Se$; which shows that Se is 0-minimal. We have thus shown that (A) implies (B).

Assume now that (B) holds. Let $x \in eSe \backslash 0$. Then $x \in Se \backslash 0$ and, since Se is 0-minimal, $Sex = Se$. Thus $eSe \cdot x = eSe$. This shows that eSe is a left 0-simple semigroup. Consequently, by Theorem 2.27, $eSe \backslash 0$ is a left simple semigroup. And further $eSe \backslash 0$ has e as an identity element, whence $eSe \backslash 0$ is a group. Hence (B) implies (A).

That (A) is equivalent to (B′) now follows by symmetry.

That (C) is equivalent to each of (A), (B) and (B′) now follows from Lemma 2.47 and Theorem 2.48.

From Theorem 2.33, Lemma 2.34, and Lemma 6.1, we have

LEMMA 8.21. *If S is a 0-simple semigroup containing a 0-minimal right ideal then S is the 0-disjoint union of its 0-minimal right ideals, each of which is non-nilpotent. The 0-minimal right ideal containing $x \neq 0$ is xS.*

As a corollary, we have a theorem due to Schwarz [1951, Theorem 7.2].

THEOREM 8.22. *Let S be a 0-simple semigroup containing a 0-minimal right ideal. Then S is completely 0-simple if and only if S contains a non-zero idempotent.*

A characterization of 0-simple semigroups containing a 0-minimal right ideal follows directly from Corollary 6.34. We now turn our attention principally to those which are not completely 0-simple, i.e. to those without non-zero idempotents.

If S is 0-simple, without non-zero idempotents and contains a 0-minimal right ideal, then it can be shown that $xy = y$ for x, y in S implies $y = 0$ (cf. Lemma 8.15). Lemma 8.16, similarly, has an analogue and we have that

$$|Ss| = |Ss \backslash Sts|$$

is independent of the choice of s, t in S. However, this does not appear to lead to a result analogous to Theorem 8.18; and so we discuss this no further.

L. M. Gluskin [1959b] considered an equivalence on a 0-simple semigroup which has no analogue for simple semigroups. We now give Gluskin's results.

A preliminary lemma will be helpful.

LEMMA 8.23. *Let S be a 0-simple semigroup containing a 0-minimal right ideal. Then S is categorical at zero* (§7.7).

PROOF. Let $a, b, c \in S$ with $ab \neq 0$ and $bc \neq 0$. We have to show that $abc \neq 0$. By Lemma 8.21, $S = \bigcup \{R_i : i \in I\}$ where the R_i are the 0-minimal right ideals of S. Suppose that $a \in R_i$, $b \in R_j$ and $c \in R_k$. Since $bc \neq 0$, therefore $bcS = bR_k = R_j$ (Lemma 8.21). Similarly, since $ab \neq 0$, $aR_j = R_i$. Hence $abcS = aR_j = R_i$; whence $abc \neq 0$.

Consider now the principal right congruence $\mathscr{R}_{\{0\}}$, determined by $\{0\}$ (§7.2). For simplicity of notation, let us denote $\mathscr{R}_{\{0\}}$ by $\mathscr{P}$. Thus (cf. Exercise 1 for § 10.4).

$$(1) \qquad \mathscr{P} = \{(x, y) \in S \times S : xa = 0 \quad \text{if and only if} \quad ya = 0 \ (a \in S)\}.$$

Lemma 8.24. *Let S be a 0-simple semigroup containing a 0-minimal right ideal. Let $a, b \in S$ with $ab \neq 0$. Then $ab \,\mathscr{P}\, b$.*

Proof. Let $abx = 0$ for x in S. Then, by the previous lemma, since $ab \neq 0$, it follows that $bx = 0$. Conversely $bx = 0$ implies $abx = 0$. Consequently, $ab \,\mathscr{P}\, b$.

Lemma 8.25. *Let S be a 0-simple semigroup containing a 0-minimal right ideal. Then $\mathscr{L} \subseteq \mathscr{P}$.*

Proof. Suppose $(a, b) \in \mathscr{L}$. If $a = b$, then $(a, b) \in \mathscr{P}$ trivially. If $a \neq b$, then $ua = b$ and $vb = a$ for some u, v in S. If $ax = 0$, then $bx = u(ax) = 0$ and if $bx = 0$, then $ax = v(bx) = 0$; so $(a, b) \in \mathscr{P}$.

For any a in S denote by P_a the $\mathscr{P}$-class containing a. As usual, R_a denotes the $\mathscr{R}$-class containing a.

Lemma 8.26. *Let S be a 0-simple semigroup containing a 0-minimal right ideal. Let $a, b \in S$ with $ab \neq 0$. Then*

$$a(R_b \cap P_b) = R_a \cap P_b.$$

Proof. Let $x \in a(R_b \cap P_b)$. Then $x = ac$, where $c \in R_b \cap P_b$. Since $\mathscr{R}$ is a left congruence (§2.1) $c \,\mathscr{R}\, b$ implies $ac \,\mathscr{R}\, ab$; whence $ac \neq 0$. Now $ac \in aS$, the 0-minimal right ideal containing a (Lemma 8.21). Hence $ac \,\mathscr{R}\, a$, i.e. $x \in R_a$. Moreover, by Lemma 8.24, since $ac \neq 0$, $ac \,\mathscr{P}\, c$; whence $ac \,\mathscr{P}\, b$, i.e. $x \in P_b$. Thus $x \in R_a \cap P_b$.

Conversely, let $x \in R_a \cap P_b$. Let $R_1 = aS$, $R_2 = bS$. Since $ab \neq 0$, $aR_2 = R_1$. From $x \in R_a = R_1 \backslash 0$ we conclude that $x = ac$ with $c \in R_2 \backslash 0 = R_b$. By Lemma 8.24, $ac \,\mathscr{P}\, c$, and since $ac = x \,\mathscr{P}\, b$, we conclude that $c \,\mathscr{P}\, b$. Thus $c = R_b \cap P_b$, and $x = ac \in a(R_b \cap P_b)$.

This completes the proof of the lemma.

We now define Gluskin's equivalence $\mathscr{K}$ to be $\mathscr{R} \cap \mathscr{P}$. Let S be a 0-simple semigroup containing a 0-minimal right ideal. Then, since $xS = 0$ if and only if $x = 0$, $\{0\}$ is both an $\mathscr{R}$-class and a $\mathscr{P}$-class. Hence $\{0\}$ is a $\mathscr{K}$-class. Let $\{P_\lambda : \lambda \in \Lambda\}$ be the non-zero $\mathscr{P}$-classes of S. By Lemma 8.21, $S = \bigcup \{R_i : i \in I\}$, where R_i are the 0-minimal right ideals of S. Then $\{R_i^* : R_i^* = R_i \backslash 0,\ i \in I\}$ are the non-zero $\mathscr{R}$-classes of S and $K_{i\lambda} = R_i^* \cap P_\lambda$, $i \in I$, $\lambda \in \Lambda$, are the non-zero $\mathscr{K}$-classes of S. For each λ in Λ define I_λ, thus:

$$(2) \qquad I_\lambda = \{i \in I : P_\lambda R_i \neq 0\}.$$

With the above notation, we then have

LEMMA 8.27. *$P_\lambda R_i \neq 0$ if and only if $xr \neq 0$ for all x in P_λ and all r in R_i^*.*

PROOF. Assume $P_\lambda R_i \neq 0$. Then there exist x_1 in P_λ and r_1 in R_i^* such that $x_1 r_1 \neq 0$. Let $x \in P_\lambda$ and $r \in R_i^*$. From $(x, x_1) \in \mathscr{P}$ and $x_1 r_1 \neq 0$, we conclude $x r_1 \neq 0$. Since $rS = R_i$, there exists s in S such that $rs = r_1$, and from $0 \neq x r_1 = xrs$ we conclude $xr \neq 0$. The converse is trivial.

LEMMA 8.28. *For each λ in Λ, $I_\lambda \neq \square$. Furthermore, $I = \bigcup\{I_\lambda \colon \lambda \in \Lambda\}$.*

PROOF. Let $a \in P_\lambda$. Then $a \neq 0$ and so $aS \neq 0$. Hence $aR_i \neq 0$, for some i in I. Thus $P_\lambda R_i \neq 0$ and $I_\lambda \neq \square$.

Now consider $a \in R_i^*$. Then $SaS = S$ and so $Sa \neq 0$. Hence $P_\lambda a \neq 0$, for some λ in Λ. Thus $P_\lambda R_i \neq 0$ and $i \in I_\lambda$. Hence $I = \bigcup\{I_\lambda \colon \lambda \in \Lambda\}$.

Combining Lemmas 8.26 and 8.27 we have

LEMMA 8.29. *Let $i, j \in I$, $\lambda, \mu \in \Lambda$. Let $k_{i\lambda} \in K_{i\lambda}$. Then*

$$k_{i\lambda} K_{j\mu} = \begin{cases} K_{i\mu}, & \text{if } j \in I_\lambda; \\ 0, & \text{otherwise.} \end{cases}$$

Whence it follows that

$$K_{i\lambda} K_{j\mu} = \begin{cases} K_{i\mu}, & \text{if } j \in I_\lambda; \\ 0, & \text{otherwise.} \end{cases}$$

For each i in I, define Λ_i thus:

$$\Lambda_i = \{\lambda \colon i \in I_\lambda\}. \tag{3}$$

Then $\lambda \in \Lambda_i$ if and only if $i \in I_\lambda$ and so, for each i in I, $\Lambda_i \neq \square$, and $\Lambda = \bigcup\{\Lambda_i \colon i \in I\}$.

For each i in I, define T_i, thus:

$$T_i = \bigcup\{K_{i\lambda} \colon \lambda \in \Lambda_i\}.$$

Then $T_i \subseteq R_i^*$. Set $A_i = R_i \backslash T_i$.

LEMMA 8.30. *For each i in I, T_i is a right simple subsemigroup of S (in particular, $T_i \neq \square$) and $A_i^2 = 0$.*

PROOF. That $T_i \neq \square$ follows because $\Lambda_i \neq \square$. Let $x \in T_i$. Then $x = k_{i\lambda} \in K_{i\lambda}$, for some λ in Λ_i. Hence

$$\begin{aligned} xT_i &= \bigcup\{k_{i\lambda} K_{i\mu} \colon \mu \in \Lambda_i\} \\ &= \bigcup\{K_{i\mu} \colon \mu \in \Lambda_i\}, \end{aligned}$$

by Lemma 8.29, i.e. $xT_i = T_i$. This shows that T_i is a right simple subsemigroup of S.

Let $x, y \in A_i$. If either of x, y is zero, then $xy = 0$. Otherwise $x \in K_{i\lambda}$ and $y \in K_{i\mu}$ where, in particular $\lambda \notin \Lambda_i$, i.e. $i \notin I_\lambda$. From Lemma 8.29 it follows that $xy = 0$. Thus $A_i^2 = 0$.

This completes the proof of the lemma.

With reference to Exercise 7(a) for §6.1, we note that

$$T_i = \{t \in R_i \colon tR_i = R_i\},$$
$$A_i = \{a \in R_i \colon aR_i = 0\}.$$

For each element x of R_i^* belongs to exactly one $K_{i\lambda} = P_\lambda \cap R_i^*$. If $P_\lambda R_i \neq 0$, then $xR_i \neq 0$ by Lemma 8.27, and hence $xR_i = R_i$. If $P_\lambda R_i = 0$, then $xR_i = 0$. But, by the definition of T_i and A_i, $K_{i\lambda} \subseteq T_i$ or $K_{i\lambda} \subseteq A_i$ according to whether $P_\lambda R_i$ is $\neq 0$ or $= 0$.

Finally, we have the following result relating the different T_i and A_i in S. We remark that, for any i, j in I, there always exists an element a of R_i such that $aR_j = R_i$. For, by Lemma 6.12, $R_iR_j = R_i$, and hence there exists a in R_i such that $aR_j \neq 0$. But then aR_j is a non-zero right ideal contained in R_i, so $aR_j = R_i$.

LEMMA 8.31. *Let $i, j \in I$ and let a be an element of R_i such that $aR_j = R_i$. Then*

$$T_i = \{ar \colon r \in R_j,\ ra \in T_j\},$$

and

$$A_i = \{ar \colon r \in R_j,\ ra = 0\}.$$

PROOF. Since $a \in R_i$, $a = k_{i\lambda} \in K_{i\lambda}$ for some λ in Λ; for, clearly, $a \neq 0$. Let $r = k_{j\mu} \in K_{j\mu}$ be any element of R_j^*. From $aR_j = R_i \subseteq P_\lambda R_j$ we conclude that $j \in I_\lambda$. Hence, by Lemma 8.29, $ar = k_{i\lambda}k_{j\mu} \in K_{i\mu}$. Thus $ar \in T_i$ if and only if $\mu \in \Lambda_i$. But $ra = k_{j\mu}k_{i\lambda} \neq 0$ if and only if $\mu \in \Lambda_i$, again by Lemma 8.29. However when $ra \neq 0$, $ra \in K_{j\lambda} \subseteq T_j$, since $j \in I_\lambda$. And this suffices to prove the first part of the lemma. As for the second part, we note that when $ra = 0$, then $\mu \notin \Lambda_i$, and $ar \in K_{i\mu} \subseteq A_i$.

The previous lemmas combine to provide a proof of the following theorem (Gluskin [1959b]).

THEOREM 8.32. *Let S be a 0-simple semigroup containing a 0-minimal right ideal. Then $S = \bigcup\{R_i \colon i \in I\}$ where the R_i are the 0-minimal right ideals of S. Let $\{P_\lambda \colon \lambda \in \Lambda\}$ be the nonzero $\mathscr{P}$-classes of S, where $\mathscr{P}$ is defined by (1). Each $\mathscr{P}$-class is a union of $\mathscr{L}$-classes of S. Set $R_i^* = R_i \backslash 0$, so that $\{R_i^* \colon i \in I\}$ is the set of non-zero $\mathscr{R}$-classes of S. Set $K_{i\lambda} = R_i^* \cap P_\lambda$, $i \in I$, $\lambda \in \Lambda$. Then the $K_{i\lambda}$ are the non-zero $\mathscr{K}$-classes of S where $\mathscr{K} = \mathscr{R} \cap \mathscr{P}$.*

Furthermore, if $k_{i\lambda} \in K_{i\lambda}$, then

$$k_{i\lambda}K_{j\mu} = \begin{cases} K_{i\mu}, & \text{if } j \in I_\lambda; \\ 0, & \text{otherwise,} \end{cases}$$

where the sets I_λ are defined by (2). Whence it follows that

$$K_{i\lambda}K_{j\mu} = \begin{cases} K_{i\mu}, & \text{if } j \in I_\lambda; \\ 0, & \text{otherwise.} \end{cases}$$

Finally, for each i in I, setting

$$T_i = \bigcup \{K_{i\lambda}\colon K_{i\lambda}^2 = K_{i\lambda}\}$$

and

$$A_i = \bigcup \{K_{i\lambda}\colon K_{i\lambda}^2 = 0\},$$

T_i is a right simple subsemigroup of S, $A_i^2 = 0$ and $A_i = R_i \backslash T_i$.

Exercises for §8.3

1. Let $S = \{0, e, a\}$ be a semigroup with multiplication table

	0	e	a
0	0	0	0
e	0	e	0
a	0	a	0.

Then eSe is a group with zero, but $Se = S$ is not 0-minimal since it contains properly the left ideal $L = \{0, a\}$. (Cf. Lemma 8.20.)

2. Let S be a 0-simple semigroup containing a 0-minimal right ideal and containing no non-zero idempotents. Then the equivalences $\mathscr{L}$ and $\mathscr{H}$ are each the identical equivalence on S and hence $\mathscr{D} = \mathscr{R}$.

3. Let S be a 0-simple semigroup containing a 0-minimal right ideal. Then, if S contains an identity element, S is a group with zero.

4. Let S be a 0-simple semigroup containing a 0-minimal right ideal. Then for some x in S, $x^2 \neq 0$. Furthermore, if $x^2 \neq 0$, then $x^n \neq 0$ for every positive integer n.

5. Let A be a zero semigroup: $A^2 = 0$. Let T be a right simple semigroup without idempotents, disjoint from A. Let $R = T \cup A$, and define a multiplication in R by the following rules.

(1) $AT = 0$.

(2) For each t in T, $ta = a\lambda_t$ $(a \in A\backslash 0)$, where λ_t is a mapping of $A\backslash 0$ upon itself with no fixed points, and such that $\lambda_{st} = \lambda_t \lambda_s$ for all s, t in T.

(3) 0 acts as a zero for R.

(4) Multiplication of two elements in $T[A]$ is as in the semigroup $T[A]$.

Then R is a semigroup with A as its maximum two-sided ideal.

Let Λ be any set. Let $\phi\colon \Lambda \to \mathscr{S}$ be a mapping of Λ upon $\mathscr{S}$, where $\mathscr{S}$ is a set of right translations of R (§1.3) with the following properties. (We do not know if such a set $\mathscr{S}$ always exists.)

(a) For each $s \in R\backslash 0$ there exists $\sigma \in \mathscr{S}$ such that $s\sigma \in T$.

(b) For each $\sigma \in \mathscr{S}$ there exists s in $R\backslash 0$ such that $s\sigma \in T$.

Now define a product in $M = \Lambda \times R$ by the rule $(\lambda, s)(\mu, t) = (\lambda, (s(\mu\phi))t)$.

Then M becomes, with this product, a semigroup with the subset $I = \Lambda \times \{0\}$ as an ideal. Furthermore, the quotient semigroup M/I is a 0-simple semigroup without non-zero idempotents containing a 0-minimal right ideal. In fact the 0-minimal right ideals of M/I are the sets $\{\lambda\} \times R$, modulo I, for λ in Λ. If $\iota_R \in \mathscr{S}$ and $\lambda_1\phi = \iota_R$, then $\{\lambda_1\} \times R \cong R$, and so R is embedded in M/I as a 0-minimal right ideal. (Cf. Exercises 7 and 8 for §6.1.)

6. Let $S = R^+ \times R^+$, where R^+ is the set of positive reals. Define a product in S by the rule $(a, b)(c, d) = (ac, bc + d)$. (Cf. regarding (a, b) as the mapping $x \to ax + b$ of R^+ into R^+.) Then S is simple and on S each of the equivalences $\mathscr{L}$, $\mathscr{R}$, $\mathscr{H}$ and $\mathscr{D}$ is equal to ι_S. (Cf. Exercises 9 and 10 for §2.1.) (O. Andersen [1952].)

7. Let S be a 0-simple semigroup containing the 0-minimal right ideal R. Define the relation $\mathscr{E}$ on S thus:

$$\mathscr{E} = \{(x, y)\colon xr = yr, \quad \text{for all } r \text{ in } R\backslash 0\}.$$

Let A be the left annihilator of R in R and let $T = R\backslash A$. Then $\mathscr{E}$ is an equivalence on S and for any t in T, $\mathscr{E} \cap (St \times St) = \iota_{St}$. Further $|S/\mathscr{E}| = |St|$.

8. Let S be a completely 0-simple semigroup with sandwich matrix $P = (p_{\lambda i})$, $i \in I$, $\lambda \in \Lambda$. Define the relation $\sim$ between the non-zero $\mathscr{L}$-classes of S by $L_\lambda \sim L_\mu$ if and only if the λth and μth rows of P have zeros in the same places, i.e. $p_{\lambda i} = 0$ if and only if $p_{\mu i} = 0$ for $i \in I$. Then $\sim$ is an equivalence relation on the non-zero $\mathscr{L}$-classes of S. The non-zero $\mathscr{P}$-classes of S are the unions of the $\sim$-classes.

9. Let S be a 0-simple semigroup containing a 0-minimal right ideal. Let $S^1 = S \cup \{1\}$ as usual. In the following we use the notation of Theorem 8.32.

(a) Define an $I \times \Lambda$-matrix $\Delta = (\delta_{i\lambda})$ as follows:

$$\delta_{i\lambda} = \begin{cases} 1 & \text{if } P_\lambda R_i \neq 0, \\ 0 & \text{if } P_\lambda R_i = 0. \end{cases}$$

Then Δ has at least one 1 in each row and in each column.

(b) If $\delta_{i\lambda} = 0$, then $K_{i\lambda} \cup 0$ is a null semigroup. If $\delta_{i\lambda} = 1$, then $K_{i\lambda}$ is a right simple subsemigroup of S. In either case, $K_{i\lambda} \cup 0$ is a left ideal of R_i.

(c) For each element $k_{i\lambda}$ of $K_{i\lambda}$, we have

$$k_{i\lambda} K_{j\mu} = \delta_{j\lambda} K_{i\mu}, \quad \text{and} \quad k_{i\lambda} R_j = \delta_{j\lambda} R_i.$$

(d) $T_i = \bigcup \{\delta_{i\lambda} K_{i\lambda}\colon \lambda \in \Lambda\}\backslash 0$.

8.4 Bisimple inverse semigroups

Since an inverse semigroup always contains idempotents a simple inverse semigroup with a minimal one-sided ideal is necessarily completely simple (Theorem 8.14). Because the idempotents of an inverse semigroup commute, a completely simple semigroup can contain only one primitive idempotent and therefore only one idempotent. A completely simple inverse semigroup is therefore a group. We can therefore restrict the discussion of simple inverse semigroups to those semigroups without minimal one-sided ideals.

We shall begin with a characterization of simple inverse semigroups. Thereafter the discussion will be restricted to simple semigroups which are also bisimple, i.e. which possess only one $\mathscr{D}$-class. The results we present are due to Clifford [1953]. In the proofs we give we have preferred to make more use of the general properties of inverse semigroups and of their representations by partial transformations.

THEOREM 8.33. *An inverse semigroup S is simple if and only if for any two idempotents e,f in S there exists an element of S with left unit e and with right unit less than or equal to f.*

PROOF. Suppose that the condition is satisfied and let a, b be any two elements of S. Let $aa^{-1} = e$, $bb^{-1} = f$. Then there exists an element u in S such that $uu^{-1} = e$ and $u^{-1}u \leqq f$. Hence $ub(b^{-1}u^{-1}a) = u(u^{-1}u)(bb^{-1})u^{-1}a = u(u^{-1}u)fu^{-1}a = uu^{-1}uu^{-1}a = ea = a$; i.e. there exist x $(= u)$ and y $(= b^{-1}u^{-1}a)$ such that $xby = a$. Thus S is simple.

Conversely, suppose S is simple. Let e,f be any two idempotents of S. Then there exist x, y in S such that $e = xfy$, and we may assume that $ex = x = xf$, $fy = y = ye$. Hence $xyx = (xfy)x = ex = x$ and $yxy = y(xfy) = ye = y$. Thus $y = x^{-1}$ and so $xx^{-1} = xy = xfy = e$. Further, $x^{-1}x \cdot f = x^{-1}x$. Thus $xx^{-1} = e$ and $x^{-1}x \leqq f$; and this completes the proof of the theorem.

In §8.5 it will be shown that there exist simple inverse semigroups containing an arbitrary cardinal number of $\mathscr{D}$-classes.

We now turn to the bisimple case and begin by collecting some facts about $\mathscr{D}$-classes in an inverse semigroup. Recall, firstly, that each $\mathscr{R}$-class and each $\mathscr{L}$-class in an inverse semigroup S contains precisely one idempotent (Corollary 2.19). Whence it follows (Lemma 2.12), that if e is an idempotent, then R_e consists of all elements of S with e as left unit. Dually, L_e consists of all elements of S with e as right unit. Whence it follows, in particular, that $L_e = R_e^{-1}$ and $R_e = L_e^{-1}$. It also follows that if e and f are two idempotents in the same $\mathscr{D}$-class D of S then there is an element a, say, in D with e as left unit and f as right unit. Conversely, if S is an inverse semigroup such that corresponding to any two idempotents e and f of S there exists an element a with e as left unit and f as right unit, then $e \mathscr{D} f$ and hence, by Lemma 2.12, S is bisimple. We thus have the following analogue of Theorem 8.33.

LEMMA 8.34. *Let S be an inverse semigroup. Then S is bisimple if and only if for any two idempotents e,f in S there exists an element of S with left unit e and right unit f.*

Part (i) of the following lemma, valid for any regular $\mathscr{D}$-class, was stated as Exercise 2 for §2.3.

Lemma 8.35. *Let S be an inverse semigroup and let e be any idempotent of S. Then*

(i) $L_eR_e = D_e$;

(ii) *let $l, l_1 \in L_e$ and $r, r_1 \in R_e$; then $lr = l_1r_1$ if and only if there exists $u \in H_e$ such that $lu = l_1$ and $ur_1 = r$.*

Proof. (i) Let $a \in D_e$ and choose $b \in R_e \cap L_a$ and $c \in R_a \cap L_e$. Then, by Theorem 2.17, $H_cb = H_a$ (H_x denotes the $\mathscr{H}$-class containing x). Thus $a \in H_cb \subseteq L_eb \subseteq L_eR_e$. This shows that $D_e \subseteq L_eR_e$. The converse containment holds by Theorem 2.4. Thus $L_eR_e = D_e$.

(ii) Let $l, l_1 \in L_e$ and $r, r_1 \in R_e$, and suppose that $lr = l_1r_1$. Then $lrr_1^{-1} = l_1r_1r_1^{-1} = l_1e = l_1$. Set $rr_1^{-1} = u$. Then $lu = l_1$. Now $lr = l_1r_1$ implies $l^{-1}lrr_1^{-1} = l^{-1}l_1r_1r_1^{-1}$, i.e. $(er)r_1^{-1} = l^{-1}(l_1e)$, i.e. $rr_1^{-1} = l^{-1}l_1$. Thus $u = l^{-1}l_1$; and hence $ur_1 = l^{-1}l_1r_1 = l^{-1}lr = er = r$, i.e. $ur_1 = r$. Finally, $uu^{-1} = l^{-1}(l_1r_1)r^{-1} = l^{-1}lrr^{-1} = e^2 = e$ and so $u \in R_e$; and $u^{-1}u = r_1r^{-1}\cdot l^{-1}l_1 = r_1(lr)^{-1}l_1 = r_1(l_1r_1)^{-1}l_1 = r_1r_1^{-1}\cdot l_1^{-1}l_1 = e^2 = e$ and so $u \in L_e$. Thus $u \in H_e$.
Conversely, if $lu = l_1$ and $ur_1 = r$, then clearly $lr = l_1r_1$.

Remark. With the notation of part (ii) of the preceding lemma, suppose that $lp = lu$ with $u \in H_e$. Then $ep = u$. In particular, the element u in (ii) is unique.

We will say that *e is an identity element for D_e* if $ea = a = ae$ for all a in D_e, without implying thereby that D_e is a subsemigroup.

Lemma 8.36. *Let S be an inverse semigroup and e an idempotent of S. Then e is an identity element for D_e if and only if R_e is a subsemigroup of S.*

Proof. Suppose that e is an identity element for D_e and let $a, b \in R_e$. Then $ab(ab)^{-1} = a(bb^{-1})a^{-1} = aea^{-1} = aa^{-1} = e$; thus $ab \in R_e$. Consequently, R_e is a subsemigroup of S.

Conversely, suppose that R_e is a subsemigroup of S. To show that e is an identity element for D_e it will clearly suffice to show that $ef\ (= fe) = f$ for any idempotent f of D_e. Consider, therefore an idempotent f in D_e and choose $a \in R_e \cap L_f$. Since R_e is a subsemigroup, $ae \in R_e$. Thus $aea^{-1} = ae(ae)^{-1} = e$. Hence $a^{-1}aea^{-1}a = a^{-1}(ea) = a^{-1}a$, i.e., $fef = f$, i.e., $ef = f$.
This completes the proof of the lemma.

Lemma 8.37. *Let S be an inverse semigroup and e an idempotent of S. Suppose that R_e is a subsemigroup of S. Then R_e is a right cancellative semigroup with e as identity element.*

Proof. That e is the identity element of R_e follows from the preceding lemma. Suppose that $ax = bx$, with $a, b, x \in R_e$. Then $a = ae = (ax)x^{-1} = bxx^{-1} = be = b$. Thus R_e is right cancellative.

Our discussion is ultimately aimed at giving a description of a bisimple inverse semigroup with an identity element. For this purpose it will be useful, as well as of independent interest, to characterize those $\mathscr{D}$-classes of an inverse semigroup which are subsemigroups. Such subsemigroups are then bisimple semigroups (Exercise 6 for §2.3).

If X is any subset of a semigroup S then we will denote by $E(X)$ the set of all idempotents in X.

LEMMA 8.38. *Let D be a $\mathscr{D}$-class of the inverse semigroup S. Then D is a subsemigroup of S if and only if $E(D)$ is a subsemigroup of S.*

Moreover, when D is a subsemigroup, D is a bisimple semigroup.

PROOF. Suppose that $E(D)$ is a subsemigroup of S. Let $a, b \in D$. Set $bb^{-1} = g$, $a^{-1}a = f$. Then, by hypothesis, $gf \in D$. Now $ag \in L_{gf}$. Hence also $ag(ag)^{-1} \in D$, i.e. $aga^{-1} = ab(ab)^{-1} \in D$; and so $ab \in D$. This shows that D is a subsemigroup. Conversely, when D is a subsemigroup, so also, clearly, is $E(D)$.

Finally, let D be a subsemigroup of S, let $a, b \in D$ and suppose that $a \mathscr{R} b$. Then $a(a^{-1}b) = b$ and $b(b^{-1}a) = a$, with $a^{-1}b, b^{-1}a \in D$. Thus $a \mathscr{R} b$ within D. Similarly, two elements of S $\mathscr{L}$-equivalent in S are also $\mathscr{L}$-equivalent in D. Hence D is bisimple.

From this lemma we have as a corollary,

COROLLARY 8.39. *Let D be a $\mathscr{D}$-class of the inverse semigroup S and let R be an $\mathscr{R}$-class of S contained in D. Then D is a subsemigroup of S if and only if for any a, b in R there exists c in R such that $Sa \cap Sb = Sc$.*

PROOF. Assume that D is a subsemigroup of S and let $a, b \in R$. Set $a^{-1}a = f$, $b^{-1}b = g$. Then $Sa = Sf$, $Sb = Sg$ and $Sa \cap Sb = Sfg$ (Lemma 1.19). By hypothesis, $fg \in D$. Choose $c \in R \cap L_{fg}$. Then $Sfg = Sc$.

Proceeding conversely, let f, g be any two idempotents of D. Then there exist $a, b \in R$ with $a^{-1}a = f$ and $b^{-1}b = g$. Let $c \in R$ be such that $Sa \cap Sb = Sc$. Then $c^{-1}c = fg$ (Lemma 1.19 and Corollary 2.19). Consequently, $fg \in D$. The required result now follows from the preceding lemma.

For $\mathscr{D}$-classes with an identity element we have the following lemma. It will be convenient to denote by f_x the right unit of the element x.

LEMMA 8.40. *Let e be an idempotent of the inverse semigroup S and suppose that e is an identity element for D_e. Write $R = R_e$.*

Then (i) *if $a \in R$, $Sa \cap R = Ra$;*

(ii) *if $a, b, c \in R$ and $f_a f_b = f_c$, then $Ra \cap Rb = Rc$.*

PROOF. (i) By Lemmas 8.36 and 8.37, R is a right cancellative semigroup with e as identity element. Hence $Ra \subseteq R \cap Sa$. Let $x \in R \cap Sa$. Then $x = sa$ $(s \in S)$ and $sa(sa)^{-1} = e$. Thus $e = saa^{-1}s^{-1} = ses^{-1} = se(se)^{-1}$. Thus $se = r$, say, belongs to R. Consequently, $x = sa = s(ea) = (se)a = ra \in Ra$; and so $R \cap Sa \subseteq Ra$. Combining the two inequalities we have $R \cap Sa = Ra$.

(ii) Let $a, b, c \in R$ and suppose that $f_a f_b = f_c$. Then $Sc = Sf_c = Sf_a \cap Sf_b = Sa \cap Sb$. Hence, by (i), $Rc = R \cap Sc = (R \cap Sa) \cap (R \cap Sb) = Ra \cap Rb$.

LEMMA 8.41. *Let D be a $\mathscr{D}$-class of the inverse semigroup S. Suppose that D contains an identity element e and that D is a subsemigroup of S. Write $R = R_e$.*

Then, for any $a, b \in R$, there exists $c \in R$ such that $Ra \cap Rb = Rc$, i.e. since $Ra = a \cup Ra$, the principal left ideals of R form a semilattice under intersection.

PROOF. If $a, b \in R$, then $f_a, f_b \in D$. Since $D^2 \subseteq D$, therefore $f_a f_b \in D$. Hence there exists c in R such that $f_c = f_a f_b$. Consequently, by Lemma 8.40, $Rc = Ra \cap Rb$.

In the preceding lemma D is in fact a bisimple semigroup (Lemma 8.38). Consequently the preceding lemmas combine to provide a proof of the greater part of the following theorem.

THEOREM 8.42. *Let S be a bisimple inverse semigroup with identity element e. Write $R = R_e$ and $L = L_e$.*

Then R is a right cancellative semigroup with an identity element and the principal left ideals of R form a semilattice under intersection.

Furthermore, $L = R^{-1}$ and $LR = S$. Each element of S can be written in the form $a^{-1}b$ with $a, b \in R$. If $a, b, c, d \in R$, then $a^{-1}b = c^{-1}d$ if and only if there exists $u \in H_e$ such that $a = uc$ and $b = ud$.

Let $a^{-1}b$, $c^{-1}d$ (a, b, c, $d \in R$) be any two elements of S. Let $Rb \cap Rc = Rx$ with $x \in R$. Set $p = xb^{-1}$, $q = xc^{-1}$. Then $p, q \in R$ and $bc^{-1} = p^{-1}q$. Consequently, $(a^{-1}b)(c^{-1}d) = (pa)^{-1}(qd)$ with $pa, qd \in R$.

PROOF. Only the assertions in the final paragraph remain to be proved.

Let $Rb \cap Rc = Rx$, with $b, c, x \in R$. By Lemma 8.40, $Sb \cap Sc \cap R = Sx \cap R$, i.e. $Sf_b f_c \cap R = Sx \cap R$. Now S is bisimple and hence the left ideal $Sf_b f_c$ has a generator z, say, in R. Then $f_z = f_b f_c$. Since $z \in Sx$, therefore $f_z f_x = f_z$. However, $x \in Sx \cap R$, and so $x \in Sf_b f_c$. Thus $f_x(f_b f_c) = f_x$, i.e. $f_x f_z = f_x$. It follows that $f_b f_c = f_z = f_x$.

Since $x \in Rx = Rb \cap Rc$, there exist $p, q \in R$ such that $x = pb = qc$. Then $p = pe = pbb^{-1} = xb^{-1}$ and, similarly, $q = xc^{-1}$. Further, $p^{-1}q = bx^{-1}xc^{-1} = bf_x c^{-1} = bf_b f_c c^{-1} = bc^{-1}$, since $f_x = f_b f_c$.

This suffices to prove the theorem.

The following corollary makes explicit the fact that S is determined by R.

COROLLARY 8.43. *Let S be a bisimple semigroup with identity element e. Write $R = R_e$ and $H = H_e$.*

On the set $R \times R$ define the equivalence relation τ thus:
$(a, b)\,\tau\,(a', b')$ if and only if $a = ua'$ and $b = ub'$ for some $u \in H$.

Define a product on the quotient set $(R \times R)/\tau$ thus:

$$(a, b)\tau(c, d)\tau = (pa, qd)\tau, \tag{1}$$

where p, q are such that $pb = qc = x$, say, and $Rb \cap Rc = Rx$.

Then, with this product, $(R \times R)/\tau$ becomes a semigroup isomorphic to S.

If R is any semigroup with an identity and such that its set of principal left ideals is closed under intersection, then, denoting by H its group of units, the relation τ may be defined on $R \times R$ as in Corollary 8.43 and τ is then an equivalence relation on $R \times R$. However, if we then attempt to define a product in $(R \times R)/\tau$ by equation (1) the product is not in general well defined. For example, this is the case if we take for R a left group which is not a group and to which an identity has been adjoined. When R is also right cancellative then the construction is always possible and, moreover, $(R \times R)/\tau$ is then a bisimple inverse semigroup with an identity and such that the $\mathscr{R}$-class containing the identity is isomorphic to R. This result of Clifford [1953] follows from the preceding corollary when combined with the next theorem.

THEOREM 8.44. *Let R be a right cancellative semigroup with an identity element 1 in which the intersection of any two principal left ideals is a principal left ideal. Let $\Sigma = \Sigma(R)$ be the inverse hull of R.*

Then Σ is a bisimple inverse semigroup with an identity element. Let P be the $\mathscr{R}$-class of Σ containing the identity element. Then P is isomorphic to R.

PROOF. Let $\rho: a \to \rho_a$ $(a \in R)$ be the right regular representation of R. Then Σ is the inverse subsemigroup generated by $R\rho$ of the symmetric inverse semigroup $\mathscr{I}_R$. Set $P = R\rho$. Then P contains $\rho_1 = \iota$, say, the identical transformation of R. Clearly ι is the identity element of Σ. We proceed to show that $P = R_\iota$, the $\mathscr{R}$-class in Σ containing ι.

Let $\xi \in \Sigma$ and $\xi\xi^{-1} = \iota$. This equation implies that ξ is a mapping of the whole of R into R. Thus, in particular, the element 1 of R is mapped by ξ into the element t, say, of R. By Lemma 1.21(i), ξ is a partial right translation of R. Hence, since it is defined on the whole of R, $r\xi = (r1)\xi = r(1\xi) = rt$ for all r in R. Thus $\xi = \rho_t$; and so $\xi \in P$. Thus $R_\iota \subseteq P$. However, it is immediate that when $\xi \in P$ then $\xi\xi^{-1} = \iota$. Hence $P = R_\iota$.

Note further that, since R contains an identity element, ρ is faithful. Hence P is isomorphic to R.

It remains to show that Σ is bisimple. Since P^{-1} is an $\mathscr{L}$-class of Σ (the $\mathscr{L}$-class containing ι) it will suffice to show that $P^{-1}P = \Sigma$ (Theorem 2.4). Since $\Sigma = \langle P \cup P^{-1} \rangle$ it will follow that $\Sigma = P^{-1}P$ if we show that for any $\xi, \eta \in P$ the element $\xi\eta^{-1}$ can be written in the form $\alpha^{-1}\beta$ for some $\alpha, \beta \in P$.

Since P is isomorphic to R, for any $\xi, \eta \in P$ there exists $\zeta \in P$ such that $P\xi \cap P\eta = P\zeta$. There exist $a, b, c \in R$ such that $\xi = \rho_a$, $\eta = \rho_b$, $\zeta = \rho_c$. Then $\epsilon_\xi = \xi^{-1}\xi$, $\epsilon_\eta = \eta^{-1}\eta$ and $\epsilon_\zeta = \zeta^{-1}\zeta$ are the identical mappings on Ra, Rb and Rc, respectively. Hence $\epsilon_\xi\epsilon_\eta$ is the identical mapping on $Ra \cap Rb$ But $Ra \cap Rb = Rc$; for this follows immediately from $P\xi \cap P\eta = P\zeta$ when we recall that ρ is an isomorphism of R onto P. Hence $\epsilon_\zeta = \epsilon_\xi\epsilon_\eta$.

Now $\zeta \in P\zeta = P\xi \cap P\eta$. Hence there exist α, β in P such that $\zeta = \alpha\xi = \beta\eta$. Then $\alpha = \alpha\xi\xi^{-1} = \zeta\xi^{-1}$ and $\beta = \beta\eta\eta^{-1} = \zeta\eta^{-1}$. Thus $\alpha^{-1}\beta = \xi\zeta^{-1}\zeta\eta^{-1} = \xi\epsilon_\zeta\eta^{-1} = \xi\epsilon_\xi\epsilon_\eta\eta^{-1} = \xi\eta^{-1}$.

This completes the proof of the theorem.

Theorem 8.44 affords us a means of constructing bisimple inverse semigroups Σ from more familiar semigroups R. And Corollary 8.43 provides a means of carrying out this construction by means of ordered pairs of elements of R. If, for example, R is the multiplicative semigroup of positive integers then $Rb \cap Rc = R(b \vee c)$, where $b \vee c$ denotes the L.C.M. of b and c. In the notation of Corollary 8.43, τ is here the identical relation on $R \times R$ and $R \times R$ becomes a bisimple inverse semigroup with a product defined by

$$(a, b)(c, d) = \left(\frac{b \vee c}{b} a, \frac{b \vee c}{c} d\right).$$

The same formula holds if R is the positive part of an abelian lattice ordered group and $b \vee c$ denotes the lattice join of b and c. For example, if R is the additive semigroup of non-negative integers, then $b \vee c$ is max (b, c), and

$$(a, b)(c, d) = (a - b + \max(b, c), d - c + \max(b, c)).$$

In this case Σ is the bicyclic semigroup (cf. Exercise 2 for §1.12).

In conclusion, we mention briefly some recent work on special classes of bisimple inverse semigroups by R. J. Warne [1964], [1965], [1966a], [1966b], and by N. R. Reilly [1965], [1966].

In the paper [1964], Warne describes the homomorphisms of one bisimple inverse semigroup with identity into another such. In the course of this, he shows (Theorem 2.2) that if P is a right cancellative semigroup with identity in which Green's relation $\mathscr{L}$ is a congruence, then P is a Schreier extension of its group U of units by $P/\mathscr{L}$, the latter being a right cancellative semigroup with identity and trivial group of units. This generalizes a theorem of Rees [1948]. (The notion of Schreier extension was carried over from groups to semigroups by L. Rédei [1952], with analogous results. We shall not stop to define the notion here, but mention that an instance of it occurred in Exercise 8 for §4.3.) Moreover, P satisfies the condition that the intersection of two principal left ideals is a principal left ideal (cf. Theorem 8.42), if and only if $P/\mathscr{L}$ satisfies it. By Corollary 8.43, this can easily be converted into the following theorem: if S is a bisimple inverse semigroup with identity, and if Green's relation $\mathscr{H}$ is a congruence on S, then S is a Schreier extension of its group U of units by $S/\mathscr{H}$, and the latter is a bisimple inverse semigroup with identity and with trivial group of units.

In [1965], Warne gives several characterizations of a bisimple inverse semigroup S for which the semilattice E_S of idempotents is totally ordered. Every element of such a semigroup is either left regular or right regular (§4.1), and S is a union of groups, right cancellative subsemigroups, and left cancellative subsemigroups. The structure of such a semigroup S is not, however, fully determined thereby. In [1966], less restrictive conditions are placed on E_S.

Reilly [1966] (see also [1965]) gives an elegant determination of the structure of all bisimple inverse semigroups for which E_S is a totally ordered set anti-isomorphic to the natural numbers; i.e. $E_S = \{e_0, e_1, e_2, \cdots\}$ with $e_0 > e_1 > e_2 > \cdots$. A semigroup S for which E_S has this structure is called an ω-semigroup. Let G be a group and α an endomorphism of G. Let $S = S(G, \alpha)$ be the set of all triples $(m; g; n)$ with m, n non-negative integers and g in G. Define product in S by

$$(m; g; n)(p; h; q) = (m + p - r; g\alpha^{p-r} \cdot h\alpha^{n-r}; n + q - r),$$

when $r = \min(n, p)$, and α^0 means the identity automorphism of G. Then S is a bisimple inverse ω-semigroup. Conversely, every bisimple inverse ω-semigroup S has this structure, G being isomorphic with the group of units of S. $\mathscr{H}$ is a congruence on S, and $S/\mathscr{H}$ is isomorphic with the bicyclic semigroup.

In [1966b], Warne shows how Reilly's theorem can be derived from his Theorem 2.2 of [1964].

Exercises for §8.4

1. If S is a simple inverse semigroup then $|Ss| = |tS|$ for any s, t in S.

2. Let S be a bisimple inverse semigroup with an identity element e. Write $R = R_e$. Let $a, b, c \in R$. Then $Ra \cap Rb = Rc$ if and only if $f_a f_b = f_c$ (where f_x denotes the right unit of x).

3. If e and f are idempotents of a semigroup S, then $e \in SfS$ if and only if there exists an idempotent g of S such that $e \mathscr{D} g$ and $g \leqq f$.

Hence an inverse semigroup S is simple if and only if, for any two idempotents e and f of S, there exists an idempotent g of S such that $e \mathscr{D} g$ and $g \leqq f$.

4. Let S be a bisimple inverse semigroup with an identity element e. Write $R = R_e$.

   (a) Every idempotent of S has the form $a^{-1}a$ with a in R.

   (b) The principal left ideals of R form a semilattice under intersection which is isomorphic with the semilattice of idempotents of S. (Clifford [1953]).

8.5 Any semigroup can be embedded in a simple semigroup

In this section we prove R. H. Bruck's theorem [1958] that any semigroup S can be embedded in a simple semigroup $\mathscr{C}(S)$ possessing an identity element. We then discuss some of the properties which $\mathscr{C}(S)$ shares with S^1. The latter results are new.

Theorem 8.45. *Any semigroup can be embedded in a simple semigroup with an identity.*

PROOF. It will suffice to prove the theorem for a semigroup $S = S^1$ with an identity element 1.

Let $\mathscr{C}(S)$ be the semigroup generated by $S \cup \{a, b\}$, where $a, b \notin S$, and subject to the relations $ab = 1$, $as = a$, $sb = b$, for any s in S, and the equations which hold in S. Setting $a^0 = 1$, $b^0 = 1$ we easily see that the elements of $\mathscr{C}(S)$ are $b^i s a^j$ ($s \in S$, i, j non-negative integers); and it is not difficult to show that $b^i s a^j = b^m t a^n$ if and only if $i = m$, $s = t$ and $j = n$ (cf. Exercises 1 and 2, below, for two other constructions of $\mathscr{C}(S)$).

Let $\alpha = b^i s a^j$ and $\beta = b^m t a^n$ be any two elements of $\mathscr{C}(S)$. Then $\alpha = b^i s a^{m+1} \cdot \beta \cdot b^{n+1} 1 a^j$, so that $\mathscr{C}(S)$ is simple; furthermore, 1 is an identity for $\mathscr{C}(S)$.

This completes the proof of the theorem.

We will continue to denote by $\mathscr{C}(S)$ the semigroup generated by $S = S^1$ and $\{a, b\}$ constructed in the above proof.

Consider the special case in which S is the one-element semigroup. Then the elements of $\mathscr{C}(S)$ are $b^i 1 a^j = b^i a^0 a^j = b^i a^j$ (i, j non-negative integers) and these combine subject solely to the relations $ab = 1$, $a1 = a$, $1b = b$, $a^0 = 1$, $b^0 = 1$. Consequently, $\mathscr{C}(S)$ is the bicyclic semigroup $\mathscr{C}$ (§1.12), i.e. $\mathscr{C}(\langle 1 \rangle) = \mathscr{C}$.

If S is any semigroup such that $S = S^1$ then the homomorphism of S onto $\langle 1 \rangle$ induces the homomorphism

$$\phi_1 \colon b^i s a^j \to b^i a^j \qquad (s \in S)$$

of $\mathscr{C}(S)$ onto $\mathscr{C}$. Thus each semigroup $\mathscr{C}(S)$ may be thought of as an extension of the bicyclic semigroup $\mathscr{C}$.

We now identify the $\mathscr{L}$-, $\mathscr{R}$- and $\mathscr{D}$-classes of $\mathscr{C}(S)$ in terms of those of S. It will be convenient to denote by A and B the subsemigroups of $\mathscr{C}(S)$

$$A = \{a^i \colon i = 0, 1, 2, \cdots\},$$

$$B = \{b^i \colon i = 0, 1, 2, \cdots\}.$$

LEMMA 8.46. (i) *If* $\{L_\lambda \colon \lambda \in \Lambda\}$ *are the* $\mathscr{L}$*-classes of* S, *then* $\{B L_\lambda a^n \colon \lambda \in \Lambda, n = 0, 1, 2, \cdots\}$ *are the* $\mathscr{L}$*-classes of* $\mathscr{C}(S)$.

(ii) *If* $\{R_i \colon i \in I\}$ *are the* $\mathscr{R}$*-classes of* S, *then* $\{b^m R_i A \colon i \in I, m = 0, 1, 2, \cdots\}$ *are the* $\mathscr{R}$*-classes of* $\mathscr{C}(S)$.

(iii) *If* $\{D_\delta \colon \delta \in \Delta\}$ *are the* $\mathscr{D}$*-classes of* S, *then* $\{B D_\delta A \colon \delta \in \Delta\}$ *are the* $\mathscr{D}$*-classes of* $\mathscr{C}(S)$.

PROOF. (i) The elements $b^i s a^j$ and $b^m t a^n$ are $\mathscr{L}$-equivalent in $\mathscr{C}(S)$ if and only if there exist $b^p x a^q$ and $b^u y a^v$ in $\mathscr{C}(S)$ such that

$$b^p x a^q b^i s a^j = b^m t a^n, \tag{1}$$

$$b^u y a^v b^m t a^n = b^i s a^j. \tag{2}$$

There are several possibilities. Listing them we have

$$b^p x a^q b^i s a^j = \begin{cases} b^p x a^{j+q-i}, & \text{if } q > i, \\ b^p x s a^j, & \text{if } q = i, \\ b^{p+i-q} s a^j, & \text{if } q < i. \end{cases}$$

And also

$$b^u y a^v b^m t a^n = \begin{cases} b^u y a^{n+v-m}, & \text{if } v > m, \\ b^u y t a^n, & \text{if } v = m, \\ b^{u+m-v} t a^n, & \text{if } v < m. \end{cases}$$

Suppose that $q > i$. Then from equation (1) we have $j + (q - i) = n$, and from equation (2) $n \leqq j$. This is impossible. Hence $q \leqq i$; and similarly, $v \leqq m$. And each of these implies further that $j = n$.

Since $q \leqq i$, from equation (1) we have either $p = m$ and $xs = t$ or $p + (i - q) = m$ and $s = t$. Since $v \leqq m$, from equation (2) we have either $u = i$ and $yt = s$ or $u + (m - v) = i$ and $t = s$. For any non-negative integers i, m we can find non-negative integers p, q, u, v satisfying these conditions. Hence we have shown that $b^i s a^j$ and $b^m t a^n$ are $\mathscr{L}$-equivalent in $\mathscr{C}(S)$ if and only if $n = j$ and $s \mathscr{L} t$ in S.

(ii) This is the left-right dual of (i).

(iii) The elements $b^i s a^j$ and $b^m t a^n$ are $\mathscr{D}$-equivalent in $\mathscr{C}(S)$ if and only if there exists $b^p x a^q$ such that $b^i s a^j \mathscr{L} b^p x a^q \mathscr{R} b^m t a^n$. By (i) and (ii) this holds if and only if $j = q$, $p = m$ and $s \mathscr{L} x \mathscr{R} t$ in S. Hence $b^i s a^j \mathscr{D} b^m t a^n$ in $\mathscr{C}(S)$ if and only if $s \mathscr{D} t$ in S. From this, assertion (iii) follows.

We have the following corollary (Preston [1959]).

Corollary 8.47. *$\mathscr{C}(S)$ is bisimple if and only if $S = S^1$ is bisimple.*

In the next section we will show how to embed any semigroup in a bisimple semigroup.

Let $b^i s a^j$ and $b^m t a^n$ be two elements of $\mathscr{C}(S)$ with $s, t \in S$. Then we easily verify that

$$(b^i s a^j)(b^m t a^n)(b^i s a^j) = \begin{cases} b^i s^2 a^j, & \text{if } j > m,\ n + (j - m) = i, \\ b^i s t s a^j, & \text{if } j = m,\ n = i; \end{cases}$$

and that these are the only cases for which the product on the left is equal to $b^i x a^j$ for some x in S. It follows that the inverses of $b^i s a^j$ in $\mathscr{C}(S)$ are the elements $b^j t a^i$ where t is an inverse of s in S. Moreover, this implies that $b^i s a^j$ has a unique inverse in $\mathscr{C}(S)$ if and only if s has a unique inverse in S. Thus we have proved the following theorem.

Theorem 8.48. *Let $S = S^1$ be a semigroup. Then $\mathscr{C}(S)$ is regular if and only if S is regular. Moreover, $\mathscr{C}(S)$ is an inverse semigroup if and only if S is an inverse semigroup.*

Since S^1 is a regular [inverse] semigroup if and only if S is a regular [inverse] semigroup we have the following corollaries. (See Appendix.)

COROLLARY 8.49. *Any regular [inverse] semigroup can be embedded in a simple regular [inverse] semigroup with identity.*

COROLLARY 8.50. *There exist simple inverse (and hence regular) semigroups with an identity with an arbitrary cardinal number of $\mathscr{D}$-classes.*

PROOF. Because of Theorem 8.48 and Lemma 8.46 (iii) it suffices to observe that in an inverse semigroup of idempotents each $\mathscr{D}$-class consists of a single idempotent.

EXERCISES FOR §8.5

1. Let N be the set of non-negative integers. Let $U = N \times S^1 \times N$, where S is a semigroup. Define a multiplication in U by the rules:

$$(m, s, n)(m', s', n') = (m + [m' - n], f(n - m'; s, s'), n' + [n - m'])$$

where

$$[x] = x \text{ for } x \geqq 0 \quad \text{and} \quad [x] = 0 \text{ for } x < 0,$$

$$f(x; s, s') = s,\ ss' \text{ or } s' \text{ according as } x > 0,\ x = 0 \text{ or } x < 0.$$

Then U is a semigroup isomorphic to the semigroup $\mathscr{C}(S)$ constructed in the proof of Theorem 8.45. For any fixed integer n the mapping $s \to (n, s, n)$ is an isomorphism of S into U. (To show that U is associative observe that

$$[x] + [y - [-x]] = [x + [y]],$$

and

$$f(x + [y]; f(y; s, s'), s'') = f(y - [-x]; s, f(x; s', s'')).)$$

(Bruck [1958].)

2. (a) Let T be a semigroup with identity element 1. Let S be a subsemigroup of T containing 1, and let a and b be elements of T such that (i) T is generated by $S \cup \{a, b\}$, (ii) $ab = 1$, (iii) $ba \notin S$, (iv) $as = a$ and $sb = b$ for all s in S. Then every element of T is uniquely expressible in the form $b^i s a^j$ with i and j non-negative integers, and s in S.

(b) Let S be any semigroup. Let M be a set disjoint from S^1 and such that $|M \cup S^1| = |M|$. Let $X = M \cup S^1$. Let α be a one-to-one mapping of X upon M. Let β be the mapping of X upon X defined as follows: $\beta | M$ is the inverse of α, and $s\beta = 1$ for every s in S^1. For each s in S^1 define a transformation τ_s of X as follows:

$$x\tau_s = \begin{cases} xs & \text{if } x \in S^1 \\ x & \text{if } x \in M. \end{cases}$$

Let T be the subsemigroup of $\mathscr{T}_X$ generated by $\Sigma \cup \{\alpha, \beta\}$, where $\Sigma = \{\tau_s : s \in S^1\}$. Then conditions (i)–(iv) of part (a) are satisfied with S, a, b replaced by Σ, α, β. Moreover, $\Sigma \cong S$ and $T \cong \mathscr{C}(S)$.

3. Let $\phi: S \rightarrow T$ be a homomorphism of $S = S^1$ upon $T = T^1$. Then

$$\phi^*: b^i s a^j \rightarrow b^i (s\phi) a^j \quad (s \in S)$$

is a homomorphism of $\mathscr{C}(S)$ upon $\mathscr{C}(T)$.

4. Let e be an idempotent of $S = S^1$. Then, in the notation introduced just before Lemma 8.46, BeA is a subsemigroup of $\mathscr{C}(S)$ isomorphic to $\mathscr{C}$.

5. The idempotents $\mathscr{C}(S)$ all have the form $b^m e a^m$ with e an idempotent of $S = S^1$.

6. The idempotents of $\mathscr{C}(S)$ commute if and only if the idempotents of $S = S^1$ commute.

8.6 Any semigroup can be embedded in a bisimple semigroup with identity

We first give a construction, due to M. P. Schützenberger (see Preston [1959]), of a class of bisimple semigroups with identity.

Let A be any infinite set and let $\mathscr{M}(A)$ denote the set of all mappings ξ of A into A such that (i) $|A\xi| = |A|$ and (ii) for any $b \in A\xi$, $|b\xi^{-1}| < |A|$. Here $b\xi^{-1}$ denotes the set of all elements of A mapped onto b by ξ. The following theorem due to Preston [1962] gives necessary and sufficient conditions for $\mathscr{M}(A)$ to be a semigroup. It can also be regarded as providing a characterization of regular cardinals. The cardinal $|A|$ of the set A is said to be *regular* if there does not exist a disjoint cover $\{A_i: i \in I\}$ of A such that $|A_i| < |A|$ for each i in I and with $|I| < |A|$. Given any cardinal p there always exists a regular cardinal $q > p$ (cf. Bachmann [1955], Chapter 7).

Theorem 8.51. *Let A be an infinite set. Then $\mathscr{M}(A)$ forms a semigroup under the operation of composition of mappings if and only if $|A|$ is a regular cardinal.*

Proof. Since composition of mappings is associative therefore $\mathscr{M}(A)$ forms a semigroup under composition if and only if it is closed under composition.

Firstly, suppose that $|A|$ is regular. Let $\xi, \eta \in \mathscr{M}(A)$. Then $|A(\xi\eta)| = |A\xi| = |A|$. For let $A\xi\eta = B$ and suppose that $|B| < |A|$. Then $A\xi = B\eta^{-1} \cap A\xi = \bigcup\{b\eta^{-1} \cap A\xi: b \in B\}$. This conflicts with the assumption that $|A|$ is regular; for $|A\xi| = |A|$, but on the other hand $\eta \in \mathscr{M}(A)$ implies that $|b\eta^{-1} \cap A\xi| < |A|$ and by hypothesis $|B| < |A|$. Hence we must have $|B| = |A|$. Further, for $b \in B$, $b(\xi\eta)^{-1} = (b\eta^{-1})\xi^{-1}$. Let $b\eta^{-1} = C$ so that $|C| < |A|$. Then $b(\xi\eta)^{-1} = \bigcup\{c\xi^{-1}: c \in C\}$; and hence since each $|c\xi^{-1}| < |A|$, the regularity of $|A|$ implies that $|b(\xi\eta)^{-1}| < |A|$. Thus $\xi\eta$ satisfies both of the defining conditions (i) and (ii) of $\mathscr{M}(A)$ above, i.e. $\xi\eta \in \mathscr{M}(A)$ so that $\mathscr{M}(A)$ is a semigroup.

Conversely, suppose that $|A|$ is not regular. Then there exists a set $\{B_i: i \in I\}$ of disjoint subsets of A such that $|B_i| < |A|$ and $|I| < |A|$, while $|B| = |A|$, where $B = \bigcup\{B_i: i \in I\}$; and furthermore we may suppose that

$|A\backslash B| = |A|$. Then there exists an element ξ, say, in $\mathscr{M}(A)$ which maps $A\backslash B$ onto B (one-to-one for example) and which for each i in I maps the elements of the set B_i onto a single element of B_i. Then $\xi^2 \notin \mathscr{M}(A)$. For $A\xi^2 = B\xi$ and $|B\xi| = |I| < |A|$. Thus it follows that if $\mathscr{M}(A)$ is a semigroup then $|A|$ is regular.

This completes the proof of the theorem.

We now assume in what follows that $|A|$ is regular. We will show that $\mathscr{M}(A)$ is bisimple (with an identity, the identical mapping of A upon A). We need two preliminary lemmas.

LEMMA 8.52. *If $\xi, \eta \in \mathscr{M}(A)$, then $\xi \mathscr{L} \eta$ if and only if $A\xi = A\eta$.*

LEMMA 8.53. *If $\xi, \eta \in \mathscr{M}(A)$, then $\xi \mathscr{R} \eta$ if and only if $\xi \circ \xi^{-1} = \eta \circ \eta^{-1}$.*

We give no proofs of these lemmas for they are merely restatements for $\mathscr{M}(A)$ of Lemmas 2.5 and 2.6, respectively, which applied to $\mathscr{T}_A$. The proofs given in Lemmas 2.5 and 2.6 were designed to apply also to the semigroup $\mathscr{M}(A)$.

We can now easily show that $\mathscr{M}(A)$ is a bisimple semigroup when $|A|$ is an infinite regular cardinal. Let $\xi, \eta \in \mathscr{M}(A)$. We will find a mapping α in $\mathscr{M}(A)$ such that $\xi \mathscr{L} \alpha \mathscr{R} \eta$. Denote by ρ the equivalence $\eta \circ \eta^{-1}$. Then $|A/\rho| = |A\eta| = |A| = |A\xi|$. Let θ be any one-to-one mapping of A/ρ onto $A\xi$, and let α be the mapping of A into itself which maps the elements in each ρ-class onto the image of the ρ-class under θ. Then $\alpha \circ \alpha^{-1} = \eta \circ \eta^{-1}$ and $A\alpha = A\xi$. Now $\alpha \circ \alpha^{-1} = \rho$ implies that $|A\alpha| = |A/\rho| = |A|$ and that each set $a\alpha^{-1}$, for $a \in A\alpha$, is a ρ-class and so is of cardinal less than $|A|$. Thus $\alpha \in \mathscr{M}(A)$. Consequently, by Lemmas 8.52 and 8.53, $\xi \mathscr{L} \alpha \mathscr{R} \eta$; and this shows that $\mathscr{M}(A)$ is bisimple. We have therefore proved the following theorem, due to Schützenberger (letter to one of the authors).

THEOREM 8.54. *If A is an infinite set such that $|A|$ is regular then $\mathscr{M}(A)$ is a bisimple semigroup with identity.*

From this, using a proof suggested by Schützenberger, we derive the next theorem (Preston [1959]). The paper cited contains a constructive proof which does not depend on Theorem 8.54.

THEOREM 8.55. *Any semigroup can be embedded in a (necessarily regular) bisimple semigroup with identity.*

PROOF. Let S be any semigroup. Choose a set A containing S such that $|A|$ is an infinite regular cardinal and $|A| > |S|$. Let a_0 be any (fixed) element in $A\backslash S$. Each element s of S then determines a mapping ρ_s of A into A defined thus:

$$x\rho_s = \begin{cases} xs & \text{if } x \in S \\ s & \text{if } x = a_0 \\ x & \text{if } x \in A\backslash\{S \cup a_0\}. \end{cases}$$

It is easily verified that $\rho_s \in \mathcal{M}(A)$. Moreover, the mapping $s \to \rho_s$ is an isomorphism of S into $\mathcal{M}(A)$ (cf. the extended right regular representation of S). And this, because of Theorem 8.54, completes the proof of the theorem.

This construction embeds a given semigroup S in another semigroup of cardinal necessarily greater than that of S. In fact when $|S|$ is an infinite cardinal then S can be embedded in a bisimple semigroup of the same cardinal as S. This we now proceed to show.

Observe that a semigroup T with identity is bisimple if and only if for any a, b in T there exist elements s, t, u, v in T such that $as = ub$, $ast = a$ and $vub = b$. Now let M be any bisimple semigroup with identity in which the semigroup S is embedded. Adjoin the identity of M to S to form the subsemigroup S^1, say, of M. Since M is bisimple with an identity, for each pair of elements a, b in S^1, there exist elements s, t, u, v in M such that $as = ub$, $ast = a$ and $vub = b$. For each pair a, b in S^1, choose such elements s, t, u, v in M and let P denote the set of all such elements selected from M. Let $S(1)$ denote the subsemigroup of M generated by $P \cup S^1$. Now construct $S(2)$ from $S(1)$ in exactly the same way as $S(1)$ was constructed from S^1. Similarly we construct $S(n+1)$ from $S(n)$ for any integer $n \geqq 1$. Let $T = \bigcup_{n=1}^{\infty} S(n)$.

Then T contains an identity, viz., the identity element of M. Further for any a, b in T there exists an integer n such that $a, b \in S(n)$, and then there necessarily exist s, t, u, v in $S(n+1)$, and hence in T, such that $as = ub$, $ast = a$ and $vub = b$. Thus T is bisimple. An easy induction shows that if $|S|$ is infinite, then $|S| = |T|$; and that if $|S|$ is finite then $|T|$ is at most countable. Hence we have shown

Corollary 8.56. *Any semigroup S can be embedded in a bisimple semigroup T with identity, such that $|T| = |S|$ if $|S|$ is infinite, and $|T|$ is countably infinite if $|S|$ is finite.*

N. R. Reilly [1965] has shown that Corollary 8.56 holds if we replace "semigroup" by "inverse semigroup" in both places.

CHAPTER 9

FINITE PRESENTATIONS OF SEMIGROUPS AND FREE PRODUCTS WITH AMALGAMATION

In this chapter we discuss a variety of constructions of semigroups, with the common feature that each construction involves explicitly the construction or determination of a congruence. §9.1 collects together some preliminary results on free semigroups. Of independent interest are the characterization of a free semigroup given in Theorem 9.6, due to Levi [1944] and Dubreil-Jacotin [1947], that of a free subsemigroup of a free semigroup, due to Schützenberger [1955], and the result of Evans [1952] that any countable semigroup can be embedded in a 2-generator semigroup.

§9.2 is concerned with showing that if a semigroup is finitely presented (in the sense that it is defined by means of a finite set of generators satisfying a finite set of relations), in terms of one set of generators and relations, then it is also finitely presented in terms of any other finite set of generators. Thus the concept of a finitely presented semigroup does not depend on the finite set of generators selected.

§9.2 provides an introduction to §9.3, which is devoted to an account of Rédei's [1963] determination of the congruences on a finitely generated free commutative semigroup F. Rédei begins by showing that each such congruence is associated with a unique *congruence pair* (M, f), where M is a subgroup of the free abelian group generated by F and f is a mapping, of a certain restricted kind, of M into the set of ideals of F (Theorem 9.17). This characterization forms the basis for the proof of the result (Theorem 9.28): all finitely generated commutative semigroups are finitely presented (Rédei, loc. cit.).

Given two semigroups S and T, with a common subsemigroup U such that $S \cap T = U$, when is it possible to embed $S \cup T$ in a semigroup R, say, such that, within R, $S \cap T$ (or the intersection of the images of S and T in the embedding) remains equal to U. That such an embedding is not always possible was shown by Kimura [1957]. J. M. Howie initiated the systematic investigation of this problem and §9.4 consists of a survey of his results. Howie's basic result, Theorem 9.44, provides a sufficient condition ensuring the possibility of such an embedding (Howie [1962]). The applications of this result indicate the extent to which the various properties of S, T, and U, in relation to one another, are preserved (Howie [1963a, b, c] and [1964a, b].

We remark that a basic tool of the discussion is the free product of a set $\{S_i : i \in I\}$ of semigroups; and observe that, when each semigroup S_i is a group, the semigroup free product does not coincide with the usual group-theoretic free product. The reason for this is a natural one: in forming the group-theoretic free product of a set of groups, it is assumed that the groups all have a common identity element. No such assumption is possible for an arbitrary set of semigroups. In our terminology, the group-theoretic free product of a set of groups G_i, $i \in I$, becomes the free (semigroup) product of the semigroups G_i with an amalgamation of their common identity element.

The final section of the chapter presents various constructions of the cancellative congruence generated by a given relation. First is presented what may be described as the analogue for cancellative congruences of the construction of a congruence by means of elementary transitions. Next we present an ingenious and well-known method of construction of a cancellative congruence which involves the introduction of formal left and right inverses. We do not know to whom this method owes its invention. Finally, in Theorem 9.54, we give a discussion of canonical forms due to Croisot [1954].

9.1 Free semigroups

The concept of the free semigroup $\mathscr{F}_X$ on X was introduced in §1.12. $\mathscr{F}_X$ consists of all finite non-empty words in the alphabet X, products being formed by juxtaposition. $\mathscr{F}_X^1$ may then be regarded as $\mathscr{F}_X$ together with the "empty word". An isomorphic image of $\mathscr{F}_X$ will also be called a free semigroup (on the image of X under the isomorphism).

We now give some alternative characterizations of free semigroups and, in particular, we determine a necessary and sufficient condition for a subsemigroup of a free semigroup to be free. The results we give are due to F. W. Levi [1944; 1946], M. L. Dubreil-Jacotin [1947], and M. P. Schützenberger [1955/6].

Theorem 9.1. *The semigroup S is a free semigroup on X if and only if $X \subseteq S$ and every element of S can be expressed uniquely as a product of elements of X.*

Proof. If $S = \mathscr{F}_X$, then since words of length one in the elements of X are identified with elements of X (cf. §1.12), it follows that $X \subseteq S$. Further, since two words are distinct unless they are identical, it follows that any element of S is a unique product of elements of X.

Conversely, suppose that every element of S can be expressed uniquely as a product of elements of its subset X. By Lemma 1.28, the inclusion mapping of X in S can be extended to a homomorphism ϕ of $\mathscr{F}_X$ onto S. That ϕ is one-to-one, and hence an isomorphism, is scarcely more than a precise statement of what we mean by stating that an element of S expressed as a product of elements of X is uniquely so expressed.

If S is a free semigroup on X, then by the *length* of an element $w = x_1x_2 \cdots x_n$ ($x_i \in X$) of S, we mean the number n of elements of X occurring in the (unique) expression of w as a product of elements of X.

Corollary 9.2. *If S is a free semigroup on X then $X = S \backslash S^2$.*

Proof. Since each element of X is a word of length one in the elements of X and each element of S^2 is of length at least two, it follows that $X \subseteq S \backslash S^2$. Conversely, since any element of S, which is a product of two or more elements of X, belongs to S^2, therefore $S \backslash S^2 \subseteq X$.

Corollary 9.3. *Let S be a free semigroup on X and T a free semigroup on Y. Let ϕ be an isomorphism of S onto T. Then $X\phi = Y$.*

Proof. By the previous corollary, $X = S \backslash S^2$ and $Y = T \backslash T^2$. Clearly, however, $S^2\phi = S\phi \cdot S\phi \subseteq T^2$, and so $X\phi = T \backslash S^2\phi \supseteq Y$. By symmetry it follows that $X\phi = Y$.

The next theorem gives an important characterization of free semigroups. We note that Lemma 1.28 implies that, when S is free on $M\mu$, the homomorphism ϕ is uniquely determined by μ and ν.

Theorem 9.4. *Let M be a set and let $\mu: M \to S$ be a one-to-one mapping of M onto a set of generators of the semigroup S. Then S is a free semigroup on $M\mu$ if and only if for any semigroup T and mapping $\nu: M \to T$, there exists a homomorphism $\phi: S \to T$ such that $\mu\phi = \nu$.*

Proof. Suppose that S is the free semigroup on $M\mu$. Since μ is one-to-one, a mapping $\nu: M \to T$ determines a mapping $\mu^{-1}\nu$ of $M\mu$ into the semigroup T. By Lemma 1.28, $\mu^{-1}\nu$ can be extended to a (unique) homomorphism ϕ, say, of S into T. We then have $\mu\phi = \nu$.

Conversely, suppose that S has the property of the theorem, i.e. suppose that any mapping of $M\mu$ into a semigroup T can be extended to a homomorphism of S into T. Take T as the free semigroup $\mathscr{F}_X$ on $X = M\mu$ and choose $\nu: M \to \mathscr{F}_X$ so that $\mu^{-1}\nu$ is the inclusion mapping of X into $\mathscr{F}_X$. Then ν is also a one-to-one mapping, and by Lemma 1.28 the (inclusion) mapping $\nu^{-1}\mu$ of X into S can be extended to a homomorphism of $\mathscr{F}_X$ into S. Thus there exist homomorphisms $\phi: S \to \mathscr{F}_X$ and $\psi: \mathscr{F}_X \to S$, extending the identity mapping on X, so that, when restricted to X, both $\phi\psi: S \to S$ and $\psi\phi: \mathscr{F}_X \to \mathscr{F}_X$ reduce to the identity mapping on X. Since X generates both S and $\mathscr{F}_X$, therefore $\phi\psi$ is the identity on S and $\psi\phi$ is the identity on $\mathscr{F}_X$. Hence S and $\mathscr{F}_X$ are isomorphic. Thus S is a free semigroup on $X\psi$, i.e., on $X = M\mu$.

Corollary 9.5. *Let S be any semigroup and let M be a set of generators of S. Let X be any set such that $|X| \geqq |M|$. Then there exists a congruence ρ on $\mathscr{F}_X$ such that $\mathscr{F}_X/\rho \cong S$.*

Proof. Since $|X| \geqq |M|$, there exists a mapping ν, say, of X onto M. Since ν is then a mapping of X into S it follows from the theorem (or from Lemma 1.28), that there exists a homomorphism ϕ, say, of $\mathscr{F}_X$ into S which extends ν. Since ν maps X onto a set of generators of S, therefore ϕ is a homomorphism of $\mathscr{F}_X$ onto S. Taking $\rho = \phi \circ \phi^{-1}$, we have $\mathscr{F}_X/\rho \cong S$.

Let a, b, c be elements of a semigroup and let $a = bc$. Then b is said to be a *left divisor* of a and c a *right divisor* of a. Each of b and c are said to be *divisors* of a.

The next theorem is due to Levi and to Dubreil-Jacotin (loc. cit.).

THEOREM 9.6. *The semigroup S is a free semigroup if and only if it satisfies each of the following conditions*:

(1) *the left and right cancellation laws hold in S*;
(2) *S contains no two-sided identity element*;
(3) *if $ax = by$ for a, b, x, y in S, then $a = b$ or $a = bu$ or $b = av$ for u, v in S*;
(4) *each element of S has only a finite number of left divisors.*

PROOF. The necessity of the conditions is easy to verify and follows directly from Theorem 9.1.

Suppose then that the conditions are satisfied. Define X to be the set $S \backslash S^2$ i.e. as the set of elements of S without any divisors. We will show that X freely generates S.

Firstly, X is non-empty and generates S. For let a be any element of S. If a has no divisors, then $a \in X$. Otherwise $a = bc$ and then both b and c belong to X or $a = xyz$; and so on. Either the process terminates in an expression for a as a product of elements of X or, for any n, as large as we please, there exist $a_1, a_2, \cdots, a_n$ in S with $a = a_1a_2 \cdots a_n$. When $a = a_1a_2 \cdots a_n$ then $a_1, a_1a_2, \cdots, a_1a_2 \cdots a_{n-1}$ are left divisors of a. They are all distinct because, if $x = xy$ in S, then $xy = xy^2$ and so by the left cancellation law (condition (1)), $y = y^2$. But any idempotent of a cancellative semigroup must be the identity element of that semigroup (cf. Exercise 1(b) for §1.1). Hence, by condition (2), S has no idempotents. Thus $a_1, a_1a_2, \cdots, a_1a_2 \cdots a_{n-1}$ are $n-1$ distinct left divisors of a. That there should be no upper bound to n is contrary to condition (4). Hence X generates S.

Suppose that $x_1x_2 \cdots x_r = x_1'x_2' \cdots x_s'$ where the x_i and x_j' belong to X. Let $x_2 \cdots x_r = x$ and $x_2' \cdots x_s' = x'$; then $x_1x = x_1'x'$. Hence, by condition (3), $x_1 = x_1'$ or one of x_1, x_1' can be factorized. The latter possibility is excluded by the definition of X. Hence $x_1 = x_1'$, and by condition (1) therefore $x = x'$. It similarly follows now that $x_2 = x_2'$; and proceeding step by step we finally obtain $r = s$ and $x_i = x_i'$ for $i = 1, 2, \cdots, r$. Thus every element of S can be expressed uniquely as a product of elements of X. By Theorem 9.1, S is therefore a free semigroup on X.

REMARK. Conditions (3) and (4) of the preceding theorem can be replaced by their left-right duals, when the conclusion of the theorem will continue to hold.

Since conditions (1), (2) and (4) are automatically satisfied for subsemigroups of free semigroups, we have as a corollary,

Corollary 9.7. *Let T be a subsemigroup of a free semigroup. Then T is itself a free semigroup if and only if when $ax = by$ with a, b, x, y in T, then $a = b$ or $a = bu$ or $b = av$ with u, v in T.*

This corollary, like the theorem, lacks symmetry as we commented in the Remark above. For a symmetrical characterization of free subsemigroups of free semigroups we have the following useful result due to Schützenberger (loc. cit.).

Corollary 9.8. *Let T be a subsemigroup of the free semigroup S. Then T is itself a free semigroup if and only if, for any w in S, $Tw \cap T \neq \square$ and $wT \cap T \neq \square$ together imply that $w \in T$.*

Proof. Suppose that T is a free semigroup and let aw and wb both belong to T, for some a, b in T, w in S. Then $a(wb) = (aw)b \in T$, and so, by the preceding corollary, $a = aw$ or $a = (aw)u$ or $av = aw$ for u, v in T. Since these factorizations all take place within the free semigroup S, the only possibility (Theorem 9.1) is that $av = aw$ and that $v = w$. Thus $w \in T$; and we have shown that the condition is necessary.

Assume conversely that, for w in S, $Tw \cap T \neq \square$ and $wT \cap T \neq \square$ together imply that $w \in T$. Let $ax = by$ for any a, b, x, y in T. Since a, b, x, y belong to S, therefore $a = b$ or $a = bu$ or $b = av$ with u, v in S, by the theorem. Suppose that $a = bu$. Then $ax = bux = by$; and so, by left cancellation, $ux = y \in T$. Hence $ux \in uT \cap T$ and $bu \in Tu \cap T$. From the hypothesis it follows that $u \in T$. Similarly, if $b = av$, then $v \in T$. From the preceding corollary it now follows that T is free.

As an application let A be the free semigroup on two generators: $A = \mathscr{F}_X$ with $X = \{x, y\}$. Put $a_i = yx^iy$, $i = 1, 2, \cdots$. Let B be the subsemigroup of A generated by all the a_i. Then an element of A belongs to B if and only if it can be expressed in the form $yx^iy^2x^jy^2\cdots y^2x^ky$, where $i > 0, j > 0, \cdots$, $k > 0$. Hence, if $u \in B$ and $wu \in B$ then $w \in B$ and, if $v \in B$ and $vw \in B$ then $w \in B$. Consequently, by the preceding corollary, B is a free semigroup. (It is also obvious, by directly applying Theorem 9.1, that B is a free semigroup on the a_i, $i = 1, 2, \cdots$.) The a_i, $i = 1, 2, \cdots$, are easily seen to be a minimal set of generators of B.

This example was used by T. Evans [1952] to prove that any countable semigroup can be embedded in a semigroup generated by two elements. Evans' proof of this result now follows.

We need a preliminary result on the embedding of one semigroup in another.

Lemma 9.9. *Let T be a subsemigroup of the semigroup S. Let τ be a congruence on T and σ a congruence on S. Then T/τ is naturally embedded (by the mapping $t\tau \to t\sigma$ $(t \in T)$) in S/σ if and only if $\tau = \sigma \cap (T \times T)$.*

Proof. Suppose that $t\tau \to t\sigma$ $(t \in T)$ does embed T/τ in S/σ. Then

$(t_1, t_2) \in \tau$, i.e. $t_1\tau = t_2\tau$, if and only if $t_1, t_2 \in T$ and $t_1\sigma = t_2\sigma$, i.e. if and only if $(t_1, t_2) \in \sigma \cap (T \times T)$. Thus $\tau = \sigma \cap (T \times T)$.

Conversely, suppose that $\sigma \cap (T \times T) = \tau$. In particular, then $\tau \subseteq \sigma$ and so $t_1\tau = t_2\tau$ implies $t_1\sigma = t_2\sigma$. Thus $t\tau \to t\sigma$ $(t \in T)$ is a well-defined mapping of T/τ into T/σ. Further, it is easily verified that this mapping is always a homomorphism.

Suppose that $t_1\sigma = t_2\sigma$ with $t_1, t_2 \in T$. Then $(t_1, t_2) \in \sigma \cap (T \times T)$ which, by assumption, equals τ. Thus $t_1\tau = t_2\tau$. Hence the mapping $t\tau \to t\sigma$ is one-to-one. This completes the proof of the lemma.

Now consider again the semigroup A freely generated by $\{x, y\}$ containing B as a free subsemigroup freely generated by $\{a_i = yx^iy\colon i = 1, 2, \cdots\} = Y$, say. Let S be any countable semigroup. Since S is countable it contains a set of generators M, say, which is at most countable. Since $|Y| \geqq |M|$, by Corollary 9.5, there exists a congruence β, say, on B such that $B/\beta \cong S$. Let β generate the congruence α on A.

Suppose that $(w_1, w_2) \in \alpha \cap (B \times B)$. Since α is generated by β, w_2 is obtained from w_1 by a finite sequence of elementary β-transitions in A (§1.5, p. 18). We wish to show that $(w_1, w_2) \in \beta$. It will suffice, for the result then follows by induction, to show that $(w_1, w_2) \in \beta$ when w_2 (in A) is obtained from w_1 (in B) by a single elementary β-transition. To this end assume that $w_1 = upv$ and $w_2 = uqv$ where $u, v \in A^1$ and $(p, q) \in \beta$. We then have $w_1 \in B$, $p \in B$ and $w_1 = upv$. Recalling that an element of A belongs to B if and only if it is of the form $yx^iy^2x^jy^2 \cdots y^2x^ky$, we easily see that $u, v \in B^1$. Hence, since β is a congruence on B, $(w_1, w_2) \in \beta$.

We have thus shown, since clearly $\beta \subseteq \alpha \cap (B \times B)$, that $\beta = \alpha \cap (B \times B)$. An application of Lemma 9.9 now gives that B/β is embedded naturally in A/α. Since S is isomorphic to B/β, S can also be embedded in A/α. Since A is generated by $\{x, y\}$, therefore A/α is generated by $\{x\alpha, y\alpha\}$. Hence we have proved Evans' theorem.

THEOREM 9.10. *Any countable semigroup can be embedded in a semigroup generated by two elements.*

For an alternative approach to this result the paper of B. H. Neumann [1960] may be consulted. Adapting a notion from the theory of groups, Neumann defines wreath products of two semigroups as follows.

We need two preliminary ideas. Firstly, suppose that A and B are two semigroups and that $\phi : b \to b\phi$ is an anti-representation of B in the semigroup of endomorphisms of A. Then $A \times B$ becomes a semigroup if we define a product by

$$(a, b)(a', b') = (a(a'(b\phi)), bb').$$

Secondly, let S be a semigroup and Y a non-empty set and denote by S^Y the semigroup of all mappings of Y into S combined under the operation (of component-wise multiplication), written as juxtaposition, defined thus:

$$y(fg) = (yf)(yg), \text{ for } y \in Y, f, g \in S^Y.$$

Let T be (or have a faithful representation as) a subsemigroup of $\mathscr{T}_Y$. Then $\phi : t \to t\phi$, $t \in T$, is an anti-representation of T in the semigroup of endomorphisms of S^Y, if we define $t\phi$ thus:

$$y(f(t\phi)) = (yt)f, \ y \in Y, f \in S^Y.$$

Here yt is the image of y under t.

Now take S^Y and T as A and B in the earlier construction and the semigroup $A \times B$ becomes a *wreath product* of S and T.

Neumann's proof of Evans' theorem may now be outlined. Let Q be a countable semigroup. Let S be a semigroup obtained from Q by adjoining, if necessary, both a zero element and an identity element ($S = (Q^0)^1$). Let T be a suitable cyclic group (of order $\geq 3d$, where d is the number of generators of S, if d is finite, of infinite order, otherwise). Let $Y = T$. Then S, and so also Q, is embedded in a subsemigroup of the wreath product of S and T which has three generators if T is infinite and two generators if T is finite. Evans' theorem follows.

For finite semigroups Neumann's method allows the stronger conclusion that a finite semigroup can be embedded in a finite two-generator semigroup. It is possible to preserve other properties of Q in the embedding. For example, if Q is finitely generated and periodic then Q can be embedded in a periodic two-generator semigroup. For further refinements the reader is referred to Neumann's paper (*loc. cit.*).

Exercises for §9.1

1. $\mathscr{F}_X$ is isomorphic to $\mathscr{F}_Y$ if and only if $|X| = |Y|$.

2. The automorphism group of $\mathscr{F}_X$ is isomorphic to $\mathscr{G}_X$, the symmetric group on X.

3. Let $T = \langle a, ab, ba \rangle$ be a subsemigroup of the free semigroup $\mathscr{F}_{\{a,b\}}$. T is not a free semigroup.

4. Let T be a subsemigroup of S having the property that $xTy \cap T \neq \square$, for any x, y in S, implies that $x, y \in T$. Let τ be a congruence on T and let σ be the congruence on S generated by τ regarded as a relation on S. Then $\tau = \sigma \cap (T \times T)$; and, consequently, T/τ is naturally embedded in S/σ.

5. Let T be a subsemigroup of the semigroup S. Let τ be a congruence on T and σ a congruence on S. Then $t\tau \to t\sigma$ ($t \in T$) defines a homomorphism of T/τ into S/σ if and only if $\tau \subseteq \sigma \cap (T \times T)$.

9.2 Finitely presented semigroups

The results of this section are probably well known. We formulate and prove them since they are fundamental for our development.

It will be convenient to introduce the following notation. We will denote by ρ^* the congruence on a semigroup generated by the binary relation ρ on that semigroup (§1.5, p. 18).

When for a semigroup S, we find a set X and a relation ρ on $\mathscr{F}_X$ such that $\mathscr{F}_X/\rho^* \cong S$, then $\mathscr{F}_X/\rho^*$ will be said to *present* S *by generators and relations.* X will be called *the set of generators* and ρ *the set of relations* of S in this presentation (cf. §1.12). If X can be chosen to be finite then S is said to be *finitely generated* (in fact the image of $X\rho^*$ is then a finite set of generators of S). If ρ is a finite set then the presentation $\mathscr{F}_X/\rho^*$ will be said to be *finitely related.* If X is finite and ρ is finite then $\mathscr{F}_X/\rho^*$ is said to *present* S *finitely.* Our object in this section is to show that if S has a finite presentation and $S \cong \mathscr{F}_Y/\sigma$, where Y is finite, then σ contains a finite subset τ such that $\tau^* = \sigma$. Thus a semigroup may be described, simply, as *finitely presented* without specifying the finite set of generators involved.

The following lemma, which provides an essential tool for our argument, is of some interest in itself.

LEMMA 9.11. *Let y be an element not in X and set $Y = X \cup y$. Let ρ be a binary relation on $\mathscr{F}_X$, let $w \in \mathscr{F}_X$ and set $\sigma = \rho \cup (w, y)$. Then $\mathscr{F}_X/\rho^* \cong \mathscr{F}_Y/\sigma^*$ under the homomorphism generated by the mapping $x\rho^* \to x\sigma^*$ $(x \in X)$.*

PROOF. We may regard $\mathscr{F}_X$ as a subsemigroup of $\mathscr{F}_Y$ and apply Lemma 9.9. Since $\rho \subseteq \sigma \cap (\mathscr{F}_X \times \mathscr{F}_X)$, therefore $\rho^* \subseteq \sigma^* \cap (\mathscr{F}_X \times \mathscr{F}_X)$. Conversely, if $(a, b) \in \sigma^* \cap (\mathscr{F}_X \times \mathscr{F}_X)$, then $a, b \in \mathscr{F}_X$ and b is obtained from a by a finite sequence of elementary σ-transitions $a = a_0 \to a_1 \to \cdots \to a_n = b$, say. Suppose that the element y is involved in these transitions and let $a_i = uwv \to uyv = a_{i+1}$, where $u, v \in \mathscr{F}_X^1$, be the first transition at which y is introduced. Since $b \in \mathscr{F}_X$, this occurrence of y must be removed by a later σ-transition $a_j \to a_{j+1}$, say, and this σ-transition can only replace y by w. From the definition of σ, this occurrence of y cannot be involved in any of the σ-transitions $a_{i+1} \to \cdots \to a_j$. It follows that, if we simply delete the σ-transitions $a_i \to a_{i+1}$ and $a_j \to a_{j+1}$, the remaining σ-transitions can be applied to derive a_{j+1} from a_i. By applying this argument as many times as is necessary, it follows that b can be derived from a by σ-transitions that do not involve y, i.e. b can be derived from a by ρ-transitions. Hence, $(a, b) \in \rho^*$, and we have shown that $\sigma^* \cap (\mathscr{F}_X \times \mathscr{F}_X) \subseteq \rho^*$. Consequently, $\sigma^* \cap (\mathscr{F}_X \times \mathscr{F}_X) = \rho^*$. By Lemma 9.9, therefore $\mathscr{F}_X/\rho^*$ is naturally embedded in $\mathscr{F}_Y/\sigma^*$ by the mapping $a\rho^* \to a\sigma^*$ $(a \in \mathscr{F}_X)$, i.e. by the homomorphism generated by $x\rho^* \to x\sigma^*$ $(x \in X)$. Since $y \in w\sigma^*$ this mapping is onto $\mathscr{F}_Y/\sigma^*$.

This completes the proof of the lemma.

We shall need the following special case of Lemma 9.9.

LEMMA 9.12. *Let σ be a congruence on $\mathscr{F}_Y$ and let $\tau \subseteq \sigma$. Suppose that*

$\mathscr{F}_Y/\tau^*$ is naturally embedded in $\mathscr{F}_Y/\sigma$ by the mapping $b\tau^* \to b\sigma$ ($b \in \mathscr{F}_Y$). Then $\tau^* = \sigma$.

It will be convenient to introduce some further notation. Let $X = \{x_1, x_2, \cdots, x_n\}$ be a finite subset of the semigroup S, say, and let b be an element of S which can be expressed as a product of the elements of X. Choose some definite expression of b as such a product. In what follows either S will be the free semigroup $\mathscr{F}_X$ on X, in which event there will be a unique expression of b as a product of the elements of X, or the factors involved will be clear from the context. We shall denote such a product by $b(x)$. Let $x_i \to v_i$, $i = 1, 2, \cdots, n$, be a mapping of X into some semigroup T. Replace each occurrence of x_i in the given factorization $b(x)$ of b by v_i for $i = 1, 2, \cdots, n$. The resulting product of the v_i will be denoted by $b(v)$.

The following lemma deals with a detail of technique that we require for the proof of our theorem.

LEMMA 9.13. *Let σ be a binary relation on the semigroup S. Let x_i, v_i, $i = 1, 2, \cdots, n$, be elements of S. Let $a_k(x)$, $b_k(x)$, $k = 1, 2, \cdots, t$, be elements of S expressed in some given manner as products of the x_i. Set*

$$\mu = \sigma \cup \{(x_i, v_i): i = 1, 2, \cdots, n\} \cup \{(a_k(x), b_k(x)): k = 1, 2, \cdots, t\},$$

$$\nu = \sigma \cup \{(x_i, v_i): i = 1, 2, \cdots, n\} \cup \{(a_k(v), b_k(v)): k = 1, 2, \cdots, t\}.$$

Then $\mu^* = \nu^*$.

PROOF. The replacement of a single occurrence of x_j by v_j in the given expression of $a_k(x)$ as a product of the x_i is an elementary μ-transition of $a_k(x)$. A finite sequence of elementary μ-transitions thus derives $a_k(v)$ from $a_k(x)$. Consequently, $(a_k(x), a_k(v)) \in \mu^*$. Similarly, $(b_k(x), b_k(v)) \in \mu^*$. Since $(a_k(x), b_k(x)) \in \mu^*$, therefore $(a_k(v), b_k(v)) \in \mu^*$. It follows that $\nu^* \subseteq \mu^*$. Similarly, $\mu^* \subseteq \nu^*$; whence $\mu^* = \nu^*$, which completes the proof.

THEOREM 9.14. *Let $X = \{x_i: i = 1, 2, \cdots, n\}$ and $Y = \{y_j: j = 1,2, \cdots, m\}$. Let ρ be the finite binary relation on $\mathscr{F}_X$,*

$$\rho = \{(a_k(x), b_k(x)): k = 1, 2, \cdots, t\}.$$

Let σ be a congruence on $\mathscr{F}_Y$ and let

$$\alpha: \mathscr{F}_X/\rho^* \to \mathscr{F}_Y/\sigma$$

be an isomorphism onto $\mathscr{F}_Y/\sigma$.

Then there exists a finite subset τ, say, of σ such that $\tau^* = \sigma$.

In fact, we may take τ to be the subset

$$\tau = \{(a_k(v), b_k(v)): k = 1, 2, \cdots, t\} \cup \{(y_j, u_j(v)): j = 1, 2, \cdots, m\},$$

where v_i, $i = 1, 2, \cdots, n$ are elements of $\mathscr{F}_Y$ such that

$$(x_i\rho^*)\alpha = v_i\sigma$$

and $u_j(x)$, $j = 1, 2, \cdots, m$, *are elements of* $\mathscr{F}_X$ *such that*

$$(u_j(x)\rho^*)\alpha = y_j\sigma.$$

PROOF. We begin by adjoining, step by step, the elements of Y to X. Set $Y_1 = X \cup y_1$, $Y_2 = Y_1 \cup y_2, \cdots, Y_m = Y_{m-1} \cup y_m = X \cup Y$. Set $\rho_1 = \rho \cup (y_1, u_1(x))$, $\rho_2 = \rho_1 \cup (y_2, u_2(x)), \cdots, \rho_m = \rho_{m-1} \cup (y_m, u_m(x))$, where the $u_j(x)$, $j = 1, 2, \cdots, m$, are elements of $\mathscr{F}_X$ such that $(u_j(x)\rho^*)\alpha = y_j\sigma$. Successive applications of Lemma 9.11 show that $\mathscr{F}_X/\rho^*$ is isomorphic, in turn, to $\mathscr{F}_{Y_1}/\rho_1^*$, $\mathscr{F}_{Y_2}/\rho_2^*, \cdots, \mathscr{F}_{Y_m}/\rho_m^*$. And the isomorphism β, say, of $\mathscr{F}_X/\rho^*$ onto $\mathscr{F}_{Y_m}/\rho_m^*$ is generated by the mapping $x_i\rho^* \rightarrow x_i\rho_m^*$, $i = 1, 2, \cdots, n$.

Choose elements $v_i(y)$ of $\mathscr{F}_Y$ such that $(x_i\rho^*)\alpha = v_i(y)\sigma$, $i = 1, 2, \cdots, n$. Note that the $v_i(y)\sigma$, $i = 1, 2, \cdots, n$, are then a set of generators of $\mathscr{F}_Y/\sigma$ because α is onto $\mathscr{F}_Y/\sigma$. Then $(x_i, v_i(y)) \in \rho_m^*$, $i = 1, 2, \cdots, n$. For, since $(y_j, u_j(x)) \in \rho_m^*$,

$$\begin{aligned}(v_i(y)\rho_m^*)\beta^{-1}\alpha &= (v_i(u)\rho_m^*)\beta^{-1}\alpha \\ &= (v_i(u(x))\rho^*)\alpha \\ &= v_i(y)\sigma \\ &= (x_i\rho_m^*)\beta^{-1}\alpha.\end{aligned}$$

Since $\beta^{-1}\alpha$ is one-to-one, therefore $x_i\rho_m^* = v_i(y)\rho_m^*$, i.e. $(x_i, v_i(y)) \in \rho_m^*$.

Set

$$\begin{aligned}\mu = \{(a_k(x), b_k(x)) \colon k = 1, 2, \cdots, t\} &\cup \{(x_i, v_i(y)) \colon i = 1, 2, \cdots, n\} \\ &\cup \{(y_j, u_j(x)) \colon j = 1, 2, \cdots, m\},\end{aligned}$$

so that what we have shown gives $\mu^* = \rho_m^*$; and set

$$\begin{aligned}\nu = \{(a_k(v), b_k(v)) \colon k = 1, 2, \cdots, t\} &\cup \{(x_i, v_i(y)) \colon i = 1, 2, \cdots, n\} \\ &\cup \{(y_j, u_j(v)) \colon j = 1, 2, \cdots, m\}.\end{aligned}$$

An application of Lemma 9.13 gives $\nu^* = \mu^*$ $(= \rho_m^*)$. Thus $\mathscr{F}_X/\rho^*$ is isomorphic to $\mathscr{F}_{X \cup Y}/\nu^*$ under the homomorphism β generated by $x_i\rho^* \rightarrow v_i(y)\nu^*$.

Now apply Lemma 9.11 to remove the x_i from $X \cup Y$. Successive applications enable us to remove each x_i in turn, simultaneously dropping $(x_i, v_i(y))$ from the set of congruence generators. We finish up with an isomorphism δ, say, of $\mathscr{F}_{X \cup Y}/\nu^*$ onto $\mathscr{F}_Y/\tau^*$, where

$$\tau = \{(a_k(v), b_k(v)) \colon k = 1, 2, \cdots, t) \cup \{(y_i, u_j(v)) \colon j = 1, 2, \cdots, m\}$$

and where δ is generated by the mapping $v_j(y)\nu^* \rightarrow v_j(y)\tau^*$.

Consider now the isomorphism $\delta^{-1}\beta^{-1}\alpha$ of $\mathscr{F}_Y/\tau^*$ onto $\mathscr{F}_Y/\sigma$. It is generated by the mapping $v_j(y)\tau^* \rightarrow v_j(y)\sigma$ and so is the natural embedding of $\mathscr{F}_Y/\tau^*$ in $\mathscr{F}_Y/\sigma$. Further, $\tau \subseteq \sigma$. For $y_j\sigma = (u_j(x)\rho^*)\alpha = u_j(v)\sigma$, so that $(y_j, u_j(v)) \in \sigma$; and $a_k(v)\sigma = (a_k(x)\rho^*)\alpha = (b_k(x)\rho^*)\alpha = b_k(v)\sigma$, so that $(a_k(v), b_k(v)) \in \sigma$. Hence $\tau \subseteq \sigma$. Lemma 9.12 now gives $\tau^* = \sigma$.

This completes the proof of the theorem.

Exercises for §9.2

1. A subsemigroup of a finitely presented semigroup is not necessarily finitely presented.

2. Let U be a subsemigroup of the semigroup S. Let $\mathscr{F}_Y/\sigma^*$ present U by generators and relations. Then there exists a set X, containing Y, and a relation ρ on $\mathscr{F}_X$, such that $\sigma = \rho \cap (\mathscr{F}_Y \times \mathscr{F}_Y)$ and such that $\mathscr{F}_X/\rho^*$ is a presentation of S by generators and relations.

Furthermore, if S is a finitely presented semigroup and $\mathscr{F}_Y/\sigma^*$ presents U finitely, then X and ρ can be chosen so that $\mathscr{F}_X/\rho^*$ presents S finitely.

3. Let S and T be two semigroups such that $S \cap T = U$. Let $\mathscr{F}_Y/\sigma^*$ present U by generators and relations. Then there exist sets X and Z, such that $X \cap Z = Y$, and relations ρ on $\mathscr{F}_X$ and τ on $\mathscr{F}_Z$ such that

$$\sigma = \rho \cap (\mathscr{F}_Y \times \mathscr{F}_Y) = \tau \cap (\mathscr{F}_Y \times \mathscr{F}_Y)$$

and such that $\mathscr{F}_X/\rho^*$ and $\mathscr{F}_Z/\tau^*$ present S and T, respectively, by generators and relations.

9.3 Finitely generated commutative semigroups are finitely presented

The results we give in this section are due to L. Rédei [1963]. Let F be a free commutative semigroup with n generators and let G be the free abelian group generated by F. We give first Rédei's characterization (Theorem 9.17) of any congruence ρ on F in terms of an associated subgroup M, say, of G and a mapping of M into the set of ideals of F. Following Rédei (loc. cit., §33) we then use this characterization to show that any congruence on F is finitely generated. For further details, including explicit constructions for various classes of congruences on F, the reader is referred to Rédei's book [1963].

The free commutative semigroup on the set $X = \{x_1, x_2, \cdots, x_n\}$ may be defined as the set of all words $x_1^{a_1} x_2^{a_2} \cdots x_n^{a_n}$, where the a_i are non-negative integers and where the product of $x_1^{a_1} x_2^{a_2} \cdots x_n^{a_n}$ and $x_1^{b_1} x_2^{b_2} \cdots x_n^{b_n}$ is defined to be $x_1^{a_1+b_1} x_2^{a_2+b_2} \cdots x_n^{a_n+b_n}$. This is isomorphic to the set of all sequences $(a_1, a_2, \cdots, a_n)$, where the a_i are non-negative integers combined under the addition

$$(a_1, a_2, \cdots, a_n) + (b_1, b_2, \cdots, b_n) = (a_1 + b_1, a_2 + b_2, \cdots, a_n + b_n).$$

This is the free commutative semigroup on n generators, with identity element $0 = (0, 0, \cdots, 0)$. It is isomorphic to the direct product (§1.11) of n infinite cyclic semigroups each with identity. Alternatively, $\mathscr{F}_X^1/\rho^*$, where $\rho = \{(x_i x_j, x_j x_i) : i, j = 1, 2, \cdots, n\}$, could be taken as the definition of the free commutative semigroup on n generators; and this form exhibits the fact that finitely generated free commutative semigroups are finitely presented.

Throughout this section we will denote by F the free commutative semigroup, defined above, consisting of all finite sequences of length n, $(a_1, a_2, \cdots, a_n)$, where the a_i are non-negative integers. F is contained in the free additive abelian group on n generators consisting of all finite sequences of integers $(a_1, a_2, \cdots, a_n)$. Throughout this section this group will be denoted by G.

Define the relation $\leqq$ on F thus:

$$(a_1, a_2, \cdots, a_n) \leqq (b_1, b_2, \cdots, b_n)$$

if and only if $a_i \leqq b_i$, for $i = 1, 2, \cdots, n$. This is a partial order on F. In fact, relative to this partial order, F is a lattice. If $\alpha = (a_1, a_2, \cdots, a_n)$ and $\beta = (b_1, b_2, \cdots, b_n)$, then

$$\alpha \vee \beta = (a_1 \vee b_1, a_2 \vee b_2, \cdots, a_n \vee b_n)$$

and

$$\alpha \wedge \beta = (a_1 \wedge b_1, a_2 \wedge b_2, \cdots, a_n \wedge b_n)$$

are the supremum and infimum, respectively, of α and β, where $a_i \vee b_i$ denotes the supremum of a_i and b_i and $a_i \wedge b_i$ denotes the infimum of a_i and b_i.

This partial order on F extends naturally to G by defining $\alpha \geqq \beta$ in G if and only if $\alpha - \beta \in F$. G is then a lattice under this partial order. Furthermore, if $\alpha \geqq \beta$ in G and $\gamma \in G$, then $\alpha + \gamma \geqq \beta + \gamma$. Thus G is a lattice-ordered abelian group and $F \backslash 0$ is the strictly positive part thereof. Rédei's characterization of the congruences on F which we now proceed to give (Theorem 9.17) applies in the more general situation where we take for G any lattice-ordered abelian group and for $F \backslash 0$ the strictly positive part of this group.

Some further notation will be useful. For any μ in G, define μ^+ and μ^- thus:

$$\mu^+ = \mu \vee 0,$$

$$\mu^- = (-\mu) \vee 0.$$

Then $\mu = \mu^+ - \mu^-$. Further, in this notation, we have, for any μ, ν in G,

$$\mu \vee \nu = (\mu - \nu)^+ + \nu = (\mu - \nu)^- + \mu,$$

$$\mu \wedge \nu = \mu - (\mu - \nu)^+ = \nu - (\mu - \nu)^-.$$

Again, observe that, if $\mu, \nu \in G$ and $\pi \in F$, then

$$(\pi + \mu) \vee (\pi + \nu) = \pi + \mu \vee \nu,$$

$$(\pi + \mu) \wedge (\pi + \nu) = \pi + \mu \wedge \nu.$$

Consider now any congruence ρ on F. Define M_ρ thus:

$$M_\rho = \{\alpha - \beta \in G: (\alpha, \beta) \in \rho\}. \tag{1}$$

We easily see that M_ρ is a subgroup of G. For let $\mu = \alpha - \beta$ and $\nu = \gamma - \delta$, where $(\alpha, \beta) \in \rho$ and $(\gamma, \delta) \in \rho$, be any two elements of M_ρ. Then $\mu - \nu = \alpha + \delta - (\beta + \gamma)$ and $(\alpha + \delta, \beta + \gamma) \in \rho$ since ρ is a congruence. Hence $\mu - \nu \in M_\rho$, which shows that M_ρ is a subgroup of G.

Each element μ of M_ρ determines an ideal μf_ρ of F defined thus:

$$\mu f_\rho = \{\xi \in F : (\xi + \mu^+, \xi + \mu^-) \in \rho\}. \tag{2}$$

From the definition (1) of M_ρ, it follows that there exist $\alpha, \beta \in F$ with $\mu = \alpha - \beta$ and $(\alpha, \beta) \in \rho$. Then $\alpha \wedge \beta \in \mu f_\rho$; for $\alpha \wedge \beta + (\alpha - \beta)^+ = \alpha \wedge \beta + \mu^+ = \alpha$ and $\alpha \wedge \beta + (\alpha - \beta)^- = \alpha \wedge \beta + \mu^- = \beta$. Hence μf_ρ is non-empty. Furthermore, since ρ is a congruence, $(\xi + \mu^+, \xi + \mu^-) \in \rho$ implies $(\eta + \xi + \mu^+, \eta + \xi + \mu^-) \in \rho$ for any $\eta \in F$. Thus $\xi \in \mu f_\rho$ implies $\xi + \eta \in \mu f_\rho$ for any $\eta \in F$. This shows that, as asserted, μf_ρ is an ideal of F.

Lemma 9.15. *Let ρ be a congruence on F. Define the subgroup M_ρ of G by equation* (1) *above and the mapping $f = f_\rho$ of M_ρ into the set of ideals of F by equation* (2) *above. Then, f has the properties*:

C(i) $0f = F$;
C(ii) $\mu f = (-\mu)f$, *for any μ in M_ρ*;
C(iii) $(\mu^+ + (\mu f)) \cap (\nu^+ + (\nu f)) \subseteq (\mu \vee \nu) + (\mu - \nu)f$, *for any μ, ν in M_ρ.*

Proof. Properties C(i) and C(ii) follow immediately from the fact that ρ is reflexive and symmetric.

To prove C(iii), consider an element ξ of $(\mu^+ + (\mu f)) \cap (\nu^+ + (\nu f))$. By the definition of μf, $\xi \in \mu^+ + (\mu f)$ if and only if $(\xi, \xi - \mu^+ + \mu^-) = (\xi, \xi - \mu)$ belongs to ρ. Similarly, $(\xi, \xi - \nu) \in \rho$. Set $\eta = \xi - \mu \vee \nu$. Then

$$\begin{aligned} \eta + (\mu - \nu)^+ &= \xi - \mu \vee \nu + (\mu - \nu)^+ \\ &= \xi - \nu; \end{aligned}$$

and

$$\begin{aligned} \eta + (\mu - \nu)^- &= \xi - \mu \vee \nu + (\mu - \nu)^- \\ &= \xi - \mu. \end{aligned}$$

Now, since ρ is transitive, $(\xi, \xi - \mu) \in \rho$ and $(\xi, \xi - \nu) \in \rho$ imply that $(\xi - \nu, \xi - \mu) \in \rho$. Hence $(\eta + (\mu - \nu)^+, \eta + (\mu - \nu)^-) \in \rho$, i.e. $\eta \in (\mu - \nu)f$, i.e. $\xi \in \mu \vee \nu + (\mu - \nu)f$; which is what we had to prove.

Let M be any subgroup of G. Let f be a mapping of M into the set of ideals of F which has the properties C(i)-C(iii) of the previous lemma. Then the pair (f, M) will be said to be a *congruence pair on* F.

Any congruence pair $\mathscr{P} = (f, M)$ determines a relation $\rho(\mathscr{P})$ on F defined thus:

$$\rho(\mathscr{P}) = \{(\alpha, \beta) \in F \times F : \alpha - \beta \in M \quad \text{and} \quad \alpha \wedge \beta \in (\alpha - \beta)f\}. \tag{3}$$

LEMMA 9.16. *If $\mathscr{P} = (f, M)$ is a congruence pair, then $\rho = \rho(\mathscr{P})$, defined by* (3), *is a congruence on F.*

PROOF. If $\alpha \in F$, then $\alpha \wedge \alpha = \alpha \in (\alpha - \alpha)f = 0f$, since $0f = F$ by (Ci). Thus ρ is reflexive. Property C(ii) similarly ensures that ρ is symmetric. Suppose that $(\alpha, \beta) \in \rho$ and that $\xi \in F$. Then $(\alpha + \xi) \wedge (\beta + \xi) = \xi + \alpha \wedge \beta$; and, hence, since $(\alpha - \beta)f$ is an ideal of F, $\xi + \alpha \wedge \beta \in (\alpha - \beta)f$. Consequently, $(\alpha + \xi, \beta + \xi) \in \rho$. This shows that ρ is compatible with the operation of F.

It remains to prove that ρ is transitive. Consider (α, β) and (β, γ) both elements of ρ. Then $\alpha - \beta \in M$ and $\beta - \gamma \in M$ give that $\alpha - \gamma \in M$. And $\alpha \wedge \beta \in (\alpha - \beta)f$ and $\beta \wedge \gamma \in (\beta - \gamma)f$ give that

$$\beta - (\beta - \alpha)^+ \in (\alpha - \beta)f = (\beta - \alpha)f,$$

and

$$\beta - (\beta - \gamma)^+ \in (\beta - \gamma)f.$$

Hence

$$\beta \in ((\beta - \alpha)^+ + (\beta - \alpha)f) \cap ((\beta - \gamma)^+ + (\beta - \gamma)f).$$

Applying C(iii) we therefore have that

$$\beta \in (\beta - \alpha) \vee (\beta - \gamma) + (\alpha - \gamma)f,$$

i.e.,

$$\beta \in \beta + (-\alpha) \vee (-\gamma) + (\alpha - \gamma)f.$$

Thus

$$0 \in (-\alpha) \vee (-\gamma) + (\alpha - \gamma)f,$$

and since $(-\alpha) \vee (-\gamma) = -(\alpha \wedge \gamma)$, we conclude that $\alpha \wedge \gamma \in (\alpha - \gamma)f$. Thus ρ is transitive.

This completes the proof of the lemma.

We have now shown that any congruence ρ on F determines a congruence pair $(f_\rho, M_\rho) = \mathscr{P}(\rho)$ and that, conversely, any pair $\mathscr{P} = (f, M)$ determines a congruence $\rho(\mathscr{P})$ on F. In fact this relationship between congruences and congruence pairs is one-to-one. The following is Rédei's "Fundamental Theorem" (loc. cit., p. 20). Rédei calls f_ρ the "kernel function" associated with ρ.

THEOREM 9.17. *The mapping $\rho \to (f_\rho, M_\rho) = \mathscr{P}(\rho)$ defined by* (1) *and* (2) *above, is a one-to-one mapping of the set of all congruences on F onto the set of all congruence pairs associated with F. The reciprocal of this mapping is the mapping $(f, M) = \mathscr{P} \to \rho(\mathscr{P})$, defined by* (3) *above.*

PROOF. Let ρ be a congruence on F and set $\mathscr{P} = (f_\rho, M_\rho)$. We will show that $\rho(\mathscr{P}) = \rho$. Since $(\alpha, \beta) \in \rho$ and $(\alpha, \beta) \in \rho(\mathscr{P})$ each imply that $(\alpha - \beta) \in M_\rho$, it suffices to show that, for $\alpha - \beta$ in M_ρ, $(\alpha, \beta) \in \rho$ if and only if $\alpha \wedge \beta \in (\alpha - \beta)f_\rho$, i.e. if and only if $(\alpha \wedge \beta + (\alpha - \beta)^+, \alpha \wedge \beta + (\alpha - \beta)^-) \in \rho$. But this is true because $\alpha \wedge \beta + (\alpha - \beta)^+ = \alpha$ and $\alpha \wedge \beta + (\alpha - \beta)^- = \beta$.

Conversely, let $\mathscr{P} = (f, M)$ be any congruence pair associated with F and set $\rho = \rho(\mathscr{P})$. We will show that $f_\rho = f$ and $M_\rho = M$. Firstly, let $\mu \in M_\rho$. Then, by (1), $\mu = \alpha - \beta$, for some α, β such that $(\alpha, \beta) \in \rho$. But, by (3), $(\alpha, \beta) \in \rho$ implies $\alpha - \beta \in M$. Thus $M_\rho \subseteq M$. Conversely, let $\mu \in M$. Choose $\xi \in \mu f$. Then $\xi = \xi + 0 = \xi + (\mu^+ \wedge \mu^-) = (\xi + \mu^+) \wedge (\xi + \mu^-) \in \mu f$ and $\xi + \mu^+ - (\xi + \mu^-) = \mu$. Hence, by (3), $(\xi + \mu^+, \xi + \mu^-) \in \rho$; whence, by (1), $\mu \in M_\rho$. Hence $M \subseteq M_\rho$. Thus $M = M_\rho$.

If $\mu \in M$, then $\xi \in \mu f_\rho$ if and only if $(\xi + \mu^+, \xi + \mu^-) \in \rho$, i.e. if and only if

$$\xi = \xi + 0 = \xi + (\mu^+ \wedge \mu^-) = (\xi + \mu^+) \wedge (\xi + \mu^-) \in \mu f.$$

Thus $f = f_\rho$.

This completes the proof of the theorem.

The rest of this section is devoted to proving Rédei's theorem, that any congruence on F is finitely generated. The next theorem, which is fundamental for this proof, is attributed by Rédei to Dickson [1913]. He states it (loc. cit., p. 52) in an equivalent form: *any subset of F, no two distinct elements of which are comparable, is finite.* It was also proved by S. G. Bourne [1949] for the slightly more general semigroup consisting of a group G to which n commuting indeterminates have been adjoined, in the form of an analogue of the Hilbert Basis Theorem for polynomial rings: *every ideal in F has a finite basis* (Corollary 9.20 below). (See Appendix.)

Let A be a subset of F. Then $\alpha \in A$ will be said to be a *minimal element* of A if, for β in F, $\beta \leqq \alpha$ implies either $\beta = \alpha$ or $\beta \notin A$.

THEOREM 9.18. *Let A be a subset of F. Then the set N of all minimal elements of A is finite. Furthermore, if $\alpha \in A$, then there exists $\nu \in N$ such that $\nu \leqq \alpha$.*

PROOF. We proceed by induction on n, the number of generators of F. The result clearly holds for $n = 1$. Assume therefore that it has already been proved for free commutative semigroups with $(n - 1)$ generators. Consider the set of integers which occur as jth components of the elements of A and let l_j be the least integer in this set. Denote by A_j the set of all elements of A with jth component equal to l_j. By the inductive hypothesis, the set M_j, say, of all minimal elements of A_j, is finite. Set $M = M_1 \cup M_2 \cup \cdots \cup M_n$. Let m_j be the largest jth component of the elements of the (finite) set M, and write $\mu = (m_1, m_2, \cdots, m_n)$. Then it is clear that every element of M is less than or equal to μ.

Denote by N_j the set of all minimal elements of A with jth component p_j satisfying $l_j \leqq p_j \leqq m_j$. From the inductive hypothesis, it follows that each set N_j is finite. Write $N = N_1 \cup N_2 \cup \cdots \cup N_n$. Then N is a finite set. Further, N is the set of all minimal elements of A.

For let $\gamma = (c_1, c_2, \cdots, c_n)$ be any minimal element of A. If $\gamma \notin N$, then $\gamma \notin N_j$ and so, by the definition of N_j and of l_j, $c_j > m_j$, for $j = 1, 2, \cdots, n$.

Thus $\gamma > \mu$ and each element of M is consequently less than γ, contrary to the choice of γ as a minimal element of A.

The final assertion of the theorem follows because it is clear that any strictly descending chain of elements of A is finite.

The unique set of minimal elements of a subset A of F will be called the *basis* of A.

COROLLARY 9.19. *Let M be a subgroup of G and suppose that $S = M \cap (F\backslash 0)$ contains at least one (non-zero) element. Then S is a finitely generated subsemigroup of F; in fact, the basis of S finitely generates S.*

PROOF. Let B be the basis of S. Let $\alpha \in S$. Then there exists β in B such that $\alpha \geqq \beta$. Then $\alpha - \beta \geqq 0$ and so $\alpha - \beta \in F$. Also $\alpha, \beta \in M$; hence $\alpha - \beta \in M$. Consequently, either $\alpha = \beta$, or $\alpha - \beta \in M \cap (F\backslash 0) = S$. Hence, if $\alpha \neq \beta$, there exists β' in B such that $\alpha - \beta \geqq \beta'$. It follows, as before, that $\alpha - \beta - \beta' \in S$ or $\alpha = \beta + \beta'$. This process must terminate in a finite number of steps, giving $\alpha = \beta + \beta' + \cdots + \beta^{(k)}$; and this shows that B generates the semigroup S.

COROLLARY 9.20. *Let A be an ideal of F. Let B be the basis of A. Then the ideal of F generated by B is A. Furthermore, any set of generators of A must contain B.*

Thus any ideal of F contains a unique minimal finite set of generators.

PROOF. Let $\alpha \in A$. Then there exists β in B such that $\alpha \geqq \beta$. Since $\alpha - \beta \geqq 0$, $\alpha - \beta = \eta$, say, belongs to F. Hence, since $\alpha = \eta + \beta$, α belongs to the ideal of F generated by B.

Let C be any set of generators of A. Then each element of A can be written in the form $\eta + \gamma$, where $\eta \in F$ and γ is a (non-empty) sum of elements of C. In particular, if $\beta \in B$, then β is such a sum, $\beta = \eta + \gamma$. This implies that β is greater than or equal to each element of C used to form the sum γ. Since β is a minimal element of A, therefore $\beta \in C$.

This completes the proof of the corollary.

Two elements $\alpha = (a_1, a_2, \cdots, a_n)$ and $\beta = (b_1, b_2, \cdots, b_n)$ of G will be said to be *compatible* if $a_i \geqq 0$ if and only if $b_i \geqq 0$ for $i = 1, 2, \cdots, n$. A set of elements of G compatible in pairs will be called a *compatible set.*

LEMMA 9.21. *Let $\{\alpha_j : j = 1, 2, \cdots, k\}$ be a compatible set of elements of G. Let $\mu = \sum\{c_j\alpha_j : j = 1, 2, \cdots, k\}$, where each $c_j \geqq 0$. Then*

$$\mu^+ = \sum\{c_j\alpha_j^+ : j = 1, 2, \cdots, k\},$$

$$\mu^- = \sum\{c_j\alpha_j^- : j = 1, 2, \cdots, k\}.$$

PROOF. The assertion in the lemma follows from the observation that, for any μ in G, μ^+ is formed from μ by deleting the negative components of μ, and that $\mu^- = (-\mu)^+$.

Denote by Φ the subgroup of the automorphism group of G consisting of all $\phi: G \to G$ which merely reverse the sign of some fixed subset of the components of each element of G. Let $\alpha = (a_1, a_2, \cdots, a_n)$. Then

$$\alpha^\phi = ((-1)^{\phi_1}a_1, (-1)^{\phi_2}a_2, \cdots, (-1)^{\phi_n}a_n,)$$

where $(\phi_1, \phi_2, \cdots, \phi_n)$ is a sequence of 0's and 1's determining ϕ. The group Φ contains 2^n elements. For each ϕ in Φ, ϕ^2 is the identical transformation of G.

LEMMA 9.22. *Let M be a subgroup of G. Then there exists a finite subset C, say, of M such that each element of M can be expressed as a sum of non-negative multiples of the elements of some compatible subset of C.*

PROOF. Let $\phi \in \Phi$ and denote by M^ϕ the subgroup $M^\phi = \{\alpha^\phi: \alpha \in M\}$ of G. Then, when $M^\phi \cap (F\backslash 0)$ is non-empty, it possesses a finite basis (Theorem 9.18). Denote this finite basis by M_ϕ. When $M^\phi \cap (F\backslash 0) = \square$, take M_ϕ to be $\square$.

Write $M_\phi^\phi = \{\alpha^\phi: \alpha \in M_\phi\}$. Then M_ϕ^ϕ is a compatible subset of G: because each element of M_ϕ belongs to F. Set $C = \bigcup\{M_\phi^\phi: \phi \in \Phi\}$. Then C is finite because each M_ϕ^ϕ is finite and Φ is finite. Evidently $C \subseteq M$.

Now consider any element $\mu \neq 0$ of M. Clearly there exists ϕ in Φ such that $\mu^\phi \in F\backslash 0$. Then, for such a ϕ, $\mu^\phi \in M^\phi \cap (F\backslash 0)$. Applying Corollary 9.19 gives that μ^ϕ is a sum of non-negative multiples of the elements of M_ϕ,

$$\mu^\phi = \sum\{c_i\alpha_i: i = 1, 2, \cdots, k\},$$

say, where $\alpha_i \in M_\phi$ and $c_i \geqq 0$, $i = 1, 2, \cdots, k$. Thus, since $(\mu^\phi)^\phi = \mu$,

$$\mu = \sum\{c_i\alpha_i^\phi: i = 1, 2, \cdots, k\},$$

where $\alpha_i^\phi \in M_\phi^\phi \subseteq C$, $i = 1, 2, \cdots, k$.

This completes the proof of the lemma.

REMARK. As an immediate corollary to Lemma 9.22 it follows that any subgroup of a finitely generated free abelian group is finitely generated; from which it follows that any subgroup of a finitely generated abelian group is finitely generated.

LEMMA 9.23. *Let ρ be a congruence on F and let $\mathscr{P} = \mathscr{P}(\rho) = (f, M)$. Let $\{\mu_j: j = 1, 2, \cdots, k\}$ be a compatible subset of M. Then, if $c_j \geqq 0$, $j = 1, 2, \cdots, k$,*

$$\bigcap\{\mu_j f: j = 1, 2, \cdots, k\} \subseteq (c_1\mu_1 + c_2\mu_2 + \cdots + c_k\mu_k)f.$$

PROOF. We prove the result first for $k = 1$. We have to show that $\mu f \subseteq (m\mu)f$ for any μ and for any non-negative integer m. We proceed by induction on m. The inequality clearly holds for $m = 0$, because $0f = F$. Assume therefore that it is true for $m - 1$. Let $\xi \in \mu f$, i.e. $(\xi + \mu^+, \xi + \mu^-) \in \rho$. By the inductive hypothesis, $\xi \in ((m - 1)\mu)f$, i.e. $(\xi + ((m - 1)\mu)^+, \xi +$

$((m-1)\mu)^-)\in\rho$. But $((m-1)\mu)^+ = (m-1)\mu^+$ and $((m-1)\mu)^- = (m-1)\mu^-$. Thus $(\xi + (m-1)\mu^+, \xi + (m-1)\mu^-)\in\rho$. Using the compatibility of ρ we have $((\xi + (m-1)\mu^+) + \mu^+, (\xi + (m-1)\mu^-) + \mu^+)\in\rho$, i.e. $(\xi + (m\mu)^+, \xi + (m-1)\mu^- + \mu^+)\in\rho$, and again $((\xi + \mu^+) + (m-1)\mu^-, (\xi + \mu^-) + (m-1)\mu^-)\in\rho$, i.e. $(\xi + (m-1)\mu^- + \mu^+, \xi + (m\mu)^-)\in\rho$. Thus, using transitivity, $(\xi + (m\mu)^+, \xi + (m\mu)^-)\in\rho$, i.e. $\xi\in(m\mu)f$. Thus $\mu f \subseteq ((m-1)\mu)f$ implies $\mu f\subseteq (m\mu)f$. By induction this inequality therefore holds for all non-negative integers m.

Now let μ, ν be any two compatible elements. Then $(\xi + \mu^+, \xi + \mu^-)\in\rho$ implies $(\xi + \mu^+ + \nu^+, \xi + \mu^- + \nu^+)\in\rho$, and $(\xi + \nu^+, \xi + \nu^-)\in\rho$ implies $(\xi + \nu^+ + \mu^-, \xi + \mu^- + \nu^-)\in\rho$. Hence $\xi\in\mu f\cap\nu f$ implies $(\xi + \mu^+ + \nu^+, \xi + \mu^- + \nu^-)\in\rho$, i.e., since μ and ν are compatible and using Lemma 9.21, $(\xi + (\mu+\nu)^+, \xi + (\mu+\nu)^-)\in\rho$. Hence $\mu f\cap\nu f\subseteq(\mu+\nu)f$.

We can now carry out an induction on k. The result of the lemma has already been shown for $k = 1$. Assume that it holds for any set of $k-1$ compatible elements. Set $\mu = c_1\mu_1 + c_2\mu_2 + \cdots + c_{k-1}\mu_{k-1}$ and $\nu = c_k\mu_k$. Then μ and ν are compatible and so, by what we have just proved, $\mu f\cap\nu f\subseteq(\mu+\nu)f$. By our inductive assumption on k,

$$\bigcap\{\mu_j f : j = 1, 2, \cdots, k-1\}\subseteq \mu f.$$

The case of $k = 1$ gives $\mu_k f\subseteq \nu f$. Hence

$$\bigcap\{\mu_j f : j = 1, 2, \cdots, k\}\subseteq \mu f\cap\nu f\subseteq(\mu+\nu)f;$$

which is what had to be shown.

Let ρ be a congruence on F and let $\mathscr{P}(\rho) = \mathscr{P} = (f_\rho, M_\rho) = (f, M)$ be its associated congruence pair. Then the *core* of ρ, or of $\mathscr{P}$, written $k(\rho)$, or $k(\mathscr{P})$, is defined thus:

$$k(\rho) = k(\mathscr{P}) = \bigcap\{\mu f : \mu\in M\}.$$

If non-empty the core of ρ is an ideal of F. We proceed to show that $k(\rho)\neq\square$ for any congruence ρ on F.

THEOREM 9.24. *Let ρ be a congruence on F. Then $k(\rho)$, the core of ρ, is non-empty.*

PROOF. Let $\mathscr{P} = \mathscr{P}(\rho) = (f, M)$. Let $\mu\in M$. By Lemma 9.22, M contains a finite subset C, independent of μ, such that

$$\mu = \sum\{c_j\mu_j : j = 1, 2, \cdots, k\}$$

where the c_j are non-negative integers and $\{\mu_j : j = 1, 2, \cdots, k\}$ is a compatible subset of C. By Lemma 9.23,

$$\bigcap\{\mu_j f : j = 1, 2, \cdots, k\}\subseteq\mu f.$$

Consequently, for all μ in M,

$$\bigcap\{\nu f : \nu\in C\}\subseteq\mu f.$$

Hence

$$\bigcap\{\nu f\colon \nu \in C\} = \bigcap\{\mu f\colon \mu \in M\} = k(\rho).$$

Since C is finite, $k(\rho)$ is the intersection of a finite set of ideals. Thus $k(\rho)$ is non-empty.

Let A be any ideal of F. Let $B = \{\beta_1, \beta_2, \cdots, \beta_k\}$, say, be the basis of A. The *norm* of A, written $\|A\|$, is defined thus:

$$\|A\| = \sum\{\|\beta_i\|\colon i = 1, 2, \cdots, k\},$$

where the norm $\|\alpha\|$ of an element $\alpha = (a_1, a_2, \cdots, a_n)$ of F is defined to be

$$\|\alpha\| = a_1 + a_2 + \cdots + a_n.$$

(We prefer "norm" to Rédei's term "height.")

The core of the congruence ρ on F is an ideal of F and the *norm* of ρ, or of $\mathscr{P} = \mathscr{P}(\rho)$, written $\|\rho\|$, or $\|\mathscr{P}\|$, is defined thus:

$$\|\rho\| = \|\mathscr{P}\| = \|k(\rho)\|.$$

Let M be a subgroup of G. Then M determines a congruence of F, which we denote by ρ_M, defined thus:

$$\rho_M = \{(\alpha, \beta) \in F \times F\colon \alpha - \beta \in M\}.$$

Then ρ_M is just the restriction to F of the congruence on G determined by the (normal) subgroup M.

The proof that congruences on F are finitely generated will proceed by means of a double induction on their norms and on the number of generators of F. We start with the case of norm zero, and such congruences are characterized in the following lemma.

LEMMA 9.25. *Let ρ be a congruence on F. Then the following assertions are equivalent.*

(i) $\rho = \rho_M$ *for some subgroup M of G.*
(ii) $\|\rho\| = 0$.
(iii) $k(\rho) = F$.

In fact, when (i), (ii) *and* (iii) *hold,* $\rho = \rho_M$ *with* $M = M_\rho$.

PROOF. Clearly $k(\rho) = F$ if and only if $\|\rho\| = 0$. Thus (ii) and (iii) are equivalent.

Let $\rho = \rho_M$. Then $M_\rho = \{\alpha - \beta\colon (\alpha, \beta) \in \rho\}$; from which the definition of ρ_M gives immediately $M_\rho = M$. Let $\mu \in M$. Then $\mu f_\rho = \{\xi \in F\colon (\xi + \mu^+, \xi + \mu^-) \in \rho\}$. Since $\xi + \mu^+ - (\xi + \mu^-) = \mu \in M$, therefore $\mu f_\rho = F$. Hence $k(\rho) = F$ and $\|\rho\| = 0$, i.e. (i) implies (ii) and (iii) and that $M = M_\rho$.

Assume that (iii) holds for ρ. Let $\mathscr{P} = \mathscr{P}(\rho) = (f_\rho, M_\rho) = (f, M)$. If $(\alpha, \beta) \in \rho$, then by the definition of M, $\alpha - \beta \in M$. Conversely, assume that $\alpha - \beta = \mu \in M$. Because $k(\rho) = F$, it follows that $\mu f = F$. Hence, in particular, $\alpha \wedge \beta \in \mu f$, i.e. $(\alpha \wedge \beta + (\alpha - \beta)^+, \; \alpha \wedge \beta + (\alpha - \beta)^-) \in \rho$, i.e. $(\alpha, \beta) \in \rho$. Thus (iii) implies (i).

This completes the proof of the lemma.

The fact that congruences of norm zero are finitely generated now follows directly from Lemma 9.22, as we now show.

LEMMA 9.26. *Let ρ be a congruence on F and suppose that $\|\rho\| = 0$. Then ρ is finitely generated.*

PROOF. From the previous lemma it follows that $\rho = \rho_M$ where $M = M_\rho$. By Lemma 9.22, M possesses a finite subset C such that each element of M can be expressed as a sum of non-negative multiples of the elements of some compatible subset of C. Thus, if $\mu \in M$,

$$\mu = \sum \{c_i \nu_i : i = 1, 2, \cdots, k\},$$

say, where $c_1, c_2, \cdots, c_k$ are non-negative integers and $\{\nu_i : i = 1, 2, \cdots, k\}$ is a compatible subset of C. By Lemma 9.21, therefore

$$\mu^+ = \sum \{c_i \nu_i^+ : i = 1, 2, \cdots, k\},$$
$$\mu^- = \sum \{c_i \nu_i^- : i = 1, 2, \cdots, k\}.$$

Since $\rho = \rho_M$ and each $\nu_i \in M$, therefore, $(\nu_i^+, \nu_i^-) \in \rho$. From the above expressions for μ^+ and μ^- it therefore follows that μ^- can be derived from μ^+ by a sequence of elementary ρ-transitions, each ρ-transition involving the replacement of an occurrence of a ν_i^+ by ν_i^-.

Set

$$\sigma = \{(\nu^+, \nu^-) : \nu \in C\}.$$

Then σ is finite and $\sigma \subseteq \rho$ and we have shown that, for any μ in M, $(\mu^+, \mu^-) \in \sigma^*$. Now let (α, β) be any element of ρ, i.e. let $\alpha - \beta \in M$. Set $\alpha - \beta = \mu$. Then $\alpha = \alpha \wedge \beta + \mu^+$ and $\beta = \alpha \wedge \beta + \mu^-$. Since $(\mu^+, \mu^-) \in \sigma^*$, clearly also $(\alpha \wedge \beta + \mu^+, \alpha \wedge \beta + \mu^-) \in \sigma^*$, i.e. $(\alpha, \beta) \in \sigma^*$. Thus we have shown that $\rho \subseteq \sigma^*$; whence $\rho = \sigma^*$. Consequently ρ is finitely generated; and the proof of the lemma is complete.

To start our induction we must also deal with the case of $n = 1$, i.e. where F is the commutative free semigroup with one generator. Here it is reasonably straight-forward to verify that each congruence is finitely generated, in fact there always exists a one-element subset of any congruence which generates it. The details are stated in Exercise 5 below.

As a step in our proof we now show how to construct from certain congruences on F other congruences with a smaller norm.

LEMMA 9.27. *Let ρ be a congruence on F such that some element in the basis of $k(\rho)$ has non-zero first component. Let λ denote the element $\lambda = (1, 0, 0, \cdots, 0)$ of F. Define ρ_1, thus:*

$$\rho_1 = \{(\alpha, \beta) \in F \times F : (\alpha + \lambda, \beta + \lambda) \in \rho\}.$$

Then ρ_1 is a congruence on F and $\|\rho_1\| < \|\rho\|$.

PROOF. Clearly ρ_1 inherits the properties of reflexivity, symmetry, transitivity and compatibility from ρ. Thus ρ_1 is a congruence on F. Observe further that evidently $M_\rho = M_{\rho_1} = M$, say.

Firstly, let us show that $k(\rho) \subseteq k(\rho_1)$ and that $k(\rho_1) + \lambda \subseteq k(\rho)$. For, if $\beta \in k(\rho)$, i.e. $(\beta + \mu^+, \beta + \mu^-) \in \rho$ for all μ in M, then $(\beta + \lambda + \mu^+, \beta + \lambda + \mu^-) \in \rho$, i.e. $(\beta + \mu^+, \beta + \mu^-) \in \rho_1$, for all μ in M; thus $\beta \in k(\rho_1)$. Consequently, $k(\rho) \subseteq k(\rho_1)$; and the second assertion follows similarly. We will use these two relations without comment in what follows.

Let $\beta_1, \beta_2, \cdots, \beta_k$ be the basis of $k(\rho)$, and suppose that $\beta_1, \cdots, \beta_t$ are those that have positive first component. By hypothesis, $t \geqq 1$. Set $\gamma_j = \beta_j - \lambda$ for $j \leqq t$. Then the γ_j, $j = 1, 2, \cdots, t$ are distinct basis elements of $k(\rho_1)$. For, firstly, the γ_j are elements of F. Further, $\gamma_j \in k(\rho_1)$. For $(\beta_j + \mu^+, \beta_j + \mu^-) \in \rho$, i.e. $(\gamma_j + \lambda + \mu^+, \gamma_j + \lambda + \mu^-) \in \rho$, i.e. $(\gamma_j + \mu^+, \gamma_j + \mu^-) \in \rho_1$ for all μ in M. Suppose that $\alpha \in k(\rho_1)$ and that $\alpha \leqq \gamma_j$. Then $\alpha + \lambda \leqq \gamma_j + \lambda = \beta_j$. Now $\alpha + \lambda \in k(\rho)$; hence, since β_j is a basis element of $k(\rho)$, $\alpha + \lambda = \beta_j$. Whence $\alpha = \gamma_j$; which shows that γ_j is a basis element of $k(\rho_1)$.

Consider now any basis element γ of $k(\rho_1)$. Then $\gamma + \lambda \in k(\rho)$ and so there exists a basis element β_i of $k(\rho)$ such that $\beta_i \leqq \gamma + \lambda$. If $i \leqq t$, then $\gamma_i = \beta_i - \lambda \leqq \gamma$; whence, since γ_i is a basis element of $k(\rho_1)$, $\gamma_i = \gamma$. If $i > t$, then β_i has first component zero and hence $\beta_i \leqq \gamma$. But $\beta_i \in k(\rho_1)$; hence, since γ is a minimal element of $k(\rho_1)$, $\beta_i = \gamma$.

We have thus shown that the basis elements of $k(\rho_1)$ divide into two subsets. One subset consists of the γ_j, $j = 1, 2, \cdots, t$, corresponding to which are basis elements $\beta_j = \gamma_j + \lambda$, $j = 1, 2, \cdots, t$, of $k(\rho)$. This subset is non-empty and, for each j, $\|\gamma_j\| < \|\beta_j\|$. The other subset, possibly empty, consists of basis elements of $k(\rho_1)$ which are also basis elements of $k(\rho)$. It follows immediately that $\|\rho_1\| < \|\rho\|$. This completes the proof of the lemma.

We are now in a position to complete the proof of Rédei's theorem. It has been shown that a congruence ρ on F is finitely generated if either $\|\rho\| = 0$ (Lemma 9.26) or if n, the number of generators of F, equals one (remarks following Lemma 9.26). Suppose then that $n > 1$ and that $\|\rho\| = s > 0$ and that the result has been proved both for congruences on F with norm smaller than s and for any congruence on a free commutative semigroup with less than n generators.

Since $\|\rho\| = s > 0$, there exists a non-zero basis element of $k(\rho)$. Without loss of generality we may assume that $k(\rho)$ has a basis element with non-zero first component. Then the construction of Lemma 9.27 gives a congruence ρ_1 on F with $\|\rho_1\| < s$. By our inductive assumption ρ_1 is finitely generated. Let σ_1 be a finite set of generators of ρ_1 and set

$$\sigma = \{(\alpha + \lambda, \beta + \lambda) \colon (\alpha, \beta) \in \sigma_1\}.$$

From the definition of ρ_1 it follows that $\sigma \subseteq \rho$.

Denote by H_{n-1} the elements of F with first component zero. H_{n-1} is then isomorphic to the free commutative semigroup with $(n-1)$ generators. Clearly, $\rho \cap (H_{n-1} \times H_{n-1})$ is a congruence on H_{n-1}, and by our inductive assumptions it follows that $\rho \cap (H_{n-1} \times H_{n-1})$ is generated by a finite subset τ, say.

Consider now the set, possibly empty, of all (α, β) in ρ for which $\alpha \in H_{n-1}$ and $\beta \notin H_{n-1}$ and denote this set by W. Set

$$A = \{\alpha: (\alpha, \beta) \in W, \quad \text{for some } \beta\}.$$

By Theorem 9.18, the set of all minimal elements of A is finite. Let $\alpha_1, \alpha_2, \cdots, \alpha_k$ be the minimal elements of A. Choose $\beta_1, \beta_2, \cdots, \beta_k$ such that $(\alpha_i, \beta_i) \in W$, $i = 1, 2, \cdots, k$. Set

$$\pi = \{(\alpha_i, \beta_i): i = 1, 2, \cdots, k\}.$$

Then, as we now proceed to show, $(\pi \cup \tau \cup \sigma)^* = \rho$.

Consider (α, β) in ρ. If $\alpha = \beta$, then $(\alpha, \beta) \in (\pi \cup \tau \cup \sigma)^*$. The remaining pairs (α, β) are such that (a) α and β each have positive first component, i.e. $\alpha \notin H_{n-1}$ and $\beta \notin H_{n-1}$ or (b) $(\alpha, \beta) \in H_{n-1} \times H_{n-1}$ or (c) either (α, β) or (β, α) belongs to W. We deal with each case separately.

(a) Here $\alpha - \lambda$ and $\beta - \lambda$ belong to F and so $(\alpha - \lambda, \beta - \lambda) \in \rho_1$. Consequently $\beta - \lambda$ can be derived from $\alpha - \lambda$ by a finite sequence of elementary σ_1-transitions. Adding λ throughout it follows that β can be derived from α by a finite sequence of elementary σ-transitions. Thus $(\alpha, \beta) \in \sigma^*$ and, *a fortiori*, $(\alpha, \beta) \in (\pi \cup \tau \cup \sigma)^*$.

(b) Here $(\alpha, \beta) \in \rho \cap (H_{n-1} \times H_{n-1})$. Consequently, $(\alpha, \beta) \in \tau^*$, whence $(\alpha, \beta) \in (\pi \cup \tau \cup \sigma)^*$.

(c) It suffices to consider the case in which $(\alpha, \beta) \in W$. Let α_i be a minimal element of A such that $\alpha_i \leqq \alpha$. Then $\alpha - \alpha_i \in F$. Adding $\alpha - \alpha_i$ to both members of (α_i, β_i), we obtain, since $(\alpha_i, \beta_i) \in \pi$, $(\alpha, \alpha - \alpha_i + \beta_i) \in \pi^*$. Since $\pi^* \subseteq \rho$ and $(\alpha, \beta) \in \rho$, we conclude that $(\alpha - \alpha_i + \beta_i, \beta) \in \rho$. Since $\beta_i, \beta \notin H_{n-1}$, it follows from case (a) that $(\alpha - \alpha_i + \beta_i, \beta) \in \sigma^*$. Hence $(\alpha, \beta) \in (\pi \cup \sigma)^* \subseteq (\pi \cup \tau \cup \sigma)^*$.

Observe, finally, that if ρ is a congruence on the semigroup S, then $\rho^1 = \rho \cup \{(1, 1)\}$ is a congruence on S^1 and ρ is finitely generated if and only if ρ^1 is finitely generated. We have therefore proved Rédei's theorem (loc. cit., p. 124). (See Appendix.)

THEOREM 9.28. *All finitely generated commutative semigroups are finitely presented.*

EXERCISES FOR §9.3

1. Let ρ be a congruence on the semigroup S. Then ρ is a subsemigroup of the direct product semigroup $S \times S$. Suppose that ρ is finitely generated, *qua* subsemigroup, by its finite subset σ. Then, also, $\sigma^* = \rho$.

The converse does not hold. For example take S to be a zero semigroup with $|S|$ infinite. The identical relation ι_S on S is then a finitely generated congruence on S. But ι_S is not a finitely generated subsemigroup of $S \times S$.

2. Let F be the free commutative semigroup with identity on n generators consisting of all ordered n-tuples $(a_1, a_2, \cdots, a_n)$ of non-negative integers. Let S be the subsemigroup of F consisting of all ordered n-tuples of positive integers. Then, if $n > 1$, S is not finitely generated: any set of generators of S must include all n-tuples $(a_1, a_2, \cdots, a_n)$ for which $a_i = 1$ for at least one of the $i = 1, 2, \cdots, n$.

3. Let F be the free commutative semigroup with identity on one generator, so that we may take F as the additive semigroup of non-negative integers. Then any subsemigroup of F is finitely generated. (Rédei [1963], p. 137.)

4. Let S be a semigroup such that every subsemigroup of $S \times S$ is finitely generated. Then every congruence on S is finitely generated.

5. Let F be the semigroup of non-negative integers. Let ρ be a congruence on F. Then either ρ is the identical relation on F or there exist integers $r \geqq 0$ and $m > 0$ such that $(r, r+m) \in \rho$. Choose r and m so that $r + m$ is as small as possible. Then $\{(r, r+m)\}^* = \rho$.

The different types of quotient image F/ρ may be classified in terms of r and m as follows. If $r = 0$, then F/ρ is a cyclic group of order m; if $r = 1$, then F/ρ is a cyclic group of order m to which a further identity has been adjoined; if $r > 1$, then F/ρ is a cyclic finite semigroup of index r and period m to which an identity has been adjoined. (Cf. Theorem 1.9.)

6. Let F be the semigroup of non-negative integers. Let ρ be a congruence, not the identity, on F and choose r, $r + m$, as in Exercise 5, so that $\{(r, r+m)\}^* = \rho$. When ρ is the identity congruence we may include it in our scheme by taking $r = \infty$, $m = 0$, where ∞ is adjoined to F to form F^0, and where we regard $m < \infty$ for any m in F. We then formally write $\iota_F = \{(\infty, \infty + 0)\}^*$.

With this notation, let $\rho = \{(r, r+m)\}^*$ and $\sigma = \{(s, s+n)\}^*$ be any two congruences on F. Then $\rho \subseteq \sigma$ if and only if (i) $r \geqq s$ and (ii) $n | m$ (i.e. n divides m). Thus the lattice of congruences on F is isomorphic to the order dual of the direct product of the lattices L and M, where L denotes F^0 with the natural ordering, and M denotes the lattice F of non-negative integers ordered by the usual division relation. L and M are then each distributive lattices. Whence it follows that $L \times M$, and so also the lattice of congruences on F, is distributive.

7. Let F be the semigroup of non-negative integers. Let ρ be a congruence on F distinct from ι_F. In the notation of Exercise 5, ρ may be written $\rho = \{(r, r+m)\}^*$ for some integers $r \geqq 0$ and $m > 0$. Then M_ρ is the (cyclic) subgroup of G generated by m, and if $\mu \in M_\rho \backslash 0$, μf_ρ is the ideal of F generated by r. Thus, in particular, $\mu f_\rho = F$ for all μ in M_ρ if and only if $r = 0$, i.e. if and only if F/ρ is a group, i.e. if and only if $\rho = \rho_{M_\rho}$.

8. Let $A = \mu^+ + \mu f$, $B = \nu^+ + \nu f$ and $C = \mu \vee \nu + (\mu - \nu)f$, where $\mu, \nu \in M$ and (f, M) is a congruence pair on F. Then

$$A \cap B = B \cap C = C \cap A.$$

(Hint: apply C(iii) of Lemma 9.15 to the pair $(-\nu, \mu - \nu)$, and add ν to both sides.) (Rédei, [1963], p. 30.)

9. Let ρ be a congruence on F, and let $M = M_\rho = \{\alpha - \beta\colon (\alpha, \beta) \in \rho\}$. Then F/ρ is cancellative if and only if $\rho = \rho_M$, in the notation of Lemma 9.25. (Rédei [1963], p. 121.)

10. Let ρ be a congruence on F, and let (f, M) be the associated congruence pair. Then the kernel $k(\rho) = \bigcap\{\mu f\colon \mu \in M\}$ of ρ consists of all elements ξ of F having the following property: if α and β are elements of F such that $\alpha \geqq \xi$, $\beta \geqq \xi$, and $\alpha - \beta \in M$, then $(\alpha, \beta) \in \rho$. (Rédei [1963], p. 98.)

9.4 Embedding semigroup amalgams: free products with amalgamations

All the results presented in this section, with the sole exception of an example of Kimura [1957], are due to Howie ([1962], [1963a, b, c, d], [1964a, b]). We have simplified Howie's original treatment by introducing the concept of "proper words".

Let $\{S_i\colon i \in I\}$ be a set of semigroups, indexed by some set I, and suppose that each S_i contains an isomorphic copy of the semigroup U; thus there exists $\phi_i\colon U \to S_i$, an isomorphic mapping of U into S_i, for each i in I. Such a system of semigroups and associated isomorphisms will be denoted by $[\{S_i\colon i \in I\};\ U;\ \{\phi_i\colon i \in I\}]$, or alternatively, by one of the following briefer notations, $[\{S_i\};\ U;\ \{\phi_i\}]$, $[S_i;\ U;\ \phi_i]$, $[\{S_i\};\ U]$ or $[S_i;\ U]$. The system $[S_i;\ U;\ \phi_i]$ will be called a *semigroup amalgam.*

The semigroup amalgam $[S_i;\ U;\ \phi_i]$ determines a partial groupoid $\mathscr{G} = \mathscr{G}[S_i;\ U;\ \phi_i]$ defined as follows.

Firstly, for each i in I, set $U_i = U\phi_i$, and

$$S_i' = (S_i \backslash U_i) \cup U,$$

and define an operation $*$ in S_i' so that the mapping

$$\chi_i\colon S_i' \to S_i \text{ defined by}$$

$$a\chi_i = \begin{cases} a\ , & \text{if } a \in S_i \backslash U_i, \\ a\phi_i, & \text{if } a \in U, \end{cases}$$

is an isomorphism of S_i' onto S_i. This defines $*$ uniquely. Secondly, assume, and this is essential to the argument, that $S_i' \cap S_j' = U$, if $i \neq j$. The operation $*$ defined on S_i' coincides, when restricted to U, with the operation of the semigroup U. Thus $U = (U, *)$ is a subsemigroup of $(S_i', *)$ for all i in I. Now define $\mathscr{G}$ thus. The set of the partial groupoid $\mathscr{G}$ is

$$\mathscr{G} = \bigcup\{S_i'\colon i \in I\}.$$

The operation of $\mathscr{G}$ is the operation $*$ wherever it is defined. Throughout this section the symbols S_i' and χ_i will, when used, have the above defined meanings in relation to $\mathscr{G}$.

It will be convenient to include within our discussion the case when U is empty. The partial groupoid $\mathscr{G}[S_i; \square]$ is then merely the disjoint union of the semigroups S_i.

The problem that concerns us is to ask, given a semigroup amalgam $[S_i; U; \phi_i]$, when can its partial groupoid $\mathscr{G}[S_i; U; \phi_i]$ be embedded in a semigroup? The embedding, of course, is to be performed in such a manner that the products defined in $\mathscr{G}$ coincide with those in the semigroup in which it is embedded. If $\mathscr{G}[S_i; U; \phi_i]$ can be embedded in the semigroup T, then we will also say that the *semigroup amalgam* $[S_i; U; \phi_i]$ can be embedded in T. We begin with an example of Kimura [1957] to show that a semigroup amalgam cannot always be embedded in a semigroup.

Let U be the zero semigroup $U = \{u, v, w, z\}$ in which all products are equal to z. Let $S_1 = U \cup \{a\}$, where $a \notin U$, $au = ua = v$ and all other products in S_1 are set equal to z. Let $S_2 = U \cup \{b\}$, where $b \notin S_1$, $bv = vb = w$ and all other products in S_2 are set equal to z. Then $S_1 \cap S_2 = U$ and a straightforward verification shows that, with the defined products, S_1 and S_2 are each semigroups. Suppose that there exists a semigroup T such that $S = S_1 \cup S_2$ can be embedded in T. Then, since T is associative, the following equations hold in T:

$$w = bv = bua = za = z.$$

Thus the elements w and z must coincide in T. Thus $[\{S_1, S_2\}; U]$ cannot be embedded in any semigroup.

Let us dispose of two simple cases first. Firstly, observe that $\mathscr{G}[S_i; U; \phi_i]$ is a semigroup if and only if U_i is a proper subsemigroup of S_i, i.e. $U_i \neq S_i$, for at most one i in I. When $\mathscr{G}$ is not a semigroup, i.e. when $*$ is not defined for all pairs of elements in $\mathscr{G}$, then $[S_i; U; \phi_i]$ is called a *proper* amalgam.

Secondly, we ask: *can the proper amalgam* $[S_i; U]$ *be embedded in a semigroup simply by adjoining a zero to its partial groupoid and setting all undefined products equal to zero?* The answer to this question is given in the following theorem.

We need a preliminary definition. The subset C, possibly empty, of the semigroup T is said to be a *consistent subset* of T if $ab \in C$, for a, b in T, implies both $a \in C$ and $b \in C$. In particular, the empty set is then a consistent subset of T. (Dubreil [1941].)

THEOREM 9.29. *Let* $[S_i; U; \phi_i]$ *be a proper semigroup amalgam, so that* U_i *is properly contained in* S_i *for at least two indices* i *in* I. *Let* $\mathscr{G}^0$ *be the groupoid derived from* $\mathscr{G}$ *by adjoining a zero element,* 0, *to* $\mathscr{G}$ *and setting all products of elements in* $\mathscr{G}^0$, *not defined in* $\mathscr{G}$, *equal to* 0 *in* $\mathscr{G}^0$. *Other products in* $\mathscr{G}^0$ *are to be evaluated as in* $\mathscr{G}$.

Then $\mathscr{G}^0$ is a semigroup if and only if U_i is a consistent subsemigroup of S_i for each i in I.

PROOF. Recall that, by Lemma 3.7, $\mathscr{G}^0$ is a semigroup if and only if the following conditions hold for any a, b, c in $\mathscr{G}$.

(i) If $a * b$ and $(a * b) * c$ are defined, so are $b * c$ and $a * (b * c)$, and

$$a * (b * c) = (a * b) * c.$$

(ii) If $b * c$ and $a * (b * c)$ are defined, so are $a * b$ and $(a * b) * c$, and

$$(a * b) * c = a * (b * c).$$

Suppose that U_i is consistent in S_i for each i in I. Let a, b, c be elements of $\mathscr{G}$ such that $a * b$ and $(a * b) * c$ are both defined. Since $a * b$ is defined, there exists i in I such that a, $b \in S_i'$. Similarly, there exists j in I such that $a * b$, $c \in S_j'$. If $i = j$ then $b * c$ and $a * (b * c)$ are both defined in S_i' and $a * (b * c) = (a * b) * c$. If $i \neq j$, then $a * b \in U$, whence, since U_i is consistent in S_i, a, $b \in U$. Hence a, $b \in S_j'$; and the required conclusion follows as before. This shows that condition (i) above holds. Condition (ii) follows similarly. Thus $\mathscr{G}^0$ is a semigroup.

Conversely, suppose $\mathscr{G}^0$ is a semigroup and suppose that, if possible, U_i is not a consistent subset of S_i. Then there exist a, b in S_i', with a, say, not in U, such that $ab \in U$. Let c be an element of $S_j \backslash U_j$ with $j \neq i$. By hypothesis on $[S_i; U; \phi_i]$, there exists $j \neq i$ with $S_j \backslash U_j \neq \square$. Then $c * (a * b)$ is defined. However, $c * a$ is not defined in $\mathscr{G}$. Thus condition (ii) above is not satisfied by $\mathscr{G}$. Consequently, contrary to assumption, $\mathscr{G}^0$ is not a semigroup. This is a contradiction. Hence, when $\mathscr{G}^0$ is a semigroup, U_i is a consistent subsemigroup of S_i for each i in I.

This completes the proof of the theorem.

Observe that, when U is empty, the above theorem applies to show that $\mathscr{G}^0$ is then a semigroup.

We now turn to what is the key concept for our problem, the concept of a free product of semigroups with an amalgamated subsemigroup. We shall see (Theorem 9.31) that a semigroup amalgam can be embedded in some semigroup if and only if it can be embedded in its associated amalgamated free product.

We first define the *free product* of the semigroups S_i, $i \in I$. For this construction we assume that $S_i \cap S_j = \square$, if $i \neq j$. Form first the set W consisting of all finite, non-empty, sequences $(a_1, a_2, \cdots, a_k)$ such that if a_j belongs to $S_{i(j)}$, say, for $j = 1, 2, \cdots, k$ then for $j = 1, 2, \cdots, k-1$, $i(j) \neq i(j+1)$. Now define a product $\circ$ on the set of "words" W as follows:

$$(a_1, a_2, \cdots, a_k) \circ (b_1, b_2, \cdots, b_s) = (a_1, a_2, \cdots, a_k, b_1, b_2, \cdots, b_s),$$

if $a_k \in S_i$, $b_1 \in S_j$ and $i \neq j$;

$$(a_1, a_2, \cdots, a_k) \circ (b_1, b_2, \cdots, b_s) = (a_1, \cdots, a_{k-1}, a_k b_1, b_2, \cdots, b_s),$$

if $a_k \in S_i$ and $b_1 \in S_i$ for some $i \in I$. A short and straightforward verification shows that $\circ$ is an associative product on W, i.e. that $(W, \circ)$ is a semigroup. We define $(W, \circ)$ to be the *free product of the semigroups* S_i, $i \in I$, and denote this free product by $\prod^*\{S_i: i \in I\}$, or, more simply, by $\prod^*\{S_i\}$, or by $\prod^* S_i$.

A special case of a free product of semigroups is the free semigroup $\mathscr{F}_X$ on the set X. If $X = \{x_i: i \in I\}$, then $\mathscr{F}_X$ is the free product of the infinite cyclic semigroups $\langle x_i \rangle$, $i \in I$ (cf. §1.12).

For each i in I, define the mapping $\kappa_i: S_i \to \prod^* S_i$, thus:

$$\kappa_i: a \to a\kappa_i = (a), \qquad a \in S_i.$$

Then κ_i is an isomorphic embedding, called the *canonical isomorphism or embedding*, of S_i into $\prod^* S_i$. We shall frequently, when no ambiguity results, identify S_i with its canonical image $S_i\kappa_i$ in $\prod^* S_i$. Then $\prod^* S_i$ is to be regarded as generated by its subsemigroups S_i, $i \in I$. When this identification is made, then the element $(a_1, a_2, \cdots, a_k)$ of $\prod^* S_i$ becomes identified with $(a_1, a_2, \cdots, a_k) = (a_1) \circ (a_2) \circ \cdots \circ (a_k) = a_1 \circ a_2 \circ \cdots \circ a_k$. The latter word we then write simply as $a_1 a_2 \cdots a_k$. The elements of $\prod^* S_i$ are thus the finite non-empty words $a_1 a_2 \cdots a_k$ such that, for each $j = 1, 2, \cdots, k$, $a_j \in S_{i(j)}$ and such that $i(j) \neq i(j+1)$ for $j = 1, 2, \cdots, k-1$. A word $a_1 a_2 \cdots a_k$ satisfying these conditions will be said to be *irreducible*.

For later use, observe one property of $\prod^*\{S_i: i \in I\}$. Let $\psi_i: S_i \to T$ be a homomorphism of S_i into T for each i in I. Define $\phi: \prod^* S_i \to T$ as follows. Let w be any element of $\prod^* S_i$. Let $w = a_1 a_2 \cdots a_k$ as an irreducible product of elements of the S_i's. Let $a_j \in S_{i(j)}$, $j = 1, 2, \cdots, k$. Define $w\phi$ thus:

$$w\phi = a_1\psi_{i(1)} \cdots a_k\psi_{i(k)}.$$

This defines $w\phi$ uniquely, because, by the definition of $\prod^* S_i$, distinct irreducible products are distinct elements of $\prod^* S_i$. It is easily verified that ϕ is a homomorphism of $\prod^* S_i$ into T. Furthermore, ϕ coincides with ψ_i when restricted to S_i, for each i in I, and since the S_i generate their free product, ϕ is the unique homomorphism of $\prod^* S_i$ which extends each ψ_i.

Alternative definitions that could be given, to determine $\prod^* S_i$ to within isomorphism, occur in Exercises 1 and 2 below.

Consider now the semigroup amalgam $[\{S_i: i \in I\}; U; \{\phi_i: i \in I\}]$ where again, for simplicity in the general discussion, we assume that $S_i \cap S_j = \square$, if $i \neq j$. Define the relation v on $\prod^* S_i$, thus:

$$v = \{((u_i), (u_j)): u_i = u\phi_i,\ u_j = u\phi_j \text{ for some } i, j \text{ in } I \text{ and for some } u \text{ in } U\}.$$

More simply, identifying each S_i with its canonical image in $\prod^* S_i$, we have:

$$v = \{(u_i, u_j): u_i = u\phi_i,\ u_j = u\phi_j \text{ for some } i, j \text{ in } I \text{ and for some } u \text{ in } U\}.$$

Then we define *the amalgamated free product of the* $\{S_i, i \in I\}$, *amalgamating* U, *determined by the semigroup amalgam* $[S_i; U; \phi_i]$, to be the semigroup $(\prod^*\{S_i: i \in I\})/v^*$. We continue to use the notation (cf. §9.2) where ρ^*

denotes the congruence generated by the relation ρ. This amalgamated free product will also be written $\prod^*_U\{S_i: i \in I\}$, or, simply, $\prod^*_U S_i$. Briefly, $\prod^*_U S_i$ may be called the *free product of the amalgam* $[S_i; U]$.

Some alternative definitions, equivalent to within isomorphism, of amalgamated free products are given in Exercises 3–5 below.

Regarding each S_i as identified with its canonical image in $\prod^* S_i$, we define the *canonical homomorphisms* $\mu_i: S_i \to \prod^*_U S_i$, thus:

$$\mu_i: a \to a\mu_i = av^*, \quad a \in S_i.$$

It then follows, that for any $u \in U$,

$$u\phi_i\mu_i = u\phi_j\mu_j,$$

since $(u\phi_i, u\phi_j) \in v^*$. Set $\phi_i\mu_i = \phi_j\mu_j = \mu$, for i, j in I. Then μ is a homomorphism of U into $\prod^*_U S_i$.

The canonical homomorphisms μ_i induce a homomorphism γ of the partial groupoid $\mathcal{G} = \mathcal{G}[S_i; U; \phi_i]$ into $\prod^*_U S_i$ defined thus:

$$\gamma: a \to a\gamma = a\chi_i\mu_i, \quad \text{if } a \in S_i'.$$

Then it is readily verified that γ is a *homomorphism of* $\mathcal{G}$ in the sense that if $a, b \in \mathcal{G}$ and $a * b$ is defined in $\mathcal{G}$, then $(a * b)\gamma = a\gamma b\gamma$. The mapping γ will be called the *canonical homomorphism of* $\mathcal{G}$ *into* $\prod^*_U S_i$. The homomorphisms μ_i determine γ and conversely γ determines the set of μ_i, $i \in I$. More generally we have the following lemma.

Lemma 9.30. *Let* $[S_i; U; \phi_i]$ *be a semigroup amalgam and let* $\mathcal{G}$ *be its partial groupoid. Let* T *be a semigroup.*

Let $\delta: \mathcal{G} \to T$ *be a homomorphism. Then* $\psi_i = \chi_i^{-1}\delta$ *is a homomorphism of* S_i *into* T *and* $\phi_i\psi_i = \phi_j\psi_j$ *for all* i, j *in* I.

Conversely, let $\psi_i: S_i \to T$ *be a family of homomorphisms such that* $\phi_i\psi_i = \phi_j\psi_j$ *for all* i, j *in* I. *Define* $\delta: \mathcal{G} \to T$, *thus:*

$$\delta: a \to a\delta = a\chi_i\psi_i, \quad \text{if } a \in S_i'.$$

Then δ *is a homomorphism.*

Moreover, the relationship established in the preceding two paragraphs between homomorphisms δ *of* $\mathcal{G}$ *and families* $\{\psi_i\}$ *of homomorphisms of* $\{S_i\}$ *is one-to-one.*

Proof. Since δ restricted to S_i' is a homomorphism of the semigroup S_i', it is clear that ψ_i is a homomorphism of S_i. From the definition of the χ_i it follows that $\phi_i\psi_i = \delta | U$.

Conversely, given the ψ_i, to show that δ is a homomorphism, it suffices to show that δ is well defined as a mapping, i.e. that δ is defined in a unique manner on U. But this is ensured by the condition that $\phi_i\psi_i = \phi_j\psi_j$ for all i, j.

The remaining assertion of the lemma is clear.

The central position, for our embedding problem, held by the concept of the amalgamated free product is shown by the following theorem.

THEOREM 9.31. 1. *The semigroup amalgam* $[\{S_i: i \in I\}; U; \{\phi_i: i \in I\}]$ *can be embedded in a semigroup if and only if the canonical homomorphism* γ *of the partial groupoid* $\mathscr{G} = \mathscr{G}[S_i; U; \phi_i]$ *is an embedding of* $\mathscr{G}$ *into* $\prod^*{}_U S_i$, *i.e. if and only if* γ *is one-to-one.*

2. *The homomorphism* γ *is one-to-one if and only if both of the following conditions* (i) *and* (ii) *hold.*

(i) *Each canonical homomorphism* $\mu_i: S_i \to \prod^*{}_U S_i$ *is one-to-one.*

(ii) $S_i\mu_i \cap S_j\mu_j = U\mu$, *for* $i \neq j$, *where* $\mu = \phi_i\mu_i$.

PROOF. Consider, firstly, the assertion of the second paragraph. Suppose that γ is one-to-one. Let $a \in S_i$, $b \in S_j$ and suppose that $a\mu_i = b\mu_j$. Choose $x \in S_i'$, $y \in S_j'$ such that $x\chi_i = a$, $y\chi_j = b$. Then $x\chi_i\mu_i = y\chi_j\mu_j$, i.e. $x\gamma = y\gamma$. Since γ is one-to-one, therefore $x = y$. Taking $i = j$ this shows that μ_i is one-to-one. Taking $i \neq j$, we infer that $S_i\mu_i \cap S_j\mu_j \subseteq U\mu$; whence, since $U\mu \subseteq S_i\mu_i$ for all i, we have $S_i\mu_i \cap S_j\mu_j = U\mu$. Thus both conditions (i) and (ii) hold.

Conversely, suppose that both (i) and (ii) hold. Let $a, b \in \mathscr{G}$, so that $a \in S_i'$, $b \in S_j'$, say. If $a\gamma = b\gamma$, then $a\chi_i\mu_i = b\chi_j\mu_j$. If $i = j$, then $a = b$ follows, because μ_i and χ_i are both one-to-one. If $i \neq j$, then condition (ii) implies that $a\chi_i\mu_i = u\mu = u\phi_i\mu_i$ for some u in U. By (i), $a\chi_i = u\phi_i$, whence $a = u \in U$, and $a\chi_i\mu_i = a\mu$. Similarly, $b \in U$, and $b\chi_j\mu_j = b\mu$, whence $a\mu = b\mu$. Since ϕ_i and μ_i are both one-to-one, so is μ, and we conclude that $a = b$. Hence γ is one-to-one.

Consider the first assertion of the theorem. Clearly, if $\mathscr{G}$ can be embedded in $\prod^*{}_U S_i$ by γ, then $[S_i; U; \phi_i]$ is embeddable in a semigroup. Conversely, assume that $[S_i; U; \phi_i]$ is embedded in a semigroup T, i.e. that $\mathscr{G}$ is embedded in T, by the mapping δ, say. Then, by Lemma 9.30, $\psi_i = \chi_i^{-1}\delta$ is a homomorphism of S_i into T. Since both χ_i and δ are one-to-one, so also is ψ_i.

Now define $\psi: \prod^* S_i \to T$ as the unique homomorphism of $\prod^* S_i$ into T such that, for each i, $\psi | S_i = \psi_i$. Let (u_i, u_j) be any element of v, so that $u_i = u\phi_i$, $u_j = u\phi_j$, for some $u \in U$. Then $u_i\psi = u\phi_i\psi = u\chi_i\psi_i = u\delta = u_j\psi$. Hence $v \subseteq \psi \circ \psi^{-1}$, so that $v^* \subseteq \psi \circ \psi^{-1}$. Thus the homomorphism ψ can be factored through $\prod^*{}_U S_i$: $\psi = (v^*)^\natural\phi$, say, where $\phi: \prod^*{}_U S_i \to T$. Let $a \in \mathscr{G}$ and let $a \in S_i'$. Then $a\gamma\phi = a\chi_i\mu_i\phi = a\chi_i v^{*\natural}\phi = a\chi_i\psi = a\chi_i\psi_i = a\delta$. Thus $\gamma\phi = \delta$; whence, since δ is one-to-one, so also is γ. This completes the proof.

We now turn to Howie's [1962] principal result giving sufficient conditions under which an amalgam can be embedded in a semigroup. We need a new concept.

Recall that the subset U of the semigroup S is *left unitary* in S if, when $u \in U$ and $s \in S$, then $us \in U$ implies $s \in U$. *Right unitariness* is defined dually. And U is *unitary* if it is both left and right unitary (§7.3). Howie generalized the concept of unitariness to that of almost unitariness. The subset U of S is said to be *almost unitary in S (with associated mappings λ and ρ)*, if there exist mappings λ and ρ of S into S with the following properties.

(a) λ is a left translation, ρ is a right translation of S and λ commutes with ρ.

(b) λ and ρ are each idempotent, i.e. are projections of S.

(c) λ and ρ are linked: i.e. $s(\lambda t) = (s\rho)t$ for all s, t in S. (Cf. §1.3.)

(d) When restricted to U, λ and ρ each become the identity mapping on U.

(e) U is unitary in $\lambda S\rho$.

NOTATION. It will be convenient, whenever dealing with almost unitary subsets, to write the mapping λ on the left and the mapping ρ on the right, as has been done in (c) and (e) above. The elements $(\lambda a)\rho$ and $\lambda(a\rho)$, equal by (a), will both be written $\lambda a\rho$ $(a \in S)$. Regarding (e), it follows from (d) that $U \subseteq \lambda S\rho$. That $\lambda S\rho$ is a subsemigroup of S follows from the identity $\lambda a\rho \cdot \lambda b\rho = \lambda(a\rho \cdot \lambda b)\rho$, $(a, b \in S)$.

Let U be any unitary subset of S. Then U is almost unitary in S taking the associated mappings to be each the identity mapping on S. The connexion between unitary and almost unitary is explored more fully in the following two theorems.

THEOREM 9.32. *Let U be a subsemigroup of the semigroup S and suppose that U has an identity element e. Then U is almost unitary in S if and only if U is unitary in eSe.*

PROOF. Let U be unitary in eSe. Take for λ and ρ the inner left and inner right translations of S, λ_e and ρ_e, respectively (§1.3). Then it is readily verified that U is almost unitary in S with these as associated mappings.

Conversely, suppose that U is almost unitary in S with associated mappings λ and ρ. Now $e \in U$. Hence (defining condition (d)) $\lambda e = e$ and $e\rho = e$. Consequently, $eSe \subseteq \lambda S\rho$; whence it follows immediately that U is unitary in eSe (defining condition (e)).

REMARK. In general, if U is almost unitary in S with associated mappings λ and ρ, then there will be many other pairs of mappings that could play the part of λ and ρ in ensuring that U is almost unitary in S. The preceding theorem effectively shows that, when U has an identity element e, then the pair λ and ρ may be replaced by λ_e and ρ_e.

COROLLARY 9.33. *Let U be a subgroup of the semigroup S. Then U is almost unitary in S.*

PROOF. Let e be the identity of U. By the theorem it suffices to show that U is unitary in eSe. Suppose that $u \in U$ and $eseu \in U$. Then $u^{-1} \in U$ and so $eseuu^{-1} = ese^2 = ese \in U$. This shows that U is right unitary in eSe. That U is left unitary in eSe follows similarly.

THEOREM 9.34. *Let U be a subsemigroup of the semigroup S. Then U is almost unitary in S if and only if it is possible to adjoin an idempotent e to S, forming the semigroup $S^e = S \cup \{e\}$, with the following properties*:

(i) $eS \subseteq S$ *and* $Se \subseteq S$;

(ii) *e is an identity element for* $U^e = U \cup \{e\}$;

(iii) *U^e is unitary in eSe.*

PROOF. Suppose that an idempotent e can be adjoined to S with the asserted properties. Take for λ and ρ the restrictions to S of λ_e and ρ_e, respectively, as defined on S^e. Since $se \neq e$ and $es \neq e$, if $s \in S$, it then readily follows that U is almost unitary in S with λ and ρ as associated mappings.

Conversely, let U be almost unitary in S with λ and ρ as associated mappings. Set $e = (\lambda, \rho)$ and define $se = s\rho$ and $es = \lambda s$, for s in S, and $e^2 = e$ (cf. the translational hull of §1.3). Then it is straightforward to verify that e has the required properties.

This completes the proof of the theorem.

Let S be a group. Let U be a subsemigroup of S which is unitary in S. Then, if e denotes the identity element of S, $eu = u \in U$ implies $e \in U$; whence $uu^{-1} = e \in U$ implies $u^{-1} \in U$ when $u \in U$. Consequently U is a subgroup of S. Conversely any subgroup is clearly unitary. Thus *a subsemigroup of a group is unitary if and only if it is a subgroup.* (Cf. Exercise 3 for §7.3.) From Theorem 9.32 it follows directly that, for subsemigroups of a group, the concepts of unitariness and almost unitariness coincide.

Howie's basic result is that the semigroup amalgam $[S_i; U; \phi_i]$ can be embedded in a semigroup when $U_i (= U\phi_i)$ is almost unitary in S_i for each i. This result rests on certain combinatorial lemmas which we now give.

Suppose that $[\{S_i: i \in I\}; U; \{\phi_i; i \in I\}]$ is a semigroup amalgam and that $U_i = U\phi_i$ is almost unitary in S_i with associated mappings λ_i and ρ_i, $i \in I$. Let the word $w = a_1 a_2 \cdots a_k$ be an (irreducible) element of $\prod^* S_i$. Then $a_1, a_2, \cdots, a_k$ will be called the *syllables* of w. The syllable a_1 is called the *left end* syllable, and a_k is called the *right end* syllable of w. If $k > 2$, then $a_2, \cdots, a_{k-1}$ are called the *internal* syllables of w. The word w is said to be a *proper* word if (i) all its internal syllables belong to $\bigcup \lambda_i S_i \rho_i$, (ii) its left end syllable, when $k \geqq 2$, belongs to $\bigcup S_i \rho_i$ and (iii) its right end syllable, when $k \geqq 2$, belongs to $\bigcup \lambda_i S_i$. A word w will be called *right* [*left*] *proper* if it is proper and if its right [left] end syllable also belongs to $\bigcup S_i \rho_i$ [$\bigcup \lambda_i S_i$], and *bi-proper* if it is both right and left proper. Thus, for example, any single syllable word is proper, but not in general right or left proper.

Let us examine more closely elementary v-transitions. An elementary v-transition takes a factor $u_i \in U_i$, of a word w, and replaces it by $u_j \in U_j$, where $u_i = u\phi_i$ and $u_j = u\phi_j$ for some u in U. Now such a change can be related to the syllables of the word w in three ways: (a) u_i is a syllable of w; (b) s_i is a syllable of w and $s_i = u_ib_i$ or $s_i = a_iu_i$; (c) s_i is a syllable of w and $s_i = a_iu_ib_i$. Transitions of types (a), (b) and (c) will be called *S-steps*, *E-steps*, and *M-steps*, respectively (syllable, end, and middle steps). It is understood that the subsequent reduction of the new word to irreducible form is to be considered as part of the elementary v-transition.

A sequence of elementary v-transitions is thus a sequence of S-steps, E-steps and M-steps. Our first purpose is to establish the basic combinatorial result that, under certain general conditions, any S-steps in such a sequence may be presumed to be at the beginning.

A conventional notation for elementary v-transitions will be convenient. Let x be any element of U. Then x_i will denote the element $x\phi_i$ of U_i, for i in I.

First a lemma we shall need,

LEMMA 9.35. *Let w' be derived from w by an M-step followed by an S-step. Then w' can be derived from w by an alternative sequence of elementary v-transitions, of length at most two, and such that, if the length is two, then the second transition is not an S-step.*

PROOF. Let $s_i = a_iu_ib_i \in S_i$ be the syllable of w which is transformed to $a_iu_jb_i$, where $j \neq i$, by the M-step. Then a_i, u_j and b_i are syllables of w'', the word obtained from w by this M-step. If the syllable of w'' which is changed by the S-step to give w' is not adjacent to either a_i or b_i then, clearly, the given M-step and S-step could be performed in reverse order to give w'. If the syllable u'_k, say, of w'' involved immediately precedes the syllable a_i (and similar remarks apply if it immediately follows b_i) and is transformed to u'_m by the S-step then again, if $m \neq i$, the M-step and S-step may be interchanged. If $m = i$, then almost as simple an interchange is again possible. For w' may be derived from w by an S-step followed by an M-step and the M-step transforms the syllable $u'_ia_iu_ib_i$ to $u'_ia_iu_jb_i$.

There remain three cases to consider. Firstly, the S-step changes $a_i = u'_i$ to u'_m. In this case $s_i = a_iu_ib_i = u'_iu_ib_i$ and an E-step applied to w replaces s_i by $u'_mu_ib_i$; a further E-step then replaces $u'_mu_ib_i$ by $u'_mu_jb_i$. Thus two E-steps suffice to derive w' from w. Similar remarks apply to the second case in which b_i is the syllable changed by the S-step. Finally, suppose that the S-step changes u_j to u_m. If $m = i$, then $w' = w$. If $m \neq i$, then a single M-step replacing u_i by u_m suffices to derive w' from w.

This completes the proof of the lemma.

The result stated in the next lemma contains some immediate implications of the definition of an almost unitary subsemigroup. It will be used where necessary in computations without further comment.

LEMMA 9.36. *Let U be an almost unitary subsemigroup of the semi-*

group S with λ and ρ as associated mappings. Then, if $u \in U$ and s, $t \in S$,

$$\lambda(sut)\rho = (\lambda s\rho)\, u(\lambda t\rho);$$
$$\lambda(su) = \lambda(su)\rho = (\lambda s\rho)u;$$
$$(us)\rho = \lambda(us)\rho = u(\lambda s\rho).$$

A sequence $w = w_0 \to w_1 \to \cdots \to w_n = w'$ of elementary v-transitions will be said to be [*left, right, bi-*] *proper* if w_i is [left, right, bi-] proper for $i = 0, 1, 2, \cdots, n$. We now show that any sequence of transitions between two proper words may be assumed to be proper. A further definition will be convenient.

Let $w = a_1a_2\cdots a_k$, where $a_i \in S_{j(i)}$, be a word in irreducible form. If $k > 1$, set $b_i = \lambda_{j(i)}a_i\rho_{j(i)}$, if $i \notin \{1, k\}$, $b_1 = a_1\rho_{j(1)}$ and $b_k = \lambda_{j(k)}a_k$; if $k = 1$, set $b_1 = a_1$. Then $w^* = b_1b_2\cdots b_k$ is a proper word called the *associated proper word* of w.

If we modify the foregoing by defining $b_1 = \lambda_{j(1)}a_1\rho_{j(1)}$, or $b_k = \lambda_{j(k)}a_k\rho_{j(k)}$, or both, we obtain the associated left proper, or right proper, or bi-proper word of w, respectively. Although we state and prove the following lemma only for (unmodified) proper words and sequences, it is easy to see that it remains valid if we change "proper" to "left proper", "right proper", or "bi-proper". We shall have occasion to call upon these "modifications of Lemma 9.37" in the sequel.

Lemma 9.37. *Let $w = w_0 \to w_1 \to \cdots \to w_n = w'$ be a sequence of elementary v-transitions from the proper word w to the proper word w'. Let w_t^* be the associated proper word of w_t, for $t = 0, 1, 2, \cdots, n$.*

Then

$$w = w_0^* \to w_1^* \to \cdots \to w_n^* = w'$$

is a proper sequence of elementary v-transitions. Furthermore, $w_t^* \to w_{t+1}^*$ is an S-step, M-step, or E-step according as $w_t \to w_{t+1}$ is an S-step, M-step, or E-step.

Proof. Let $w_t = a_1a_1\cdots a_k$ in irreducible form, with $a_i \in S_{j(i)}$; thus $w_t^* = b_1b_2\cdots b_k$, with $b_i \in S_{j(i)}$. Suppose that $w_t \to w_{t+1}$ is an S-step. Then, for some i, $a_i = u_{j(i)} \in U_{j(i)}$ and w_{t+1} is obtained from w_t by replacing $u_{j(i)}$ by u_m, say. In this event, $a_i = u_{j(i)} = \lambda_{j(i)}u_{j(i)}\rho_{j(i)} = b_i$. If u_m is a syllable of w_{t+1}, then it is also a syllable of w_{t+1}^*. If u_m is not a syllable of w_{t+1}, then it combines with a syllable of w_t to form a syllable of w_{t+1}. Suppose $m = j(i+1)$. Then u_ma_{i+1} is a syllable of w_{t+1}, and the associated syllable of w_{t+1}^* is $\lambda_mu_ma_{i+1}\rho_m$, or $\lambda_mu_ma_{i+1}$, if $i + 1 = k$. But $\lambda_mu_ma_{i+1} = u_m\lambda_ma_{i+1}$ and $\lambda_mu_ma_{i+1}\rho_m = u_m\lambda_ma_{i+1}\rho_m$ and $u_m = \lambda_mu_m\rho_m$. Similar comments apply if $m = j(i-1)$. Hence, in all cases, $w_t^* \to w_{t+1}^*$ is the elementary v-transition resulting from the S-step which replaces $a_i = u_{j(i)}$ by u_m.

Suppose that $w_t \to w_{t+1}$ is an M-step. Then, for some i, $a_i = c_iu_{j(i)}d_i$, where $u_{j(i)} \in U_{j(i)}$, and w_{t+1} is obtained from w_t by replacing $u_{j(i)}$ by u_m, say, where $m \neq j(i)$. It is evident, since $a_i = (c_i\rho_{j(i)})u_{j(i)}(\lambda_{j(i)}d_i)$, that w_{t+1}^* is also obtained from w_t^* by an M-step.

Suppose that $w_t \to w_{t+1}$ is an E-step. Then, for some i, $a_i = c_iu_{j(i)}$, where $u_{j(i)} \in U_{j(i)}$, (or alternatively, and this case is entirely similar, $a_i = u_{j(i)}d_i$) and w_{t+1} is obtained from w_t by replacing $u_{j(i)}$ by u_m, say. If u_m is a syllable of w_{t+1}, then, since $a_i = (c_i\rho_{j(i)})u_{j(i)}$, w_{t+1}^* is obtained from w_t^* by the E-step which, if $i > 1$, replaces $b_i = \lambda_{j(i)}c_i\rho_{j(i)}u_{j(i)}$ by $\lambda_{j(i)}c_i\rho_{j(i)}u_m$, or, if $i = 1$, replaces $b_1 = c_1\rho_{j(1)}u_{j(1)}$ by $c_1\rho_{j(1)}u_m$. If u_m is not a syllable of w_{t+1}, then $m = j(i+1)$ and because $\lambda_m u_m a_{i+1}\rho_m = u_m\lambda_m a_{i+1}\rho_m$ and $\lambda_m u_m a_{i+1} = u_m\lambda_m a_{i+1}$, it similarly follows that w_{t+1}^* is derived from w_t^* by an E-step.

This completes the proof of the lemma.

A word w of $\prod^* S_i$, each of whose syllables is an element of $\bigcup\{U_i\}$, will be called a *U-word*. A word w of $\prod^* S_i$, each of whose syllables is either an element of S_i or is an element of $\bigcup\{U_i\}$, will be called a *(U, i)-word*. Thus, a U-word is a (U, i)-word for any i. Let w be a (U, i)-word. Replace each syllable $u_m \in U_m$ of w by u_i, for all $m \neq i$. The resulting word is then denoted by $w(i)$.

LEMMA 9.38. *Let w be a left proper (U, k)-word [U-word], and let w' be a bi-proper word obtained from w by a left proper sequence consisting wholly of E-steps. Then w' is also a (U, k)-word [U-word], and $w(k) = w'(k)$.*

PROOF. Let the given left proper sequence be

$$w = w_0 \to w_1 \to \cdots \to w_{n-1} \to w_n = w'.$$

We prove the lemma by induction on n. Assume therefore that the lemma holds for sequences of length $n - 1$. The proof of the assertion for a single E-step can be inferred from that of the inductive step from $n - 1$ to n, by taking $w' = w_1$.

If w_1 is a (U, k)-word, then we conclude from the hypothesis for induction that w' is a (U, k)-word, and that $w'(k) = w_1(k)$. Clearly $w_1(k) = w_0(k) = w(k)$, and hence this case is concluded. We may therefore assume that w_1 is not a (U, k)-word.

The transition $w_0 \to w_1$ cannot take place on a syllable in S_k, since then w_1 would be a (U, k)-word. Hence it must have the form, with $i \neq k$,

$$w_0 = \cdots u_i \cdots = \cdots u_i'a_i \cdots \to \cdots u_j'a_i \cdots = w_1,$$

or with a_iu_i' instead of $u_i'a_i$. Since we can evidently assume $j \neq i$, it is clear that a_i is a syllable of w_1. Suppose that a_i were not the right end syllable of w_1. Then, since w_1 is left proper, $\lambda_i a_i \rho_i = a_i$. Hence, using Lemma 9.36,

$$u_i = \lambda_i u_i \rho_i = \lambda_i(u_i'a_i)\rho_i = u_i'(\lambda_i a_i \rho_i) = u_i'a_i.$$

Since $a_i \in \lambda_i S_i \rho_i$, and U_i is unitary in $\lambda_i S_i \rho_i$, we conclude that $a_i \in U_i$. Since this would imply that w_1 is a (U, k)-word, contrary to our assumption, we conclude that a_i must be the right end syllable of w_1.

But then u_i is the right end syllable of $w_0 = w$, and so w is bi-proper. We may now apply one of the modifications of Lemma 9.37 to convert the given sequence into a new one,

$$w = w_0 \to w_1^* \to \cdots \to w_{n-1}^* \to w_n = w'$$

which is bi-proper. The two end terms, w and w', are unaffected, since they are already bi-proper.

The above argument showing that $a_i \in U_i$ now holds with a_i in any position, and so w_1^* is a (U, k)-word. That $w_n = w'$ is also a (U, k)-word, and that $w'(k) = w_1^*(k) = w(k)$ now follows by induction.

The above proof also covers the bracketed version of the lemma.

We now investigate the possibility of bringing an S-step forward when it follows an E-step. It will turn out, as we shall see in the next lemma, that the argument is straightforward except in the following situation. Let $w = \cdots s_i s_j$, with final two syllables $s_i \in S_i$ and $s_j \in S_j$. Suppose that $s_i = a_i u_i$, $u_i \in U_i$. Suppose that $u_j s_j = u_j' \in U_j$. Let the E-step and S-step be as follows:

$$\begin{aligned} w = \cdots s_i s_j &= \cdots a_i u_i s_j \\ &\xrightarrow{E} \cdots a_i u_j s_j = \cdots a_i u_j' \\ &\xrightarrow{S} \cdots a_i u_m'. \end{aligned}$$

When $s_j \notin \lambda_j S_j \rho_j$, following Howie, we shall call this the *irreducible ES-configuration*. There is similarly an irreducible ES-configuration which involves the first two syllables of w.

LEMMA 9.39. *Let w' be derived from w by an E-step followed by an S-step: $w \xrightarrow{E} w'' \xrightarrow{S} w'$; and suppose that this sequence is proper.*

Then, except possibly when the E-step and S-step are in irreducible ES-configuration, w' can be derived from w by an alternative proper sequence of elementary v-transitions, of length at most two, and such that, if the length is two, the second transition is not an S-step.

PROOF. Let $s_i \in S_i$ be the syllable of w involved in the E-step. Suppose that $s_i = a_i u_i$ and that the E-step transforms s_i to $(a_i \rho_i) u_j$. Observe that we must have $a_i \rho_i$ and not simply a_i after transforming s_i because, by assumption, w'' is proper. Strictly speaking, we first make the "internal transition" $\cdots a_i u_i \cdots \to \cdots (a_i \rho_i) u_i \cdots$ on w, which does not alter the irreducible word associated with w, since $a_i u_i = a_i(\lambda_i u_i) = (a_i \rho_i) u_i$, and then we make the E-step $\cdots (a_i \rho_i) u_i \cdots \to \cdots (a_i \rho_i) u_j \cdots$. No confusion will result from our referring to $\cdots a_i u_i \cdots \to \cdots (a_i \rho_i) u_j \cdots$ as an E-step.

Entirely similar considerations will hold throughout if we begin with a left E-step, instead of a right E-step, for the transformation $w \to w''$. There are several possibilities for the S-step $w'' \to w'$ which we now consider in turn.

Firstly, if the S-step involves a syllable of w'' which is also a syllable of w, then the S-step and E-step may simply be performed in reverse order to derive w' from w. If the S-step changes u_j to u_m, then a single E-step (or no step at all if $m = i$) replacing $s_i = a_i u_i$ by $(a_i\rho_i)u_m$ serves to derive w' from w. If the S-step changes the syllable $a_i\rho_i$ of w'', then $a_i\rho_i = u_i'$, say, and the S-step replaces u_i' by u_m', say. Then $s_i = u_i'u_i$ may be replaced firstly by $u_m'u_m$ and then by $u_m'u_j$ (the latter step not being necessary, if $j = m$). This derives w' from w either by a single S-step or by an S-step followed by an E-step. Observe that in each case the alternative sequence which derives w' from w is also proper.

Finally, there remains the case when the syllable adjacent on the right to s_i in w is an element of S_j, s_j, say, so that $u_j s_j$ is a syllable of w'', and $u_j s_j$ is the syllable of w'' involved in the S-step. Then $u_j s_j = u_j'$, say, and the S-step transforms u_j' to u_m', say. Suppose that s_j is an internal syllable of w. Since w is proper, $s_j \in \lambda_j S_j \rho_j$. Hence, since U_j is unitary in $\lambda_j S_j \rho_j$, the equation $u_j s_j = u_j'$ implies that $s_j = u_j''$, say, an element of U_j. Then

$$w = \cdots a_i u_i u_j'' \cdots \xrightarrow{S} \cdots a_i u_i u_m'' \cdots$$

$$\xrightarrow{E} \cdots (a_i\rho_i)u_m u_m'' \cdots = \cdots (a_i\rho_i)u_m' \cdots = w'$$

derives w' from w by an S-step followed by an E-step. Further this sequence is proper.

Suppose now that s_j is the right end syllable of w. This is then either the irreducible ES-configuration, which by hypothesis we exclude, or $s_j \in \lambda_j S_j \rho_j$. In the latter event, the argument just given for an internal syllable holds without change. And this completes the proof of the lemma.

If, in the foregoing lemma, we replace "proper" by "bi-proper", then the irreducible ES-configuration cannot arise. This is, in particular, the case when U_i is unitary in S_i for each i, and each λ_i and ρ_i is taken as the identity mapping of S_i, so that $\lambda_i S_i \rho_i = S_i$ for each i. From Lemmas 9.35 and 9.39 we obtain the following corollary.

Corollary 9.40. *Any sequence of elementary v-transitions from one bi-proper word w to another w' can be replaced by a bi-proper sequence from w to w' in which all the S-steps, if any, come first. In particular, this holds for arbitrary words w and w' if U_i is unitary in S_i for all i in I.*

To handle the general case, we develop a technique for bypassing the irreducible ES-configurations.

Observe that if $w \xrightarrow{E} w'' \xrightarrow{S} w'$ is an irreducible ES-configuration, then the final syllable of w'', u_j' in our earlier notation, satisfies $u_j' = u_j'\rho_j$. The following lemma provides what we require.

Lemma 9.41. *Let $s_k \in S_k$ and let $s_k \to \cdots \to w$ be a proper sequence of elementary v-transitions from s_k to w. Suppose that t_j is the final syllable of w and that $t_j = t_j\rho_j$. Then $s_k = s_k\rho_k$.*

PROOF. Suppose there are S-steps in the sequence from s_k to w. Consider the first such S-step. If it follows an E-step in irreducible ES-configuration, then the word to which the S-step is applied has its right-most syllable satisfying the hypotheses imposed on w in the lemma. We may then take this word as w for the remainder of the proof. If the first S-step immediately follows an M-step or an E-step not in the irreducible ES-configuration, then, by Lemmas 9.35 and 9.39, we can find a proper sequence of elementary v-transitions from s_k to w either with one less S-step or with its first S-step occurring earlier than in the given sequence from s_k to w. Continuing in this way with the first S-step, then with the second S-step, and so on, we finally obtain a proper sequence from s_k to w in which either all the S-steps are performed first or the first S-step follows an E-step in irreducible ES-configuration. In the latter event we proceed as has been stated. In the former event, if there is an initial S-step, then $s_k \in U_k$, and so $s_k = s_k \rho_k$. Alternatively, there are no S-steps in the sequence from s_k to w.

The proof of the lemma is thus reduced to the case when there are no S-steps in the sequence from s_k to w. We argue now by induction on the number of M-steps in the sequence.

Suppose firstly that w is derived from s_k entirely by E-steps. The first of these must have one of the following two forms: $s_k = u_k s'_k \to u_j s'_k$, or $s_k = s'_k u_k \to s'_k u_j$. The second of these gives the conclusion of the lemma immediately, and hence we may assume the first. This implies that $\lambda_k s_k = s_k$, i.e., that the word s_k is left proper. By one of the modifications of Lemma 9.37, we can pass from the given sequence $s_k \to \cdots \to w$ to a sequence $s_k \to \cdots \to w^*$ that is left proper. Since w^* is obtained from w merely by replacing its left end syllable b_j, say, by $\lambda_j b_j$, it is clear that w^* is also right proper, hence bi-proper. We now apply Lemma 9.38 to conclude that w^* is a (U, k)-word, and that $w^*(k) = w(k) = s_k$. Since w^* is bi-proper, it is clear that $w^*(k)\rho_k = w^*(k)$, and hence $s_k \rho_k = s_k$.

Assume, therefore, as an inductive hypothesis, that the result of the lemma holds if the sequence from s_k to w contains fewer than p M-steps. Suppose that the sequence from s_k to w contains p M-steps and let us examine the portion of the sequence from the final M-step to w.

Let $w_1 = \cdots a_m u_m b_m \cdots \xrightarrow{M} \cdots a_m u_n b_m \cdots = w_2$, where $a_m u_m b_m \in S_m$ is a syllable of w_1, be the final M-step. Let $w_2 \to w_3 \to \cdots \to w_q = w$ be the remaining transitions from w_2 to w. By assumption, these transitions are all E-steps. Observe that an E-step applied to a syllable $s_m \in S_m$ leaves behind it a non-empty syllable s'_m, say, also in S_m, which we will call a *descendant* of s_m; we shall also include s_m itself among the descendants of s_m. Furthermore, descendants of s'_m, or of its descendants, will be called descendants of s_m. Each word in the sequence $w_2 \to w_3 \to \cdots \to w_q$ will thus contain syllables, belonging to S_m, which are the descendants of the syllables a_m and b_m of w_2. Let $a_m^{(r)}$ and $b_m^{(r)}$ be the descendants of a_m and b_m, respectively, in w_r, $r = 3, \cdots, q$. Interpret $a_m^{(2)}$ as a_m and $b_m^{(2)}$ as b_m. Let $w_r = \cdots a_m^{(r)} y_r b_m^{(r)} \cdots$.

For $r = 2$, $y_2 = u_n$ is a U-word. Assume that we have already shown that y_r is a U-word. If the E-step $w_r \to w_{r+1}$ acts on a syllable of y_r, then by Lemma 9.38, y_{r+1} is also a U-word. If the E-step acts on a syllable of w_r to the left of $a_m^{(r)}$ or to the right of $b_m^{(r)}$, then $y_r = y_{r+1}$. If the E-step acts on $a_m^{(r)}$ and peels off a syllable from its left or, similarly, peels off a syllable from the right of $b_m^{(r)}$, then again $y_r = y_{r+1}$. In the only remaining possibilities y_r is extended by either a U-syllable on its left or a U-syllable on its right. Thus in all cases, y_{r+1} is a U-word. By induction on r, each y_r is therefore a U-word.

For $r = 2, 3, \cdots, q$, replace y_r in w_r by $y_r(m)$ to give the word $w'_r = \cdots a_m^{(r)} y_r(m) b_m^{(r)} \cdots$. Then as is easily seen, each transformation $w'_r \to w'_{r+1}$, $r = 2, 3, \cdots, q-1$, is either an elementary v-transition or leaves w'_r unchanged. Since y_q is an internal syllable of w_q, w'_q has the same final syllable as $w_q = w$. Since $w'_2 = w_1$, the sequence from s_k to w'_q, that has been derived from the sequence from s_k to w has only $p-1$ M-steps. By our inductive hypothesis, therefore $s_k \rho_k = s_k$.

This completes the proof of the lemma.

A detail of the procedure used in the final two paragraphs of the proof of Lemma 9.41 we will require again. For convenience of reference we state it formally as a lemma.

LEMMA 9.42. *Let* $w_1 = \cdots a_m u_m b_m \cdots \xrightarrow{M} \cdots a_m u_n b_m \cdots = w_2$, *where* $a_m u_m b_m \in S_m$ *is a syllable of* w_1, *be an* M-*step. Let* $w_2 \to w_3 \to \cdots \to w_q$ *be* E-*steps; and suppose that* $w_1 \to w_2 \to \cdots \to w_q$ *is a proper sequence.*

Then w_q *may be written as an irreducible word in the form* $w_q = xyz$, *where* x, z *are non-empty words, where* y *is a* U-*word, and such that there exists a proper sequence* $w_1 \to w'_3 \to \cdots \to w'_q$, *consisting entirely of* E-*steps from* w_1 *to* $w'_q = xy(m)z$.

We now gather together the results we have proved to derive

LEMMA 9.43. *Let* $s_k \in S_k$ *and* $s_h \in S_h$ *and suppose that* $(s_k, s_h) \in v^*$. *Then there is a sequence of elementary* v-*transitions from* s_k *to* s_h *in which the* S-*steps, if there are any, are all performed first.*

PROOF. By assumption, there is a sequence of elementary v-transitions from s_k to s_h. By Lemma 9.37, we can assume that this sequence is proper. If there are no S-steps in the sequence then there is nothing to prove. In the contrary case, consider the first S-step in the sequence. If this immediately follows either an M-step or an E-step not in irreducible ES-configuration, then, by Lemmas 9.35 and 9.39, there is a proper sequence from s_k to s_h, either with one less S-step, or with its first S-step occurring nearer the beginning, than in the given sequence from s_k to s_h.

Continue in this way until we have a proper sequence from s_k to s_h in which the first S-step is the first transition, or in which there is one less S-step than in the initial sequence from s_k to s_h, or in which the first S-step follows an E-step in irreducible ES-configuration. In the latter event, recalling the comment preceding Lemma 9.41, we may apply Lemma 9.41 to deduce that $s_k\rho_k = s_k$ (or, in the dual situation, that $\lambda_k s_k = s_k$).

Suppose now that $w_1 \to w_2$ is any elementary υ-transition between any two words w_1 and w_2 and that the final syllable $t_j \in S_j$, say, of w_1, satisfies $t_j\rho_j = t_j$. Let $s_n \in S_n$, say, be the final syllable of w_2. Let w_2' be formed from w_2 by replacing its final syllable by $s_n\rho_n$. Then $w_1 \to w_2'$ is also an elementary υ-transition. This is easily seen by considering in turn each of the possibilities. For example, let $t_j = a_j u_j b_j$, where $u_j \in U_j$, and suppose that u_j is replaced by u_m by the transition $w_1 \to w_2$. Then $b_j = s_n$; and since $t_j = t_j\rho_j$, we also have $t_j = a_j u_j(b_j\rho_j)$. Thus $w_1 \to w_2'$ is an elementary υ-transition. The other cases are similar. Furthermore, it is clear that $w_1 \to w_2'$ is an E-step, M-step or S-step according as $w_1 \to w_2$ is an E-step, M-step or S-step.

Suppose now that, in our earlier notation, the E-step and S-step in irreducible ES-configuration being considered are:

$$\begin{aligned} w = \cdots s_i s_j &= \cdots a_i u_i s_j \\ &\xrightarrow{E} \cdots a_i u_j s_j = \cdots a_i u_j' = w'' \\ &\xrightarrow{S} \cdots a_i u_m' = w'. \end{aligned}$$

Since $s_k\rho_k = s_k$, the argument we have just given applies to show that the sequence from s_k to w can be replaced by a proper sequence, of the same length and containing no S-steps, from s_k to $w_1 = \cdots s_i(s_j\rho_j)$, where w_1 is derived from w by replacing its final syllable s_j by $s_j\rho_j$. Since $u_j' = u_j'\rho_j = u_j(s_j\rho_j)$, $w_1 \to w''$ is an E-step.

The S-step $w'' \to w'$ is now the first S-step in the proper sequence $s_k \to \cdots \to w_1 \xrightarrow{E} w'' \xrightarrow{S} w'$; but, since $s_j\rho_j \in \lambda_j S_j \rho_j$, the final E-step and S-step are now no longer in irreducible ES-configuration. Consequently we may apply Lemma 9.39, as before, to this pair of transitions.

Thus, in all cases, the original sequence from s_k to s_h may be replaced by a proper sequence from s_k to s_h, either with a smaller number of S-steps than the original sequence, or with the same number of S-steps, but with the first elementary υ-transition an S-step.

Repetitions of this argument to deal, in turn, with any remaining S-steps not at the beginning, complete the proof of the lemma.

We can now quickly prove Howie's basic result on embedding semigroup amalgams.

THEOREM 9.44. *Let* $[\{S_i : i \in I\}; U; \{\phi_i : i \in I\}] = [S_i; U; \phi_i]$, *be a semigroup amalgam. Let* U_i $(= U\phi_i)$ *be almost unitary in* S_i, *with associated mappings* λ_i *and* ρ_i, *for* $i \in I$.

Then the semigroup amalgam $[S_i; U; \phi_i]$ *can be embedded in a semigroup.*

PROOF. By Theorem 9.31, part 1, it suffices to show that $[S_i; U; \phi_i]$ can be embedded in $\prod^* {}_U S_i$. By Theorem 9.31, part 2, it therefore suffices to show that:

(i) each canonical homomorphism $\mu_k: S_k \to \prod^* {}_U S_k$ is one-to-one; and
(ii) $S_k\mu_k \cap S_h\mu_h = U\mu$, for $h \neq k$, where $\mu = \phi_h\mu_h$.

Let $s_k \in S_k$ and $s'_h \in S_h$. Suppose that $s_k\mu_k = s'_h\mu_h$, i.e. that $(s_k, s'_h) \in v^*$. From Lemma 9.43, there is a sequence of elementary v-transitions from s_k to s'_h in which the S-steps, if any, are performed first. If there are initial S-steps then their combined effect can clearly be achieved by a single S-step, $s_k \to s_l$, say. Consider the sequence from s_l to s'_h. Observe that each E-step or M-step increases the number of syllables in a one-syllable word and cannot decrease the number of syllables when applied to any word. Consequently, the sequence from s_l to s'_h, which, by hypothesis, contains no S-steps, and which terminates on a word of one syllable, can contain no E-steps or M-steps either. Thus $s_l = s'_h$.

Hence there are only two possibilities. Firstly, there are no S-steps in the sequence from s_k to s'_h, when we infer $s_k = s'_h$. Secondly, the transformation from s_k to s'_h can be achieved in a single S-step. In this event, $s_k = u_k$, say, belongs to U_k, and $s'_h = u_h$. Thus we have shown that if $s_k\mu_k = s'_h\mu_h$, then either $s_k = s'_h$ or else $h \neq k$ and $s_k \in U_k$. To prove (i), we assume at the outset that $h = k$, and hence we are driven to the first conclusion. To prove (ii), since we always have $U\mu \subseteq S_k\mu_k \cap S_h\mu_h$, it suffices to show that if $x \in S_k\mu_k \cap S_h\mu_h$ for $k \neq h$, then $x \in U\mu$. Hence assume $x = s_k\mu_k = s'_h\mu_h$. Since $k \neq h$, we are driven to the second conclusion, that $s_k \in U_k$. Hence $s_k = u\phi_k$ for some u in U, and $x = s_k\mu_k = u\phi_k\mu_k = u\mu \in U\mu$. This completes the proof of the theorem.

The combinatorial apparatus built up to prove the preceding theorem may be used to derive further information on amalgamated free products. We give some further results of Howie.

THEOREM 9.45. *Let $[S_i; U; \phi_i]$ be a semigroup amalgam in which U_i is a unitary subsemigroup of S_i for each i in I. Then $S_i\mu_i$ is a unitary subsemigroup of $\prod^* {}_U S_i$, for each i in I. Furthermore, $U\mu$ is a unitary subsemigroup of $\prod^* {}_U S_i$.*

PROOF. From the preceding theorem it follows that $S_i\mu_i \cap S_j\mu_j = U\mu$, if $i \neq j$. Consequently, since the intersection of any two unitary subsemigroups is itself a unitary subsemigroup, it will suffice to prove that each $S_i\mu_i$ is unitary in $\prod^* {}_U S_i$.

We will show that $S_i\mu_i$ is right unitary in $\prod^* {}_U S_i$. Left unitariness will follow similarly. Suppose then that $s_i, t_i \in S_i$, that $w \in \prod^* S_i$ and that $(ws_i, t_i) \in v^*$. We must show that this implies that $wv^* \in S_i\mu_i$.

By Corollary 9.40, there is a sequence of elementary v-transitions from t_i to ws_i in which the S-steps, if any, are performed first. Any initial S-steps can be combined into a single S-step; and hence, in both cases, there is a sequence, containing no S-steps, from $t_j \in S_j$, say, to ws_i, where t_j is a (U, i)-word. The proof is by induction on the number of M-steps in this sequence.

Suppose, firstly, the sequence consists entirely of E-steps. Then, by Lemma 9.38, since, under the assumption of this theorem, all words are bi-proper, ws_i is a (U, i)-word. Hence w is a (U, i)-word, whence $wv^* \in S_i\mu_i$.

Assume now that the required conclusion can be drawn for sequences which contain fewer than p M-steps. Suppose the sequence from t_j to ws_i contains p M-steps and let us examine the portion of the sequence from the final M-step to ws_i. Applying Lemma 9.42 it follows that ws_i may be written as an irreducible word in the form xyz, where x, z are non-empty words and y is a non-empty U-word; and, further, that there exists a sequence of elementary v-transitions, containing no S-steps, and containing only $p-1$ M-steps, from t_j to $xy(m)z = w'$, say, for some m in I.

Consider the various possibilities for w'. We have that $ws_i = xyz$ and that, in this form, xyz is irreducible, i.e. that the final syllable of x and the first syllable of y lie in distinct S_i and, similarly, the final syllable of y and the first syllable of z lie in distinct S_i. Observe also that s_i need not be a syllable of ws_i. It follows from this that the final syllable of z is a_is_i, where a_i either belongs to S_i or is the empty word. Consequently $w' = xy(m)z$ may be written $w' = w''s_i$, where $w'' = xy(m)b$ and where $bs_i = z$. It is possible here that b is the empty word. Applying our inductive hypothesis to the sequence from t_j to $w''s_i$, we infer that $w''v^* \in S_i\mu_i$. However, $w = xyb$ and so $(w, w'') \in v^*$. Consequently, $wv^* \in S_i\mu_i$. This suffices to complete the proof of the theorem.

Strengthening the previous theorem, by reading "almost unitary" for "unitary" throughout, is not possible. For a counter example, see Exercise 7, below.

We now give a theorem on the generation of subsemigroups of $\prod^*{}_U S_i$. For conditions which ensure that the hypotheses of this theorem are satisfied, see Exercise 8, below. Exercise 9 shows that the theorem would be false if we omitted the hypothesis that T_i is almost unitary in S_i for each i.

THEOREM 9.46. *Let $[S_i; U; \phi_i]$ be a semigroup amalgam in which U_i is an almost unitary subsemigroup of S_i, with associated mappings λ_i and ρ_i, for each i in I. For each i, let T_i be an almost unitary subsemigroup of S_i, containing U_i and with the same associated mappings λ_i and ρ_i. Then the semigroups $T_i\mu_i$ generate in $\prod^*{}_U S_i$ a subsemigroup isomorphic to $\prod^*{}_U T_i$.*

PROOF. Since $T_i \subseteq S_i$, $\prod^* T_i \subseteq \prod^* S_i$. Since $[T_i; U; \phi_i]$ is a semigroup amalgam, the relation $v = \{(u_i, u_j) \colon i, j \in I,\ u\phi_i = u_i,\ u\phi_j = u_j$, for some u in $U\}$ is a relation on $\prod^* T_i$. Denote by $v^\dagger$ the congruence on $\prod^* T_i$ generated by v. Then $\prod^* T_i/v^\dagger$ is the amalgamated product $\prod^*{}_U T_i$. By Lemma 9.9, to show that $\prod^*{}_U T_i$ is embedded naturally in $\prod^*{}_U S_i$, i.e. that $\prod^*{}_U T_i$ is isomorphic to the subsemigroup of $\prod^*{}_U S_i$ generated by the semigroups

$T_i\mu_i$, it suffices to show that

$$\upsilon^\dagger = \upsilon^* \cap (\prod{}^* T_i \times \prod{}^* T_i).$$

And to show this, all that remains to be proved is that

$$\upsilon^* \cap (\prod{}^* T_i \times \prod{}^* T_i) \subseteq \upsilon^\dagger.$$

To this end, consider words w, w' in $\prod^* T_i$ such that $(w, w') \in \upsilon^*$. Since, by hypothesis, $T_i \subseteq \lambda_i S_i \rho_i$ for each i, w and w' are bi-proper words. By Corollary 9.40, there exists a bi-proper sequence of υ-transitions

$$w = w_0 \to w_1 \to \cdots \to w_s \to w_{s+1} \to \cdots \to w_n = w'$$

from w to w' such that the first s transitions $(0 \leq s \leq n)$ are all S-steps, while the remaining $n - s$ are either E-steps or M-steps. We proceed to show that $w_r \in \prod^* T_i$ for all $r = 0, 1, \cdots, n$; and this will complete the proof.

We first make two observations about an υ-transition $w_r \to w_{r+1}$. A syllable which does not belong to $\bigcup T_i$ will be called a *non-T-syllable*.

(1) An S-step cannot increase the number of non-T-syllables.
(2) An E- or M-step cannot decrease the number of non-T-syllables.

To see (1), let $w_r = \cdots s_i u_j s_k' \cdots \to w_{r+1} = \cdots s_i u_l s_k' \cdots$ be an S-step. If $l \neq i$ and $l \neq k$, there is no change in the number of non-T-syllables, since u_j and u_l both belong to $\bigcup T_i$. If $l = i \neq k$, and $s_i \in T_i$, then $s_i u_i \in T_i$ also. If $l = i = k$, and both s_i and s_i' belong to T_i, then $s_i u_i s_i' \in T_i$. The other cases are clear.

To see (2), let $w_r = \cdots s_i \cdots = \cdots a_i u_i b_i \to w_{r+1} = \cdots a_i u_k b_i \cdots$ be an M-step. If $s_i \notin T_i$, then either $a_i \notin T_i$ or $b_i \notin T_i$. For $a_i \in T_i$ and $b_i \in T_i$ would imply $s_i = a_i u_i b_i \in T_i$.

Let $w_r = \cdots s_i s_k \cdots = \cdots a_i u_i s_k \cdots \to w_{r+1} = \cdots a_i u_l s_k \cdots$ be an E-step. If $s_i \notin T_i$, then $a_i \notin T_i$, as above. Suppose that $s_k \notin T_k$ and that $l = k$. Since w_r is bi-proper, $s_k \in \lambda_k S_k \rho_k$. Hence, since T_k is unitary in $\lambda_k S_k \rho_k$, $u_k s_k \notin T_k$. If $l \neq k$, then s_k itself is a non-T-syllable of w_{r+1}.

Now $w = w_0$ has no non-T-syllables, and, by (1), the same must be true of w_s. But no non-T-syllables can appear in the sequence $w_s \to \cdots \to w_r = w'$; for, if any did, their number could, by (2), never be thereafter reduced, and w' would not be in $\prod^* T_i$. Hence every $w_r \in \prod^* T_i$, which concludes the proof.

We complete this section by giving a result of Howie's on amalgamated direct sums of semigroups.

Let $\{S_i : i \in I\}$ be a disjoint family of semigroups. Let σ be the relation on $\prod^* S_i$ defined thus:

$$\sigma = \{(s_i s_j, s_j s_i); s_i \in S_i, s_j \in S_j, i, j \in I, i \neq j\}.$$

Then the *direct sum* of the S_i, written $\sum\{S_i : i \in I\}$, is defined by

$$\sum S_i = (\prod{}^* S_i)/\sigma^*.$$

Observe that, if each S_i has an identity element and if $|I|$ is finite, then $\sum S_i$ need not be isomorphic with the direct product $\prod S_i$. (For this reason, some other term, such as *free sum*, might be preferable. For further comments on direct products and direct sums, see Exercises 10–13 below.)

Consider now a semigroup amalgam $[S_i; U; \phi_i]$. Define the relation v on $\prod^* S_i$, as usual. Then the *direct sum of the amalgam* $[S_i; U; \phi_i]$, or, the *direct sum of the* S_i, *amalgamating* U, denoted by $\sum_U S_i$, is defined thus:

$$\textstyle\sum_U S_i = (\prod^* S_i)/(\sigma \cup v)^*.$$

Our concern is to give sufficient conditions under which a semigroup amalgam can be embedded naturally in its direct sum.

Regarding each S_i as identified with its canonical image in $\prod^* S_i$, we define the canonical homomorphisms, $\nu_i \colon S_i \to \sum_U S_i$, thus:

$$\nu_i \colon a \to a\nu_i = a(\sigma \cup v)^*, \quad a \in S_i.$$

It then follows, that for any $u \in U$,

$$u\phi_i\nu_i = u\phi_j\nu_j,$$

since $(u\phi_i, u\phi_j) \in v \subseteq (\sigma \cup v)^*$. Consequently, we may apply Lemma 9.30, to deduce that π defined by

$$\pi \colon a \to a\pi = a\chi_i\nu_i, \quad \text{if } a \in S_i',$$

is a homomorphism of the partial groupoid $\mathscr{G} = \mathscr{G}[S_i; U; \phi_i]$ into $\sum_U S_i$. The amalgam $[S_i; U; \phi_i]$ will be said to be *embedded naturally in its direct sum* if π is one-to-one.

First, a necessary condition. A subsemigroup U of a semigroup S is said to be *central* in S if $us = su$ for any $u \in U$ and any $s \in S$.

Lemma 9.47. *Let π be a one-to-one mapping of $\mathscr{G}$ into $\sum_U S_i$. Then, provided $|I| > 1$, U_i is central in S_i, for each i in I.*

Proof. Let $u \in U$, $u_i = u\phi_i$, $s_i \in S_i$ and choose $j \neq i$, $j \in I$. Let $u_j = u\phi_j$. From the definition of σ and v,

$$(s_i u_j, u_j s_i) \in \sigma,$$
$$(s_i u_j, s_i u_i) \in v^*,$$
$$(u_j s_i, u_i s_i) \in v^*;$$

whence, by transitivity,

$$(s_i u_i, u_i s_i) \in (\sigma \cup v)^*.$$

Hence, $(s_i u_i)\pi = (u_i s_i)\pi$. It follows immediately, that, if π is one-to-one, then $s_i u_i = u_i s_i$. This shows that U_i is central in S_i.

When U_i is almost unitary in S_i, for each i in I, then the condition of Lemma 9.47 is also sufficient.

THEOREM 9.48. *Let* $[\{S_i; i\in I\}; U; \{\phi_i: i\in I\}] = [S_i; U; \phi_i]$ *be a semigroup amalgam for which* $|I| > 1$. *Let* U_i $(= U\phi_i)$ *be almost unitary in* S_i, *with associated mappings* λ_i *and* ρ_i *for each* $i\in I$.

Then $[S_i; U; \phi_i]$ *can be embedded naturally in its direct sum if and only if* U_i *is central in* S_i *for each* $i \in I$.

PROOF. Because of Lemma 9.47 it remains only to prove the "if" part of the theorem. We know that $[S_i; U; \phi_i]$ is naturally embedded in $\prod^*{}_U S_i$ (Theorem 9.44 and Theorem 9.31) and we could proceed to show that if U_i is central in S_i, for each i, then the natural homomorphism of $\prod^*{}_U S_i$ onto $\sum_U S_i$ does not collapse $\mathscr{G}$. However, we prefer a direct approach.

We easily see that π is one-to-one if and only if both the following conditions hold.

(i) Each canonical homomorphism $\nu_i: S_i \to \sum_U S_i$ is one-to-one.

(ii) $S_i\nu_i \cap S_j\nu_j = U\nu$, for $i \neq j$, where $\nu = \phi_i\mu_i$.

The analogy with Theorem 9.31, part 2, is clear; and the analogy holds for the proof also. We proceed to show that (i) and (ii) hold under our hypotheses.

Observe that an elementary $(\sigma \cup v)$-transition, either is an elementary v-transition and so is an E-step, M-step or S-step, or is an elementary σ-transition.

Suppose that $s_i \in S_i$ and that $(s_i, w) \in (\sigma \cup v)^*$. Then there exists a sequence

$$s_i = w_0 \to w_1 \to \cdots \to w_n = w,$$

say, of elementary $(\sigma \cup v)$-transitions from s_i to w. By applying σ-transitions only, each word w_r can be transformed to a word v_r, say, with the property that no two of its syllables belong to the same semigroup S_i. There is, in general, more than one word v_r which can be obtained from w_r in this way. Select one such corresponding to each w_r, for $r = 0, 1, 2, \cdots, n$.

Observe now that either there are no transitions in the given sequence and $s_i = w$, or $s_i \in \lambda_i S_i \rho_i$. The latter conclusion follows because $w_0 \to w_1$ cannot be a σ-transition and so $s_i = u_i \in U_i$, or $s_i = a_i u_i$, or $s_i = u_i b_i$, or $s_i = a_i u_i b_i$, where $a_i, b_i \in S_i$. Consider the case $s_i = a_i u_i b_i$. Since U_i is central in S_i,

$$s_i = a_i b_i u_i = a_i b_i (u_i \rho_i) = (a_i b_i u_i)\rho_i = s_i \rho_i$$

and

$$s_i \rho_i = a_i b_i u_i = u_i a_i b_i = (\lambda_i u_i) a_i b_i = \lambda_i (u_i a_i b_i) = \lambda_i s_i \rho_i.$$

Thus $s_i \in \lambda_i S_i \rho_i$. A similar argument applies to each of the other cases. For later use we also remark that if $w_n = w$ is a single syllable word (and $n > 0$) in S_j, say, then $w_{n-1} \to w_n$ must be an v-transition; whence we infer that $w \in \lambda_j S_j \rho_j$.

Assume now that $n > 0$ and, for each r, define v_r^* as follows. Suppose that $v_r = s_{i(1)} s_{i(2)} \cdots$, where $s_{i(j)} \in S_{i(j)}$; then we set $v_r^* = \lambda_{i(1)} s_{i(1)} \rho_{i(1)} \cdot \lambda_{i(2)} s_{i(2)} \rho_{i(2)} \cdots$.

We then have $s_i = w_0 = v_0 = v_0^*$, since, as has been shown, $s_i \in \lambda_i S_i \rho_i$; furthermore, v_0^* is a (U, i)-word and $s_i = v_0^*(i)$. Suppose that it has already been shown that v_r^* is a (U, i)-word and that $s_i = v_r^*(i)$. We will show that it then follows that v_{r+1}^* is a (U, i)-word and that $v_{r+1}^*(i) = s_i$.

Consider the possibilities for $w_r \to w_{r+1}$, in turn. If this is a σ-transition, then clearly the set of syllables of v_{r+1} coincides with that of v_r. Hence v_{r+1}^* is a (U, i)-word and, since U_i is central in S_i, $v_{r+1}^*(i) = v_r^*(i) = s_i$.

Suppose that $w_r \to w_{r+1}$ is an E-step and let $s_m = a_m u_m$, say, where $u_m \in U_m$, be the syllable of w_r transformed to $a_m u_k$, say, in this transition. In v_r, $a_m u_m$ appears in a syllable $t_m = x a_m u_m y$, say, where x and y either are empty words or belong to S_m. Thus $\lambda_m t_m \rho_m$ is a syllable of v_r^* (in fact $t_m = \lambda_m t_m \rho_m$). Now, v_r^* is a (U, i)-word. Hence, either $m = i$, or $\lambda_m t_m \rho_m \in U_m$. In the former event, $\lambda_m x a_m y \rho_m \in S_i$. In the latter event, since $\lambda_m t_m \rho_m = \lambda_m x a_m y \rho_m u_m$ and U_m is unitary in $\lambda_m S_m \rho_m$, $\lambda_m x a_m y \rho_m \in U_m$. Thus, in both cases, $\lambda_m x a_m y \rho_m$, the syllable of v_{r+1}^* in S_m, is a (U, i)-word. Straightforwardly, we see that the syllable of v_{r+1}^* in S_k, is also a (U, i)-word. Thus v_{r+1}^* is a (U, i)-word. Since U_i is central in S_i, evaluating $v_{r+1}^*(i)$ gives $v_{r+1}^*(i) = v_r^*(i) = s_i$.

Because U_m is central in S_m for each $m \in I$, a very similar argument can be applied when $w_r \to w_{r+1}$ is either an M-step or an S-step. Thus we conclude, in all cases, $v_{r+1}^*(i) = s_i$. By induction on r, it follows that $v_n^*(i) = s_i$.

Suppose now that w is a single syllable word, $w_n = w = t_j \in S_j$, say. Then either $n = 0$ or, as remarked earlier, $t_j \in \lambda_j S_j \rho_j$. Thus, if $n > 0$, $t_j = w_n = v_n = v_n^*$; and so t_j is a (U, i)-word and $t_j(i) = s_i$. There are thus only two possibilities. Firstly, $j = i$ and $t_j = s_i$; secondly, $t_j = u_j \in U_j$ and $s_i = u_i$. These two possibilities combine to show that conditions (i) and (ii) hold for the mappings ν_i, $i \in I$. And this completes the proof of the theorem.

Exercises for §9.4

1. Let $\{S_i : i \in I\}$ be an indexed set of semigroups. Suppose that, for each i, the semigroup S_i is presented by generators X_i and relations σ_i: thus $S_i \cong \mathscr{F}_{X_i}/\sigma_i^*$. Suppose that $X_i \cap X_j = \square$ if $i \neq j$. Set $X = \bigcup\{X_i : i \in I\}$ and $\sigma = \bigcup\{\sigma_i : i \in I\}$. Then $\mathscr{F}_X/\sigma^*$ is, to within isomorphism, the free product $\prod^*\{S_i : i \in I\}$.

2. Let $\{S_i : i \in I\}$ be an indexed set of semigroups. Let S be a semigroup generated by the isomorphic images $S_i \nu_i$, $i \in I$, of the S_i, where $\nu_i : S_i \to S$ is an isomorphic embedding, $i \in I$. Suppose that S has the following property: let T be any semigroup such that there exist homomorphisms $\psi_i : S_i \to T$ for each i in I; then there exists a homomorphism $\phi : S \to T$ such that $\nu_i \phi = \psi_i$ for each i in I. Then S is isomorphic to $\prod^*\{S_i : i \in I\}$.

3. Let $[S_i; U; \phi_i]$ be a semigroup amalgam. Suppose that, for each i, the semigroup S_i is presented by generators X_i and relations σ_i, where $X_i \cap X_j = \square$, if $i \neq j$: thus $S_i \cong \mathscr{F}_{X_i}/\sigma_i^*$. Set $V_i = (U\phi_i)(\sigma_i^{*\natural})^{-1}$. Set $\tau = \{(v_i, v_j) \in$

$V_i \times V_j$: $v_i\sigma_i^* = u\phi_i$, $v_j\sigma_j^* = u\phi_j$ for some i, j and for some $u \in U\}$. Set $\sigma = (\bigcup\{\sigma_i\}) \cup \tau$ and $X = \bigcup\{X_i\}$. Then $\mathscr{F}_X/\sigma^*$ is isomorphic to the free product $\prod^*{}_U S_i$ of the amalgam $[S_i; U; \phi_i]$. (Howie [1962].)

4. Let $[S_i; U; \phi_i]$ be a semigroup amalgam. Let U be presented by generators Y and relations π: thus $U \cong \mathscr{F}_Y/\pi^*$. Then there exist sets X_i, such that $X_i \supseteq Y$ and $X_i \cap X_j = Y$ if $i \neq j$, and relations σ_i on $\mathscr{F}_{X_i}$ such that $S_i \cong \mathscr{F}_{X_i}/\sigma_i^*$ and $\pi^* = \sigma_i^* \cap (\mathscr{F}_Y \times \mathscr{F}_Y)$ for each i. (Cf. Exercises 2 and 3 for §9.2.) Set $X = \bigcup\{X_i\}$ and $\sigma = \bigcup\{\sigma_i\}$. Then $\mathscr{F}_X/\sigma^*$ is isomorphic to the free product $\prod^*{}_U S_i$ of the amalgam $[S_i; U; \phi_i]$.

5. Let $[\{S_i: i \in I\}; U; \{\phi_i: i \in I\}]$ be a semigroup amalgam. Let S be a semigroup generated by the homomorphic images $S_i\nu_i$, $i \in I$, of the S_i, where $\nu_i: S_i \to S$ is a homomorphism, and where $\phi_i\nu_i = \phi_j\nu_j$ for any i, j in I. Suppose further that S and the family of homomorphisms ν_i have the following property: let T be any semigroup such that there exist homomorphisms $\psi_i: S_i \to T$ for each i in I, such that $\phi_i\psi_i = \phi_j\psi_j$ for any i, j in I; then there exists a homomorphism $\phi: S \to T$ such that $\nu_i\phi = \psi_i$ for each i in I. Then S is isomorphic to the free product $\prod^*{}_U S_i$ of the amalgam $[S_i; U; \phi_i]$.

6. Let $[S_i; U; \phi_i]$ be a proper semigroup amalgam and suppose that each semigroup S_i is a group. Suppose that $[S_i; U; \phi_i]$ can be embedded in a semigroup. Then U is a group. (Cf. Theorem 1.11.) (Howie [1962].)

7. Let U be presented by generators $\Gamma = \{e, u_1, \cdots, u_5\}$ and relations

$$\rho = \{(e^2, e), (u_1u_3, u_2), (u_4u_5, u_1)\} \cup \{(eu_r, u_r), (u_re, u_r): 1 \leqq r \leqq 5\}.$$

Let S be presented by generators $\Gamma \cup \{s\}$ and relations

$$\rho' = \rho \cup \{(u_1s, u_2), (ese, u_3), (s^2, ses)\}.$$

Let T be presented by generators $\Gamma \cup \{t\}$ and relations

$$\rho'' = \rho \cup \{(u_4t, u_1), (ete, u_5), (t^2, tet)\}.$$

Then U is a subsemigroup of S, unitary in eSe and hence (Theorem 9.32) almost unitary in S. Similarly U is an almost unitary subsemigroup of T. Consequently, the amalgam $[\{S, T\}; U]$ is embedded in $\prod^*{}_U\{S, T\}$.

Denote (cf. Theorem 9.31) the canonical isomorphisms of S, T and U into $\prod^*{}_U\{S, T\}$ by μ_1, μ_2 and μ, respectively. Then, it may be shown that $(e\mu)(t\mu_2)(s\mu_1)(e\mu) \notin U\mu$, whereas $(u_4\mu)\cdot(e\mu)(t\mu_2)(s\mu_1)(e\mu) = u_2\mu \in U\mu$. Consequently, $U\mu$ is not unitary in $(e\mu)\cdot\prod^*{}_U\{S, T\}\cdot(e\mu)$; whence, by Theorem 9.32, $U\mu$ is not almost unitary in $\prod^*{}_U\{S, T\}$. (Howie [1963b].)

8. Let $[S_i; U; \phi_i]$ be a semigroup amalgam. For each i in I, let T_i be a subsemigroup of S_i containing U_i. Suppose further that T_i is almost unitary in S_i with associated mappings λ_i and ρ_i.

Then the hypotheses of Theorem 9.46 are satisfied if and only if

(i) U_i is a unitary subsemigroup of $\lambda_iS_i\rho_i$.

Suppose U_i is an almost unitary subsemigroup of S_i with associated mappings γ_i and δ_i. Then each of the following conditions suffices to ensure that (i) holds.

(ii) $\lambda_i S_i \rho_i \subseteq \gamma_i S_i \delta_i$.

(iii) U_i and T_i are unitary in S_i.

(iv) T_i contains an identity element e_i which belongs to U_i.

(v) U is a group and T_i is a subgroup of S_i. (Howie [1963b].)

9. Let U be the free semigroup on the two generators u, v. Let T_1 be the free semigroup on the four generators u, v, t_1, t_1'. Let S_1 be presented by generators u, v, t_1, t_1', s_1 and relations $\{(s_1 u, t_1), (s_1 v, t_1')\}$. Let $T_2 = S_2$ be presented by generators u, v, t_2, t_2' and the relation (ut_2, vt_2').

Then U is a unitary subsemigroup of S_1 and S_2 (i.e. its natural image in S_1 and S_2 is unitary). Furthermore, T_1 is a subsemigroup of S_1. This may be shown by observing that S_1 is isomorphic (Lemma 9.11) to the free semigroup on u, v, s_1 and then using Corollary 9.8 to show that the subsemigroup of this free semigroup generated by u, v, $s_1 u$ and $s_1 v$ is a free semigroup on these generators.

The semigroup amalgam $[\{S_1, S_2\}; U]$ is embedded in $\prod^*{}_U\{S_1, S_2\}$ and the semigroup amalgam $[\{T_1, T_2\}; U]$ is embedded in $\prod^*{}_U\{T_1, T_2\}$.

Let μ_i be the canonical embedding (Theorem 9.31) of S_i into $\prod^*{}_U\{S_1, S_2\}$, $i = 1, 2$. Then the subsemigroup of $\prod^*{}_U\{S_1, S_2\}$, generated by $T_1\mu_1$ and $T_2\mu_2$ is not naturally isomorphic to $\prod^*{}_U\{T_1, T_2\}$. For $t_1\mu_1 t_2\mu_2 = t_1'\mu_1 t_2'\mu_2$ in $\prod^*{}_U\{S_1, S_2\}$; whereas $t_1\mu_1 t_2\mu_2 \neq t_1'\mu_1 t_2'\mu_2$ in $\prod^*{}_U\{T_1, T_2\}$. (Howie [1963b].)

Let $\{S_i : i \in I\}$ be a family of semigroups. $S = \prod\{S_i : i \in I\} = \prod S_i$ denotes the *direct product* of the S_i and is defined thus. The set S is the cartesian product of the sets S_i and thus consists of all sets (or sequences) $\{a_i : i \in I\}$ where $a_i \in S_i$ for each i. The product of elements of S is the component-wise product:

$$\{a_i\}\cdot\{b_i\} = \{a_i b_i\}.$$

The element a_i is said to be the *i-component* of $\{a_i\}$. If $x \in \prod S_i$, then the i-component of x is denoted also by x_i.

A subsemigroup B of S is called a *subdirect product of the* S_i if for each i in I and for each $a_i \in S_i$ there is an element of B with a_i as its i-component.

10. Let $\{S_i : i \in I\}$ be a family of semigroups. Adjoin to S_i an identity element e_i (whether S_i already has an identity element or not) to form the semigroup T_i. Let B be the subdirect product of the T_i consisting of all elements t of $\prod T_i$ with at most a finite number of components $t_i \neq e_i$. Denote by e the identity of $\prod T_i$ ($e = \{e_i\}$).

Then $A = B\backslash\{e\}$ is a subsemigroup of $\prod T_i$ isomorphic to the direct sum $\sum S_i$. A is a subdirect product of the T_i.

Denote by $\bar{S}_j$ the set of all elements t belonging to A for which $t_i = e_i$ when $i \neq j$. Then $\bar{S}_j$ is a subsemigroup of A isomorphic to S_j. If $j \neq k$, then each element of $\bar{S}_j$ commutes with each element of $\bar{S}_k$. The semigroup A is generated by its subsemigroups $\bar{S}_j$, $j \in I$.

11. Let $\{S_i : i \in I\}$ be a family of semigroups.

(a) Let S be the set of all mappings f of I into $\bigcup\{S_i\}$ satisfying $if \in S_i$ for each i in I and endow S with a product by defining, for $f, g \in S$,

$$i(fg) = (if)(ig), \; i \in I.$$

Then S is the direct product $\prod S_i$.

(b) Let Σ be the set of all mappings f of non-empty finite subsets of I into $\bigcup\{S_i\}$ satisfying $if \in S_i$ for each i in I for which if is defined. Endow Σ with a product by agreeing that for f, g in Σ and i in I, $i(fg)$ is defined whenever if or ig is defined and in this event

$$i(fg) = \begin{cases} (if)(ig), & \text{if both } if \text{ and } ig \text{ are defined,} \\ if, & \text{if } ig \text{ is not defined,} \\ ig, & \text{if } if \text{ is not defined.} \end{cases}$$

Then Σ is a semigroup isomorphic to the direct sum $\sum S_i$.

12. Let $\{S_i : i \in I\}$ be a family of semigroups. Let $\mathscr{I}$ be a family of subsets of I. Let D be the set of all mappings f of elements of $\mathscr{I}$ into $\bigcup S_i$ such that $if \in S_i$ when if is defined.

Let $f, g \in D$ and suppose that f and g are mappings of the elements J and K of $\mathscr{I}$, respectively, into $\bigcup S_i$. Define $fg \colon J \cup K \to \bigcup\{S_i\}$, thus:

$$i(fg) = \begin{cases} (if)(ig), & \text{if } i \in J \cap K, \\ if, & \text{if } i \in J \backslash K, \\ ig, & \text{if } i \in K \backslash J. \end{cases}$$

Then, with respect to this product, D is a semigroup if and only if $\mathscr{I}$ is closed under the operation of union, i.e. if and only if $\mathscr{I}$ is a semilattice under the operation of union.

For each i in I, adjoin to S_i an identity element e_i (whether S_i already has an identity element or not) to form the semigroup T_i. Denote by $e = \{e_i\}$ the identity element of $\prod T_i$. Then, when D is a semigroup, it is (isomorphic to) a subsemigroup of $\prod T_i$.

If $\mathscr{I}$ is the set of all subsets of I, then D is isomorphic to $\prod T_i$. If $\mathscr{I}$ has I as its sole element then D is isomorphic to $\prod S_i$.

13. Let $\{S_i : i \in I\}$ be a family of semigroups. For each i in I, adjoin to S_i an identity element e_i (whether S_i already has an identity element or not) to form the semigroup T_i. Denote by θ_{ij} the identical mapping of S_i, if $i = j$, and the mapping of S_i onto e_j if $i \neq j$. Let S be a semigroup.

Then S is isomorphic to the direct sum $\sum S_i$ if and only if there exist, for each i in I, homomorphisms

$$f_i: S_i \to S,$$

$$g_i: S \to T_i,$$

such that $f_i g_j = \theta_{ij}$, for all i, j in I, and, for each s in S,

$$s = \prod\{sg_i f_i: i \in I \quad \text{and} \quad sg_i \in S_i\},$$

the value of this product being independent of the order in which the elements $sg_i f_i$ are taken. Observe that, since only finite products of elements are defined in a semigroup, the above expression of s as a product implies that $sg_i \in S_i$ for only a finite number of i in I.

9.5 Construction of cancellative congruences

By a *presentation of a cancellative semigroup* S we mean the specification of a set X and a relation ρ on the free semigroup $\mathscr{F}_X$ such that $\mathscr{F}_X/\rho^c \cong S$, where ρ^c is the smallest congruence containing ρ such that $\mathscr{F}_X/\rho^c$ is cancellative. In general, ρ^c is larger than the congruence ρ^* generated by ρ, and hence this notion of presentation of S differs from that defined in §9.2. Perhaps we should call it a "cancellative presentation", but this precaution is scarcely necessary.

In the case of groups, one constructs first the free group $\mathscr{F}\mathscr{G}_X$ on a suitable set X (Vol. 1, p. 43). Then, to present a given group G, one needs only find a relation ρ on $\mathscr{F}\mathscr{G}_X$ such that $\mathscr{F}\mathscr{G}_X/\rho^* \cong G$. Similarly, in the case of commutative semigroups S, one constructs first the free commutative F_X (§9.3), and then looks for a relation ρ on F_X such that $F_X/\rho^* \cong S$. Thus, in these two important cases, thanks to the "master systems" $\mathscr{F}\mathscr{G}_X$ and F_X, we find it possible to deal simply with the congruence ρ^* generated by ρ. But no such "master system" exists for the class of cancellative semigroups. The reason for this is that this class is not closed under the operation of taking homomorphic images. We must content ourselves—unless the future produces a better method—with $\mathscr{F}_X$, and develop techniques for obtaining ρ^c from ρ.

To increase the scope of the present section, we discuss the derivation of ρ^c from ρ, not just on $\mathscr{F}_X$, but on any semigroup S. The important special case $S = \mathscr{F}_X$ we keep always in mind. The results for general S (Theorems 9.50 and 9.53) are new.

The latter part of this section is devoted to an expanded account of an ingenious method due to Croisot [1954b] for obtaining a set Y of "canonical forms" in a semigroup S with respect to a congruence ρ on S, i.e. a set Y such that each element of S is congruent modulo ρ to exactly one member of Y. This method applies to arbitrary, as well as cancellative, congruences.

Applied to the important special case $S = \mathscr{F}_X$, it gives us a method for describing the structure of a [cancellative] semigroup directly from a [cancellative] presentation thereof, without recourse to representations of various sorts (such as by transformations, matrices, ordered n-tuples, etc.).

A congruence σ on a semigroup S is said to be *cancellative* if S/σ is a cancellative semigroup. Thus, if σ is a congruence on S, then σ is cancellative if and only if $(ab, ac) \in \sigma$ and $(ba, ca) \in \sigma$ each imply that $(b, c) \in \sigma$, for any a, b, c in S. The following lemma is easily proved.

LEMMA 9.49. *Let* $\{\sigma_i : i \in I\}$ *be a family of cancellative congruences on the semigroup* S. *Then* $\sigma = \bigcap\{\sigma_i : i \in I\}$ *is a cancellative congruence on* S.

Since the universal relation $S \times S$ on S is a cancellative congruence, the foregoing lemma shows that, given any binary relation ρ on S, there is a unique minimal cancellative congruence on S which contains ρ, namely, the intersection of all cancellative congruences on S which contain ρ. We shall denote this congruence by ρ^c and we shall say that ρ^c is the cancellative congruence on S generated by ρ. Our concern in this section is to present various constructions for ρ^c.

The following notation will simplify the description of our first construction. Denote by $\mathscr{B} = \mathscr{B}_S$, the set of all binary relations on the semigroup S and define the mappings C^*, D, G and θ of $\mathscr{B}$ into $\mathscr{B}$ as follows. For any ρ in $\mathscr{B}$,

$$\begin{aligned}
C^*\colon \rho \to \rho C^* &= \{(x, y) \colon x = sut,\ y = svt;\ s, t \in S^1;\ (u, v) \in \rho\},\\
D\colon \rho \to \rho D &= \{(x, y) \colon u = xa,\ v = ya;\ a \in S^1;\ (u, v) \in \rho\},\\
G\colon \rho \to \rho G &= \{(x, y) \colon u = ax,\ v = ay;\ a \in S^1;\ (u, v) \in \rho\},\\
\theta\colon \rho \to \rho^\theta &= (\rho \circ \rho) \cup \rho C^* \cup \rho D \cup \rho G.
\end{aligned}$$

The following theorem may be regarded as an analogue of Theorem 1.8.

THEOREM 9.50. *Let* ρ *be a binary relation on the semigroup* S *and set* $\tau = \rho \cup \rho^{-1} \cup \iota_S$. *Then*

$$\rho^c = \bigcup\{\tau^{\theta^n} \colon n = 1, 2, 3, \cdots\}.$$

PROOF. Set $\sigma = \bigcup\{\tau^{\theta^n} \colon n = 1, 2, 3, \cdots\}$. We begin by showing that σ is a cancellative congruence containing ρ.

Since τ is reflexive, $\tau \subseteq \tau \circ \tau$ and so $\rho \subseteq \tau \subseteq \tau \circ \tau \subseteq \tau^\theta \subseteq \sigma$. Since $\iota_S \subseteq \tau$, also $\iota_S \subseteq \sigma$. Hence σ is a reflexive relation containing ρ.

The relation τ is symmetric by construction and, for a reflexive, symmetric relation, each of the mappings, $\circ$, $\bigcup$, C^*, D and G used to define θ, preserves symmetry. Thus θ, and, by induction on n, each of its iterates θ^n, preserves symmetry. Consequently, σ is symmetric.

It has already been observed that $\tau \subseteq \tau^\theta$; consequently, $\tau^{\theta^n} \subseteq \tau^{\theta^{n+1}}$, for each n. We use this relation to show that σ is transitive. For, consider (a, b) and (b, c) in σ. By the definition of σ, for some integers m, n, $(a, b) \in \tau^{\theta^n}$

and $(b, c) \in \tau^{\theta^m}$. Thus if p denotes the larger of m, n both (a, b) and (b, c) are in τ^{θ^p}; whence $(a, c) \in \tau^{\theta^p} \circ \tau^{\theta^p} \subseteq \tau^{\theta^{p+1}} \subseteq \sigma$.

To show that σ is compatible, consider (a, b) in σ; so that, for some integer n, $(a, b) \in \tau^{\theta^n}$. Then, for any c in S, $(ac, bc) \in \tau^{\theta^n} C^* \subset \tau^{\theta^{n+1}} \subseteq \sigma$. Thus σ is right compatible; and left compatibility follows similarly.

Finally, to show that σ is cancellative, consider (ac, bc) in σ so that, for some integer n, $(ac, bc) \in \tau^{\theta^n}$. Then, $(a, b) \in \tau^{\theta^n} D \subseteq \sigma$. Thus σ is right cancellative; and that σ is left cancellative follows similarly.

To complete the proof it will suffice to show that $\sigma \subseteq \rho^c$. To this end, consider any relation β contained in ρ^c. Then, it is easily verified that $\beta^\theta \subseteq \rho^c$. For example, let $(x, y) \in \beta D$; then, for some $(u, v) \in \beta$ and a in S^1, $u = xa$ and $v = ya$. Thus, $(xa, ya) \in \beta \subseteq \rho^c$. Since ρ^c is right cancellative, therefore $(x, y) \in \rho^c$. The three other possibilities are disposed of similarly. Consequently, since $\tau \subseteq \rho^c$, it follows, by induction, that $\tau^{\theta^n} \subseteq \rho^c$, for $n = 1, 2, \cdots$; whence, $\sigma \subseteq \rho^c$.

This completes the proof of the theorem.

Our second construction for ρ^c we give first for a relation ρ on a free semigroup $\mathscr{F}_X$. The construction makes use of two copies of X,

$$X^L = \{x^L : x \in X\} \quad \text{and} \quad X^R = \{x^R : x \in X\},$$

say, where $x \to x^L$ and $x \to x^R$ are each one-to-one mappings, and where we suppose that X^L and X^R are disjoint from each other and from X. Set $Y = X \cup X^L \cup X^R$, and consider the following kinds of transformation of elements of $\mathscr{F}_Y$.

(i) An elementary ρ-transition.

(ii) An *insertion* of $x^L x$ or of $x x^R$ $(x \in X)$: $w \to w'$, where $w = w_1 w_2$, $w' = w_1 x^L x w_2$ or $w' = w_1 x x^R w_2$, where $w_1, w_2 \in \mathscr{F}_Y^1$.

(iii) A *deletion* of $x^L x$ or of $x x^R$ $(x \in X)$; a deletion being the reverse of a transformation of type (ii).

A chain

$$w = w_1 \to w_2 \to \cdots \to w_{n+1} = w'$$

of transformations $w_i \to w_{i+1}$, each of type (i), (ii) or (iii), will be called *ρ-allowable* (where ρ is a relation on $\mathscr{F}_X$) if w and w' belong to $\mathscr{F}_X$ and if—a more precise statement follows—any elements x^L or x^R which are inserted are deleted in an order which is the reverse of their order of insertion. More precisely, suppose that the step $w_i \to w_{i+1}$ is an insertion, introducing the word $x^L x$, say. Then, since w' belongs to $\mathscr{F}_X$, this occurrence of x^L must be deleted at some step in the chain from w to w'. For the chain to be ρ-allowable we require that this occurrence of x^L in w_{i+1} be deleted before the deletion of any other indexed elements, y^R or y^L, say, already present in w_{i+1}. This requirement also applies to occurrences, if any, already present of x^L itself in w_{i+1}. Precisely similar conditions are imposed upon the insertion and deletion of elements of X^R. To ensure that these conditions are satisfied we must keep track of each indexed symbol as it appears in the chain, distinguishing between repetitions of the same symbol by placing, for example, markers, 1, 2, 3, $\cdots$, successively under its occurrences. The following example of an allowable sequence will make this process clear.

Let $X = \{p, q, r, s, t, u, v, x, y, z\}$. Let $\rho = \{(qt, rs), (rt, xy), (px, zq), (qy, ut), (us, vt)\}$. Then

$$\begin{aligned} pq &\to pqt\underset{1}{t^R} \to prs\underset{1}{t^R} \to prt\underset{2}{t^R}s\underset{1}{t^R} \\ &\to pxy\underset{2}{t^R}s\underset{1}{t^R} \to zqy\underset{2}{t^R}s\underset{1}{t^R} \\ &\to zut\underset{2}{t^R}s\underset{1}{t^R} \to zus\underset{1}{t^R} \to zvt\underset{1}{t^R} \\ &\to zv, \end{aligned}$$

is a ρ-allowable chain from pq to zv.

THEOREM 9.51. *Let ρ be a relation on $\mathscr{F}_X$. Then, for w, w' in $\mathscr{F}_X$, $(w, w') \in \rho^c$ if and only if there exists a ρ-allowable chain from w to w'.*

PROOF. Denote by $\bar{\rho}$ the relation $\bar{\rho} = \{(w, w')$: there exists a ρ-allowable chain from w to $w'\}$. Then $\bar{\rho}$ is clearly reflexive. Since a ρ-allowable chain traversed in the reverse direction is also ρ-allowable, $\bar{\rho}$ is symmetric. Suppose that (w_1, w_2) and (w_2, w_3) belong to $\bar{\rho}$; so that there exist ρ-allowable chains

$$w_1 \to \cdots \to w_2, \quad \text{and} \quad w_2 \to \cdots \to w_3.$$

Then, the chain from w_1 to w_3, obtained by following the first of these chains by the second, is ρ-allowable. Thus $(w_1, w_3) \in \bar{\rho}$; and this shows that $\bar{\rho}$ is transitive.

Let $(w_1, w_2) \in \bar{\rho}$ and $w \in \mathscr{F}_X$. Then the ρ-allowable chain $w_1 \to \cdots \to w_2$ gives another ρ-allowable chain $ww_1 \to \cdots \to ww_2$. Thus $(ww_1, ww_2) \in \bar{\rho}$. Hence $\bar{\rho}$ is left, and similarly right, compatible. Consequently we have shown that $\bar{\rho}$ is a congruence on $\mathscr{F}_X$.

To show that $\bar{\rho}$ is cancellative, consider firstly a pair (xw_1, xw_2) belonging to $\bar{\rho}$, where $x \in X$. Thus, there exists a ρ-allowable chain,

$$xw_1 \to \cdots \to xw_2;$$

and so the chain

$$w_1 \to x^L x w_1 \to \cdots \to x^L x w_2 \to w_2$$

is also ρ-allowable. Thus, $(w_1, w_2) \in \bar{\rho}$. This shows that we can cancel, on the left, an element of X. Hence, by cancelling one generator at a time, we can infer that, for any w in $\mathscr{F}_X$, $(ww_1, ww_2) \in \bar{\rho}$ implies $(w_1, w_2) \in \bar{\rho}$. Similarly, we show that $\bar{\rho}$ is right cancellative.

It follows, since clearly $\rho \subseteq \bar{\rho}$, that $\rho^c \subseteq \bar{\rho}$. We now proceed to show that the reverse inequality also holds.

To this end, consider $(w, w') \in \bar{\rho}$. Then there exists a ρ-allowable chain from w to w'. If only elementary ρ-transitions occur in this chain, then $(w, w') \in \rho^* \subseteq \rho^c$. Otherwise, at some stage in the chain, there is an insertion. We will deal with the case where the first insertion is an insertion of an element of X^L. A similar treatment will apply if the first insertion is an insertion of an element of X^R. The argument will proceed by induction on the length of the chain from w to w'. Since chains of length 1 are merely elementary ρ-transitions there is no difficulty in starting the induction.

Assume therefore that there is a ρ-allowable chain

$$w \xrightarrow{(a)} w_1w_2 \xrightarrow{(b)} w_1x^Lxw_2 \xrightarrow{(c)} w_3x^Lxw_4 \xrightarrow{(d)} w_3w_4 \xrightarrow{(e)} w',$$

say, where

(a) consists of a chain of elementary ρ-transitions,
(b) is an insertion of x^Lx,
(c) is an allowable chain in which x^L is not involved,
(d) is a deletion of x^Lx, and
(e) is a ρ-allowable chain from the word $w_3w_4 \in \mathscr{F}_X$ to w'.

Observe, since the chain from w to w' is ρ-allowable, that, on the removal of x^L at stage (d), the resulting word w_3w_4 belongs to $\mathscr{F}_X$; consequently, the assumption about the chain (e) is justified.

The assumption that (c) consists of a chain of transitions in which x^L is not involved is justified by the definition of a ρ-allowable chain. Hence, throughout the chain (c), the symbol x^L, introduced at (b), acts as a barrier, so that, effectively, the transitions of (c) take place independently either to the left or to the right of this x^L. If taken in sequence, the transitions to the left of x^L form a chain, which is easily seen to be ρ-allowable, from w_1 to w_3; and, similarly, the transitions to the right of x^L, form a ρ-allowable chain from xw_2 to xw_4. Each of these chains, from w_1 to w_3 and from xw_2 to xw_4, is of length less than the given chain from w to w'. Hence, by our inductive assumption, $(w_1, w_3) \in \rho^c$ and $(xw_2, xw_4) \in \rho^c$.

Since ρ^c is cancellative, therefore $(w_1, w_3) \in \rho^c$ and $(w_2, w_4) \in \rho^c$; whence, $(w_1w_2, w_3w_4) \in \rho^c$. The chain from w to w_1w_2 is of length less than the given chain, as is also the chain from w_3w_4 to w'. Hence, since both these chains are ρ-allowable, our inductive assumption gives $(w, w_1w_2) \in \rho^c$ and $(w_3w_4, w') \in \rho^c$. From the transitivity of ρ^c we infer $(w, w') \in \rho^c$. This completes the proof of the theorem.

As a preliminary to extending the construction of Theorem 9.51 from a free semigroup to an arbitrary semigroup we state the following lemma. First, a definition.

Let $\alpha: A \to B$ be a mapping of the set A into the set B. Then the mapping $(\alpha, \alpha): A \times A \to B \times B$ is defined thus:

$$(a, a')(\alpha, \alpha) = (a\alpha, a'\alpha), \quad \text{for } (a, a') \in A \times A.$$

LEMMA 9.52. *Let $\phi: S \to T$ be a homomorphism of the semigroup S onto the semigroup T.*

(i) *The mappings $\sigma \to \sigma(\phi, \phi)$ and $\tau \to \tau(\phi, \phi)^{-1}$ are mutually inverse, one-to-one mappings of the set of all congruences σ on S containing $\phi \circ \phi^{-1}$ onto the set of all congruences τ on T, and vice-versa. Furthermore, this mapping is such that the corresponding quotient semigroups are isomorphic; more precisely, if τ is a congruence on T, then the natural mapping,*

$$\phi_\tau^*: a(\tau(\phi, \phi)^{-1}) \to (a\phi)\tau \quad (a \in S),$$

is an isomorphism of $S/\tau(\phi, \phi)^{-1}$ onto T/τ.

(ii) *If τ is a congruence on T, then τ is cancellative if and only if $\tau(\phi, \phi)^{-1}$ is cancellative.*

(iii) *Let ρ be any reflexive binary relation on T. Then $(\rho^*)(\phi, \phi)^{-1} = (\rho(\phi, \phi)^{-1})^*$ and $(\rho^c)(\phi, \phi)^{-1} = (\rho(\phi, \phi)^{-1})^c$.*

PROOF. (i) Let τ be a congruence on T, and let $(a, b) \in S \times S$. Since $(a, b) \in \tau(\phi, \phi)^{-1}$ if and only if $(a\phi, b\phi) \in \tau$, and this in turn if and only if $a\phi\tau^\natural = b\phi\tau^\natural$, it follows that $\tau(\phi, \phi)^{-1} = \phi\tau^\natural \circ (\phi\tau^\natural)^{-1}$. Since $\phi\tau^\natural$ is a homomorphism of S onto T/τ, this shows that $\tau(\phi, \phi)^{-1}$ is a congruence on S. Since $\tau \supseteq \iota_T$, we have

$$\tau(\phi, \phi)^{-1} \supseteq \iota_T(\phi, \phi)^{-1} = \phi \circ \phi^{-1}.$$

Conversely, let σ be any congruence on S which contains $\phi \circ \phi^{-1}$. Set $\tau = \sigma(\phi, \phi)$. Then τ is a congruence on T. That τ is reflexive follows because ϕ is a mapping onto T. It is immediate that τ is symmetric. Transitivity follows because $\sigma \supseteq \phi \circ \phi^{-1}$: for let $(a, b), (b, c) \in \tau$, so that there exist x, y, z, u such that $x\phi = a$, $y\phi = b$, $z\phi = b$, $u\phi = c$ and $(x, y), (z, u) \in \sigma$. Since $y\phi = z\phi$ and $\phi \circ \phi^{-1} \subseteq \sigma$, therefore $(y, z) \in \sigma$. Hence $(x, u) \in \sigma$; whence $(a, c) \in \tau$. To show that τ is compatible, consider $(a, b) \in \tau$, $c \in T$. Choose $x, y, z \in S$ so that $x\phi = a$, $y\phi = b$, $(x, y) \in \sigma$ and $z\phi = c$. Then $(zx, zy), (xz, yz) \in \sigma$; whence $(ca, cb), (ac, bc) \in \tau$. Thus we have shown that $\tau = \sigma(\phi, \phi)$ is a congruence on T.

It is not difficult to see that the two mappings are inverse to each other, and we omit the details.

To complete the proof of part (i), let τ be any congruence on T and let us show first that ϕ_τ^* is well-defined. To this end consider a, b in S such that $a(\tau(\phi, \phi)^{-1}) = b(\tau(\phi, \phi)^{-1})$. We have to show that then $(a\phi)\tau = (b\phi)\tau$; but this follows directly from the definition of $\tau(\phi, \phi)^{-1}$. That ϕ_τ^* is a mapping onto T/τ is clear. Suppose further that $(a(\tau(\phi, \phi)^{-1}))\phi_\tau^* = (b(\tau(\phi, \phi)^{-1}))\phi_\tau^*$ for some $a, b \in S$. Then $(a\phi)\tau = (b\phi)\tau$, i.e. $(a, b)(\phi, \phi) \in \tau$, whence $(a, b) \in \tau(\phi, \phi)^{-1}$. This shows that ϕ_τ^* is one-to-one; and, since ϕ_τ^* is clearly also a homomorphism, this completes the proof that ϕ_τ^* is an isomorphism of $S/\tau(\phi, \phi)^{-1}$ onto T/τ.

(ii) This is an immediate corollary of the fact that T/τ and $S/\tau(\phi, \phi)^{-1}$ are isomorphic.

(iii) Let ρ be a reflexive binary relation on T. As seen at the beginning of the proof of part (i), that ρ is reflexive enables us to infer that $\phi \circ \phi^{-1} \subseteq \rho(\phi, \phi)^{-1}$. Consequently, also $(\rho(\phi, \phi)^{-1})^*$ and $(\rho(\phi, \phi)^{-1})^c$ both contain $\phi \circ \phi^{-1}$.

Since $\rho \subseteq \rho^*$, therefore $\rho(\phi, \phi)^{-1} \subset \rho^*(\phi, \phi)^{-1}$; and hence, by (i), $(\rho(\phi, \phi)^{-1})^* \subseteq \rho^*(\phi, \phi)^{-1}$. Conversely, we have $\rho(\phi, \phi)^{-1} \subseteq (\rho(\phi, \phi)^{-1})^*$, whence $\rho \subseteq (\rho(\phi, \phi)^{-1})^*(\phi, \phi)$. By (i) it follows that $\rho^* \subseteq (\rho(\phi, \phi)^{-1})^*(\phi, \phi)$; whence $\rho^*(\phi, \phi)^{-1} \subseteq (\rho(\phi, \phi)^{-1})^*$. Combining this with the earlier inequality gives $\rho^*(\phi, \phi)^{-1} = (\rho(\phi, \phi)^{-1})^*$.

Using the result of (ii) an exactly analogous argument enables us to infer that $\rho^c(\phi, \phi)^{-1} = (\rho(\phi, \phi)^{-1})^c$.

The preceding lemma, combined with Theorem 9.51, gives as an immediate corollary

THEOREM 9.53. *Let ρ be a binary relation on the semigroup S. Choose a set X and a mapping ϕ, so that $\phi\colon \mathscr{F}_X \to S$ is a homomorphism of the free semigroup $\mathscr{F}_X$ onto the semigroup S. Set $\tau = (\rho \cup \iota_S)(\phi, \phi)^{-1}$.*

Then, for a, b in S, $(a, b) \in \rho^c$ if and only if, for w, w' in $\mathscr{F}_X$ such that $w\phi = a$ and $w'\phi = b$, there exists a τ-allowable chain of transformations starting at w and terminating at w'.

We complete this section by giving an account of R. Croisot's discussion [1954b], of canonical forms associated with congruences and cancellative congruences on a semigroup. If σ is a congruence on the semigroup S, then by a *set of canonical forms in S for σ* is meant a cross-section (cf. Theorem 2.10(i)) of the partition of S determined by σ. Thus $Y \subseteq S$ is a set of canonical forms in S for σ if and only if Y meets each σ-class in precisely one element of S.

Let ρ be a binary relation on the semigroup S. A mapping $a \to \bar{a}$ of S into S is called a *ρ^*-canonical projection* if

(i) $(a, \bar{a}) \in \rho^*$;

(ii) if $(a, b) \in \rho \cup \rho^{-1} \cup \iota_S$, then $\bar{a} = \bar{b}$;

(iii) $\overline{\bar{a}\bar{b}} = \overline{ab}$;

for all a, b in S. It is called a *ρ^c-canonical projection* if (ii) and (iii) hold and also

(i′) $(a, \bar{a}) \in \rho^c$;

(iv) if $\overline{ac} = \overline{bc}$ or $\overline{ca} = \overline{cb}$, then $\bar{a} = \bar{b}$;

for all a, b, c in S.

THEOREM 9.54. *Let ρ be a binary relation on the semigroup S.*

Let Y be a subset of S and let $\phi\colon S \to Y$ be a mapping of S onto Y. Then Y is a set of canonical forms for $\rho^*[\rho^c]$ in S if $a \to a\phi$ is a $\rho^*[\rho^c]$-canonical projection of S.

Conversely, let Y *be a set of canonical forms for* $\rho^*[\rho^c]$ *in* S. *Define* $\phi: S \to Y$, *onto* Y, *thus*:

$$\{a\phi\} = a\rho^* \cap Y \ [= a\rho^c \cap Y], \quad a \in S.$$

Then ϕ *is a* $\rho^*[\rho^c]$*-canonical projection of* S.

PROOF. We shall prove only the first reading of the theorem and leave to the reader the similar proof of the alternative reading for cancellative congruences.

The converse part of the theorem is clear. To show the first part suppose therefore that $\phi: S \to Y$ is a ρ^*-canonical projection of S onto Y. Since, by (i), $(a, a\phi) \in \rho^*$ for all a in S, there is an element of Y in each ρ^*-class. Suppose that $x, y \in Y$ and that $(x, y) \in \rho^*$. Then, since ϕ is onto Y, there exist a, b, say, in S, such that $a\phi = x$ and $b\phi = y$. Since $(a, a\phi) \in \rho^*$ and $(b, b\phi) \in \rho^*$, therefore also $(a, b) \in \rho^*$. Thus (Theorem 1.8) b can be obtained from a by a finite sequence of elementary ρ-transitions. Let $upv \to uqv$ be one of these transitions, so that $u, v \in S^1$ and $(p, q) \in \rho \cup \rho^{-1} \cup \iota_S$. Using (ii) gives that $p\phi = q\phi$. Then, by several applications of property (iii),

$$\begin{aligned}(upv)\phi &= ((up)\phi(v\phi))\phi = ((((u\phi)(p\phi))\phi)(v\phi))\phi \\ &= ((((u\phi)(q\phi))\phi)(v\phi))\phi \\ &= (uqv)\phi.\end{aligned}$$

Applying this result to each of the transitions from a to b, we infer that $a\phi = b\phi$. Thus $x = y$; and this completes the proof of the theorem.

In applying Theorem 9.54, it is often convenient to set up the mapping $w \to \overline{w} = w\phi$ by means of an auxiliary transformation ψ of S having the following properties:

(1) For each w in S, $w \to w\psi$ is an elementary ρ-transition.

(2) For each w in S, there exists a positive integer n such that

$$w\psi^{n+1} = w\psi^n.$$

(3) For each pair w_1, w_2 of elements of S, there exist positive integers k and l such that

$$\begin{aligned}(w_1w_2)\psi^{k+1} &= ((w_1\psi)w_2)\psi^k, \\ (w_1w_2)\psi^{l+1} &= (w_1(w_2\psi))\psi^l.\end{aligned}$$

The meaning of (2) is that, after n applications of ψ to w, we arrive at a word $w\psi^n$ which is unaffected by ψ. We take $\overline{w}$ $(= w\phi) = w\psi^n$. (i) is clear from (1), while (iii) follows from (3). For (3) evidently implies

$$((w_1\psi)w_2)\phi = (w_1w_2)\phi = (w_1(w_2\psi))\phi.$$

This implies

$$((w_1\psi)(w_2\psi))\phi = (w_1w_2)\phi,$$

and we obtain (iii), $((w_1\phi)(w_2\phi))\phi = (w_1w_2)\phi$, by iteration. (ii), and (iv) if appropriate, must be checked separately.

As an example, let $X = \{x, y, z\}$, and let

$$\rho = \{(zx, xy), (yx, xz), (zy, yz)\},$$

a relation on the free semigroup $\mathscr{F}_X$. For any w in $\mathscr{F}_X$, let $w\psi$ be obtained from w as follows.

If some z in w is followed by an x or a y, select the right-most such z in w, and replace zx by xy, or zy by yz, as the case may be. If no such z exists, and if some y in w is followed by an x, select the right-most such y in w, and replace yx by xz. If no such z and no such y exist, then define $w\psi$ to be w.

(1) is immediate, and (2) is fairly easy to check, with $\bar{w} = w\psi^n$ having the form $x^\alpha y^\beta z^\gamma$. We prove the first equality in (3) as follows; the proof of the second is similar. Suppose

$$w_1 = w_3pw_4 \quad \text{and} \quad w_1\psi = w_3qw_4, \quad \text{where } (p, q) \in \rho.$$

From the nature of ρ and ψ it is clear that there exists a non-negative integer k such that the first k applications of ψ to either w_1w_2 or $(w_1\psi)w_2$ affect only w_4w_2, while the next application of ψ to $(w_1w_2)\psi^k$ is just $p \to q$. If $k > 0$, we have

$$\begin{aligned}(w_1w_2)\psi^{k+1} &= [w_3p((w_4w_2)\psi^k)]\psi = w_3q((w_4w_2)\psi^k) \\ &= (w_3qw_4w_2)\psi^k = ((w_1\psi)w_2)\psi^k;\end{aligned}$$

while, if $k = 0$,

$$(w_1w_2)\psi = (w_1\psi)w_2,$$

which implies, for any $k > 0$, that

$$(w_1w_2)\psi^{k+1} = ((w_1\psi)w_2)\psi^k.$$

For example, suppose $w_1 = w_3zxw_4$ and $w_1\psi = w_3xyw_4$. After, say, k applications of ψ to w_1w_2, we drive all the z's in w_4w_2 to the extreme right end:

$$(w_1w_2)\psi^k = w_3zx((w_4w_2)\psi^k).$$

The next application of ψ must replace zx by xy:

$$(w_1w_2)\psi^{k+1} = w_3xy((w_4w_2)\psi^k).$$

But this is what we get after k applications of ψ to

$$w_3xyw_4w_2 = (w_1\psi)w_2.$$

The direct check of (ii) is quite trivial:

$$\overline{zx} = xy = \overline{xy}, \quad \overline{yx} = xz = \overline{xz}, \quad \overline{zy} = yz = \overline{yz}.$$

We conclude that

$$Y = \{x^\alpha y^\beta z^\gamma \colon \alpha \geqq 0, \beta \geqq 0, \gamma \geqq 0, \alpha + \beta + \gamma > 0\}$$

is a set of ρ^*-canonical forms.

This method is not necessarily the easiest one to establish that Y is a set of ρ^*-canonical forms. It is easy to see directly that any word in $\mathscr{F}_X$ is ρ^*-equivalent to one in Y; the difficult thing is to see that no two apparently distinct members of Y are ρ^*-equivalent. The latter may also be shown as follows.

It is a matter of routine calculation to show that

$$(1) \qquad (x^\alpha y^\beta z^\gamma \cdot x^{\alpha'} y^{\beta'} z^{\gamma'})\phi = \begin{cases} x^{\alpha+\alpha'} y^{\beta+\beta'} z^{\gamma+\gamma'} & \text{if } \alpha' \text{ is even,} \\ x^{\alpha+\alpha'} y^{\gamma+\beta'} z^{\beta+\gamma'} & \text{if } \alpha' \text{ is odd.} \end{cases}$$

Let T be the set of ordered triples (α, β, γ) of non-negative integers, excluding $(0, 0, 0)$, and define a product in T as suggested by (1). It is laborious but routine to check that this product is associative. Furthermore we note that $(1, 0, 0)$, $(0, 1, 0)$, $(0, 0, 1)$ satisfy the generating relations ρ for x, y, z, respectively. Hence the mapping $x \to (1, 0, 0)$, $y \to (0, 1, 0)$, $z \to (0, 0, 1)$ induces a homomorphism of $R = \mathscr{F}_X/\rho^*$ onto T, and if two members of Y were equal in R, they would be equal in T.

The point is that Croisot's method gives us an additional way of establishing that a certain set Y of words is ρ^*-canonical. Moreover, it is one that deals with the words in $\mathscr{F}_X$ themselves, without recourse to a representation of R by transformations of a set, by matrices, or by ordered n-tuples.

Exercises for §9.5

1. Let $X = \{p, q, e\}$ and ρ be the relation defined on $\mathscr{F}_X$ thus:

$$\rho = \{(pq, e), (ep, p), (pe, p), (eq, q), (qe, q)\}.$$

Let $Y = \{q^m p^n \colon m, n \geqq 0\}$, where we set $q^0 = p^0 = e$.

Then Y is a set of canonical forms for ρ^* in $\mathscr{F}_X$. $\mathscr{F}_X/\rho^*$ is the bicyclic semigroup. (Cf. Exercises 1 and 2 for §1.12.)

2. Let $X = \{l, m, n\}$ and ρ be the relation defined on $\mathscr{F}_X$ thus:

$$\rho = \{(ml, l^2m), (nl^2, ln), (mn, nm)\}.$$

Let $Y = \{n^\nu l^\lambda m^\mu \colon \nu \geqq 0, \lambda \geqq 0, \mu \geqq 0, \lambda + \mu + \nu > 0\}$, where any factor raised to a zero power is suppressed.

Then Y is a set of canonical forms for ρ^* in $\mathscr{F}_X$. Furthermore, if we define a product on Y by the rule

$$n^\nu l^\lambda m^\mu \cdot n^{\nu'} l^{\lambda'} m^{\mu'} = n^{\nu+\nu'} l^{2^{\nu'}\lambda + 2^\mu \lambda'} m^{\mu+\mu'},$$

then Y becomes a semigroup isomorphic to $\mathscr{F}_X/\rho^*$. In other words, the right-hand side of this equation is the element of Y in the ρ^*-class containing the element of $\mathscr{F}_X$ on the left-hand side. (Croisot [1954b].)

3. Let ϕ be a homomorphism of a semigroup S onto a semigroup T. Let (ϕ, ϕ) and $(\phi, \phi)^{-1}$ be the mappings of Lemma 9.52. Regarding ϕ as the

relation $\{(x, a) \in S \times T : x\phi = a\}$, and defining products as usual for binary relations, then (in the notation of Lemma 9.52)

$$\tau = \sigma(\phi, \phi) \quad \text{if and only if } \tau = \phi^{-1} \circ \sigma \circ \phi,$$
$$\sigma = \tau(\phi, \phi)^{-1} \quad \text{if and only if } \sigma = \phi \circ \tau \circ \phi^{-1}.$$

From $\tau \supseteq \iota_T$ we have $\sigma \supseteq \phi \circ \iota_T \circ \phi^{-1} = \phi \circ \phi^{-1}$. From this and $\phi^{-1} \circ \phi = \iota_T$ we can easily show that

$$\phi^{-1} \circ (\phi \circ \tau \circ \phi^{-1}) \circ \phi = \tau$$
$$\phi \circ (\phi^{-1} \circ \sigma \circ \phi) \circ \phi^{-1} = \sigma.$$

This gives an alternative proof of part (i) of Lemma 9.52.

CHAPTER 10

CONGRUENCES

In the preceding chapter congruences were essentially involved in the constructions discussed. We here continue our discussion of congruences, but now with our attention more explicitly directed at the congruences themselves.

The chapter begins with a section discussing the extent to which a congruence is determined by a subset of its congruence classes. Theorem 7.38 and its corollaries provided some special results, for regular semigroups and for inverse semigroups, in this direction. The next two sections, §§10.2–3, discuss two important classes of (in general, one-sided), congruences which can be defined on an arbitrary semigroup, first introduced in the fundamental and ground-breaking paper of Dubreil [1941]. Our account is confined essentially to the general theory of these congruences and the reader is referred to the extensive literature, especially of the French school, for more detailed investigations. The importance of the Dubreil principal equivalences of §10.2 has already been seen in their application to the representations, by one-to-one partial transformations, of an inverse semigroup (§§7.2–3). These equivalences play a similar basic rôle in the representations of an arbitrary semigroup by means of one-to-one partial transformations, as will be seen in §11.4. The key result for this later application is Theorem 10.22. The concluding result of §10.2 provides a characterization of congruences determining quotient semigroups which are either groups or groups with zero. The similar results of Levi [1944], [1946] on this problem are related to those of Dubreil. §10.3 includes a further discussion of the same problem in terms of reversible equivalences.

The principal equivalences and the reversible equivalences of Dubreil are each one-sided congruences. Congruences, analogous to Dubreil's principal equivalences, called bilateral equivalences by Croisot, are treated in §10.4. The results here are due to Croisot [1957]. The account culminates in the constructions, embodied in Theorems 10.37 and 10.39, of all homomorphisms of a semigroup onto cancellative semigroups with a kernel.

There are several references in the literature to possible analogues for semigroups of the Jordan-Hölder theorems for groups (e.g. see Rees [1940] and Preston [1954a]). These are essentially theorems about the congruences on a semigroup and its subsemigroups. One possible way of generalizing the Jordan-Hölder theorems, including an application to semigroups as a special case, may be found in Birkhoff's book [1948]. In §§10.5–10.6, we present, for the special case of semigroups, a general theory of theorems of the Jordan-Hölder kind, due to Goldie [1950]. An account of Goldie's theory for general algebras in which any two congruences commute has been given by P. M. Cohn in his recent book [1965].

The remaining two sections of the chapter are concerned with congruences on two special kinds of semigroups. §10.7 deals with completely 0-simple semigroups and begins with a description of their congruences following the results of Gluskin [1956], Tamura [1960] and Preston [1961]. The account includes the interesting example (Theorem 10.52) of double coset decompositions of a completely simple semigroup used by Schwarz [1962] to describe homomorphisms of such a semigroup onto a group. The section concludes with a result (Theorem 10.58) of Malcev [1952], for use in the next section, on congruences on completely 0-simple semigroups of the form I_{n+1}/I_n, where I_m is the subsemigroup of $\mathscr{T}_X$ consisting of all elements of rank less than m.

The final section, §10.8, presents Malcev's [1952] results on the congruences on a semigroup $\mathscr{T}_X$. For $|X|$ finite, the lattice of congruences on $\mathscr{T}_X$ is seen to form a chain and each congruence is either a Rees ideal congruence or a congruence determined by a congruence on one of the completely 0-simple semigroups I_{n+1}/I_n (Theorem 10.68). For $|X|$ infinite the situation is more complicated. A third type of congruence is defined and it is shown that the congruences on $\mathscr{T}_X$ of these three types form a set of generators of the lattice of congruences on $\mathscr{T}_X$. Each congruence on $\mathscr{T}_X$, not of the first two kinds, is associated with a finite sequence of (infinite) cardinals and, using this finite sequence, an explicit construction of the congruence is given in terms of the above described generators of the lattice.

10.1 Admissible and normal sets

In §1.5 (Theorem 1.8), we showed how to construct the congruence generated by an arbitrary relation on a semigroup. In this section we discuss the related constructions of one-sided congruences and the dual fact that, in addition to generating a unique minimal congruence containing it, any equivalence on a semigroup also contains a unique maximal congruence. Also discussed are sets of congruences with a given set of congruence classes in common. The ideas developed in this section derive from the papers of E. S. Lyapin [1950] and Marianne Teissier [1951]. We follow the more general account of G. B. Preston [1961]. A considerable number of new results have been included, notably the first three lemmas and the one-sided formulation of Theorem 10.4 and its corollaries.

Let S be a semigroup and let $\mathscr{S}$ denote the set of all symmetric reflexive relations on S, $\mathscr{E}$ the set of all equivalences on S and $\mathscr{P}[\mathscr{Q}, \mathscr{C}]$ the set of all left [right, two-sided] congruences on S. Let $\mathscr{F}$ be any one of these sets. If $\rho_i \in \mathscr{F}$ $(i \in I)$, then it is easily verified that also $\bigcap\{\rho_i \colon i \in I\} \in \mathscr{F}$.

Further $S \times S \in \mathscr{F}$. Hence, defining the operation $\vee$ in the usual way by

$$\bigvee \{\rho_i : i \in I\} = \bigcap \{\rho : \rho \in \mathscr{F}, \rho \supseteq \rho_i, \text{ for all } i \in I\},$$

$\mathscr{F}$ becomes a complete lattice with respect to $\bigcap$ and $\bigvee$.

LEMMA 10.1. *Let $\mathscr{F}$ be any one of $\mathscr{E}$, $\mathscr{P}$, $\mathscr{Q}$, $\mathscr{C}$. Let $\rho, \sigma \in \mathscr{F}$, and put $\tau = \{(a, b): a\rho x_1 \sigma x_2 \rho \cdots \sigma x_m \rho b$, for some $x_1, x_2, \cdots, x_m$ in $S\}$.*
Then $\tau = \rho \vee \sigma$.

Note. Here, and similarly elsewhere, we write $a\rho x_1 \sigma x_2 \rho \cdots \sigma x_m \rho b$ as a shorthand for $(a, x_1) \in \rho$, $(x_1, x_2) \in \sigma, \cdots, (x_m, b) \in \rho$.

PROOF. Since $\rho \vee \sigma$ is an equivalence relation, and hence a transitive relation, containing both σ and ρ, we must have $\tau \subseteq \rho \vee \sigma$. It will suffice therefore to show that $\tau \in \mathscr{F}$. It is clear that τ is reflexive and symmetric. Suppose that $(a, b) \in \tau$ and $(b, c) \in \tau$, so that there exist $x_1, x_2, \cdots, x_m$, $y_1, y_2, \cdots, y_n$ in S such that $a\rho x_1 \sigma x_2 \rho \cdots \sigma x_m \rho b$ and $b\rho y_1 \sigma y_2 \rho \cdots \sigma y_n \rho c$. If we insert between these two chains the link $b\sigma b$ we obtain a chain of the desired kind from a to c. Thus $(a, c) \in \tau$; and this shows that τ is transitive. That τ is also a left [right, two-sided] congruence when ρ and σ are such is clear. Thus $\tau \in \mathscr{F}$.

Since the construction for $\rho \vee \sigma$ is the same for each choice of $\mathscr{F}$, we have immediately the

COROLLARY 10.2. *$\mathscr{P}$, $\mathscr{Q}$ and $\mathscr{C}$ are each sublattices of $\mathscr{E}$. $\mathscr{C}$ is a sublattice of $\mathscr{P}$ and of $\mathscr{Q}$; in fact $\mathscr{C} = \mathscr{P} \cap \mathscr{Q}$.*

We now define six mappings, L, L^*, R, R^* and C, C^* of $\mathscr{B}$ into $\mathscr{B}$, where $\mathscr{B}$ denotes the set of all (binary) relations on S. C^* has already been defined in §9.5. Let $\rho \in \mathscr{B}$; then we define

$$\begin{aligned}
\rho L &= \{(x, y): (sx, sy) \in \rho \text{ for all } s \in S^1\}, \\
\rho L^* &= \{(x, y): x = su, y = sv \text{ for some } (u, v) \in \rho, s \in S^1\}, \\
\rho C &= \{(x, y): (sxt, syt) \in \rho \text{ for all } s, t \in S^1\}, \\
\rho C^* &= \{(x, y): x = sut, y = svt \text{ for some } (u, v) \in \rho, s, t \in S^1\},
\end{aligned}$$

while ρR and ρR^* are defined to be the left-right duals of ρL and ρL^*, respectively.

It is clear, for any ρ, that

$$\begin{aligned}
\rho L &\subseteq \rho \subseteq \rho L^*, \\
\rho R &\subseteq \rho \subseteq \rho R^*, \\
\rho C &\subseteq \rho \subseteq \rho C^*.
\end{aligned}$$

More precisely, denoting by ρT the transitive closure of ρ (in Volume 1, p. 14, the transitive closure $\bigcup \{\rho^n : n = 1, 2, \cdots\}$ was denoted by ρ^t), we have

LEMMA 10.3. *$L[R, C]$ is an idempotent, intersection preserving mapping of $\mathscr{E}$ onto $\mathscr{P}[\mathscr{Q}, \mathscr{C}]$, and*

$$LR = RL = C.$$

$L^*T[R^*T, C^*T]$ is an idempotent mapping of $\mathscr{S}$ onto $\mathscr{P}[\mathscr{Q}, \mathscr{C}]$, and

$$L^*TR^*T = R^*TL^*T = C^*T.$$

If $\rho \in \mathscr{E}$ then $\rho L[\rho R, \rho C]$ is the unique maximal left [right, two-sided] congruence contained in ρ. If $\rho \in \mathscr{S}$ then $\rho L^*T[\rho R^*T, \rho C^*T]$ is the unique minimal left [right, two-sided] congruence containing ρ.

PROOF. We deal with the unstarred mappings first.

Let $(x, y) \in \rho L$. Then $(sx, sy) \in \rho$ for all s in S^1. Hence, for any z in S,

$$(szx, szy) \in \rho$$

for all s in S^1. Thus $(zx, zy) \in \rho L$. Hence ρL is left compatible. Now, if $\rho \in \mathscr{E}$, it is clear that ρL is reflexive and symmetric. Further ρL is transitive; for from $(sx, sy) \in \rho$ and $(sy, sz) \in \rho$, for all s in S^1, we have, since ρ is transitive, $(sx, sz) \in \rho$. Hence ρL is a left congruence, i.e. $\rho L \in \mathscr{P}$.

Now consider ρ in $\mathscr{P}$. Since ρ is then left compatible $(x, y) \in \rho$ implies that $(sx, sy) \in \rho$ for all s in S^1. Thus $\rho \subseteq \rho L$. Since, as already observed, we always have $\rho L \subseteq \rho$, it follows that $\rho = \rho L$. Thus L is an idempotent mapping of $\mathscr{E}$ onto $\mathscr{P}$.

Let $\rho_i \in \mathscr{E}$, $i \in I$. Then

$$(x, y) \in (\bigcap \rho_i)L,$$

if and only if, $(sx, sy) \in \bigcap \rho_i$ for all s in S^1, i.e. if and only if, for each i, $(sx, sy) \in \rho_i$, i.e. if and only if

$$(x, y) \in \bigcap (\rho_i L).$$

Hence

$$(\bigcap \rho_i)L = \bigcap (\rho_i L);$$

i.e. L is intersection preserving.

Finally, consider ρ in $\mathscr{E}$ and let σ be any left congruence contained in ρ. Then $\sigma \cap \rho = \sigma$, and so, since L is intersection preserving,

$$\sigma L \cap \rho L = \sigma L.$$

But, since $\sigma \in \mathscr{P}$, $\sigma L = \sigma$. Thus

$$\sigma \cap \rho L = \sigma,$$

i.e. $\sigma \subseteq \rho L$. Thus ρL is the unique maximal left congruence contained in ρ.

The corresponding assertions for R and C follow similarly. That $LR = RL = C$ follows immediately from the definitions.

To deal with the starred mappings note firstly that the assertion that C^*T maps $\mathscr{S}$ into $\mathscr{C}$ is merely a restatement of part of Theorem 1.8. That

C^*T is idempotent is clear—from Theorem 1.8 or directly from the definition—and from this it follows that C^*T maps $\mathscr{S}$ onto $\mathscr{C}$. The similar assertions about L^*T and R^*T follow in the same way. It is a straightforward computation to verify that L^*T and R^*T commute and that their product is C^*T.

Of the remaining assertions in the lemma about the starred mappings, that about C^*T is part of Theorem 1.8. For L^*T and R^*T the assertions follow similarly. For example, consider L^*T. It is clear from the definitions that if $\rho \in \mathscr{S}$ then

$$\rho \subseteq \rho L^* \subseteq \rho L^*T.$$

Let σ be any left congruence containing ρ. Let $(x, y) \in \rho L^*$. Then there exist s, u, v such that $x = su$, $y = sv$, $(u, v) \in \rho$ and $s \in S^1$. Now $\rho \subseteq \sigma$, and so $(u, v) \in \sigma$. Since σ is a left congruence, therefore $(su, sv) \in \sigma$, i.e. $(x, y) \in \sigma$. Hence $\rho L^* \subseteq \sigma$; whence $\rho L^*T \subseteq \sigma T = \sigma$. And so ρL^*T is the unique minimal left congruence containing ρ.

This completes the proof of the lemma.

We now apply these constructions to obtain a generalization of the result of Lyapin [1950] and Teissier [1951], mentioned earlier, determining necessary and sufficient conditions for a subset of a semigroup to be a congruence class relative to some congruence.

Let $\mathscr{A}$ be a set of disjoint subsets of a semigroup S. We recall (§7.4) that $\mathscr{A}$ is said to be [*left, right*] *admissible* if $\mathscr{A}$ is a subset of the set of congruence classes of some [left, right] congruence on S; and that if ρ be an equivalence, such that the elements of $\mathscr{A}$ are ρ-classes, then ρ is said to *admit* $\mathscr{A}$.

It is to be observed that if $\mathscr{A}$ is both left and right admissible then $\mathscr{A}$ is not necessarily admissible. For example, when S is a group and $\mathscr{A}$ consists of a single subset H of S, then if H is a subgroup it (or $\mathscr{A}$) is both left and right admissible. However H is only admissible (when a subgroup) if it is a normal subgroup of S.

In general, if $\mathscr{A}$ is [left, right] admissible, then there will be several [left, right] congruences on S which admit $\mathscr{A}$. We easily see that the intersection and the union ($\vee$, as defined earlier) of any set of equivalences which admit $\mathscr{A}$ also admit $\mathscr{A}$. Thus the set of [left, right] congruences which admit $\mathscr{A}$ forms a complete sublattice of the lattice of [left, right] congruences on S. We proceed to find the 0-elements and 1-elements of these sublattices. We will work primarily in terms of left congruences.

We start from the 0-element and 1-element of the lattice of equivalences on S which admit $\mathscr{A}$. Suppose that $\mathscr{A} = \{A_i: i \in I\}$, a disjoint set of subsets of S. Write

$$(1) \qquad \begin{cases} A = \bigcup\{A_i: i \in I\}, \\ \alpha(\mathscr{A}) = \alpha = \bigcup\{A_i \times A_i: i \in I\} \cup \{(x, x): x \in S \backslash A\}, \\ \beta(\mathscr{A}) = \beta = \bigcup\{A_i \times A_i: i \in I\} \cup (S \backslash A \times S \backslash A). \end{cases}$$

Then, clearly, α is the minimal and β is the maximal equivalence admitting $\mathscr{A}$. Further, an equivalence ρ has the property that each A_i, $i \in I$, is a union of ρ-classes if and only if $\rho \subseteq \beta$. Hence, from the preceding lemma, βL is the maximal left congruence on S such that each A_i is a union of congruence classes. Similarly, again by the preceding lemma, since $\alpha \subseteq \rho$ if and only if each A_i is contained in some ρ-class, the left congruence $\alpha L^* T$ is the minimal left congruence on S with the property that each set A_i is contained in a single congruence class.

Suppose that $\mathscr{A}$ is left admissible and that ρ is a left congruence on S. Now the set A_i is a ρ-class if and only if it is both a union of ρ-classes and is contained in a ρ-class. Thus it is clear from the above remarks that the left congruence ρ admits $\mathscr{A}$ if and only if

$$(2) \qquad \alpha L^* T \subseteq \rho \subseteq \beta L.$$

It is further evident that $\mathscr{A}$ is left admissible if and only if

$$(3) \qquad \alpha L^* T \subseteq \beta L.$$

We have thus proved part (i) of the following theorem.

THEOREM 10.4. *Let $\mathscr{A} = \{A_i : i \in I\}$ be a disjoint set of subsets of S. Define A, $\alpha(\mathscr{A}) = \alpha$ and $\beta(\mathscr{A}) = \beta$ by equations* (1) *above.*

(i) *Then $\mathscr{A}$ is a left admissible set of subsets of S if and only if*

$$(3) \qquad \alpha L^* T \subseteq \beta L;$$

and then the left congruence ρ admits $\mathscr{A}$ if and only if

$$(2) \qquad \alpha L^* T \subseteq \rho \subseteq \beta L.$$

(ii) *$\mathscr{A}$ is a left admissible set of subsets of S if and only if*

$$(4) \qquad xA_i \cap A_j \neq \square \text{ implies that } xA_i \subseteq A_j$$

for $x \in S^1$ and $i, j \in I$.

PROOF. It remains only to prove part (ii) of the theorem. The "only if" half is clear, for the condition immediately follows from the fact that the elements of $\mathscr{A}$ are congruence classes of some left congruence on S.

Conversely, let us assume that $\mathscr{A}$ satisfies condition (4). Then each A_i is a βL-class. For let $a, a' \in A_i$ and let $x \in S^1$. Then either $xa \in A_j$, for some j in I, or $xa \in S \backslash A$. In the former case $xA_i \cap A_j \neq \square$, so that by (4), $xA_i \subseteq A_j$. Hence $xa' \in A_j$. In the latter case $xa' \in S \backslash A$; for otherwise, by the argument just used, $xa \in A$. Thus, for any x in S^1, $(xa, xa') \in \beta$. Recalling the definition of the operation L, we deduce that $(a, a') \in \beta L$. Hence each class A_i is contained in a single βL-class.

Conversely, let $a \in A_i$ and suppose that $(x, a) \in \beta L$. Then, since $\beta L \subseteq \beta$, $(x, a) \in \beta$. Hence, immediately, $x \in A_i$. Consequently, each A_i is a βL-class, i.e. βL admits $\mathscr{A}$.

Since βL is a left congruence on S this completes the proof of the theorem.

As a corollary we prove an assertion made earlier about partial congruences (in the remarks preceding Theorem 7.10).

COROLLARY 10.5. *Let ρ be a partial right congruence with domain T on the semigroup S. Then there exists a right congruence τ, say, on S, such that ρ is the restriction of τ to T.*

PROOF. The assertion of the corollary amounts to saying that the set of ρ-classes is a right admissible set of subsets of S.

Let $\mathscr{A} = \{A_i : i \in I\}$ be the set of ρ-classes. Let $x \in S^1$ and suppose that $A_i x \cap A_j \neq \square$ for some $i, j \in I$. Then, $ax \in A_j$, for some $a \in A_i$. Suppose $a' \in A_i$. Then $(a, a') \in \rho$; and hence, since ρ is a partial right congruence on S, $(ax, a'x) \in \rho$. Thus $a'x \in A_j$; whence $A_i x \subseteq A_j$. From the left-right dual of part (ii) of the theorem it therefore follows that $\mathscr{A}$ is right admissible.

Since a set of subsets of S which is both left admissible and right admissible is not necessarily admissible, we state as a separate theorem the two-sided analogue of Theorem 10.4. Its proof proceeds on similar lines.

THEOREM 10.6. *Let $\mathscr{A} = \{A_i : i \in I\}$ be a disjoint set of subsets of S. Define A, $\alpha(\mathscr{A}) = \alpha$ and $\beta(\mathscr{A}) = \beta$ by equations (1) above.*

(i) *Then $\mathscr{A}$ is an admissible set of subsets of S if and only if*

$$\alpha C^* T \subseteq \beta C; \tag{5}$$

the congruence ρ admits $\mathscr{A}$ if and only if

$$\alpha C^* T \subseteq \rho \subseteq \beta C. \tag{6}$$

(ii) *$\mathscr{A}$ is an admissible set of subsets of S if and only if*

$$xA_i y \cap A_j \neq \square \text{ implies that } xA_i y \subseteq A_j \tag{7}$$

for $x, y \in S^1$ and $i, j \in I$.

Recall, further, (§7.4), that a [left, right] admissible set $\mathscr{A}$ is said to be *[left, right] normal* in S if there is only one [left, right] congruence on S which admits $\mathscr{A}$. If S is a group then any subgroup is both left and right normal in this sense. Any normal subgroup, or coset of a normal subgroup, where "normal" is used in its usual group-theoretic sense, is normal in the above sense. As an immediate corollary to Theorem 10.4, and in the notation of the theorem we have

COROLLARY 10.7. *$\mathscr{A}$ is a left normal set of subsets of S if and only if*

$$\alpha L^* T = \beta L.$$

What is virtually merely another way of putting this result is contained in the following

COROLLARY 10.8. *Let $\mathscr{A}$ be a left admissible set of subsets of S. Then $\mathscr{A}$ is left normal if and only if, when ρ is a left congruence which admits $\mathscr{A}$, then*

$$\beta L \subseteq \rho.$$

This latter corollary was effectively what was used in the proofs of normality given at various places in §7.4.

We end this section by observing that it follows from the definitions of $\beta = \beta(\mathscr{A})$ and of R that

(8) $\beta R = \{(x, y)\colon xs \in A_i$ if and only if $ys \in A_i$, for all $s \in S^1$, $i \in I\}$.

If we write $\beta_i = \beta(\{A_i\}) = \{(A_i \times A_i) \cup (S\backslash A_i \times S\backslash A_i)\}$, for each $i \in I$, then, from (8) it immediately follows that

(9) $$\beta R = \bigcap \{\beta_i R\colon i \in I\}.$$

We also note, for further use, that similarly,

(10) $\beta C = \{(x, y)\colon sxt \in A_i$ if and only if $syt \in A_i$, for all $s, t \in S^1$, $i \in I\}$;

and

(11) $$\beta C = \bigcap \{\beta_i C\colon i \in I\}.$$

$\beta_i R$ is essentially the principal right congruence of Dubreil [1941] associated with A_i, and $\beta_i C$ the principal congruence of Croisot [1957]. We shall deal with these in §§10.2 and 10.4, respectively.

EXERCISES FOR §10.1

1. Let S be either a zero semigroup or a left zero semigroup (Vol. I, p. 4). Then every equivalence on S is a congruence.
2. If ρ is a reflexive, symmetric relation on a set S, then, in general, there is not a unique maximal equivalence relation contained in ρ.
3. When restricted to $\mathscr{E}$ (in the notation of Lemma 10.3), $TL^* = L^*$. Thus, in general, $TL^* \neq L^*T$.
4. Let S be a cyclic group of order 12: $S = \langle a \rangle$, $a^{12} = e$. Let ρ_1 be the equivalence on S with equivalence classes $\{e, a, a^6, a^7\}$, $\{a^2, a^8\}$, $\{a^3, a^4, a^9, a^{10}\}$ and $\{a^5, a^{11}\}$. Then $\rho_1 L$ is the congruence on S determined by the subgroup $\langle a^6 \rangle$. Let ρ_2 be the equivalence on S with the equivalence classes $\{e, a^2, a^3, a^5, a^6, a^8, a^9, a^{11}\}$ and $\{a, a^4, a^7, a^{10}\}$. Then $\rho_2 L$ is the congruence on S determined by the subgroup $\langle a^3 \rangle$. Thus $\rho_1 L \vee \rho_2 L = \rho_2 L$. Hence, in general, L is not a lattice homomorphism of $\mathscr{E}$ onto $\mathscr{P}$ (cf. Lemma 10.3).
5. Using the notation of Lemma 10.3, considered as mappings defined on $\mathscr{S}$, $L^*R^* = R^*L^* = C^*$.
6. Let S be the direct product of the group G and the left zero semigroup E: $S = G \times E$. Then S is a left group, hence completely simple and regular. Let ρ be the equivalence on S whose equivalence classes are the one-element

sets $\{(\epsilon, e)\}$, where ϵ is the identity of G and $e \in E$, together with the sets $\{(\alpha, e): e \in E\}$, one for each $\alpha \in G \backslash \epsilon$. Then ρ is a left congruence on S. The equivalence classes which contain idempotents are thus the same for both ρ and the identity congruence on S. Thus the analogue for completely simple semigroups (and hence for regular semigroups) of Theorem 7.39 for inverse semigroups does not hold.

7. Let S be a group with zero $G^0 = G \cup \{0\}$, and adjoin a new identity element e to S to form the semigroup S^*. Let $\rho[\sigma]$ be the congruence on S^* whose equivalence classes are the sets $\{e\}$ and S [the sets $\{e\}$, $\{0\}$, and G]. Then ρ and σ both admit $\{\{e\}\}$. Hence S^* is an inverse semigroup on which congruences are not determined by the nonzero idempotent classes. (Cf. Theorem 7.38 and Exercise 4 for §10.7.)

10.2 The principal equivalences of Dubreil

Let S be a group and H a subgroup of S. The relation $\mathscr{R}_H$ on S defined thus

$$\mathscr{R}_H = \{(a, b) \in S \times S: Ha = Hb\}$$

is a right congruence on S, and the equivalence classes of $\mathscr{R}_H$ are the right cosets $\{Ha: a \in S\}$ of H in S. Conversely, if ρ is any right congruence on S then the ρ-class containing the identity of S is a subgroup H, say, of S and $\rho = \mathscr{R}_H$. P. Dubreil [1941] considered two ways of extending to semigroups the construction of right congruences on groups. We give an account of Dubreil's constructions in this section and the next section.

For the first extension, note firstly that, when H is a subgroup of the group S, we may write $\mathscr{R}_H$ thus:

$$\mathscr{R}_H = \{(a, b) \in S \times S: ax \in H \text{ if and only if } bx \in H, \text{ for } x \in S\}.$$

This alternative form of $\mathscr{R}_H$ for a group, Dubreil takes as his definition of $\mathscr{R}_H$ for a semigroup. The definition has already been given in §7.2. This differs from the relation $\beta_H = \beta(\{H\})R$ (see end of §10.1) only in that x ranges over S instead of S^1. The two relations coincide in many important cases, even when $S \neq S^1$, e.g. when H is a right unitary subsemigroup of S. Thus we could use β_H instead of $\mathscr{R}_H$ in Theorems 10.22 and 10.24, the most significant results of this section. The theory of β_H in general is a little simpler than that of $\mathscr{R}_H$ because H is always a union of β_H-classes, but is not always a union of $\mathscr{R}_H$-classes. We have deemed it better nonetheless to follow Dubreil's original account.

Let S be a semigroup and let H be a subset of S. We recall that, for any a in S, $a^{[-1]}H$ is defined thus:

$$a^{[-1]}H = \{x \in S: ax \in H\}.$$

With this notation, as in §7.2, we define

$$\mathscr{R}_H = \{(a, b) \in S \times S : a^{[-1]}H = b^{[-1]}H\}$$

and

$$\mathscr{R}_H^* = \{(a, b) \in S \times S : a^{[-1]}H = b^{[-1]}H \neq \square\}.$$

From Lemma 7.13, we have that $\mathscr{R}_H$ is a right congruence on S, called the *principal right congruence* on S determined by H, and that $\mathscr{R}_H^*$ is a partial right congruence on S, called the *principal partial right congruence* on S determined by H. Setting $W_H = \{x \in S : x^{[-1]}H = \square\}$, then W_H is the contradomain of $\mathscr{R}_H^*$, and $\mathscr{R}_H^*$ coincides with the restriction of $\mathscr{R}_H$ to $S \backslash W_H$. W_H is called the *right residue* of H. Dually, the *principal left congruence* on S determined by H is defined by

$${}_H\mathscr{R} = \{(a, b) \in S \times S : Ha^{[-1]} = Hb^{[-1]}\}.$$

The set ${}_HW = \{x \in S : Hx^{[-1]} = \square\}$ is called the *left residue* of H.

LEMMA 10.9. *If H is a subset of the semigroup S, and $W_H \neq \square$, then W_H is an $\mathscr{R}_H$-class and a right ideal of S.*

PROOF. That W_H is an $\mathscr{R}_H$-class is part of Lemma 7.13.

Let $w \in W_H$, $a \in S$. Suppose that $x \in (wa)^{[-1]}H$, i.e. $wax \in H$. Then $ax \in w^{[-1]}H$. This conflicts with the assumption that $w \in W_H$, i.e. that $w^{[-1]}H = \square$. Hence $(wa)^{[-1]}H = \square$, i.e. $wa \in W_H$. Thus W_H is a right ideal of S.

The subset H of S is said to be *strong* (in S) if, for any a, b in S,

$$a^{[-1]}H \cap b^{[-1]}H \neq \square \text{ implies } a^{[-1]}H = b^{[-1]}H.$$

The next lemma shows that this apparently one-sided concept coincides with its left-right dual.

LEMMA 10.10. *Let H be a subset of the semigroup S. Then the following assertions are equivalent.*

(A) *H is strong.*

(B) *For any a, b, x, y in S, when any three of ax, bx, ay, by belong to H then so also does the fourth.*

(C) *For any a, b in S*

$$Ha^{[-1]} \cap Hb^{[-1]} \neq \square \textit{ implies } Ha^{[-1]} = Hb^{[-1]}.$$

PROOF. (A) *implies* (B). Assume that H is strong. Consider the four elements ax, bx, ay and by. If we select three of these elements then they are of the form pu, pv, qu and the fourth is then qv, where $\{p, q\} = \{a, b\}$ and $\{x, y\} = \{u, v\}$. If $pu, pv, qu \in H$, then $u \in p^{[-1]}H \cap q^{[-1]}H \neq \square$; whence, since H is strong, $p^{[-1]}H = q^{[-1]}H$. Hence, from $pv \in H$, i.e. $v \in p^{[-1]}H$, we infer $v \in q^{[-1]}H$, i.e., $qv \in H$.

(B) *implies* (A). Suppose that (B) holds and that $a^{[-1]}H \cap b^{[-1]}H \neq \square$.

Suppose that $x \in a^{[-1]}H \cap b^{[-1]}H$. Then $ax \in H$ and $bx \in H$. Let y be any element of $a^{[-1]}H$, so that $ay \in H$. From (B), it then follows that $by \in H$, i.e. that $y \in b^{[-1]}H$. Thus $a^{[-1]}H \subseteq b^{[-1]}H$; the reverse inequality follows similarly. Consequently H is strong.

Since (B) is left-right symmetric, its equivalence with (C) follows at once from its equivalence with (A).

When H is strong we have the following analogue of Lemma 7.14.

LEMMA 10.11. *Let H be a strong subset of the semigroup S. Then the equivalence classes of $\mathscr{R}_H$ are the nonempty members of*

$$\{Hx^{[-1]}: x \in S\} \cup \{W_H\}.$$

PROOF. Let $(a, b) \in \mathscr{R}_H$, so that $a^{[-1]}H = b^{[-1]}H$. If $a^{[-1]}H = \square$, then $a, b \in W_H$, and it has already been shown, Lemma 7.13, that W_H is, when nonempty, an $\mathscr{R}_H$-class. Suppose then that $x \in a^{[-1]}H = b^{[-1]}H$, so that $ax \in H$ and $bx \in H$. Then $a \in Hx^{[-1]}$ and $b \in Hx^{[-1]}$.

Conversely, let $a, b \in Hx^{[-1]}$. Then $x \in a^{[-1]}H \cap b^{[-1]}H$. Since H is strong, $a^{[-1]}H = b^{[-1]}H$, and $(a, b) \in \mathscr{R}_H$. This completes the proof of the lemma.

If H is a subgroup of the group S then $\mathscr{R}_H = \mathscr{R}_X$ when X is any $\mathscr{R}_H$-class. In general, when H is strong, we have the following theorem.

THEOREM 10.12. *Let H be a strong subset of the semigroup S. Let X be an $\mathscr{R}_H$-class and suppose that $X \neq W_H$. Then X is strong and*

$$W_H \subseteq W_X \quad \textit{and} \quad \mathscr{R}_H \subseteq \mathscr{R}_X.$$

The restriction of $\mathscr{R}_H$ to $S \backslash W_X$ is equal to $\mathscr{R}_X^*$. Furthermore, if $H \subseteq X$, then $\mathscr{R}_H = \mathscr{R}_X$.

PROOF. By Lemma 10.11, since $X \neq W_H$, there exists x in S such that $X = Hx^{[-1]}$. Suppose that $y, z \in S$ and that $y^{[-1]}X \cap z^{[-1]}X \neq \square$. Let $a \in y^{[-1]}X \cap z^{[-1]}X$. Then $ya \in X$, i.e. $yax \in H$, and, similarly, $zax \in H$. Thus $ax \in y^{[-1]}H \cap z^{[-1]}H \neq \square$; consequently, $y^{[-1]}H = z^{[-1]}H$. Let b be any element of $y^{[-1]}X$. Then $ybx \in H$, i.e., $bx \in y^{[-1]}H = z^{[-1]}H$; whence $zbx \in H$, i.e. $b \in z^{[-1]}X$. Thus $y^{[-1]}X \subseteq z^{[-1]}X$; the reverse inequality follows similarly. And this shows that X is strong.

Suppose that $(a, b) \in \mathscr{R}_H$, i.e. that $a^{[-1]}H = b^{[-1]}H$. Then $(a^{[-1]}H)x^{[-1]} = (b^{[-1]}H)x^{[-1]}$. But, as is easily verified, this is the same as $a^{[-1]}(Hx^{[-1]}) = b^{[-1]}(Hx^{[-1]})$, i.e. $a^{[-1]}X = b^{[-1]}X$, i.e. $(a, b) \in \mathscr{R}_X$. Consequently, $\mathscr{R}_H \subseteq \mathscr{R}_X$. In particular, W_H is contained in an $\mathscr{R}_X$-class. Since $a^{[-1]}H = \square$ implies $a^{[-1]}X = a^{[-1]}(Hx^{[-1]}) = (a^{[-1]}H)x^{[-1]} = \square$, $W_H \subseteq W_X$.

Since $\mathscr{R}_H \subseteq \mathscr{R}_X$, W_X is a union of $\mathscr{R}_H$-classes. We now show, that, outside W_X, $\mathscr{R}_H$ coincides with $\mathscr{R}_X$. To show this it will suffice to show that, when $a, b \in S \backslash W_X$, then $(a, b) \in \mathscr{R}_X$ implies $(a, b) \in \mathscr{R}_H$. To this end, therefore, consider $a, b \in S$ such that $a^{[-1]}X = b^{[-1]}X \neq \square$. Then

$$(a^{[-1]}H)x^{[-1]} = (b^{[-1]}H)x^{[-1]} \neq \square;$$

and so there exists z such that $zx \in a^{[-1]}H \cap b^{[-1]}H$. Since H is strong, it follows that $(a, b) \in \mathscr{R}_H$.

It remains to show that $W_H = W_X$, if $H \subseteq X$. And for this it suffices to show that $W_X \subseteq W_H$. Let $a \in W_X$, i.e. suppose that $a^{[-1]}X = \square$. Now, since $H \subseteq X$, therefore $a^{[-1]}H \subseteq a^{[-1]}X$. Hence also $a^{[-1]}H = \square$. Consequently, $a \in W_H$; this shows that $\mathscr{R}_H = \mathscr{R}_X$, if $H \subseteq X$, and completes the proof of the theorem.

Corollary 10.13. *Let H be a strong subset of the semigroup S. Then, for $a \in S$, $a^{[-1]}H$ and $Ha^{[-1]}$ are each strong subsets of S.*

Proof. The empty subset of S is strong. For nonempty sets the corollary follows immediately from Lemma 10.11, Theorem 10.12, and their left-right duals, which are valid by Lemma 10.10.

A nonempty subset H of a semigroup S is said to be *right perfect* if it is strong and if also it is contained in an $\mathscr{R}_H$-class different from W_H. Thus, if H is strong, H is right perfect if and only if $h^{[-1]}H \cap h_1^{[-1]}H \neq \square$ for all $h, h_1 \in H$. We shall denote by U_H the $\mathscr{R}_H$-class containing the right perfect subset H. By the preceding theorem, U_H is strong and $\mathscr{R}_H = \mathscr{R}_{U_H}$.

Lemma 10.14. *Let H be a right perfect subset of the semigroup S. Then H is right admissible.*

Conversely, if H is right admissible, then H is contained in an $\mathscr{R}_H$-class. Furthermore, if also H is strong and $h^{[-1]}H \neq \square$ for some $h \in H$, then H is right perfect.

Proof. Suppose that H is right perfect and suppose that $Hx \cap H \neq \square$, where $x \in S^1$. If $x = 1$, then $Hx = H$ and $Hx \subseteq H$. If $x \in S$, then there exists $h \in H$ such that $hx \in H$, i.e. $x \in h^{[-1]}H$. Since H is right perfect, therefore $x \in h_1^{[-1]}H$ for any $h_1 \in H$, i.e. $h_1 x \in H$ for any $h_1 \in H$. Thus $Hx \subseteq H$. Consequently, by the left-right dual of Theorem 10.4 (ii), H is right admissible.

If H is right admissible, then by the left-right dual of Theorem 10.4(i), H is a congruence class of the right congruence β_H, where we write β_H as a shorthand for $\beta(\{H\})R$ (in the notation of §10.1). By equation (8) of §10.1,

$$\beta_H = \{(x, y) \colon xs \in H \text{ if and only if } ys \in H \text{ for all } s \in S^1\}.$$

Thus, immediately from the definition of $\mathscr{R}_H$, $\beta_H \subseteq \mathscr{R}_H$. Hence H is contained in an $\mathscr{R}_H$-class.

The remaining assertion of the lemma follows immediately if we observe that the condition $h^{[-1]}H \neq \square$, for some $h \in H$, ensures that $H \nsubseteq W_H$.

Of particular interest in the sequel is the case when H is a subsemigroup.

Lemma 10.15. *A strong subsemigroup is right perfect.*

Proof. Let H be a strong subsemigroup of the semigroup S. Then, for any $h \in H$, $hH \subseteq H$; hence $H \subseteq h^{[-1]}H$, for all $h \in H$. Whence it follows that H is right perfect.

LEMMA 10.16. *Let H be a strong subsemigroup of the semigroup S. Then the $\mathscr{R}_H$-class, U_H, containing H is a right unitary subsemigroup of S.*

Furthermore $H = U_H$ if and only if H is right unitary in S.

PROOF. By Lemma 10.15, $H \subseteq U_H$ and $h^{[-1]}H \neq \square$ for $h \in H$. Since H is a semigroup, $H \subseteq h^{[-1]}H$, for $h \in H$. Since H is strong, therefore $u \in U_H$ if and only if $H \subseteq u^{[-1]}H$, i.e. if and only if $uH \subseteq H$. Suppose that $u_1, u_2 \in U_H$. Then $u_2H \subseteq H$ implies $u_1u_2H \subseteq u_1H \subseteq H$; thus $u_1u_2 \in U_H$. Hence U_H is a subsemigroup.

To show that U_H is right unitary, consider $u \in U_H$, $x \in S$ and suppose that $xu \in U_H$. Then $xuh \in H$, for some $h \in H$. Since $u \in U_H$, $uh \in H$, $uh = h_1$, say. Hence $xh_1 \in H$, i.e. $h_1 \in x^{[-1]}H$. Since H is strong, therefore $x \in U_H$. This shows that U_H is right unitary.

Suppose now that H is right unitary. Let $u \in U_H$, so that $uH \subseteq H$. Since H is right unitary, it immediately follows that $u \in H$.

This completes the proof of the lemma.

For a strong subsemigroup the result of Theorem 10.12 may be strengthened.

THEOREM 10.17. *Let H be a strong subsemigroup of the semigroup S. Let X be an $\mathscr{R}_H$-class and suppose that $X \neq W_H$. Then $\mathscr{R}_H = \mathscr{R}_X$.*

PROOF. By Theorem 10.12, it suffices to prove that $W_H = W_X$; and for this, again by Theorem 10.12, it suffices to prove that $W_X \subseteq W_H$.

Suppose that $X = Hx^{[-1]}$, as we may by Lemma 10.11. Let $a \in W_X$ and $b \in X$. Then $a^{[-1]}(Hx^{[-1]}) = \square$ and $bx \in H$. We have to show that $a^{[-1]}H = \square$. Suppose that, on the contrary, $z \in a^{[-1]}H$. Then $az \in H$; and because $bx \in H$ so also $a(zb)x \in H$. Thus $zb \in a^{[-1]}(Hx^{[-1]})$, contrary to assumption. Hence, in fact, $a^{[-1]}H = \square$; and this completes the proof of the theorem.

We now look at the connexion between a principal right congruence $\mathscr{R}_H$ and an arbitrary right congruence on a semigroup.

LEMMA 10.18. *Let ρ be a right congruence on the semigroup S and let H be a ρ-class. Then $\rho \subseteq \mathscr{R}_H$.*

PROOF. Let $(a, b) \in \rho$. Suppose $ax \in H$. Since ρ is right compatible, $(ax, bx) \in \rho$; whence, since H is a ρ-class, $bx \in H$. Similarly $bx \in H$ implies $ax \in H$. Hence $(a, b) \in \mathscr{R}_H$.

A right congruence ρ on a semigroup S is said to be *partially right cancellative with residue* W if W is a ρ-class which is a right ideal of S and if $ac, bc \in S\backslash W$ and $(ac, bc) \in \rho$ $(a, b, c \in S)$ imply $(a, b) \in \rho$ (i.e., if the partial right congruence $\rho \cap (S\backslash W \times S\backslash W)$ is right cancellative). A right congruence will also be said to be partially right cancellative when it is right cancellative (i.e., when $W = \square$).

THEOREM 10.19. *If H is a strong subset of the semigroup S, then $\mathscr{R}_H$ is partially right cancellative with residue W_H.*

Conversely, let ρ be a partially right cancellative right congruence on S with residue W. Let H be a ρ-class $\neq W$. Then $\rho \subseteq \mathscr{R}_H$, and the restriction of ρ to $S \backslash W_H$ coincides with $\mathscr{R}_H^*$.

PROOF. If $ac, bc \in S \backslash W_H$ and $(ac, bc) \in \mathscr{R}_H^*$, then $(ac)^{[-1]}H = (bc)^{[-1]}H \neq \square$. Let $x \in (ac)^{[-1]}H$. Then $acx \in H$ and $bcx \in H$. Thus $cx \in a^{[-1]}H \cap b^{[-1]}H$; since H is strong therefore $a^{[-1]}H = b^{[-1]}H \neq \square$, i.e. $(a, b) \in \mathscr{R}_H$ and $a, b \notin W_H$. Thus $(a, b) \in \mathscr{R}_H^*$.

Conversely, let ρ be a partially right cancellative right congruence on S with residue W, and let H be a ρ-class $\neq W$. By Lemma 10.18, $\rho \subseteq \mathscr{R}_H$. Let $(a, b) \in \mathscr{R}_H^*$, so that there exists x in $a^{[-1]}H = b^{[-1]}H$. Then $ax, bx \in H$, whence, since H is a ρ-class, $(ax, bx) \in \rho$. Since ρ is partially right cancellative with residue W, and $ax, bx \in S \backslash W$, we conclude that $(a, b) \in \rho$. Hence $\mathscr{R}_H^* \subseteq \rho$, and this completes the proof of the theorem.

The following corollary gives necessary and sufficient conditions for a principal right congruence $\mathscr{R}_H$ to be right cancellative. We need a preliminary definition.

The subset C, possibly empty, of the semigroup S is said to be a *left consistent subset* of S if $ab \in C$, for a, b in S, implies $a \in C$. (Cf. §9.4.)

COROLLARY 10.20. *Let H be a strong subset of the semigroup S. Then $\mathscr{R}_H$ is right cancellative if and only if W_H is a left consistent subset of S.*

PROOF. Let $a, b, c \in S$ and $(ac, bc) \in \mathscr{R}_H$. If $ac, bc \in S \backslash W_H$, then $(a, b) \in \mathscr{R}_H$, by the theorem. Suppose then that $ac, bc \in W_H$. If W_H is left consistent, then we infer that $a, b \in W_H$, so that $(a, b) \in \mathscr{R}_H$.

Conversely, if $\mathscr{R}_H$ is right cancellative, suppose that $ac \in W_H$. Thus $acS \cap H = \square$ and hence $accS \cap H = \square$. Consequently, $acc \in W_H$. Thus $(ac, acc) \in \mathscr{R}_H$, whence we infer that $(a, ac) \in \mathscr{R}_H$, i.e. that $a \in W_H$. This shows that W_H is left consistent and completes the proof of the corollary.

When H is a subsemigroup we can obtain more detailed results.

LEMMA 10.21. *Let H be a left unitary subsemigroup of the semigroup S. Let $h \in H$ and $a \in S$. Then $(ha, a) \in \mathscr{R}_H$.*

Furthermore, if $a \in S \backslash W_H$ and $b \in S$, then there exists $c \in S$ such that $(ac, b) \in \mathscr{R}_H$.

PROOF. If $ax \in H$ then $hax \in H$ since H is a subsemigroup. If $(ha)x = h(ax) \in H$, then $ax \in H$ since H is left unitary. Thus, for x in S, $ax \in H$ if and only if $(ha)x \in H$; in other words, $(ha, a) \in \mathscr{R}_H$.

If $a \in S \backslash W_H$, then there exists s in S such that $as \in H$; whence, for any b in S, $((as)b, b) \in \mathscr{R}_H$. Write $sb = c$; then $(ac, b) \in \mathscr{R}_H$.

If H is a strong subsemigroup of the semigroup S then, by Lemma 10.16, H is contained in an $\mathscr{R}_H$-class $U_H = U$, say, which is right unitary. Also $H = U$ if and only if H is right unitary. Moreover, from Theorem 10.17, it follows that $\mathscr{R}_H = \mathscr{R}_U$. If U is also left unitary, and hence unitary, Lemma 10.21 gives that $(ua, a) \in \mathscr{R}_U$ for any u in U and a in S. We now obtain a characterization of such principal right congruences $\mathscr{R}_U$.

THEOREM 10.22. *Let U be a strong, unitary subsemigroup of the semigroup S. Then $\rho = \mathscr{R}_U$ is a right congruence on S. Let $W = W_U$ denote the right residue of U. Then either $W = \square$ or W is a proper right ideal of S which is one of the congruence classes of ρ. Moreover*

(i) *if $ac, bc \in S \backslash W$ and $(ac, bc) \in \rho$ then $(a, b) \in \rho$;*
(ii) *there exists u in S such that $(ua, a) \in \rho$ for all a in S;*
(iii) *if $a \in S \backslash W$ and $b \in S$ then there exists $c \in S$ such that $(ac, b) \in \rho$.*

Conversely, if ρ is a right congruence on S which satisfies (i), (ii) *and* (iii), *either with $W = \square$, or with W as a ρ-class which is a proper right ideal of S, then the set U of all $u \in S$ for which* (ii) *holds is a strong, unitary subsemigroup of S and $\rho = \mathscr{R}_U$. Moreover W is then the right residue of U and U is a single ρ-class.*

The correspondence, described above, between U and $\mathscr{R}_U$ is one-to-one.

PROOF. The direct half of the theorem follows from Lemma 10.9, Theorem 10.19 and Lemma 10.21.

For the converse, suppose ρ is a right congruence on S satisfying the stated conditions and let U denote the set of all $u \in S$ for which (ii) holds. Let $u_1, u_2 \in U$. Then $(u_1(u_2a), u_2a) \in \rho$ and $(u_2a, a) \in \rho$ for all $a \in S$; whence $((u_1u_2)a, a) \in \rho$ for all $a \in S$. Hence $u_1u_2 \in U$, and so U is a subsemigroup of S.

Let $u \in U$ and $ux \in U$. Since $ux \in U$, $(uxa, a) \in \rho$ for all $a \in S$; since $u \in U$, $(u(xa), xa) \in \rho$ for all $a \in S$. Whence it follows that $(xa, a) \in \rho$ for all $a \in S$, i.e. that $x \in U$. Thus U is left unitary.

We now show that $U \cap W = \square$. Suppose, to the contrary, that $u \in U \cap W$. Then $(ua, a) \in \rho$ for all $a \in S$. By assumption, $W \neq S$; choose $b \in S \backslash W$. Since W is a right ideal, $ub \in W$. Since W is a ρ-class, $(ub, b) \in \rho$ implies that $b \in W$. This is in contradiction to the choice of b. Hence, in fact, $U \cap W = \square$.

Let $u \in U$ and suppose that $(x, u) \in \rho$. Then, since ρ is a right congruence, $(xa, ua) \in \rho$ for all $a \in S$. Since $u \in U$, $(ua, a) \in \rho$ for all $a \in S$. Hence $(xa, a) \in \rho$ for all a in S, i.e. $x \in U$. Furthermore, let $u_1, u_2 \in U$ and choose $c \in S \backslash W$. Then $(u_1c, c) \in \rho$ and $(u_2c, c) \in \rho$; whence $(u_1c, u_2c) \in \rho$. Since $c \in S \backslash W$ and W, if nonempty, is a ρ-class, therefore $u_1c, u_2c \in S \backslash W$. Hence, by property (i), $(u_1, u_2) \in \rho$. Thus we have shown that U is a ρ-class.

Denote by W_U the right residue of U. Suppose that $a \in S \backslash W$. Let $u \in U$. By property (iii), there exists x in S such that $(ax, u) \in \rho$. Since U is a ρ-class, therefore $ax \in U$; whence $a \notin W_U$, and thus we have shown that $W_U \subseteq W$. Conversely, suppose that $a \in W$. If $a \notin W_U$, then $ax \in U$, for some x in S. However, since W is a right ideal, also $ax \in W$. This conflicts with the fact that $U \cap W = \square$. Hence $W \subseteq W_U$. Combining this with the earlier inequality, we have that $W = W_U$.

Since U is a ρ-class, from Lemma 10.18 we conclude that $\rho \subseteq \mathscr{R}_U$. If

$(a, b) \in \mathscr{R}_U$ and $a, b \in W_U$, then $(a, b) \in \rho$ as we have just shown. Otherwise $(a, b) \in \mathscr{R}_U$ implies that $ax, bx \in U$ for some x in S. Choose $c \in S \setminus W$. Then $(axc, c) \in \rho$ and $(bxc, c) \in \rho$; whence $(axc, bxc) \in \rho$ and, since $c \notin W$ and W, if nonempty, is a ρ-class, $axc, bxc \in S \setminus W$. From property (i) we infer that $(a, b) \in \rho$. Thus $\mathscr{R}_U \subseteq \rho$; whence $\rho = \mathscr{R}_U$.

To see that U is strong, consider ax, bx, by in U. Since U is a ρ-class, $(ax, bx) \in \rho$; since $U \cap W = \square$, by property (i), this implies that $(a, b) \in \rho$. Since ρ is a right congruence, therefore $(ay, by) \in \rho$. Whence it follows, again since U is a ρ-class, that $ay \in U$. Thus, by Lemma 10.10, U is strong.

The final assertion of the theorem about U, viz., that U is right unitary, now follows immediately from Lemma 10.16.

The remaining assertion of the theorem, that the correspondence between strong, unitary subsemigroups U and their associated principal right congruences $\mathscr{R}_U$ is one-to-one will follow if we show that, given U, then U consists of all elements x in S such that $(xa, a) \in \mathscr{R}_U$ for all a in S. By Lemma 10.21 we know that each element of U has this property. If $(xa, a) \in \mathscr{R}_U$ for all a in S, then, in particular, $(xu, u) \in \mathscr{R}_U$ for u in U. By Lemma 10.16, U is an $\mathscr{R}_U$-class; whence $xu \in U$. Since U is right unitary, $x \in U$.

This completes the proof of the theorem.

We now turn to the consideration of congruences. A subset H of a semigroup S is said to be *symmetric* if $W_H = {}_H W$ and $\mathscr{R}_H = {}_H\mathscr{R}$. In this case $\mathscr{R}_H$ is a congruence on S.

A subset H of a semigroup S satisfying

$$ab \in H \text{ if and only if } ba \in H, \text{ for all } a, b \in S,$$

i.e. $a^{[-1]}H = Ha^{[-1]}$ for all $a \in S$, is said to be *reflexive*.

THEOREM 10.23. *Let H be a strong subsemigroup of a semigroup S. Then H is symmetric if and only if it is reflexive.*

PROOF. It is clear that any reflexive subset H of S is symmetric.

Conversely, suppose that H is a symmetric, strong subsemigroup of S. Suppose that $a^{[-1]}H = \square$. Then $a \in W_H = {}_H W$; whence $Ha^{[-1]} = \square$. Thus $a^{[-1]}H = Ha^{[-1]}$. Suppose that $x \in a^{[-1]}H$, i.e. that $ax \in H$. Then $a \in H x^{[-1]} = X$, say, and X is an $\mathscr{R}_H$-class, by Lemma 10.11. Since H is symmetric, X is an ${}_H\mathscr{R}$-class, distinct from ${}_H W = W_H$, and so, by the left-right dual of Lemma 10.11, $X = y^{[-1]}H$ for some $y \in S$. Thus $a \in y^{[-1]}H$, i.e. $ya \in H$.

By Lemma 10.15, H is contained in an $\mathscr{R}_H$-class, U, say. By Lemma 10.16 and its left-right dual, U is a unitary subsemigroup of S. By Theorem 10.17 and its left-right dual, $\mathscr{R}_H = \mathscr{R}_U = {}_U\mathscr{R}$. Since $ya \in U$, $ax \in U$ and U is unitary, therefore, for z in S, $zy \in U$ if and only if $zy(ax) \in U$, i.e. $(y, y(ax)) \in {}_U\mathscr{R}$. Similarly, $(x, (ya)x) \in \mathscr{R}_U$. Consequently, $(x, y) \in \mathscr{R}_U = {}_H\mathscr{R}$; whence it follows that $ya \in H$ implies that $xa \in H$, i.e. that $x \in Ha^{[-1]}$.

Thus we have shown that $a^{[-1]}H \subseteq Ha^{[-1]}$; the reverse inequality follows by symmetry. This completes the proof of the theorem.

THEOREM 10.24. *Let H be a strong and reflexive (hence symmetric) subsemigroup of the semigroup S. Set $\rho = \mathscr{R}_H = {}_H\mathscr{R}$. Then either $W_H \neq \square$, when S/ρ is a group with zero, with U_H as identity and W_H as zero, or $W_H = \square$, when S/ρ is a group with U_H as identity. $U = U_H$ is a strong, reflexive, unitary subsemigroup of S and $\rho = \mathscr{R}_U$.*

Conversely, let ρ be a congruence on S such that S/ρ is either a group or a group with zero. Let U be the identity element of S/ρ. Then U is a strong, reflexive, unitary subsemigroup of S and $\rho = \mathscr{R}_U$. Furthermore, $W_U \neq \square$ if and only if S/ρ has a zero. When S/ρ has a zero, this zero is W_U.

PROOF. Let H be a strong, reflexive subsemigroup of S. That H is symmetric follows from Theorem 10.23, so that $\mathscr{R}_H = {}_H\mathscr{R} = \rho$, say, and $W_H = {}_H W = W$, say.

By Lemma 10.16 and its left-right dual, H is contained in a ρ-class $U_H = {}_HU = U$, say, and U is a unitary subsemigroup of S. Moreover, by Theorem 10.12, U is strong. By Theorem 10.17 and its dual, $\rho = \mathscr{R}_U = {}_U\mathscr{R}$. Since W, if nonempty, is an ideal of S (Lemma 10.9), it is the zero element of S/ρ. Hence, by the direct part of Theorem 10.22 and its dual (part (iii)), either $W \neq \square$, when S/ρ is a group with zero, or $W = \square$, when S/ρ is a group. That U is the identity of S/ρ follows because U is a nonzero idempotent of S/ρ. That U is reflexive follows directly from Theorem 10.23.

Conversely, let ρ be a congruence on S such that S/ρ is either a group or a group with zero. Set $W = \square$, if S/ρ is a group, and, if S/ρ has a zero, let W denote this zero element. Then, if nonempty, W is an ideal of S. Denote by U the identity element of S/ρ; then, clearly, U consists of all elements u of S for which $(ua, a) \in \rho$ for all a in S, or, equivalently, for which $(au, a) \in \rho$ for all a in S. Conditions (i), (ii) and (iii) of Theorem 10.22 and their left-right duals are evidently satisfied by ρ; whence, by this theorem, $\rho = \mathscr{R}_U = {}_U\mathscr{R}$ and U is a strong unitary subsemigroup of S. Again by Theorem 10.22 and its dual, $W = W_U = {}_U W$. Hence U is symmetric; whence, from Theorem 10.23, U is reflexive. The remaining assertions of the theorem follow from the converse half of Theorem 10.22 and from Lemma 10.16.

An alternative, but closely similar, account of homomorphisms of a semigroup onto a group is contained in the two papers of F. W. Levi, [1944] and [1946]. Levi termed a subsemigroup N of a semigroup S *normal* in S if, for a, b, c in S, when any two of abc, ac, b belong to N, then all three belong to N. N is said to be *complete* (Levi) or *net à droite* (Dubreil) if it has empty residue W_N. Levi established that there is a one-to-one correspondence between complete normal subsemigroups of a semigroup S and congruences on S whose quotient semigroups are groups. The congruence on

S determined by the complete normal subsemigroup N of S is ρ_N, say, where

$$\rho_N = \{(a, b)\text{: there exists } x \text{ in } S \text{ such that } ax,\ bx \in N\}.$$

It is easily verified (see Exercise 17 below) that a subsemigroup N of S is complete normal, in Levi's sense, if and only if it is strong, reflexive, and unitary, and has empty residue. Furthermore, $\rho_N = \mathscr{R}_N$.

Exercises for §10.2

1. A semigroup S is said to be *strict* if for each nonempty strong subset H of S, $W_H = {}_HW = \square$.

Let S be a strict semigroup and let ρ be a congruence on S such that S/ρ is cancellative. Let H be any ρ-class. Then H is strong and symmetric and $\rho = \mathscr{R}_H = {}_H\mathscr{R}$. (Dubreil [1941].)

2. Let G be a group and let H be a nonempty subset of G. Then the following are equivalent assertions about H.

 (i) H is strong.
 (ii) If $h_1, h_2, h_3 \in H$, then $h_1h_2^{-1}h_3 \in H$.
 (iii) H is a right [left] coset of a subgroup of G. (Dubreil [1941].)

3. Let H be a strong subsemigroup of the semigroup S. Suppose that $W_H = {}_HW \neq \square$. Then W_H is a prime ideal of S. (Dubreil [1941].)

4. Let ρ be a congruence on the semigroup S. Suppose that S/ρ is a group or a group with zero. Let X be any ρ-class other than the zero element, if any, of S/ρ. Then $\rho = \mathscr{R}_X = {}_X\mathscr{R}$ and ${}_XW = W_X$.

5. Let $S = \{1, 2, \cdots, n, \cdots\}$ be the multiplicative semigroup of positive integers. Let $H = \{h\}$, for some $h \in S$. Then $r^{[-1]}H = \square$ if r does not divide h, $r^{[-1]}H = \{h/r\}$, if r divides h. H is strong. The equivalence $\mathscr{R}_H$ divides S into a finite number of equivalence classes, W_H consisting of all positive integers not dividing h, each divisor of h forming a one-element equivalence class. Let x divide h and let $X = \{x\}$. Then W_H is properly contained in W_X when x is a proper divisor of h. In $S \backslash W_X$, $\mathscr{R}_H = \mathscr{R}_X$. (Dubreil [1941].)

6. If X and Y are nonempty subsets of a semigroup S define $X^{[-1]}Y$ and $YX^{[-1]}$ thus:

$$X^{[-1]}Y = \{a \in S\colon Xa \subseteq Y\},$$
$$YX^{[-1]} = \{a \in S\colon aX \subseteq Y\}.$$

Let X, Y, Z be nonempty subsets of S. Then

$$(X^{[-1]}Y)Z^{[-1]} = X^{[-1]}(YZ^{[-1]}),$$
$$X^{[-1]}(Y^{[-1]}Z) = (YX)^{[-1]}Z,$$
$$(XY^{[-1]})Z^{[-1]} = X(ZY)^{[-1]}.$$

7. Let H be a subset of the semigroup S. Define the relations ρ_1 and ρ_2, depending on H, thus:

$\rho_1 = \{(a, b) \in S \times S: Ha = Hb\}$,

$\rho_2 = \{(a, b) \in S \times S: Hax \subseteq H$ if and only if $Hbx \subseteq H$ for all x in $S\}$.

Then ρ_1 and ρ_2 are right congruences on S. Moreover, if S is a group and H is a subgroup, $\rho_1 = \rho_2 = \mathscr{R}_H$.

8. The set of all right unitary subsets of a semigroup S forms a complete lattice in which intersection is set-theoretic intersection.

9. (a) A subset H of the semigroup S is right unitary in S if and only if $(S \backslash H)H \subseteq S \backslash H$.

(b) If A is a right ideal of S then $S \backslash A$ is right unitary.

(c) H is a right unitary subsemigroup of S if $S \backslash H$ is a prime, proper right ideal of S. (Chaudhuri [1959].)

10. Suppose that S is a regular, left cancellative semigroup, i.e. a right group (Exercise 4 for §1.11). Then every nonempty, left unitary subset of S is a subsemigroup of S.

In fact, S is isomorphic to the direct product $G \times E$, where G is a group and E is a right zero semigroup (§1.11). The left unitary, nonempty subsets of $G \times E$ are the subsemigroups $H \times F$, where H is a subgroup of G and F is a nonempty subset of E.

A nonempty subset of $G \times E$ is right unitary if and only if it is unitary. The unitary subsets of $G \times E$ are the subsets $H \times E$ where H is a subgroup of G. (Chaudhuri [1959].)

11. (a) Let S be a cancellative semigroup. Suppose there exists an element $h \in S$ such that $h \in Sa \cap aS$ for all a in S. Then S is a group.

(b) Let H be a strong symmetric subset of the semigroup S. Suppose that the residue $W_H = {}_HW$, of H, is empty. Then $S/\mathscr{R}_H$ is a group. (Croisot [1952].)

12. (a) Let ρ be a congruence on the semigroup S such that S/ρ is a group, and let H be a ρ-class. Then H is strong, symmetric, and with empty residue. Moreover, $\rho = \mathscr{R}_H$.

(b) Let ρ and σ be congruences on the semigroup S such that S/ρ and S/σ are groups, and such that ρ and σ have an equivalence class H in common. Then $\rho = \sigma$. (Croisot [1952].)

13. Let $\mathscr{U} = \mathscr{U}(S)$ denote the set of strong, reflexive, unitary subsemigroups of the semigroup S which have empty residue in S. Let $\mathscr{G}$ denote the set of congruences ρ on S such that S/ρ is a group.

The mapping $U \rightarrow \mathscr{R}_U$ $(U \in \mathscr{U})$ is a one-to-one mapping of $\mathscr{U}$ onto $\mathscr{G}$. Moreover, for $U_1, U_2 \in \mathscr{U}$, $U_1 \subseteq U_2$ if and only if $\mathscr{R}_{U_1} \subseteq \mathscr{R}_{U_2}$.

Let E be the set of idempotents of S, and assume $E \neq \square$. Then E is contained in every member U of $\mathscr{U}$, so the intersection M of all U in $\mathscr{U}$ is not empty. If E has empty residue, then $M \in \mathscr{U}$, and hence, by Proposition 1.7, any semigroup for which E has empty residue has a maximal group homomorphic image. The latter is, moreover, unique to within isomorphism (see §11.6). We give an explicit form of M for three important types of semigroups.

(a) For an inverse semigroup S,

$$M = \{x \in S: ex = e, \quad \text{for some } e \in E\}.$$

(Cf. Exercise 6 for §7.7.)

(b) Let S be a completely simple semigroup. Let G be any maximal subgroup of S. Let N be the normal subgroup of G generated by the set $\{ef: e = e^2, f = f^2, ef \in G\}$. Then

$$M = \bigcup \{fNe: e = e^2, f = f^2, ef \in N\}.$$

(c) Let S be a semigroup with zeroid elements, i.e. let S contain a minimal ideal which is a group. (Cf. Exercises for §2.5.) Then

$$M = \{x \in S: ex = e\}$$

where e is the identity of the group ideal of S. (Cf. Exercise 16 for §2.7.) (Stoll [1951].)

14. Let ρ be a right cancellative right congruence on a semigroup S. Denote by $\mathscr{A}$ the set of ρ-classes in S. Then

$$\rho = \bigcap \{\mathscr{R}_A: A \in \mathscr{A}\}.$$

(Thierrin [1953].)

15. Let $\phi: S \to T$ be a homomorphism of the semigroup S onto the semigroup T. Then a subset V of T is a right unitary subsemigroup of T if and only if $V\phi^{-1}$ is a right unitary subsemigroup of S. Hence, in particular, when e ($\in T$) is a right identity of T, $e\phi^{-1}$ is a right unitary subsemigroup of S.

16. (a) The lattice of right congruences on a group G is isomorphic to the lattice of subgroups of G.

(b) Let H, K be subgroups of the group G. Then $HK = KH$ if and only if $\mathscr{R}_H \circ \mathscr{R}_K = \mathscr{R}_K \circ \mathscr{R}_H$. (Dubreil [1954, pp. 153-4].)

17. For a subsemigroup N of a semigroup S, the following are equivalent.

(A) N is reflexive and unitary.

(B) N is strong, symmetric, and unitary.

(C) N is normal in the sense of Levi (see end of §10.2).

18. Let S be a cancellative, commutative semigroup. The identity relation $\Delta = \iota_S$ on S is a unitary, reflexive subsemigroup of the direct product $S \times S$, and has empty residue. The group $(S \times S)/\mathscr{R}_\Delta$ is isomorphic with the group of quotients of S. (A. J. Hulin, verbal communication.)

10.3 The reversible equivalences of Dubreil

Let S be a group and H a subgroup of S. The relation P_H on S defined thus:

$$\mathrm{P}_H = \{(a, b) \in S \times S: h_1 a = h_2 b \text{ for some } h_1, h_2 \in H\},$$

is a right congruence on S, and the equivalence classes of P_H are the right cosets $\{Ha: a \in S\}$. This is the second form in which a right congruence on a group may be given which we referred to at the beginning of §10.2, and which Dubreil [1941] uses as the starting point for investigating similar congruences on semigroups.

A semigroup S is called *right [left] reversible* if $Sa \cap Sb \neq \square$ $[aS \cap bS \neq \square]$ for all a, b in S (cf. §1.10). S is called *reversible* if it is both left and right reversible.

Let S be any semigroup and let H be a right reversible subsemigroup of S. By this we mean that H is a subsemigroup of S and that H is a right reversible semigroup. Then the *right reversible equivalence relation* P_H, corresponding to H, is defined thus:

$$\mathrm{P}_H = \{(a, b) \in S \times S: ua = vb \quad \text{for some } u, v \text{ in } H\}.$$

Then P_H is a right congruence on S. For P_H is evidently reflexive and symmetric. To show that it is transitive consider a, b, c in S such that $a \,\mathrm{P}_H\, b$ and $b \,\mathrm{P}_H\, c$. Then there exist u, v, u', v' in H such that

$$ua = vb, \qquad u'b = v'c.$$

Since H is right reversible, there exist x, y in H such that $xv = yu'$. Then

$$xua = xvb = yu'b = yv'c.$$

Since $xu, yv' \in H$, we conclude that $a \,\mathrm{P}_H\, c$. Thus P_H is an equivalence on S. That P_H is right compatible is clear. Hence P_H is a right congruence on S.

THEOREM 10.25. *Let H be a right reversible subsemigroup of the semigroup S. Then*

(i) *H is contained in a P_H-class, U, say;*
(ii) $U = \bigcup\{h^{[-1]}H: h \in H\}$;
(iii) *if $a \in S$ and $u \in U$, then $(ua, a) \in \mathrm{P}_H$;*
(iv) *U is a left unitary, right reversible subsemigroup of S.*
(v) *$H = U$ if and only if H is left unitary;*
(vi) $\mathrm{P}_U = \mathrm{P}_H$;
(vii) $\mathrm{P}_U \subseteq \mathscr{R}_U$.

PROOF. (i) Let $h, k \in H$. Since H is right reversible, there exist $u, v \in H$ such that $uh = vk$, i.e. $(h, k) \in \mathrm{P}_H$. Consequently, H is contained in a P_H-class.

(ii) Let $u \in U$. Then, by the definition of U, for h in H, $(u, h) \in \mathrm{P}_H$, whence $xu = yh$ for some x, y in H. Since H is a subsemigroup, $yh \in H$. Thus $u \in x^{[-1]}H$, where $x \in H$.

Conversely, let u be any element of $y^{[-1]}H$, where y is an element of H. Then $yu = h$, say, belongs to H. Thus $(hy)u = hh$, where $hy \in H$ since H is a subsemigroup; whence $(u, h) \in \mathrm{P}_H$ so that $u \in U$.

Consequently $U = \bigcup\{h^{[-1]}H: h \in H\}$.

(iii) Let $u \in U$, $a \in S$. Let $h \in H$. Then there exist x, y in H such that $xu = yh$. This gives $x(ua) = (yh)a$; and, since $x, yh \in H$, this implies that $(ua, a) \in \mathrm{P}_H$.

(iv) Let u_1, u_2 be arbitrary elements of U. By (iii), $(u_1u_2, u_2) \in \mathrm{P}_H$; whence, since U is a P_H-class, $u_1u_2 \in U$. Thus U is a subsemigroup of S. To see that U is left unitary, let $ua \in U$ with u in U and a in S. From $(ua, a) \in \mathrm{P}_H$ we have $a \in U$. Furthermore, if $u, v \in U$ then, because $(u, v) \in \mathrm{P}_H$, there exist x, y in H such that $xu = yv$; since $H \subseteq U$, this shows that U is right reversible.

(v) Because of (iv), when $H = U$ then H is left unitary. Conversely, suppose that H is left unitary. Let $u \in U$, so that $(u, h) \in \mathrm{P}_H$ for $h \in H$; whence $h_1u = h_2h$, for some $h_1, h_2 \in H$. Now $h_2h \in H$; hence h_1u and h_1 belong to H, whence $u \in H$ since H is left unitary. This shows that $H = U$.

(vi) That $\mathrm{P}_H \subseteq \mathrm{P}_U$ is evident from $H \subseteq U$. Conversely, consider $(a, b) \in \mathrm{P}_U$; so that there exist x, y in U such that $xa = yb$. Since $x \in U$, (ii) implies that there exist h_1, h_2 in H such that $h_1x = h_2$. Since U is a subsemigroup, therefore $h_1y \in U$, and so there exist $h_3, h_4 \in H$ such that $h_3h_1y = h_4$. Then, also $h_3h_2 = h_5$, say, is an element of H. Consequently,

$$h_5a = h_3h_2a = h_3h_1xa = h_3h_1yb = h_4b;$$

whence $(a, b) \in \mathrm{P}_H$. Thus $\mathrm{P}_U \subseteq \mathrm{P}_H$.

(vii) Let $(a, b) \in \mathrm{P}_U$, so that $ua = vb$ for some u, v in U. Suppose that $ax \in U$. Then $uax \in U$, since U is a subsemigroup. But $uax = vbx$; whence, since U is left unitary, $bx \in U$. Similarly, $bx \in U$ implies $ax \in U$. Hence $(a, b) \in \mathscr{R}_U$.

We now turn our attention to the situation when P_H coincides with its left-right dual ${}_H\mathrm{P}$:

$${}_H\mathrm{P} = \{(a, b) \in S \times S: au = bv \text{ for some } u, v \text{ in } H\},$$

${}_H\mathrm{P}$ being a left congruence when H is left reversible.

Let H be a nonempty subset of the semigroup S. Then H is said to be a *centric subset* of S if $aH = Ha$ for all a in S. H will be called a *centric subsemigroup* if it is a centric subset and also a subsemigroup of S.

LEMMA 10.26. *Let H be a centric subsemigroup of the semigroup S. Then H is reversible and ${}_H\mathrm{P} = \mathrm{P}_H$.*

PROOF. If $u, v \in H$, there exist x and y in H such that $vu = xv$ and $uv = vy$, since $vH = Hv$ by hypothesis. Hence H is both left and right reversible, i.e., H is reversible. By symmetry, we need only prove that $\mathrm{P}_H \subseteq {}_H\mathrm{P}$. Let a, b be elements of S such that $(a, b) \in \mathrm{P}_H$. This means that there exist u, v in H such that $ua = vb$. From $aH = Ha$ and $bH = Hb$ we see that there exist x, y in H such that $ax = ua$ and $by = vb$. Hence $ax = by$ with x, y in H, i.e. $(a, b) \in {}_H\mathrm{P}$.

THEOREM 10.27. *Let H be a centric subsemigroup of the semigroup S and suppose that H is without residue, i.e. that ${}_HW = W_H = \square$. Let U denote the P_H-class containing H.*

Then U is a reversible, unitary, reflexive, strong subsemigroup of S, $\mathrm{P}_H = \mathrm{P}_U = \mathscr{R}_U$, and hence S/P_U is a group.

PROOF. By the preceding lemma, H is reversible and ${}_H\mathrm{P} = \mathrm{P}_H$; hence P_H is a congruence on S. By Theorem 10.25, H is contained in a P_H-class U which is right reversible and left unitary. By the dual of Theorem 10.25, U is left reversible and right unitary. Thus U is a unitary, reversible subsemigroup of S.

Part (iii) of Theorem 10.25 and its left-right dual imply that U is the identity element of S/P_U. If $a \in S$, then there exists x in S such that $ax \in U$, since $H \subseteq U$ and $W_H = \square$. Let A denote the P_U-class containing a. Thus there exists X in S/P_U such that $AX = U$. This suffices to show that S/P_U is a group.

That $\mathrm{P}_U = \mathscr{R}_U$ and that U is strong and reflexive now follows directly from Theorem 10.24.

COROLLARY 10.28. *If H is a unitary, centric subsemigroup of S, and H is without residue, then H is reflexive and strong, and $\mathrm{P}_H = \mathscr{R}_H$.*

Let H be a centric subsemigroup of the semigroup S. By Lemma 10.26, Theorem 10.25 and its dual, H is contained in a P_H-class U, say, which is a reversible, unitary subsemigroup of S. Moreover, ${}_H\mathrm{P} = \mathrm{P}_H$, by Lemma 10.26, and so P_H is a congruence on S. U is the identity element of S/P_H, by Theorem 10.25(iii) and its dual. The question arises: what more can be said about $S' = S/\mathrm{P}_H$, beyond its possession of an identity element? The answer is: nothing. In fact nothing more can be said even if H is unitary. For if S is an arbitrary semigroup with identity element e, then $H = \{e\}$ is a unitary, centric subsemigroup of S, and $(a, b) \in \mathrm{P}_H$ if and only if $a = b$. We note, incidentally, that H need not be reflexive, for $ab = e$ need not imply $ba = e$ (take, for example, the bicyclic semigroup, §1.12).

Again, we know that $({}_H\mathrm{P} =){}_U\mathrm{P} = \mathrm{P}_U(= \mathrm{P}_H)$. But this does not necessarily imply that U is centric. So the converse of Lemma 10.26 is false. For example, let S be the semigroup on the set $\{h, t, a, b\}$ with multiplication table

	h	t	a	b
h	h	h	b	b
t	h	t	a	b
a	b	b	b	b
b	b	b	b	b.

Let $H = \{h\}$. Then H is centric. U is the set of all x in S such that $xh = h$, so that $U = \{h, t\}$. Since $aU = b$ and $Ua = \{a, b\}$, we see that U is not centric.

The foregoing example shows that, although we have $\mathrm{P}_H = \mathrm{P}_U$, we do not necessarily have $\mathscr{R}_H = \mathscr{R}_U$. The equivalence classes in the example are as follows:

$$\mathscr{R}_H\colon \{h\}, \{t\}, \{a, b\};$$

$$\mathscr{R}_U\colon \{h, t\}, \{a, b\}.$$

On the other hand we always have $W_H = W_U$; here $W_H = W_U = \{a, b\}$.

For what semigroups S do we obtain all homomorphic group images of S by means of reversible equivalence relations? Certainly not all semigroups. For if we consider the homomorphism of S onto the one-element group, we see that S must be reversible on at least one side. But the following theorem shows that we do obtain all group images of S in this fashion if S is centric (with respect to itself, i.e. $aS = Sa$ for all a in S). For S a commutative cancellative semigroup this result appears in the paper of P. Dubreil and M. L. Dubreil-Jacotin [1940].

THEOREM 10.29. *Let S be a centric semigroup.*

(i) *Let H be a unitary subsemigroup of S, and suppose that H is without residue, i.e. that ${}_HW = W_H = \square$. Then H is centric if and only if it is reflexive in S.*

(ii) *Let $\phi\colon S \to G$ be a homomorphism of S onto a group G. Let H be the subset of S mapped onto the identity of G. Then H is a unitary, centric subsemigroup of S without residue and $\phi \circ \phi^{-1} = \mathrm{P}_H$.*

PROOF. (i) We have already seen that when H is centric then it is also reflexive (Corollary 10.28). To see the converse suppose that H is reflexive. By symmetry, it suffices to show that $aH \subseteq Ha$ for all a in S. Let u be any element of H. Since, by hypothesis, S is centric, there exists c in S such that $au = ca$. Since H is without residue, there exists b in S such that $ab \in H$; whence $ba \in H$ because H is reflexive. Hence $b(ca) = b(au) = (ba)u \in H$, since H is a subsemigroup. Again, since H is reflexive, $c(ab) = (ca)b \in H$. Finally, since H is unitary, we conclude that $c \in H$, as desired.

(ii) Since H is the identity element of $S/\phi \circ \phi^{-1}$, which is isomorphic to G, that H is unitary, centric and without residue follows from Theorem 10.24 and what we have already proved in part (i). From Theorem 10.24 we also know that $\phi \circ \phi^{-1} = \mathscr{R}_H$; from Theorem 10.27 we have $\mathscr{R}_H = \mathrm{P}_H$. And this concludes the proof of the theorem.

We conclude this section with a classical example of a reversible equivalence relation (see, for example, E. Landau [1927], Part 10, Chapter 4, §4). Let S be the multiplicative semigroup of integral ideals in a finite algebraic number field, and let N be the subsemigroup of principal ideals. N is centric and reflexive since S is commutative, and is readily seen to be unitary. That

N has empty residue is the theorem (Satz 771 in Landau) that every ideal divides a principal ideal. S/P_N is thus an abelian group, in this case of finite order, the "group of ideal classes".

Exercises for §10.3

1. Let S be a cancellative semigroup. Let G be a subgroup of S. Then G is reversible and unitary in S. The equivalence classes of the right congruence P_G are the sets $\{Gx: x \in S\}$, and P_G is right cancellative.

2. Let S be a right cancellative semigroup and let H be a right reversible subsemigroup of S. Then P_H is right cancellative. (Dubreil [1941].)

3. Let S be a right cancellative semigroup. Then every right reversible, left unitary subsemigroup U of S is strong and right unitary. Moreover, if U is such a subsemigroup, $\mathrm{P}_U \subseteq \mathscr{R}_U$ and P_U coincides with $\mathscr{R}_U$ when restricted to $S \backslash W_U$. (Dubreil [1941].)

10.4 Principal congruences

Two-sided analogues of the principal equivalences of Dubreil have been considered by several authors. R. S. Pierce [1954] used them to characterize homomorphisms onto disjunctive semigroups and his result is stated as Exercise 1, below. They were used by Preston [1954c] in giving a condition for a representation of certain inverse semigroups to be faithful. A systematic study of their properties was given by Croisot [1957]. The account in this section is taken from Croisot's paper.

If H is a subset of a semigroup S and $a \in S$, then $H \,..\, a$ is defined thus:

$$H \,..\, a = \{(x, y) \in S \times S: xay \in H\}.$$

We then define

$$\mathscr{P}_H = \{(a, b) \in S \times S: H \,..\, a = H \,..\, b\}.$$

This relation differs from $\beta(\{H\})C$ (§10.1) in the same way that $\mathscr{R}_H$ does from $\beta(\{H\})R$, and the remarks made in the early part of §10.2 apply here as well. That $\mathscr{P}_H$ is an equivalence relation on S is clear. Let $(a, b) \in \mathscr{P}_H$. Then $(x, y) \in H \,..\, (ac)$ if and only if $(x, cy) \in H \,..\, a$; and $(x, y) \in H \,..\, (bc)$ if and only if $(x, cy) \in H \,..\, b$. From $H \,..\, a = H \,..\, b$ we conclude that $H \,..\, (ac) = H \,..\, (bc)$. Thus $(ac, bc) \in \mathscr{P}_H$. Similarly, $(ca, cb) \in \mathscr{P}_H$. Thus $\mathscr{P}_H$ is a congruence on S. $\mathscr{P}_H$ is called the *principal congruence* on S determined by H.

The *biresidue* W of H is the set

$$W = \{a \in S: H \,..\, a = \square\}.$$

Lemma 10.30. *If H is a subset of the semigroup S and its biresidue $W \neq \square$, then W is a (two-sided) ideal of S, and a $\mathscr{P}_H$-class.*

Proof. Let $w \in W$, $a \in S$. Suppose that $(x, y) \in H\,..\,(wa)$, i.e. $xway \in H$. Then $(x, ay) \in H\,..\,w$, contrary to the assumption that $w \in W$. Hence $H\,..\,(wa) = \square$, so that $wa \in W$. Similarly $aw \in W$. Thus W is an ideal of S.

A subset H of S is said to be *bistrong* (in S) if, for any a, b in S,

$$(H\,..\,a) \cap (H\,..\,b) \neq \square \text{ implies } H\,..\,a = H\,..\,b.$$

Theorem 10.31. *Let H be a bistrong subset of the semigroup S. Let X be a $\mathscr{P}_H$-class distinct from the biresidue W of H. Denote by W^X the biresidue of X. Then X is bistrong and*

$$W \subseteq W^X \quad \text{and} \quad \mathscr{P}_H \subseteq \mathscr{P}_X.$$

The restrictions of $\mathscr{P}_H$ and $\mathscr{P}_X$ to $S \backslash W^X$ coincide. Furthermore, if $H \subseteq X$, then $\mathscr{P}_H = \mathscr{P}_X$.

Proof. Observe that, if $x \in X$ and $(a, b) \in H\,..\,x$, then, since H is bistrong,

$$(u, v) \in X\,..\,y \text{ if and only if } (au, vb) \in H\,..\,y.$$

For $(u, v) \in X\,..\,y$ always implies $(au, vb) \in H\,..\,y$; and conversely, $(au, vb) \in H\,..\,y$ implies that $(a, b) \in (H\,..\,x) \cap (H\,..\,uyv)$, whence, since H is bistrong, $uyv \in X$, i.e. $(u, v) \in X\,..\,y$. It follows immediately that, if $y \notin W^X$, then $y \notin W$. Thus $W \subseteq W^X$.

Suppose that $(p, q) \in \mathscr{P}_H$, so that $H\,..\,p = H\,..\,q$. Then, by the above observation, $(u, v) \in X\,..\,p$ if and only if $(au, vb) \in H\,..\,p$ and $(u, v) \in X\,..\,q$ if and only if $(au, vb) \in H\,..\,q$. Since $H\,..\,p = H\,..\,q$, we conclude that $X\,..\,p = X\,..\,q$. Hence $(p, q) \in \mathscr{P}_X$; which shows that $\mathscr{P}_H \subseteq \mathscr{P}_X$.

Now suppose that $(p, q) \in \mathscr{P}_X$ and that $p, q \in S \backslash W^X$. Then there exists $(u, v) \in X\,..\,p = X\,..\,q$. Hence $(au, vb) \in (H\,..\,p) \cap (H\,..\,q)$; whence since H is bistrong, $(p, q) \in \mathscr{P}_H$. This shows that $\mathscr{P}_X$ coincides with $\mathscr{P}_H$ in $S \backslash W^X$.

Suppose now that $H \subseteq X$. To show that $\mathscr{P}_H = \mathscr{P}_X$, it will suffice to show that $W^X \subseteq W$. To this end, consider $p \in W^X$. Then $X\,..\,p = \square$. Suppose that, if possible, there exists $(u, v) \in H\,..\,p$. Then $upv \in H$. Since $H \subseteq X$, $upv \in X$ and so $(u, v) \in X\,..\,p$, contrary to assumption. Hence $H\,..\,p = \square$, i.e. $p \in W$.

This completes the proof of the theorem.

A nonempty subset H of a semigroup S is said to be *biperfect* if it is bistrong and if also it is contained in a $\mathscr{P}_H$-class distinct from the biresidue of H.

If S possesses an identity element e, say, then every bistrong subset of H is biperfect; for $(e, e) \in H\,..\,h$ for all $h \in H$. If there exist a, b in S such that $axb = x$ for all x in S, then the same conclusion follows. In fact, however, these two cases coincide, as the following lemma, of interest in itself, shows (Croisot [1957]. See also Franklin and Lindsay [1960/61].)

Lemma 10.32. *Let S be a semigroup containing elements a, b such that $axb = x$ for all x in S. Then S possesses an identity element e, say, and $e = ab = ba$.*

Proof. For any x in S we have

$$abx = a(abx)b = (aab)xb = axb = x;$$
$$xab = a(xab)b = ax(abb) = axb = x.$$

Thus ab is an identity element for S. Hence $(ab)^2 = ab$; but $(ab)^2 = a(ba)b = ba$. Thus $ab = ba$.

As for the one-sided case (Lemma 10.15) we have

Lemma 10.33. *A bistrong subsemigroup of a semigroup is biperfect.*

Proof. If H is a bistrong subsemigroup then $(h, h') \in H \mathbin{..} h$ for any $h, h' \in H$.

Bistrong subsemigroups which are without biresidue provide yet another method of describing homomorphisms of semigroups onto groups. The following theorem may be compared with Theorems 10.24 and 10.27.

Theorem 10.34. *Let H be a bistrong subsemigroup of the semigroup S and suppose that the biresidue of H is empty. Then H is contained in a $\mathscr{P}_H$-class U, say, and U is a bistrong, unitary subsemigroup of S with empty residue. $H = U$ if and only if H is unitary. Moreover, $\mathscr{P}_H = \mathscr{P}_U = \mathscr{R}_U$ and $S/\mathscr{P}_H$ is a group.*

Conversely, if ρ is a congruence on S such that S/ρ is a group and U denotes the identity element of S/ρ, then U is a bistrong, unitary subsemigroup with empty biresidue and moreover, $\rho = \mathscr{P}_U$.

The correspondence described above between U and $\mathscr{P}_U$ is one-to-one.

Proof. If H is a bistrong subsemigroup of S then, by Lemma 10.33, H is biperfect. Let U be the $\mathscr{P}_H$-class containing H.

Let $u, v \in U$. Then $(u, h) \in \mathscr{P}_H$ and $(v, h) \in \mathscr{P}_H$, for $h \in H$. Since $\mathscr{P}_H$ is a congruence, therefore $(uv, h^2) \in \mathscr{P}_H$. Since $h^2 \in H$, therefore $uv \in U$. Hence U is a subsemigroup.

Let $u \in U$, $xu \in U$. There exists $(a, b) \in H \mathbin{..} (xu)$, whence $(a, ub) \in H \mathbin{..} x$. Now $H \mathbin{..} (xu) = H \mathbin{..} (u'u)$, for $u' \in U$, since U is a semigroup and a $\mathscr{P}_H$-class. Hence $(a, b) \in H \mathbin{..} (u'u)$, i.e. $(a, ub) \in H \mathbin{..} u'$. Thus $(H \mathbin{..} x) \cap (H \mathbin{..} u') \neq \square$, and so, since H is bistrong, $x \in U$. This shows that U is right unitary. That U is left unitary follows similarly.

That U has empty biresidue follows immediately from the fact that $H \subseteq U$. That U is bistrong, follows immediately from Theorem 10.31.

Suppose that H is unitary. Let $u \in U$; so that $H \mathbin{..} u = H \mathbin{..} h \neq \square$ for h in H. Hence $huh \in H$ since $h^3 \in H$. Since H is unitary, $h(uh) \in H$ implies $uh \in H$ which, in turn, implies $u \in H$. Thus $H = U$ if and only if U is unitary.

That $\mathscr{P}_H = \mathscr{P}_U$ follows from Theorem 10.31. Consider now $S/\mathscr{P}_U$. If $a, b, \cdots$, denote elements of S, denote by $A, B, \cdots$, respectively, the elements of $S/\mathscr{P}_U$ to which they belong. Let $AB = AC$, so that $(ab, ac) \in \mathscr{P}_U$. Since U has empty biresidue, there exists $(x, y) \in U \mathbin{..} (ab) = U \mathbin{..} (ac)$. Thus

$(xa, y) \in (U \,..\, b) \cap (U \,..\, c)$; whence $(b, c) \in \mathscr{P}_U$ and $B = C$. Hence $S/\mathscr{P}_U$ is a left cancellative and, similarly, a right cancellative semigroup.

Now U is an idempotent of $S/\mathscr{P}_U$, so that, because $S/\mathscr{P}_U$ is cancellative, U is the identity element of $S/\mathscr{P}_U$. Since U has empty biresidue, given X in $S/\mathscr{P}_U$, there exist A, B in $S/\mathscr{P}_U$ such that $AXB = U$. It follows that $(XBA)^2 = XB(AXB)A = XBUA = XBA = U$, since the identity element of a cancellative semigroup is its sole idempotent. This shows that $S/\mathscr{P}_U$ is a group.

Finally, that $\mathscr{P}_U = \mathscr{R}_U$ now follows from Theorem 10.24.

To deal with the converse suppose that U is the identity element of the group S/ρ. That U is a unitary subsemigroup of S is part of Theorem 10.24. That S/ρ is a group clearly ensures that U has empty biresidue. Suppose that $(x, y) \in (U \,..\, a) \cap (U \,..\, b)$. Then $XAY = U = XBY$, with the same convention about capital letters for ρ-classes as we had above for $\mathscr{P}_U$-classes. Let $(p, q) \in U \,..\, a$. Then $PAQ = U$. But $XAY = XBY$ implies that $A = B$. Hence $PBQ = U$, whence $pbq \in U$, i.e. $(p, q) \in U \,..\, b$. This shows that $U \,..\, a \subseteq U \,..\, b$. By symmetry, it follows that U is bistrong.

Now $(a, b) \in \rho$, i.e. $A = B$, if and only if there exist X, Y such that $XAY = U = XBY$, i.e. if and only if there exist x, y such that $(x, y) \in (U \,..\, a) \cap (U \,..\, b)$, i.e. if and only if $(a, b) \in \mathscr{P}_U$. Thus $\rho = \mathscr{P}_U$.

That the correspondence we have established between bistrong, unitary subsemigroups with empty biresidue, and congruences ρ such that S/ρ is a group, is one-to-one is clear.

Corollary 10.35. *Let H be a nonempty subset of the semigroup S. Then the following are equivalent assertions.*

(A) *H is a bistrong, unitary subsemigroup of S with empty biresidue.*

(B) *H is a strong, reflexive, unitary subsemigroup of S with empty right residue.*

Proof. The corollary follows from a comparison of Theorem 10.34 with Theorem 10.24.

So far we have derived for principal congruences results directly analogous to those for principal right congruences and have, in particular, shown that they provide an alternative tool for describing homomorphisms of semigroups onto groups. We now use principal congruences to obtain a description of homomorphisms of cancellative semigroups onto cancellative semigroups with a kernel.

First we give a characterization of the kernel of a semigroup. (See Appendix.)

Theorem 10.36. *Let S be a semigroup. Then S has a kernel if and only if there exists an element in S with empty biresidue. If it exists, the kernel of S is the set of elements of S with empty biresidue.*

Proof. S has a kernel if and only if there exists an element in S which belongs to every two-sided ideal of S, i.e. if and only if there exists an element a, say, such that $a \in SxS$ for all x in S. But this is so if and only if $a \mathbin{..} x$ (where we write $a \mathbin{..} x$ for $\{a\} \mathbin{..} x$) is nonempty for all $x \in S$, i.e. if and only if a has empty biresidue.

These remarks immediately imply that the kernel, if it exists, of S is the set of elements of S which have empty biresidue.

THEOREM 10.37. *Let H be a biperfect subset of the semigroup S and suppose that H has empty biresidue in S. Then $S/\mathscr{P}_H$ is a cancellative semigroup with a kernel.*

PROOF. Let $(ad, bd) \in \mathscr{P}_H$. Since the biresidue of H is empty, there exist x, y in S such that $(x, y) \in H \mathbin{..} (ad) = H \mathbin{..} (bd)$; whence $(x, dy) \in (H \mathbin{..} a) \cap (H \mathbin{..} b)$, which, since H is bistrong implies $(a, b) \in \mathscr{P}_H$. Thus $\mathscr{P}_H$ is right cancellative. Similarly $\mathscr{P}_H$ is left cancellative. Consequently, $S/\mathscr{P}_H$ is a cancellative semigroup.

Since H is biperfect it is contained in a $\mathscr{P}_H$-class, X, say. Clearly the biresidue of X is also empty. Hence $S/\mathscr{P}_H$ contains an element with empty biresidue. From Theorem 10.36, it follows that $S/\mathscr{P}_H$ has a kernel.

LEMMA 10.38. *Let S be a semigroup and let ρ be a cancellative congruence on S. Let X be a ρ-class. Then X is bistrong in S.*

Moreover $\rho \subseteq \mathscr{P}_X$ and, if W denotes the biresidue of X in S, then ρ coincides with $\mathscr{P}_X$ in $S \backslash W$.

PROOF. Let $(a, b) \in (X \mathbin{..} p) \cap (X \mathbin{..} q)$, so that $(apb, aqb) \in \rho$. Since ρ is cancellative, therefore $(p, q) \in \rho$. Suppose that $(u, v) \in X \mathbin{..} p$. Then $upv \in X$; and $(p, q) \in \rho$ implies that $(upv, uqv) \in \rho$. Thus, since X is a ρ-class, $uqv \in X$, i.e. $(u, v) \in X \mathbin{..} q$. Consequently, $X \mathbin{..} p \subseteq X \mathbin{..} q$. By symmetry, we have $X \mathbin{..} p = X \mathbin{..} q$ and this shows that X is bistrong.

If $(a, b) \in \rho$ then, since X is a ρ-class, $paq \in X$ if and only if $pbq \in X$, i.e. $(a, b) \in \mathscr{P}_X$. If $(a, b) \in \mathscr{P}_X$ and $a, b \in S \backslash W$, then there exist p, q such that $paq \in X$ and $pbq \in X$. Then $(paq, pbq) \in \rho$ which implies, since ρ is cancellative that $(a, b) \in \rho$.

We can now prove a converse of Theorem 10.37.

THEOREM 10.39. *Let $\phi: S \to T$ be a homomorphism of the semigroup S onto the semigroup T and suppose that T is a cancellative semigroup with kernel K, say.*

Let $k \in K$ and set $H = k\phi^{-1}$. Then H is a biperfect subset of S with empty biresidue and $\mathscr{P}_H = \phi \circ \phi^{-1}$.

PROOF. By Theorem 10.36, k is an element of T with empty biresidue in T. Let $x \in S$ and set $x\phi = u$. Let $(p, q) \in k \mathbin{..} u$, so that $puq = k$. Choose a, b in S such $a\phi = p$, $b\phi = q$. This is possible because ϕ is onto T. Then it follows that $axb \in H$, i.e. that $(a, b) \in H \mathbin{..} x$. Thus $H \mathbin{..} x \neq \square$ for all x in S, i.e. H has empty biresidue.

Now $\phi \circ \phi^{-1}$ is cancellative because T is cancellative. Also H is a

$\phi\circ\phi^{-1}$-class. From the preceding lemma we conclude that H is biperfect and that $\mathscr{P}_H = \phi\circ\phi^{-1}$.

Theorems 10.37 and 10.39 combine to provide a means of constructing all homomorphisms of semigroups onto cancellative semigroups with a kernel.

We conclude this section with a theorem of Croisot's [1957].

THEOREM 10.40. *Let S be a cancellative semigroup with kernel K. Suppose that K contains an idempotent. Then $K = S$ and S is a group.*

PROOF. The theorem is readily proved directly, but we merely observe that it follows from Theorem 10.39, if we take ϕ to be the identity mapping on S and k to be the idempotent of K, and then apply Theorem 10.34.

To show that a simple cancellative semigroup is not necessarily a group it suffices to observe that the semigroup of all matrices

$$\begin{pmatrix} a & 0 \\ b & 1 \end{pmatrix},$$

where a, b are real and positive, is simple and cancellative, but contains no idempotent. This is a special case of Exercise 10(a) for §2.1. For another example, Croisot's paper [1957] may be consulted.

EXERCISES FOR §10.4

1. An ideal I of a semigroup S is said to be *inclusive* if $SaS \subseteq I$ implies $a \in I$. A semigroup S is said to be *disjunctive* if $S = S^0$ and if $\mathscr{P}_0$, the principal congruence determined by the subset $\{0\}$ of S, is the identical relation on S.

Let I be an inclusive ideal of the semigroup S. Then $S/\mathscr{P}_I$ is a disjunctive semigroup. Conversely, if $\phi: S \to T$ is a homomorphism of S onto a disjunctive semigroup T, then $I = 0\phi^{-1}$ is an inclusive ideal of S and $\phi\circ\phi^{-1} = \mathscr{P}_1$. (Pierce [1954].)

The following exercises are taken from Croisot [1957].

2. Let $\{H_i : i \in I\}$ be a family of bistrong subsets of the semigroup S. Then $\bigcap\{H_i : i \in I\}$ is a bistrong subset of S.

3. Let ρ be a congruence on the semigroup S. Let A be a ρ-class. Then $\rho \subseteq \mathscr{P}_A$.

4. Let S be a semigroup and $H \subseteq S$. Let W_H, ${}_HW$ and W be the right residue, left residue and biresidue, respectively, of H in S.

   (a) If ${}_HW \subseteq W_H$, then ${}_HW \subseteq W$.

   (b) If $H \cap S^2 \cap W_H \cap W = \square$, then $W \subseteq {}_HW$.

   (c) If H is a subsemigroup, then $W \subseteq {}_HW \cap W_H$.

5. Let S be a semigroup and let $H \subseteq S$. Let H be symmetric and strong. Then H is bistrong. Let W denote the biresidue of H. Then

$$\mathscr{R}(= \mathscr{R}_H = {}_H\mathscr{R}) \subseteq \mathscr{P}_H \quad \text{and} \quad {}_HW = W_H \subseteq W.$$

Moreover $\mathscr{R}$ and $\mathscr{P}_H$ coincide on $S\backslash W$.

6. Let S be a right simple semigroup. Then every bistrong subset of S is strong.

7. Let S be a semigroup and H a biperfect subsemigroup of S. Let U be the $\mathscr{P}_H$-class containing H. Then U is the intersection of all unitary subsemigroups of S which contain H.

8. Let H be a symmetric perfect subset of the semigroup S. Then H is biperfect and

$$_H\mathscr{R} = \mathscr{R}_H = \mathscr{P}_H, \qquad _HW = W_H = W,$$

W denoting the biresidue of H in S.

9. Let ρ be a cancellative congruence on the semigroup S. Then

$$\rho = \bigcap\{\mathscr{P}_H\colon H \text{ a } \rho\text{-class}\}.$$

10. Let G be a group. Then $H \subseteq G$ is a bistrong subset of G if and only if $xHH^{-1}x^{-1}H \subseteq H$ for all x in G. A bistrong subset of G is strong. A subgroup of G is bistrong if and only if it is normal in G.

Consequently, the bistrong subsets of a group are the cosets of normal subgroups of the group.

11. Let G be a group and let $H \subseteq G$. Then the congruence $\mathscr{P}_H$ is that determined by the greatest normal subgroup contained in $\bigcap\{Hh^{-1}\colon h \in H\}$.

12. Let S be a semigroup and let I be an ideal of S. Then I is the biresidue of some subset H of S if and only if it may be written

$$I = S^{[-1]}(AS^{[-1]})$$

for some set $A \subseteq S$. (For notation see Exercise 6 for §10.2.)

10.5 Homomorphism theorems for subsemigroups

In the treatment we present in this section of the homomorphism theorems for quotients of subsemigroups of a semigroup we follow the account of A. W. Goldie [1950]. There is little difficulty in extending the theory from semigroups to arbitrary algebras (with finitary operations); and it is for such algebras that Goldie presents his results.

Let S be a semigroup and let T_1 and T_2 be subsemigroups of S. Suppose that $T_1 \cap T_2 = P$, say, is nonempty. Let τ_i be a congruence on the semigroup T_i, $i = 1, 2$. Observe that

$$P\tau_i = \{y \in T_i\colon (p, y) \in \tau_i \quad \text{for some } p \in P\}$$

is a subsemigroup of S. If R is any subsemigroup of S, we shall write $(\tau_i)_R$ for the restriction $\tau_i \cap (R \times R)$ of τ_i to R. *We define* $\tau_1 * \tau_2$ *to be the congruence on* $P\tau_2$ *generated by the restrictions* $(\tau_1)_{P\tau_2}$ *and* $(\tau_2)_{P\tau_2}$ *of* τ_1 *and* τ_2, *respectively, to* $P\tau_2$. If $T_1 = T_2 = P$, then $\tau_1 * \tau_2$ is the join $\tau_1 \vee \tau_2$ of τ_1 and τ_2 in the lattice of congruences on P. Thus, when T_1 and T_2 are not necessarily equal,

$$(\tau_1)_P * (\tau_2)_P = (\tau_1)_P \vee (\tau_2)_P.$$

We also note that $\tau_1 * \tau_2 = (\tau_1)_P * \tau_2$.

The first part of the following lemma gives a construction for $\tau_1 * \tau_2$, and the other two parts give consequences thereof which we shall need later.

LEMMA 10.41. (i) *With the above notation,* $(a, b) \in \tau_1 * \tau_2$ *if and only if there exists a finite sequence of elements* $x_1, x_2, \cdots, x_{2m}$ *of* S *such that*

$$a\tau_2 x_1 \tau_1 x_2 \tau_2 \cdots \tau_1 x_{2m} \tau_2 b.$$

Each member x_i *of any such sequence must belong to* P.

(ii) $(\tau_1 * \tau_2)_P = (\tau_1)_P * (\tau_2)_P = (\tau_1)_P \vee (\tau_2)_P$.

(iii) *If* K *is any subsemigroup of* P,

$$K(\tau_1 * (\tau_2)_P)\tau_2 = K(\tau_1 * \tau_2).$$

REMARK. For convenience, in this section and the next, we shall frequently indicate composition of relations by juxtaposition.

PROOF. (i) Let τ denote the relation defined in the statement of the lemma. It is clear from this definition of τ that, when $(a, b) \in \tau$ then each of a and b is τ_2-equivalent to an element belonging to $T_1 \cap T_2$. Thus both a and b belong to $P\tau_2$. Consequently, τ is a binary relation on $P\tau_2$.

Let $a \in P\tau_2$. Then there exists $b \in P$ such that $(a, b) \in \tau_2$. Hence

$$a\tau_2 b \tau_1 b \tau_2 a;$$

whence $(a, a) \in \tau$, i.e. τ is reflexive. Clearly, τ is symmetric. To show that τ is transitive, consider $(a, b) \in \tau$ and $(b, c) \in \tau$, so that there exist $x_1, x_2, \cdots, x_{2m}$, say, and $y_1, y_2, \cdots, y_{2n}$, such that

$$a\tau_2 x_1 \tau_1 x_2 \cdots \tau_1 x_{2m} \tau_2 b,$$

and

$$b\tau_2 y_1 \tau_1 y_2 \cdots \tau_1 y_{2n} \tau_2 c.$$

Since $x_{2m}\tau_2 b \tau_2 y_1$, we have $x_{2m}\tau_2 y_1$; hence

$$a\tau_2 x_1 \tau_1 x_2 \cdots \tau_1 x_{2m} \tau_2 y_1 \tau_1 \cdots \tau_1 y_{2n} \tau_2 c,$$

i.e. $(a, c) \in \tau$. Thus τ is an equivalence on $P\tau_2$.

We now show that τ is a congruence. Let $(a, b) \in \tau$, and $(c, d) \in \tau$ so that

$$a\tau_2 x_1 \tau_1 x_2 \cdots \tau_1 x_{2m} \tau_2 b,$$

and

$$c\tau_2 y_1 \tau_1 y_2 \cdots \tau_1 y_{2n} \tau_2 d.$$

We may suppose that $m = n$; for, if $m < n$, say, then the first chain may be extended in length by the insertion of a suitable number of terms

$$x_{2m}\tau_2 x_{2m} \tau_1 x_{2m}.$$

Since τ_1 and τ_2 are both congruences on their respective domains, we conclude immediately

$$(ac)\tau_2(x_1 y_1)\tau_1(x_2 y_2) \cdots \tau_1(x_{2n} y_{2n})\tau_2(bd),$$

i.e. that $(ac, bd) \in \tau$.

The congruence τ clearly contains $(\tau_1)_{P\tau_2}$ and $(\tau_2)_{P\tau_2}$, and is contained in any congruence on $P\tau_2$ with this property. Hence $\tau = \tau_1 * \tau_2$.

(ii) Since, by definition, $\tau_1 * \tau_2$ contains both $(\tau_1)_{P\tau_2}$ and $(\tau_2)_{P\tau_2}$, it is clear that $(\tau_1 * \tau_2)_P$ contains both $(\tau_1)_P$ and $(\tau_2)_P$, hence also their join.

Conversely, let $(a, b) \in (\tau_1 * \tau_2)_P$. By (i),

$$a\tau_2 x_1 \tau_1 x_2 \tau_2 \cdots \tau_1 x_{2m} \tau_2 b$$

for some $x_1, x_2, \cdots, x_{2m}$ in P. Since $a, b \in P$ also, we infer that $(a, b) \in (\tau_1)_P \vee (\tau_2)_P$.

(iii) Let $a \in K(\tau_1 * \tau_2)$. Then $(k, a) \in \tau_1 * \tau_2$ for some k in K, and, by (i), there exist $x_1, \cdots, x_{2m}$ in P such that

$$k\tau_2 x_1 \tau_1 x_2 \tau_2 \cdots \tau_1 x_{2m} \tau_2 a.$$

Since k also belongs to P, we can replace each τ_2 in this, except the last, by $(\tau_2)_P$. Since

$$k(\tau_2)_P x_1 \tau_1 x_2 (\tau_2)_P \cdots \tau_1 x_{2m} (\tau_2)_P x_{2m},$$

we have $(k, x_{2m}) \in \tau_1 * (\tau_2)_P$. Hence $x_{2m} \in K(\tau_1 * (\tau_2)_P)$, and from $x_{2m}\tau_2 a$, we conclude that $a \in K(\tau_1 * (\tau_2)_P)\tau_2$.

Conversely, let $a \in K(\tau_1 * (\tau_2)_P)\tau_2$. Then $b\tau_2 a$ for some b in $K(\tau_1 * (\tau_2)_P)$, and, by (i), there exist k in K and $x_1, x_2, \cdots, x_{2m}$ in P such that

$$k(\tau_2)_P x_1 \tau_1 x_2 (\tau_2)_P \cdots \tau_1 x_{2m} (\tau_2)_P b.$$

From $x_{2m}\tau_2 b$ and $b\tau_2 a$ we have $x_{2m}\tau_2 a$. Hence

$$k\tau_2 x_1 \tau_1 x_2 \tau_2 \cdots \tau_1 x_{2m} \tau_2 a,$$

and we infer from (i) that $(k, a) \in \tau_1 * \tau_2$, or $a \in K(\tau_1 * \tau_2)$. This concludes the proof of Lemma 10.41.

Before introducing the homomorphism theorem, we shall need one more lemma.

LEMMA 10.42. *Let T be a subsemigroup of the semigroup S, and let σ be a congruence on S. Then $t\sigma_T \to t\sigma$ $(t \in T)$ is an isomorphism of T/σ_T onto $T\sigma/\sigma_{T\sigma}$.*

REMARK. The restriction $\sigma_{T\sigma}$ of σ to $T\sigma$ is the same as $\iota_T * \sigma$.

PROOF. The mapping $t\sigma_T \to t\sigma$ is single-valued and one-to-one, for $(t_1, t_2) \in \sigma_T$ if and only if $(t_1, t_2) \in \sigma$, $(t_1, t_2 \in T)$. It is evidently onto, and from

$$(t_1\sigma_T)(t_2\sigma_T) = (t_1t_2)\sigma_T \to (t_1t_2)\sigma = (t_1\sigma)(t_2\sigma),$$

it is a homomorphism.

THEOREM 10.43. *Let T be a subsemigroup of the semigroup S. Let σ be a* *congruence on S, and τ a congruence on T. Define the natural mappings α and β by*

$$(t\tau)\alpha = t(\tau * \sigma_T), \quad (t \in T),$$
$$(t(\tau * \sigma_T))\beta = t(\tau * \sigma), \quad (t \in T).$$

Then α is a homomorphism of T/τ onto $T/(\tau * \sigma_T)$, and β is an isomorphism of $T/(\tau * \sigma_T)$ onto $T\sigma/(\tau * \sigma)$.

PROOF. τ and σ_T are congruences on T, and hence $\tau * \sigma_T$ is just their join. In particular, it is a congruence on T containing τ, and hence α is a homomorphism onto.

To deal with β, we apply Lemma 10.41(ii) to the case $T_1 = T$, $T_2 = S$, $P = T_1 \cap T_2 = T$, $\tau_1 = \tau$, $\tau_2 = \sigma$. We obtain $(\tau * \sigma)_T = \tau \vee \sigma_T$. Apply Lemma 10.42, replacing S by $T\sigma$ and σ by $\tau * \sigma$. Since $T(\tau * \sigma) = T\sigma$ and $(\tau * \sigma)_{T\sigma} = \tau * \sigma$, we infer that the mapping $t(\tau * \sigma)_T \to t(\tau * \sigma)$ is an isomorphism of $T/(\tau * \sigma)_T$ onto $T\sigma/(\tau * \sigma)$. Since $(\tau * \sigma)_T = \tau \vee \sigma_T = \tau * \sigma_T$, this mapping is just β.

As a corollary, we have the following analogue of the Zassenhaus Lemma for groups.

COROLLARY 10.44. *Let T_1 and T_2 be subsemigroups of the semigroup S such that $P = T_1 \cap T_2$ is not empty. Let τ_1 and τ_2 be congruences on T_1 and T_2, respectively. Then*

$$P\tau_1/(\tau_2 * \tau_1) \cong P\tau_2/(\tau_1 * \tau_2)$$

under the natural homomorphism

$$p(\tau_2 * \tau_1) \to p(\tau_1 * \tau_2), \quad (p \in P).$$

PROOF. If in the theorem we replace S by $P\tau_1$, T by P, σ by τ_1, and τ by $(\tau_2)_P$, then the isomorphism β becomes

$$p((\tau_2)_P * (\tau_1)_P) \to p((\tau_2)_P * \tau_1), \quad (p \in P),$$

an isomorphism of $P/((\tau_2)_P * (\tau_1)_P)$ onto $P\tau_1/((\tau_2)_P * \tau_1)$. By symmetry, we also infer that the mapping

$$p((\tau_1)_P * (\tau_2)_P) \to p((\tau_1)_P * \tau_2), \quad (p \in P),$$

is an isomorphism of $P/((\tau_1)_P * (\tau_2)_P)$ onto $P\tau_2/((\tau_1)_P * \tau_2)$. As remarked earlier,

$$(\tau_1)_P * (\tau_2)_P = (\tau_1)_P \vee (\tau_2)_P = (\tau_2)_P * (\tau_1)_P,$$
$$(\tau_1)_P * \tau_2 = \tau_1 * \tau_2,$$

and

$$(\tau_2)_P * \tau_1 = \tau_2 * \tau_1.$$

Combining the two isomorphisms we have established, the assertion of the corollary follows.

Exercises for §10.5

1. Let T be a subsemigroup of the semigroup S. Let σ be the Rees-congruence, $\sigma = J \times J \cup \iota_{S \setminus J}$, on S determined by the ideal J of S. Let $\tau = \iota_T$. Then Theorem 10.43 immediately gives

$$T/(J \cap T) \cong (J \cup T)/J;$$

which is Theorem 2.36.

2. Let R and S be subsemigroups of a semigroup and let r, s be ideals of R, S, respectively. Write

$$T = r \cup (R \cap S), \quad t = r \cup (R \cap s)$$
$$U = s \cup (R \cap S), \quad u = s \cup (r \cap S).$$

Then t, u are ideals of T, U, respectively, and

$$T/t \cong U/u.$$

This result of Rees [1940] is a special case of Corollary 10.44 when $R \cap S \neq \square$, and is trivial otherwise.

10.6 Weakly permutable relations and the Jordan-Hölder Theorem*

Let S be a semigroup and let K be a subsemigroup of S. Let R and T be subsemigroups of S such that $R \cap T \supseteq K$. Let ρ be a congruence on R and let τ be a congruence on T. Then ρ and τ are said to be *weakly permutable over* K if $K\rho_T\tau_R = K\tau_R\rho_T$. (Here, as in §10.5, $\rho_T = \rho \cap (T \times T)$.) If $T = R$, so that $\rho_T = \rho$ and $\tau_R = \tau$, then if ρ and τ commute, i.e. if $\rho \circ \tau = \tau \circ \rho$, they are also weakly permutable over K. The converse does not hold. For an example, see Exercise 1, below.

In this section we shall use the concept of weak permutability, due to A. W. Goldie [1950], to establish analogues for semigroups of the Jordan-Hölder-Schreier theorems for groups. The account here is based on that of Goldie (loc. cit. for an extension to general algebras). The importance of commuting congruence relations was first emphasized by P. Dubreil and M. L. Dubreil-Jacotin [1939]. Dubreil and Dubreil-Jacotin restricted their discussion to equivalence relations on sets. Goldie extended their results in two directions, firstly the relatively straightforward extension from equivalences on sets to congruences on algebras, and secondly to bring congruences on subalgebras within the scope of the theory.

We pursue our treatment initially without introducing any weak permutability conditions of the congruences considered. The conditions will be introduced only when necessary.

---

* Throughout this section, superscripts will be reserved for use as distinguishing indices, and will not be used to indicate powers or, in the case of superscript 1, to indicate the adjunction or presence of an identity element.

A sequence of congruences ρ^i on the subsemigroups R^i of the semigroup S, $i = 1, 2, \cdots, m + 1$, will be called a *normal K-sequence of length m from S to T* if

(1) K is a subsemigroup of S;
(2) $R^i \supseteq K$, for $i = 1, 2, \cdots, m + 1$;
(3) $R^1 = S$ and $\rho^1 = S \times S$;
(4) $R^{m+1} = T$ and $\rho^{m+1} = \iota_T$;
(5) $R^{i+1} \supseteq K\rho^i \supseteq K\rho^{i+1}$, for $i = 1, 2, \cdots, m$;
(6) $K\rho^i\rho^{i+1} = K\rho^i$, for $i = 1, 2, \cdots, m$.

The normal sequence will further be said to be *reduced* if $K\rho^i = R^{i+1}$, for $i = 1, 2, \cdots, m$.

Conditions (3), (4) and (5) imply that

$$S = K\rho^1 \supseteq K\rho^2 \supseteq \cdots \supseteq K\rho^m \supseteq K\rho^{m+1} = K.$$

Condition (6) says that each $K\rho^i$, which by (5) is a subsemigroup of the domain R^{i+1} of ρ^{i+1}, is a union of ρ^{i+1}-classes.

If ρ^i, $i = 1, 2, \cdots, m + 1$, is a normal sequence from S to T, denote by $\bar{\rho}^{i+1}$ the restriction of ρ^{i+1} to $K\rho^i$, $i = 1, 2, \cdots, m$, and define $\bar{\rho}^1 = \rho^1$. Then it is easily verified that $\bar{\rho}^i$, $i = 1, 2, \cdots, m + 1$, is a reduced normal K-sequence from S to $K\rho^m$. The sequence $\bar{\rho}^i$ is said to be the reduced normal sequence associated with the sequence ρ^i.

The *factor set* of the sequence ρ^i is the set of semigroups $\{K\rho^i/\bar{\rho}^{i+1}: i = 1, 2, \cdots, m\}$. The sequence ρ^i and its associated reduced sequence $\bar{\rho}^i$ have the same factor set.

Two normal K-sequences from S to T are said to be *isomorphic* if they are of the same length and if there is a one-to-one correspondence between their factor sets such that corresponding elements are isomorphic semigroups.

A sequence of congruences $\rho^1, \cdots, \rho^{m+1}$ is said to be a *refinement* of the sequence $\sigma^1, \cdots, \sigma^{n+1}$ if $\sigma^1, \cdots, \sigma^{n+1}$ is a subsequence of $\rho^1, \cdots, \rho^{m+1}$. Note that we do not restrict the use of the term refinement to normal sequences of congruences.

Let

$$A: \rho^1, \rho^2, \cdots, \rho^{m+1},$$

and

$$B: \sigma^1, \sigma^2, \cdots, \sigma^{n+1}$$

be two reduced normal K-sequences from S to T. As before, let R^i be the subsemigroup of S on which the congruence ρ^i, $i = 1, 2, \cdots, m + 1$, is defined and also suppose that S^j is the subsemigroup of S on which the congruence σ^j is defined, $j = 1, 2, \cdots, n + 1$.

Now define

$$\rho^{j,i} = \sigma^j * \rho^{i+1}, \quad (i = 1, 2, \cdots, m; j = 1, 2, \cdots, n + 1),$$
$$\sigma^{i,j} = \rho^i * \sigma^{j+1}, \quad (i = 1, 2, \cdots, m + 1; j = 1, 2, \cdots, n).$$

The operation $*$ here is that introduced in §10.5. Then the sequences

$$C: \rho^1 \ (= \rho^{1,1}),\ \rho^{2,1}, \cdots, \rho^{n,1}, \rho^2, \rho^{2,2}, \cdots, \rho^{n,i}, \rho^{i+1}, \rho^{2,i+1}, \cdots, \rho^{m+1},$$

and

$$D: \sigma^1 \ (= \sigma^{1,1}),\ \sigma^{2,1}, \cdots, \sigma^{m,1}, \sigma^2, \sigma^{2,2}, \cdots, \sigma^{m,j}, \sigma^{j+1}, \sigma^{2,j+1}, \cdots, \sigma^{n+1},$$

are refinements of the sequences A and B respectively. The sequences C and D each contain $nm + 1$, not necessarily distinct, terms. We call C and D the *Zassenhaus refinements* of the sequences A and B.

Denote by $R^{j,i}$ the subsemigroup on which $\rho^{j,i}$ is defined and by $S^{i,j}$ the subsemigroup on which $\sigma^{i,j}$ is defined. Then, from the definition of the $*$ product, it follows that

$$R^{j+1,i} = (S^{j+1} \cap R^{i+1})\rho^{i+1},$$
$$S^{i+1,j} = (R^{i+1} \cap S^{j+1})\sigma^{j+1}.$$

Hence, by the Zassenhaus Lemma, Corollary 10.44, we have

$$R^{j+1,i}/\rho^{j+1,i} = S^{i+1,j}/\sigma^{i+1,j}$$

for $i = 1, 2, \cdots, m$ and $j = 1, 2, \cdots, n$.

Before proceeding we need a lemma on weakly permutable congruences.

LEMMA 10.45. *Let R and T be two subsemigroups of the semigroup S each of which contains the subsemigroup K. Let ρ be a congruence on R and τ be a congruence on T and suppose that ρ and τ are weakly permutable over K. Then*

$$K(\rho_T * \tau) = K(\rho * \tau) = (K\rho \cap T)\tau$$

and

$$K(\tau_R * \rho) = K(\tau * \rho) = (K\tau \cap R)\rho.$$

PROOF. Weak permutability of ρ and τ over K immediately implies that

$$K\rho_T\tau_R = K\tau_R\rho_T = K\tau_R(\rho_T\rho_T) = (K\tau_R\rho_T)\rho_T = K\rho_T\tau_R\rho_T.$$

By induction we therefore have

$$K\rho_T\tau_R = K\rho_T\tau_R\rho_T\tau_R \cdots . \tau_R\rho_T.$$

Using Lemma 10.41(i), this set of equalities implies that $K\rho_T\tau_R = K(\tau_R * \rho_T)$. Similarly, we infer $K\tau_R\rho_T = K(\rho_T * \tau_R)$. Hence, since $K\rho_T\tau_R = K\tau_R\rho_T$, we conclude

$$K\rho_T\tau_R = K(\tau_R * \rho_T) = K(\rho_T * \tau_R) = K\tau_R\rho_T.$$

From Lemma 10.41(iii), we have

$$K(\rho_T * \tau) = K(\rho * \tau) = K\rho_T\tau_R\tau = K\rho_T\tau.$$

But $K\rho_T = K\rho \cap T$; whence we have

$$K(\rho_T * \tau) = K(\rho * \tau) = (K\rho \cap T)\tau.$$

The other result asserted in the lemma now follows by symmetry.

We return to the consideration of the Zassenhaus refinements of the reduced sequences A and B. Suppose now that each congruence ρ^i in the sequence A is weakly permutable over K with each congruence σ^j in the sequence B. Then, as we shall devote the rest of this section to showing, *the Zassenhaus refinements C and D of A and B are isomorphic normal K-sequences from S to T.*

We must first show that C and D are normal K-sequences, from S to T. We shall prove this for the sequence C; a similar argument will prove it for D.

To show that C is a normal K-sequence, we have to verify that conditions (1)–(6) are satisfied. That conditions (1)–(4) are satisfied, is immediate.

Condition (5) *holds for the sequence C.*

We must show that

(a) $R^{2,i} \supseteq K\rho^i \supseteq K\rho^{2,i}$, for $i = 1, 2, \cdots, m$;

(b) $R^{i+1} \supseteq K\rho^{n,i} \supseteq K\rho^{i+1}$, for $i = 1, 2, \cdots, m$;

(c) $R^{j+1,i} \supseteq K\rho^{j,i} \supseteq K\rho^{j+1,i}$, for $i = 1, 2, \cdots, m$ and $j = 2, 3, \cdots, n-1$.

Proof of (a). By the preceding lemma,

$$\begin{aligned} K\rho^{2,i} &= K(\sigma^2 * \rho^{i+1}) \\ &= (K\sigma^2 \cap R^{i+1})\rho^{i+1}, \end{aligned}$$

since, by hypothesis, σ^2 and ρ^{i+1} are weakly permutable over K. By assumption, A is reduced; whence $R^{i+1} = K\rho^i$. Hence

$$\begin{aligned} (K\sigma^2 \cap R^{i+1})\rho^{i+1} &= (K\sigma^2 \cap K\rho^i)\rho^{i+1} \\ &\subseteq K\rho^i\rho^{i+1} = K\rho^i, \end{aligned}$$

by condition (6), which holds for the sequence A. Consequently,

$$K\rho^{2,i} \subseteq K\rho^i.$$

Moreover, as we have already seen,

$$R^{2,i} = (S^2 \cap R^{i+1})\rho^{i+1}.$$

Also, by conditions (5) and (3) on the sequence B,

$$S = K\sigma^1 \subseteq S^2$$

whence $S = S^2$; so that $S^2 \cap R^{i+1} = R^{i+1}$. Hence we have

$$R^{2,i} = (S^2 \cap R^{i+1})\rho^{i+1} = R^{i+1}\rho^{i+1} = R^{i+1}.$$

Since $K\rho^i \subseteq R^{i+1}$ (in fact $= R^{i+1}$), this completes the proof of (a). We have in fact proved slightly more, viz.,

$$R^{2,i} = K\rho^i \supseteq K\rho^{2,i}.$$

Proof of (b).

$$\begin{aligned} K\rho^{n,i} &= K(\sigma^n * \rho^{i+1}) \\ &= (K\sigma^n \cap R^{i+1})\rho^{i+1}, \end{aligned}$$

again by Lemma 10.45, since σ^n and ρ^{i+1} are, by hypothesis, weakly permutable over K. Now $K\sigma^n \supseteq K$, $R^{i+1} \supseteq K$, and so

$$K\rho^{n,i} = (K\sigma^n \cap R^{i+1})\rho^{i+1} \supseteq K\rho^{i+1}.$$

Moreover,

$$(K\sigma^n \cap R^{i+1})\rho^{i+1} \subseteq R^{i+1}\rho^{i+1} = R^{i+1};$$

and this completes the proof of (b).

PROOF OF (c).

$$K\rho^{j+1,i} = (K\sigma^{j+1} \cap R^{i+1})\rho^{i+1},$$

by Lemma 10.45, since σ^{j+1} and ρ^{i+1} are, by hypothesis, weakly permutable over K. Furthermore,

$$(K\sigma^{j+1} \cap R^{i+1})\rho^{i+1} \subseteq (K\sigma^j \cap R^{i+1})\rho^{i+1},$$

by condition (5) upon the sequence B. The above equality, for j instead of $j+1$, gives

$$(K\sigma^j \cap R^{i+1})\rho^{i+1} = K\rho^{j,i}.$$

Whence we have

$$K\rho^{j+1,i} \subseteq K\rho^{j,i}.$$

Also

$$\begin{aligned} R^{j+1,i} &= (S^{j+1} \cap R^{i+1})\rho^{i+1} \\ &= (K\sigma^j \cap R^{i+1})\rho^{i+1}, \end{aligned}$$

since B is reduced,

$$= K\rho^{j,i},$$

again by the preceding lemma. Hence we have shown that

$$R^{j+1,i} = K\rho^{j,i} \supseteq K\rho^{j+1,i};$$

and this holds for $i = 1, 2, \cdots, m$ and $j = 2, 3, \cdots, n-1$.

This completes the verification that condition (5) holds for C. Observe that we have not shown that C is reduced. However, in cases (a) and (c), we do have the equalities required for making C reduced. We will make use of these equalities later.

Condition (6) *holds for the sequence C.*

We must show that

(d) $K\rho^i\rho^{2,i} = K\rho^i$, for $i = 1, 2, \cdots, m$;
(e) $K\rho^{n,i}\rho^{i+1} = K\rho^{n,i}$, for $i = 1, 2, \cdots, m$;
(f) $K\rho^{j,i}\rho^{j+1,i} = K\rho^{j,i}$, for $i = 1, 2, \cdots, m$ and $j = 2, 3, \cdots, n-1$.

Since we proved, when dealing with condition (5) that $R^{2,i} = K\rho^i$ and $R^{j+1,i} = K\rho^{j,i}$, (d) and (f) follow immediately. (e) is an evident consequence of $K\rho^{n,i} = (K\sigma^n \cap R^{i+1})\rho^{i+1}$.

This completes the proof that C is a normal K-sequence from S to T.

We now show that C and D are isomorphic sequences. The factor set of C, making use of the equalities established in (a) and (c) above, is

$$\{R^{j+1,i}/\rho^{j+1,i}: i = 1, 2, \cdots, m; j = 1, 2, \cdots, n-1\}$$
$$\bigcup \{K\rho^{n,i}/(\rho^{i+1} \cap (K\rho^{n,i} \times K\rho^{n,i})): i = 1, 2, \cdots, m\};$$

while, similarly, the factor set of D is

$$\{S^{i+1,j}/\sigma^{i+1,j}: i = 1, 2, \cdots, m-1; j = 1, 2, \cdots, n\}$$
$$\bigcup \{K\sigma^{m,j}/(\sigma^{j+1} \cap (K\sigma^{m,j} \times K\sigma^{m,j}): j = 1, 2, \cdots, n\}.$$

Now we have already seen, using the Zassenhaus lemma, that

$$R^{j+1,i}/\rho^{j+1,i} \cong S^{i+1,j}/\sigma^{i+1,j}$$

for $i = 1, 2, \cdots, m-1$ and $j = 1, 2, \cdots, n-1$. This establishes isomorphisms between $(m-1)(n-1)$ pairs of semigroups in the factor sets of C and D. There are $m + n - 1$ further pairs to consider.

Consider the $n-1$ elements $R^{j+1,m}/\rho^{j+1,m}$, $j = 1, 2, \cdots, n-1$, of the factor set of C. We proceed to show that

$$R^{j+1,m}/\rho^{j+1,m} \cong K\sigma^{m,j}/(\sigma^{j+1} \cap (K\sigma^{m,j} \times K\sigma^{m,j})),$$

for $j = 1, 2, \cdots, n-1$.

Now $R^{j+1,m} = (S^{j+1} \cap R^{m+1})\rho^{m+1}$, whence, from condition (4) on the sequence A, $R^{j+1,m} = S^{j+1} \cap T$. Furthermore, since the sequence B is reduced, $K\sigma^j = S^{j+1}$, for $j = 1, 2, \cdots, n$, whence by condition (5) on B,

$$S^{j+1} \supseteq K\sigma^n = S^{n+1} = T,$$

for $j = 1, 2, \cdots, n$. Hence, in particular, $S^{j+1} \cap T = T$ and so $R^{j+1,m} = T$, for $j = 1, 2, \cdots, n-1$.

Furthermore,

$$\rho^{j+1,m} = \sigma^{j+1} * \rho^{m+1} = \sigma_T^{j+1} * \iota_T = \sigma_T^{j+1}.$$

By Theorem 10.43,

$$T/\sigma_T^{j+1} \cong T\sigma^{j+1}/(\iota_T * \sigma^{j+1}).$$

But $T = R^{m+1} \cap S^{j+1} = K\rho^m \cap S^{j+1}$, and so, by Lemma 10.45,

$$T\sigma^{j+1} = (K\rho^m \cap S^{j+1})\sigma^{j+1} = K\sigma^{m,j}.$$

Hence also

$$\iota_T * \sigma^{j+1} = \sigma^{j+1} \cap (T\sigma^{j+1} \times T\sigma^{j+1})$$
$$= \sigma^{j+1} \cap (K\sigma^{m,j} \times K\sigma^{m,j}).$$

Whence, finally, we have

$$R^{j+1,m}/\rho^{j+1,m} = T/\sigma_T^{j+1}$$
$$\cong K\sigma^{m,j}/(\sigma^{j+1} \cap (K\sigma^{m,j} \times K\sigma^{m,j})),$$

for $j = 1, 2, \cdots, n-1$.

By the symmetry of the situation, it follows also that

$$K\rho^{n,i}/(\rho^{i+1} \cap (K\rho^{n,i} \times K\rho^{n,i})) \cong S^{i+1,n}/\sigma^{i+1,n},$$

for $i = 1, 2, \cdots, m-1$.

To complete the proof that C is isomorphic with D it merely remains to show that the two remaining elements $K\rho^{n,m}/(\rho^{m+1} \cap (K\rho^{n,m} \times K\rho^{n,m}))$ of C and $K\sigma^{m,n}/(\sigma^{n+1} \cap (K\sigma^{m,n} \times K\sigma^{m,n}))$ of D are isomorphic. But

$$\begin{aligned} K\rho^{n,m} &= (K\sigma^n \cap R^{m+1})\rho^{m+1} \\ &= (S^{n+1} \cap R^{m+1})\rho^{m+1} \\ &= T\iota_T = T, \end{aligned}$$

and so

$$\rho^{m+1} \cap (K\rho^{n,m} \times K\rho^{n,m}) = \iota_T.$$

Thus, by symmetry, each of these two remaining elements are equal to T/ι_T; and this completes the proof.

We have proved the following theorem.

THEOREM 10.46. *Let S be a semigroup and let T and K, with $T \supseteq K$, be subsemigroups of S.*

Then any two reduced normal K-sequences from S to T, such that each congruence in one is weakly permutable over K with each congruence in the other, have isomorphic normal K-sequence refinements.

EXERCISE FOR §10.6

1. Let S be a semigroup such that there exist two congruence relations ρ and σ, say, on S which do not commute, i.e. for which $\sigma \circ \rho \neq \rho \circ \sigma$. Adjoin an (additional if $S = S^1$) identity element e, say, to S to form the semigroup S^*. Define ρ^* and σ^*, thus:

$$\begin{aligned} \rho^* &= \rho \cup \{(e, e)\} \\ \sigma^* &= \sigma \cup \{(e, e)\}. \end{aligned}$$

Then $\rho^* \circ \sigma^* \neq \sigma^* \circ \rho^*$ and ρ^* and σ^* are congruences on S^*. Let $K = \{e\}$. Then $K\rho^* \circ \sigma^* = K\sigma^* \circ \rho^* = \{e\}$. Thus ρ^* and σ^* are weakly permutable over K.

10.7 CONGRUENCES ON COMPLETELY 0-SIMPLE SEMIGROUPS

A nontrivial homomorphic image of a completely 0-simple semigroup is completely 0-simple, as was proved in Lemma 3.10. This result of L. M. Gluskin [1956], (see also G. B. Preston [1959]), when combined with Munn's theorem, Theorem 3.11 (an equivalent result was proved, independently, by Gluskin [1956]), provides the basis for the determination of congruences on a completely 0-simple semigroup. Preliminary results were obtained by

Gluskin, who gave a description of congruence classes which contain idempotents [1956]. Independent discussions by T. Tamura [1960] and G. B. Preston [1961] gave a complete description of congruences on a completely 0-simple semigroup. We follow here the account of Preston.

We make detailed use of Theorem 3.11. So we begin by restating it. In fact, we only need the special case of Theorem 3.11 for homomorphisms onto. The following theorem is easily seen to follow from applying Theorem 3.11 to this special case.

THEOREM 10.47. *Let* $S = \mathscr{M}^0(G; I, \Lambda; P)$ *be a Rees* $I \times \Lambda$ *matrix semigroup over a group with zero* G^0 *with sandwich matrix* $P = (p_{\lambda i})$. *Let* $S^* = \mathscr{M}^0(G^*; I^*, \Lambda^*; P^*)$ *be a Rees* $I^* \times \Lambda^*$ *matrix semigroup over a group with zero* $(G^*)^0$ *with sandwich matrix* $P^* = (p^*_{\lambda^* i^*})$.

Let $i \rightarrow u_i$ *and* $\lambda \rightarrow v_\lambda$ *be mappings of* I *and* Λ, *respectively, into* G^*. *Let* ϕ *and* ψ *be mappings of* I *onto* I^* *and* Λ *onto* Λ^*, *respectively. Let* ω *be a nontrivial homomorphism of* G^0 *onto* $(G^*)^0$ *such that*

$$p_{\lambda i}\omega = v_\lambda p^*_{\lambda\psi,\, i\phi} u_i \tag{1}$$

for all λ *in* Λ *and* i *in* I. *For each element* $(a; i, \lambda)$ *of* S, *define*

$$(a; i, \lambda)\theta = [u_i(a\omega)v_\lambda; i\phi, \lambda\psi], \tag{2}$$

the square bracket indicating an element of S^*. *Then* θ *is a homomorphism of* S *onto* S^*. *Conversely, if* S *is regular, then every homomorphism of* S *onto* S^* *is obtained in this fashion.*

The above theorem is apparently too special for our purpose of determining all congruences on a completely 0-simple semigroup S. The lack of generality is removed, however, by Lemma 3.10, which, as already mentioned, states that any nontrivial homomorphic image (i.e. a homomorphic image with more than one element) of a completely 0-simple semigroup is again completely 0-simple. Consequently, to within isomorphism, every nontrivial homomorphic image of S has the form of the semigroup S^* of Theorem 10.47.

It will be convenient to identify I^* with $I/\phi \circ \phi^{-1}$ and Λ^* with $\Lambda/\psi \circ \psi^{-1}$. This involves no loss of generality for I^* and Λ^* are merely indexing sets and ϕ, ψ are onto I^*, Λ^*, respectively. We shall do this and also write $i\phi = i^*$ $(i \in I)$ and $\lambda\psi = \lambda^*$ $(\lambda \in \Lambda)$; and i^* will then be regarded as the $\phi \circ \phi^{-1}$-class containing i, λ^* as the $\psi \circ \psi^{-1}$-class containing λ.

Consider now an $\mathscr{H}$-class $H^*_{i^*\lambda^*} = \{[a^*; i^*, \lambda^*]: a^* \in G^*\}$ of S^*. Its inverse image under θ is $\bigcup\{H_{j\mu}: j \in i^*, \mu \in \lambda^*\}$, where $H_{j\mu}$ denotes the $\mathscr{H}$-class of S, $H_{j\mu} = \{(a; j, \mu): a \in G\}$. Denote by Z the $I \times \Lambda$ rectangle of $\mathscr{H}$-classes $H_{i\lambda}$, $i \in I$, $\lambda \in \Lambda$. Now, by equation (1), $p^*_{\lambda^* i^*} = 0$ if and only if $p_{\mu j} = 0$, for any $j \in i^*$ and $\mu \in \lambda^*$. Thus $\{H_{j\mu}: j \in i^*, \mu \in \lambda^*\}$ consists either entirely of group $\mathscr{H}$-classes or of $\mathscr{H}$-classes none of which is a group. In the former event, we shall say that $\{H_{j\mu}: j \in i^*, \mu \in \lambda^*\}$ is a *completely simple* *subrectangle* (*with sides* i^* *and* λ^*) of Z. In the latter event, i.e. when $p^*_{\lambda^* i^*} = 0$, $\{H_{j\mu}: j \in i^*, \mu \in \lambda^*\}$ will be called a *zero subrectangle* of Z. Observe, as a justification of this terminology, that when $\{H_{j\mu}: j \in i^*, \mu \in \lambda^*\}$ is a completely simple subrectangle, then $\bigcup\{H_{j\mu}: j \in i^*, \mu \in \lambda^*\}$ is a completely simple semigroup (over G^0, with the $i^* \times \lambda^*$ submatrix of P as sandwich matrix); and that when $\{H_{j\mu}: j \in i^*, \mu \in \lambda^*\}$ is a zero subrectangle, then the product of any two elements of $\bigcup\{H_{j\mu}: j \in i^*, \lambda \in \lambda^*\}$ is zero. The subrectangles of Z just described will be called the *subrectangles of* $\theta \circ \theta^{-1}$.

A partition of Z into subrectangles is said to be *permissible* if it is induced by partitions of I and of Λ, i.e. if the sets of sides of subrectangles form partitions of I and Λ, respectively, and if also each subrectangle either consists entirely of group $\mathscr{H}$-classes (when it is called a *completely simple* partition class) or consists of $\mathscr{H}$-classes none of which is a group (when it is called a *zero* partition class). The partition of Z described in the previous paragraph is thus permissible, and will be called the partition *determined* by the congruence $\theta \circ \theta^{-1}$.

We now determine the congruence classes of $\theta \circ \theta^{-1}$. Without loss of generality, we may assume that Λ and I have a common element 1 and that H_{11} is a group. Let N be the normal subgroup of H_{11} which is the kernel of the restriction of θ to H_{11}. We shall show that there exist elements $e_i \in H_{i1}$, $i \in I$, and $f_\lambda \in H_{1\lambda}$, $\lambda \in \Lambda$, such that, except for the congruence class $\{0\}$, the $\theta \circ \theta^{-1}$-classes are the sets

$$\bigcup\{e_j N a f_\mu: j \in i^*, \mu \in \lambda^*\},$$

where $a \in H_{11}$ and i^*, λ^* are the sides of a subrectangle of the partition of Z determined by $\theta \circ \theta^{-1}$. Observe, before proceeding, that together with $\{0\}$, these classes form a disjoint cover of S.

Since ω is onto $(G^*)^0$, there exist x_i, for $i \in I$, and y_λ, for $\lambda \in \Lambda$, in G such that

$$x_i \omega = u_i$$
$$y_\lambda \omega = v_\lambda.$$

Define e_i as $(x_i^{-1} p_{11}^{-1}; i, 1)$ and f_λ as $(p_{11}^{-1} y_\lambda^{-1}; 1, \lambda)$.

Now $N \subseteq H_{11}$ and so it may be written in the form $N = (M; 1, 1)$; and if $a \in H_{11}$, we may write $a = (g; 1, 1)$. Then

$$e_j N a f_\mu = (x_j^{-1} M p_{11} g y_\mu^{-1}; j, \mu),$$

whence, by equation (2),

$$\begin{aligned}(e_j N a f_\mu)\theta &= [u_j(x_j^{-1} M p_{11} g y_\mu^{-1})\omega v_\mu; i^*, \lambda^*] \\ &= [g\omega; i^*, \lambda^*],\end{aligned}$$

since $x_j \omega = u_j$, $y_\mu \omega = v_\mu$ and, by the definition of N, $u_1(M\omega)v_1 = (p^*_{1^*1^*})^{-1}$.

Consequently, for each $a \in H_{11}$, and for each pair of sides i^*, λ^* of a subrectangle of $\theta \circ \theta^{-1}$, the set

$$\bigcup \{e_j N a f_\mu : j \in i^*, \mu \in \lambda^*\}$$

is contained in a single congruence class of $\theta \circ \theta^{-1}$.

Now it is clear that any two $\theta \circ \theta^{-1}$-congruent elements of S belong to the same subrectangle of $\theta \circ \theta^{-1}$, so that if $r\theta = s\theta$ then $r, s \in \bigcup \{H_{j\mu} : j \in i^*, \mu \in \lambda^*\}$, for some i^*, λ^*. Thus there exist $a, b \in H_{11}$ such that $r \in e_j N a f_\mu$ and $s \in e_k N b f_\kappa$ where $j, k \in i^*$ and $\mu, \kappa \in \lambda^*$; for, as already observed, the sets of the form $e_j N a f_\mu, a \in H_{11}, j \in I, \mu \in \Lambda$, form a disjoint cover of $S \backslash 0$. Let $a = (g; 1, 1)$ and $b = (h; 1, 1)$. Then, as before, $(e_j N a f_\mu)\theta = [g\omega; i^*, \lambda^*]$ and $(e_k N b f_\kappa)\theta = [h\omega; i^*, \lambda^*]$. From $r\theta = s\theta$ we have $g\omega = h\omega$; whence $a\theta = [u_1(g\omega)v_1; 1^*, 1^*] = [u_1(h\omega)v_1; 1^*, 1^*] = b\theta$, which leads to $Na = Nb$. Hence

$$r, s \in \bigcup \{e_j N a f_\mu : j \in i^*, \mu \in \lambda^*\};$$

and this completes the proof that each such set is a $\theta \circ \theta^{-1}$-congruence class.

We have associated with the congruence $\theta \circ \theta^{-1}$ a permissible partition $\mathscr{P}$, say, of Z, (the partition classes, or $\mathscr{P}$-classes, being the subrectangles of $\theta \circ \theta^{-1}$), a normal subgroup N of H_{11}, and two sets of elements of S, $\{e_i\}$, where $e_i \in H_{i1}$, $i \in I$ and $\{f_\lambda\}$, where $f_\lambda \in H_{1\lambda}$, $\lambda \in \Lambda$. And, conversely, these in turn determine $\theta \circ \theta^{-1}$. We shall write $\theta \circ \theta^{-1} = [\mathscr{P}, N, \{e_i\}, \{f_\lambda\}]$.

Suppose that $H_{i\lambda}$ belongs to a zero subrectangle of $\theta \circ \theta^{-1}$. Then $f_\lambda e_i \in H_{1\lambda} H_{i1} = 0$, since $p_{\lambda i} = 0$. Consequently, $N f_\lambda e_i = 0$, if $H_{i\lambda}$ belongs to a zero subrectangle. Suppose that $H_{i\lambda}$ and $H_{j\mu}$ belong to the same completely simple subrectangle of $\theta \circ \theta^{-1}$. Then

$$\begin{aligned}(f_\lambda e_i)\theta &= (p_{11}^{-1} y_\lambda^{-1} p_{\lambda i} x_i^{-1} p_{11}^{-1}; 1, 1)\theta \\ &= [u_1(p_{11}^{-1})\omega v_\lambda^{-1} v_\lambda p^*_{\lambda^* i^*} u_i u_i^{-1}(p_{11}^{-1})\omega v_1; 1^*, 1^*] \\ &= [u_1(p_{11}^{-1})\omega p^*_{\lambda^* i^*}(p_{11}^{-1})\omega v_1; 1^*, 1^*] \\ &= (f_\mu e_j)\theta,\end{aligned}$$

since $i^* = j^*$ and $\lambda^* = \mu^*$. Whence it follows that

$$N f_\lambda e_i = N f_\mu e_j. \tag{3}$$

We have thus shown that this equation holds whenever $H_{i\lambda}$ and $H_{j\mu}$ belong to the same subrectangle of $\theta \circ \theta^{-1}$.

Consider now the converse of the above discussion. Suppose that $\mathscr{P}$ is a permissible partition of the $I \times \Lambda$ rectangle Z of $\mathscr{H}$-classes $H_{i\lambda}$ of the completely 0-simple semigroup $S = \mathscr{M}^0(G; I, \Lambda; P)$. Suppose that $1 \in I \cap \Lambda$ and that H_{11} is a group. Let N be a normal subgroup of H_{11} and let $\{e_i\}$, $e_i \in H_{i1}$ for $i \in I$, and $\{f_\lambda\}$, $f_\lambda \in H_{1\lambda}$ for $\lambda \in \Lambda$, be two sets such that condition (3) is satisfied whenever $H_{i\lambda}$ and $H_{j\mu}$ belong to the same partition class of $\mathscr{P}$ (in fact, (3) is automatically satisfied for zero partition classes of $\mathscr{P}$). Then, as is easily seen, the set $\{0\}$ together with the sets

$$\bigcup \{e_j N a f_\mu : j \in i^*, \mu \in \lambda^*\},$$

where $a \in H_{11}$ and $i^*(\subseteq I)$, $\lambda^*(\subseteq \Lambda)$ are the sides of a partition class of $\mathscr{P}$, form a disjoint cover of S. They thus form the equivalence classes of an equivalence ρ, say, on S; and we shall write $\rho = [\mathscr{P}, N, \{e_i\}, \{f_\lambda\}]$. We shall show that ρ is in fact a congruence.

Let $(r, s) \in \rho$; so that, either $r = s = 0$, when our desired conclusion easily follows, or there exists a partition class π, say, of $\mathscr{P}$, with sides i^* and λ^*, say, and an element $a \in H_{11}$, such that $r \in e_j N a f_\mu$, $s \in e_k N a f_\kappa$ for some $j, k \in i^*$ and some $\mu, \kappa \in \lambda^*$. Let $t \in S$. If $t = 0$, then $rt = st = 0$; whence $(rt, st) \in \rho$. If $t \neq 0$, then $t \in e_l N b f_\nu$ for some b in H_{11}, where $l \in m^*$ and $\nu \in \eta^*$, say, and m^*, η^* are the sides of a $\mathscr{P}$-class. Thus

$$rt \in e_j N a f_\mu e_l N b f_\nu$$

and

$$st \in e_k N a f_\kappa e_l N b f_\nu.$$

Since $\mathscr{P}$ is permissible, m^*, λ^* are the sides of a $\mathscr{P}$-class. Hence $f_\mu e_l = 0$ if and only if $f_\kappa e_l = 0$; whence $rt = 0$ if and only if $st = 0$. When $f_\mu e_l \neq 0$, then by condition (3)

$$\begin{aligned} N a f_\mu e_l N b &= a N f_\mu e_l N b \\ &= a N f_\kappa e_l N b \\ &= N a f_\kappa e_l N b \\ &= N a f_\kappa e_l b, \end{aligned}$$

and, consequently, rt and st belong to a ρ-class. Hence, in all cases, $(rt, st) \in \rho$. The proof that ρ is left compatible follows similarly.

Observe also that there exists an element $a \in H_{11}$ such that $e_1 N a f_1 = N$; and for this element a, N is the intersection with H_{11} of the ρ-class $\bigcup \{e_i N a f_\lambda : i \in 1^*, \lambda \in 1^*\}$.

We have proved the following theorem.

THEOREM 10.48. *Let Z denote the $I \times \Lambda$ rectangle of $\mathscr{H}$-classes $H_{i\lambda}$ $(i \in I, \lambda \in \Lambda)$ containing the nonzero elements of the completely 0-simple semigroup S. Let $1 \in I \cap \Lambda$ and suppose (as we may without loss of generality) that H_{11} is a group. Let ρ be a congruence on S which is not the universal congruence and let N be the normal subgroup of H_{11} which is a congruence class of the restriction of ρ to H_{11}.*

Then there exists a permissible partition $\mathscr{P}$ of Z into subrectangles and there exist elements f_λ in $H_{1\lambda}$ $(\lambda \in \Lambda)$ and e_i in $H_{i1} (i \in I)$ with the following properties:

(3) (i) $$N f_\lambda e_i = N f_\mu e_j$$

when $H_{i\lambda}$ and $H_{j\mu}$ belong to the same $\mathscr{P}$-class;

(ii) *the sets $\bigcup \{e_j N a f_\mu : j \in i^*, \mu \in \lambda^*\}$, where $a \in H_{11}$, and i^* $(\subseteq I)$ and λ^* $(\subseteq \Lambda)$ are the sides of a subrectangle of $\mathscr{P}$, are the ρ-classes of S containing nonzero elements; i.e. ρ is the congruence $[\mathscr{P}, N, \{e_i\}, \{f_\lambda\}]$.*

Conversely, let $\mathscr{P}$ *be a permissible partition of* Z *and let* f_λ *in* $H_{1\lambda}$, e_i *in* H_{i1} *and* N, *a normal subgroup of* H_{11}, *be chosen so that* (i) *holds. Then the equivalence* $[\mathscr{P}, N, \{e_i\}, \{f_\lambda\}]$ *is a congruence on* S *such that* N *is a congruence class of its restriction to* H_{11}.

A completely simple semigroup with a zero adjoined is a completely 0-simple semigroup of a special kind viz., a completely 0-simple semigroup S such that $S\backslash 0$ is a subsemigroup or, equivalently, such that each $\mathscr{H}$-class is a group. Conversely, if S is completely 0-simple and $S\backslash 0$ is a subsemigroup, then $S\backslash 0$ is completely simple. Consequently, the above theorem can be applied to give a similar determination of the congruences on a completely simple semigroup.

We consider a special case of such congruences. Clearly, when $S\backslash 0$ is completely simple, any partition of Z, induced by partitions of I and of Λ, is permissible. In particular the partition $\mathscr{U}$, for which Z itself is the only partition class, is permissible. The congruences $[\mathscr{U}, N, \{e_i\}, \{f_\lambda\}]$ on S are those for which the quotient semigroup is a group with zero. A discussion of this situation will include, as an easy application, a discussion of group homomorphic images of completely simple semigroups. We shall derive (Theorem 10.51) R. R. Stoll's [1951] result on such homomorphisms (cf. Exercise 13 for §10.2). We first prove a general result on the ordering of congruences.

THEOREM 10.49. *Let* $\rho = [\mathscr{P}, M, \{e_i\}, \{f_\lambda\}]$ *and* $\sigma = [\mathscr{Q}, N, \{g_i\}, \{h_\lambda\}]$ *be two congruences on the completely* 0*-simple semigroup* S. *Then* $\rho \subseteq \sigma$ *if and only if*

(i) $\mathscr{P} \subseteq \mathscr{Q}$,

(ii) $M \subseteq N$,

(iii) *for each* $\mathscr{P}$*-class, with sides* i^* *and* λ^*, *say, there exist* a_{i^*} *and* b_{λ^*} *in* H_{11} *such that, for* $j \in i^*$ *and* $\mu \in \lambda^*$,

$$e_j f_1 = g_j n_j a_{i^*} h_1,$$
$$e_1 f_\mu = g_1 n_\mu b_{\lambda^*} h_\mu,$$

where n_j, $n_\mu \in N$.

PROOF. Suppose that $\rho \subseteq \sigma$. Then, clearly, $\mathscr{P} \subseteq \mathscr{Q}$ and, considering the restrictions of ρ and σ to H_{11}, $M \subseteq N$.

Consider the ρ-class $\bigcup \{e_j M f_\nu : j \in i^*, \nu \in 1^*\}$ where i^*, 1^* are the sides of the $\mathscr{P}$-class π, say, containing H_{i1}. Since $\rho \subseteq \sigma$, this is contained in a σ-class $\bigcup \{g_j N a_{i^*} h_\nu : j \in i^+, \nu \in 1^+\}$, where $a_{i^*} \in H_{11}$ and i^+, 1^+ are the sides of the $\mathscr{Q}$-class which contains π. Since $(p_{11}^{-1}; 1, 1) \in M$, $e_j f_\nu \in e_j M f_\nu$. Hence, in particular, $e_j f_1 \in g_j N a_{i^*} h_1$; and so there exists $n_j \in N$ such that $e_j f_1 = g_j n_j a_{i^*} h_1$, for $j \in i^*$. Similarly, there exists b_{λ^*} in H_{11} such that $e_1 f_\mu = g n_\mu b_{\lambda^*} h_\mu$, with $n_\mu \in N$ when $\mu \in \lambda^*$.

Assume now, conversely, that conditions (i)–(iii) are satisfied by the congruences ρ and σ. Let $\bigcup\{e_jMaf_\mu: j\in i^*, \mu\in\lambda^*\}$ be a ρ-class. Since $\mathscr{P}\subseteq\mathscr{Q}$, there exists a $\mathscr{Q}$-class, with sides i^+, λ^+, say, such that $i^*\subseteq i^+, \lambda^*\subseteq\lambda^+$. Hence, since also $M\subseteq N$,

$$\bigcup\{e_jMaf_\mu: j\in i^*, \mu\in\lambda^*\}\subseteq\bigcup\{e_jNaf_\mu: j\in i^*, \mu\in\lambda^*\},$$

$$=\bigcup\{g_jn_ja_{i^*}h_1f_1^{-1}Nae_1^{-1}g_1n_\mu b_{\lambda^*}h_\mu: j\in i^*, \mu\in\lambda^*\},$$

by condition (iii),

$$=\bigcup\{g_jNch_\mu: j\in i^*, \mu\in\lambda^*\},$$

where $c = a_{i^*}h_1f_1^{-1}ae_1^{-1}g_1b_{\lambda^*}$, since N is normal in H_{11},

$$\subseteq\bigcup\{g_jNch_\mu: j\in i^+, \mu\in\lambda^+\};$$

which latter is a σ-class. Consequently, $\rho\subseteq\sigma$; and this completes the proof of the theorem.

Conditions for the equality of two congruences, given in the form $[\mathscr{P}, M, \{e_i\}, \{f_\lambda\}]$, immediately follow: conditions (i), (ii) and (iii) of the theorem must hold symmetrically as between the two congruences. A better result is contained in the following corollary.

Corollary 10.50. *Let*

$$\rho=[\mathscr{P}, M, \{e_i\}, \{f_\lambda\}]$$

and

$$\sigma=[\mathscr{P}, M, \{g_i\}, \{h_\lambda\}].$$

Then $\rho=\sigma$ *if and only if* $\rho\subseteq\sigma$.

Proof. If $\rho\subseteq\sigma$, then by (iii) of the theorem, if i^*, λ^* are the sides of a $\mathscr{P}$-class, there exist a_{i^*}, b_{λ^*} in H_{11} and $m_j, m_\mu\in M$, such that

$$e_jf_1 = g_jm_ja_{i^*}h_1,$$
$$e_1f_\mu = g_1m_\mu b_{\lambda^*}h_\mu,$$

for $j\in i^*$ and $\mu\in\lambda^*$. Hence

$$g_jh_1 = e_jf_1h_1^{-1}a_{i^*}^{-1}m_j^{-1}h_1f_1^{-1}f_1$$
$$= e_jm_j'c_{i^*}f_1,$$

where $m_j'\in M$, and where $c_{i^*}=f_1h_1^{-1}a_{i^*}^{-1}h_1f_1^{-1}$, since M is normal in H_{11}. Similarly,

$$g_1h_\mu = e_1m_\mu'd_{\lambda^*}f_1,$$

where $m_\mu'\in M$ and $d_{\lambda^*}\in H_{11}$. By the theorem, these two sets of equations suffice to show that $\rho=\sigma$.

We now use the above theorem on ordering of congruences to prove Stoll's theorem (loc. cit.) which can be equivalently stated as follows.

Theorem 10.51. *Let T be a completely simple semigroup. Then there is a unique minimal congruence in the set of all congruences ρ on T such that T/ρ is a group.*

PROOF. We work with the completely 0-simple semigroup S obtained by adjunction of a zero to T and prove the equivalent result that there is a unique minimal congruence in the set of all congruences ρ on S such that S/ρ is a group with zero.

In our earlier notation, such congruences are congruences $[\mathscr{U}, N, \{e_i\}, \{f_\lambda\}]$, where Z is the sole subrectangle of the partition $\mathscr{U}$. By condition (i) of Theorem 10.48,

$$Nf_\lambda e_i = Nf_1 e_1,$$

for all $i \in I$, $\lambda \in \Lambda$. Thus there exist elements $n_{\lambda i} \in N$, such that

$$f_\lambda e_i n_{\lambda i} = f_1 e_1 = g^{-1},$$

say, where $g \in H_{11}$.

It follows that

$$\begin{aligned}(e_i n_{\lambda i} g f_\lambda)^2 &= e_i n_{\lambda i} g(f_\lambda e_i n_{\lambda i}) g f_\lambda \\ &= e_i n_{\lambda i} g g^{-1} g f_\lambda \\ &= e_i n_{\lambda i} g f_\lambda;\end{aligned}$$

so that $e_i n_{\lambda i} g f_\lambda = e_{i\lambda}$, say, is the idempotent of the group $H_{i\lambda}$. Then, also,

$$\begin{aligned}e_{1\lambda} e_{i1} &= e_1 n_{\lambda 1} g f_\lambda e_i n_{1i} g f_1 \\ &= e_1 n_{\lambda 1} (n_{\lambda i})^{-1} n_{1i} g f_1 \\ &\in e_1 N g f_1 = N,\end{aligned}$$

since $e_1 n_{11} g f_1 = e_{11} \in N \cap e_1 N g f_1$.

Now let E be the normal subgroup of H_{11} generated by the set $\{e_{1\lambda} e_{i1}\colon i \in I, \lambda \in \Lambda\}$. Since $E e_{1\lambda} e_{i1} = E$, $\mu = [\mathscr{U}, E, \{e_{i1}\}, \{e_{1\lambda}\}]$ is a congruence on S. In fact

$$\mu \subseteq [\mathscr{U}, N, \{e_i\}, \{f_\lambda\}];$$

for we have already seen that $E \subseteq N$, and the equations $e_{i1} = e_i n_{1i} g f_1$, $e_{1\lambda} = e_1 n_{\lambda 1} g f_\lambda$, which hold for all i and λ, suffice, by Theorem 10.49, to prove the inequality. This shows that μ is the required minimal congruence on S.

It is not difficult to show (we leave it to the reader), using Corollary 10.50, that if $[\mathscr{U}, N, \{e_i\}, \{f_\lambda\}]$ and $[\mathscr{U}, N, \{g_i\}, \{h_\lambda\}]$ are two congruences on S, where $S\backslash 0$ is completely simple, then they are equal. Furthermore, if $N \supseteq E$, then $[\mathscr{U}, N, \{e_{i1}\}, \{e_{1\lambda}\}]$ is a congruence on S. Hence the group congruences on S are in one-to-one correspondence with the normal subgroups of H_{11} which contain E. The congruence classes of $[\mathscr{U}, N, \{e_{i1}\}, \{e_{1\lambda}\}]$ are the sets $\bigcup\{e_{i1} N a e_{1\lambda}\colon i \in I, \lambda \in \Lambda\}$, together with $\{0\}$. Apart from $\{0\}$, any one of these uniquely determines a coset of N and so uniquely determines N, by its intersection with H_{11}. Consequently, within the class of congruences ρ on S such that S/ρ is a group with zero, any such congruence is uniquely determined by any one of its congruence classes other than $\{0\}$.

Returning to the completely simple case we have that, within the class of congruences ρ on the completely simple semigroup T such that T/ρ is a group, any such congruence is uniquely determined by any one of its congruence classes.

Consider again the congruence classes $\bigcup\{e_{i1}Nae_{1\lambda}: i \in I, \lambda \in \Lambda\}$ of the restriction to T of the congruence on S determined by the normal subgroup N of H_{11}. The congruence class K, say, containing all the idempotents of T is

$$K = \bigcup\{e_{i1}Ne_{1\lambda}: i \in I, \lambda \in \Lambda\}.$$

We then have, if $a \in H_{11}$,

$$\begin{aligned} KaK &= \bigcup\{e_{i1}Ne_{1\mu}ae_{j1}Ne_{1\lambda}: i, j \in I, \lambda, \mu \in \Lambda\} \\ &= \bigcup\{e_{i1}Nae_{1\lambda}: i \in I, \lambda \in \Lambda\}, \end{aligned}$$

since $e_{1\mu}a = a = ae_{j1}$ (Lemma 2.14). Thus KaK is the congruence class containing a. No special part is being played here by the $\mathscr{H}$-class H_{11}, since all the $\mathscr{H}$-classes in T are groups. Consequently, for any a in T, KaK is the congruence class containing a. Observe also that K is clearly a completely simple subsemigroup of T and that the intersection of K with any $\mathscr{H}$-class is a normal subgroup of that $\mathscr{H}$-class. In fact, as we have seen, the intersection of K with any $\mathscr{H}$-class uniquely determines K and the intersection of K with any $\mathscr{H}$-class has to satisfy the sole condition that it is a normal subgroup of that $\mathscr{H}$-class which contains all products of pairs of idempotents of T that lie in that $\mathscr{H}$-class (for the $\mathscr{H}$-class H_{11} such products are the elements $e_{1\lambda}e_{i1}$, $i \in I$, $\lambda \in \Lambda$). We have thus shown the greater part of the following (Schwarz [1962])

Theorem 10.52. *Let T be a completely simple semigroup. Then the congruences ρ on T such that T/ρ is a group are obtained in the following fashion.*

Let K be a completely simple subsemigroup of T which contains all the idempotents of T. Let H be an $\mathscr{H}$-class of T and suppose that $K \cap H$ is a normal subgroup of H which contains all products of pairs of idempotents of T which lie in H. Then the sets KaK, $a \in T$, are the congruence classes of a congruence ρ such that T/ρ is a group. Moreover, K is then the identity of T/ρ.

Proof. The proof of the theorem will be complete if we show that if K is a completely simple subsemigroup of T containing all the idempotents of T then $K \cap H$ uniquely determines K. (We already know this when K is a congruence class of a congruence ρ such that T/ρ is a group.) Reverting to our earlier notation, suppose that H is the $\mathscr{H}$-class H_{11} and let $H_{i\lambda}$ be another $\mathscr{H}$-class. Let $e_{j\mu}$ again denote the idempotent of $H_{j\mu}$, $j \in I$, $\mu \in \Lambda$. Let $N_{i\lambda} = K \cap H_{i\lambda}$. Then, since each $e_{j\mu} \in K$, $e_{i1}N_{11}e_{1\lambda} \subseteq N_{i\lambda}$ and $e_{1i}N_{i\lambda}e_{\lambda 1} \subseteq N_{11}$. Now, by Lemma 2.2, the mappings $x \to e_{i1}xe_{1\lambda}$, $x \in H_{11}$, and $y \to e_{1i}ye_{\lambda 1}$, $y \in H_{i\lambda}$, are one-to-one mappings of H_{11} onto $H_{i\lambda}$ and of $H_{i\lambda}$ onto H_{11}, respectively. Hence $e_{i1}N_{11}e_{1\lambda} = N_{i\lambda}$ and $e_{1i}N_{i\lambda}e_{\lambda 1} = N_{11}$; and this proves what is required.

An interesting comment on the above double coset decompositions of a completely simple semigroup is provided by the following result of A. H. Clifford [1963].

THEOREM 10.53. *Let S be a regular semigroup and $\phi: S \to G$ be a homomorphism of S onto a group G. Set $K = e\phi^{-1}$, where e is the identity of G.*

Then the congruence classes of $\phi \circ \phi^{-1}$ are the sets KsK, $s \in S$, if and only if K is simple.

PROOF. Assume, firstly, that K is simple. Let $s, t \in S$ be such that $s\phi = t\phi$. Let x be an inverse of t and y be an inverse of s in S. Since K contains all idempotents of S, in particular, $ys, xt \in K$. Hence

$$KsK \cdot KxK \supseteq Ks \cdot ys \cdot xt \cdot xK = KsxK = K,$$

since $sx \in K$ and K is simple. But $(KsK \cdot KxK)\phi = s\phi \cdot x\phi = e$, since $s\phi = t\phi$. Hence

$$KsK \cdot KxK \subseteq K.$$

Thus $KsK \cdot KxK = K$. Similarly, $KxK \cdot KsK = K$. In particular, it then follows (take $t = s$) that $KxK \cdot KtK = K$. Whence we have, since $K^2 = K$,

$$\begin{aligned} KsK &= Ks(KxK \cdot KtK) \\ &= (KsK \cdot KxK)KtK \\ &= K^2tK = KtK. \end{aligned}$$

Thus, in particular, if $t\phi = s\phi$, then $t \in KsK$. Since, conversely, $(KsK)\phi = s\phi$, it follows that $(s\phi)\phi^{-1} = KsK$. This proves the sufficiency of the condition.

The necessity is clear, for, if $k \in K$ and KkK is a congruence class of $\phi \circ \phi^{-1}$, then $KkK = K$. The proof of the theorem is complete.

We conclude this section with a theorem of Malcev [1952] determining the congruences on certain completely 0-simple semigroups which arise as principal factors of a full transformation semigroup $\mathscr{T}_X$.

Recall that if $\alpha \in \mathscr{T}_X$, then $|X\alpha|$ is the *rank* of α (§2.2). If $\alpha, \beta \in \mathscr{T}_X$, then

$$\text{rank } (\alpha\beta) \leqq \min (\text{rank } \alpha, \text{rank } \beta).$$

Hence, if r is any cardinal number greater than 1, the set

$$I_r = \{\alpha \in \mathscr{T}_X : \text{rank } \alpha < r\}$$

is an ideal of $\mathscr{T}_X$.

The following lemma generalizes to infinite X a remark made on p. 95 of Volume 1 for finite X.

LEMMA 10.54. *For any set X, and for any positive integer n such that $1 < n \leqq |X|$, I_{n+1}/I_n is completely 0-simple.*

PROOF. Let $\alpha, \beta \in I_{n+1} \backslash I_n$. Then $|X\alpha| = |X\beta| = n$. Thus, we can write

$$X\alpha = \{a_1, a_2, \cdots, a_n\},$$
$$X\beta = \{b_1, b_2, \cdots, b_n\}.$$

Select a_i' in $a_i\alpha^{-1}$ $(i = 1, 2, \cdots, n)$ and define γ (in $\mathscr{T}_X$) to be that mapping which, for each i, maps $b_i\beta^{-1}$ onto a_i'. Define δ as that mapping which maps each a_i onto b_i and which, further, maps $X\backslash\{a_1, a_2, \cdots, a_n\}$ onto b_1. Then, both γ and δ belong to $I_{n+1}\backslash I_n$, and $\gamma\alpha\delta = \beta$. This shows that I_{n+1}/I_n is 0-simple.

It remains to show that I_{n+1}/I_n contains a primitive idempotent. Certainly $I_{n+1}\backslash I_n$ contains idempotents. Let η, θ be two idempotents of $I_{n+1}\backslash I_n$ such that $\eta\theta = \theta\eta = \theta$. Then $X\theta = X\theta\eta \subseteq X\eta$; whence, since $|X\theta| = |X\eta|$ is finite, $X\theta = X\eta$. Hence, for any x in X, there exists a y in X such that $x\eta = y\theta$. Hence, $x\eta = y\theta = y\theta\eta = y\theta^2\eta = (y\theta)\theta\eta = x\eta\theta\eta = x\theta$, for any x in X. Thus $\eta = \theta$; and this proves that η is primitive and completes the proof of the lemma.

We now determine the congruences on I_{n+1}/I_n; and to do this we must first determine its $\mathscr{H}$-classes. The analogues of Lemmas 2.5 and 2.6 give what we require.

LEMMA 10.55. *Let $\alpha, \beta \in I_{n+1}\backslash I_n$. Then $\alpha\mathscr{L}\beta$ in I_{n+1}/I_n if and only if $X\alpha = X\beta$.*

PROOF. If $\alpha\mathscr{L}\beta$, i.e. if $\alpha = \beta$ or $\xi\alpha = \beta$ and $\eta\beta = \alpha$, for some $\xi, \eta \in I_{n+1}\backslash I_n$, then clearly $X\alpha = X\beta$. Conversely, if $X\alpha = X\beta$, then define ξ to be such that, for each y in $X\beta$, it maps all of the elements of the set $y\beta^{-1}$ upon a single element in $y\alpha^{-1}$. Then $\xi \in I_{n+1}\backslash I_n$ and $\xi\alpha = \beta$. Similarly, there exists $\eta \in I_{n+1}\backslash I_n$ such that $\eta\beta = \alpha$. Hence $\alpha\mathscr{L}\beta$.

The $\mathscr{L}$-class of I_{n+1}/I_n consisting of all $\alpha \in \mathscr{T}_X$ such that $X\alpha = Y$ (where $|Y| = n$) will be said to be *determined by* Y.

LEMMA 10.56. *Let $\alpha, \beta \in I_{n+1}\backslash I_n$. Then $\alpha\mathscr{R}\beta$ in I_{n+1}/I_n if and only if $\alpha \circ \alpha^{-1} = \beta \circ \beta^{-1}$.*

PROOF. Let $\alpha\mathscr{R}\beta$ so that either $\alpha = \beta$ or there exist ξ, η, say, in $I_{n+1}\backslash I_n$ such that $\alpha\xi = \beta$ and $\beta\eta = \alpha$ (in fact, there also exist such ξ and η when $\alpha = \beta$). In the former event, clearly $\alpha \circ \alpha^{-1} = \beta \circ \beta^{-1}$. In the latter event, $\alpha \circ \alpha^{-1} = (\beta\eta) \circ (\beta\eta)^{-1} = \beta \circ (\eta \circ \eta^{-1}) \circ \beta^{-1} \supseteq \beta \circ \beta^{-1}$; and, similarly, $\beta \circ \beta^{-1} \supseteq \alpha \circ \alpha^{-1}$. Hence $\alpha \circ \alpha^{-1} = \beta \circ \beta^{-1}$, as required.

Conversely, assume that $\alpha \circ \alpha^{-1} = \beta \circ \beta^{-1}$. Define ξ on $X\alpha$ by $x\alpha\xi = x\beta$ $(x \in X)$ and on $X\backslash X\alpha$ by $(X\backslash X\alpha)\xi = x_0\beta$, where x_0 is a fixed element of X. Now $\alpha \circ \alpha^{-1} = \beta \circ \beta^{-1}$ means that $x\alpha = y\alpha$ if and only if $x\beta = y\beta$. Hence, the restriction of ξ to $X\alpha$ is a one-to-one mapping of $X\alpha$ onto $X\beta$. Since also $X\xi \subseteq X\beta$, therefore $X\xi = X\beta$ and so $\xi \in I_{n+1}\backslash I_n$. Moreover, $\alpha\xi = \beta$. Similarly, there exists $\eta \in I_{n+1}\backslash I_n$ such that $\beta\eta = \alpha$. Thus $\alpha\mathscr{R}\beta$, as required.

The $\mathscr{R}$-class of I_{n+1}/I_n consisting of all $\alpha \in \mathscr{T}_X$ such that $\alpha \circ \alpha^{-1} = \pi$ (where $|X/\pi| = n$) will be said to be *determined by* π.

The above lemmas imply (cf. Theorem 2.9(vi)) that the $\mathscr{H}$-classes of I_{n+1}/I_n, other than $\{0\}$, can be indexed by the ordered pairs (π, Y) where π is an equivalence on X and Y is a subset of X such that $|X/\pi| = |Y| = n$. The $\mathscr{H}$-class *determined by* (π, Y) lies in the $\mathscr{R}$-class consisting of all α with $\alpha \circ \alpha^{-1} = \pi$ and in the $\mathscr{L}$-class consisting of all α with $X\alpha = Y$.

In the next lemma we shall use the following analogue of Theorem 2.10(i). *The $\mathscr{H}$-class determined by (π, Y) is a group if and only if Y contains precisely one element of each π-class.*

Lemma 10.57. (i) *Let π_1 and π_2 be two distinct equivalences on X such that $|X/\pi_1| = |X/\pi_2| = n$. Then there exists a subset Y of X such that precisely one of the $\mathscr{H}$-classes determined by (π_1, Y) and (π_2, Y), respectively, is a group.*

(ii) *Let Y_1 and Y_2 be two distinct subsets of X such that $|Y_1| = |Y_2| = n > 1$. Then there exists an equivalence π on X such that precisely one of the $\mathscr{H}$-classes determined by (π, Y_1) and (π, Y_2), respectively, is a group.*

Proof. (i) Since $\pi_1 \neq \pi_2$ there exist $a, b \in X$ such that $(a, b) \in \pi_1$ but $(a, b) \notin \pi_2$. Let Y be a subset of X containing a and b and containing precisely one element from each π_2-class. Then $|Y| = n$ and the $\mathscr{H}$-class corresponding to (π_2, Y) is a group. Since, however, $(a, b) \in \pi_1$, the $\mathscr{H}$-class corresponding to (π_1, Y) is not a group.

(ii) Since $Y_1 \neq Y_2$ and $|Y_1| = |Y_2| = n$, is finite, there exist a, b in X such that $a \in Y_1$, $a \notin Y_2$, $b \notin Y_1$, $b \in Y_2$. Since $n > 1$, there exists $c \neq b$ in Y_2. Let π be any equivalence on X which contains precisely one element of Y_1 in each π-class and which also contains b and c in a single π-class. Then the $\mathscr{H}$-class determined by (π, Y_1) is a group and that corresponding to (π, Y_2) is not a group.

This completes the proof of the lemma.

It follows from the preceding lemma that the only permissible partition of the rectangle Z of the $\mathscr{H}$-classes of nonzero elements of I_{n+1}/I_n is the identity partition. It is easy to see, from Theorem 10.49, that a congruence $[\mathscr{I}, N, \{e_i\}, \{f_\lambda\}]$, in the notation of that theorem, where $\mathscr{I}$ denotes the identity partition, is uniquely determined by N and that, indeed, e_i may be chosen as any element in H_{i1}, $i \in I$, and f_λ may be chosen as any element in $H_{1\lambda}$, $\lambda \in \Lambda$; for each partition class contains only one $\mathscr{H}$-class. Moreover, N may be chosen as any normal subgroup of the group H_{11}. We have therefore shown the following result of A. I. Malcev [1952].

Theorem 10.58. *Let X be a set and n a positive integer such that $1 < n \leqq |X|$. Then I_{n+1}/I_n is completely 0-simple and the congruences on I_{n+1}/I_n are in one-to-one correspondence with the normal subgroups of any of the group $\mathscr{H}$-classes in I_{n+1}/I_n. If N is such a normal subgroup then the congruence classes of the congruence determined by N are the sets aNb, where $a, b \in I_{n+1}/I_n$.*

Comparing Lemmas 10.55 and 10.56 with Lemmas 2.5 and 2.6 we see that the $\mathscr{H}$-classes of I_{n+1}/I_n in $I_{n+1}\backslash I_n$ coincide with $\mathscr{H}$-classes of $\mathscr{T}_X$. From Theorem 2.10(ii), it thus follows that the group $\mathscr{H}$-classes in $I_{n+1}\backslash I_n$ are isomorphic to the symmetric group $\mathscr{G}_n$ on n symbols. If, for example, $n = 4$, there are thus precisely two congruences on I_{n+1}/I_n which are distinct from the identity congruence and from the universal congruence. The identity congruence is, of course, determined by the one-element normal subgroup of $\mathscr{G}_n$, whereas the universal congruence falls outside the present description.

For further information on the lattice of congruences on a completely 0-simple semigroup we refer the reader to the paper of G. B. Preston [1965]. It is there shown if σ and ρ are two congruences on a completely 0-simple semigroup S such that there is a maximal chain of congruences on S from ρ to σ which is of finite length, then any maximal chain of congruences from ρ to σ is finite and of the same length.

Exercises for §10.7

1. Let $S = \mathscr{M}^0(G; I, \Lambda; P)$ be a completely 0-simple Rees $I \times \Lambda$ matrix semigroup for which the $\mathscr{H}$-class H_{11} is a group. Let $\rho = [\mathscr{P}, N, \{e_i\}, \{f_\lambda\}]$ be a congruence on S. Let I^* be the set of sides of $\mathscr{P}$-classes contained in I and Λ^* the set of sides of $\mathscr{P}$-classes contained in Λ. Set $G^* = H_{11}/N$ and $P^* = (p^*_{\lambda^* i^*})$, a $\Lambda^* \times I^*$ matrix, where $p^*_{\lambda^* i^*} = Nf_\mu e_j$ for $j \in i^*$, $\mu \in \lambda^*$. Then S/ρ is isomorphic to $\mathscr{M}^0(G^*; I^*, \Lambda^*; P^*)$ under the mapping which sends 0 onto 0 and each element of $e_j Na f_\mu$ onto $[Na; i^*, \lambda^*]$, where $j \in i^*$, $\mu \in \lambda^*$ and $a \in H_{11}$.

2. Let θ be a homomorphism of the completely 0-simple semigroup $S = \mathscr{M}^0(G; I, \Lambda; P)$ onto the completely 0-simple semigroup $S^* = \mathscr{M}^0(G^*; I^*, \Lambda^*; P^*)$. Suppose that θ is given as in Theorem 10.47:

$$(a; i, \lambda)\theta = [u_i(a\omega)v_\lambda; i\phi, \lambda\psi].$$

Let $x_i\omega = u_i$ and $y_\lambda\omega = v_\lambda$, and denote by α_θ the isomorphism

$$\alpha_\theta\colon (a; i, \lambda) \to (x_i a y_\lambda; i, \lambda)$$

of S onto $S_1 = \mathscr{M}^0(G; I, \Lambda; Y^{-1}PX^{-1})$, where Y^{-1} is the $\Lambda \times \Lambda$ diagonal matrix with y_λ^{-1} in the (λ, λ)-th place, and X^{-1} is the $I \times I$ diagonal matrix with x_i^{-1} in the (i, i)-th place. (Cf. Lemma 3.6.)

Set $Q = Y^{-1}PX^{-1}$, and let $S_2 = \mathscr{M}^0(G\omega; I, \Lambda; Q\omega)$. Then

$$\theta_\omega\colon (a; i, \lambda) \to (a\omega; i, \lambda)$$

is a homomorphism of S_1 onto S_2.

Let $S_3 = \mathscr{M}^0(G\omega; I^*, \Lambda; Q_3)$, where $Q_3 = (q_{\lambda i^*})$ is a $\Lambda \times I^*$ matrix with $q_{\lambda i^*} = p^*_{\lambda\psi, i^*}$. Then

$$\theta_r\colon (b; i, \lambda) \to (b; i\phi, \lambda)$$

is a homomorphism of S_2 onto S_3.

Again, let θ_l be the mapping

$$\theta_l\colon (b;\, i^*,\, \lambda) \to [b;\, i^*,\, \lambda\psi].$$

Then θ_l is a homomorphism of S_3 onto S^*.

Finally, we have $\theta = \alpha_\theta \theta_\omega \theta_r \theta_l$. (Gluskin [1956].)

3. Let X be a set and ξ any cardinal such that $1 < \xi \leqq |X|$. Let ξ' denote the successor of ξ, i.e. the least cardinal greater than ξ. Then $I_{\xi'}/I_\xi$ is 0-bisimple. (Cf. Lemma 10.54.)

4. Every congruence ρ on a completely 0-simple semigroup S is uniquely determined by the set of ρ-classes containing nonzero idempotents of S.

10.8 Congruences on a full transformation semigroup $\mathscr{T}_X$

The lattice of congruences on a full transformation semigroup $\mathscr{T}_X$ is generated, *qua* lattice, by congruences of three simple kinds. This section is devoted to an account of this result of A. I. Malcev [1952]. Malcev's theorem is given below in two parts, Theorems 10.68 and 10.72.

The first kind of congruence is a Rees congruence determined by an ideal. If I is an ideal of $\mathscr{T}_X$ then we shall denote by I^* the congruence on $\mathscr{T}_X$ whose quotient semigroup is $\mathscr{T}_X/I$.

An extension of Theorem 2.9(ii) gives a characterization of the ideals of $\mathscr{T}_X$ (Malcev [1952]). We shall use the following notation. If ξ is any cardinal, then we define

$$I_\xi = \{\alpha \in \mathscr{T}_X\colon \text{rank } \alpha < \xi\},$$

$$D_\xi = \{\alpha \in \mathscr{T}_X\colon \text{rank } \alpha = \xi\}.$$

It will be convenient, for the purposes of the present section, to denote by ξ' the successor of a cardinal number ξ, i.e. the smallest cardinal exceeding ξ. Thus

$$I_{\xi'} = I_\xi \cup D_\xi.$$

If $1 \leqq \xi \leqq |X|$ then, by Theorem 2.9(ii), $I_{\xi'}$ is the principal ideal of $\mathscr{T}_X$ generated by any element of rank ξ and, by Theorem 2.9(iii), D_ξ is the $\mathscr{D}$-class of $\mathscr{T}_X$ consisting of all elements of rank ξ.

Theorem 10.59. *Let X be a set and ξ a cardinal number such that $1 < \xi \leqq |X|'$. Then I_ξ is an ideal of $\mathscr{T}_X$. Moreover, every ideal of $\mathscr{T}_X$ is one of the ideals I_ξ, and the correspondence between the ξ and the I_ξ is one-to-one.*

Proof. That each I_ξ is an ideal of $\mathscr{T}_X$ follows immediately from the fact that for any $\alpha, \beta \in \mathscr{T}_X$, rank $(\alpha\beta) \leqq \min\{\text{rank } \alpha, \text{rank } \beta\}$.

Conversely, let I be any ideal of $\mathscr{T}_X$. Let ξ be the least cardinal exceeding the ranks of all the elements of I. Then certainly $I \subseteq I_\xi$. Conversely, if $\alpha \in I_\xi$ then, by the definition of ξ, there exists an element β, say, in I such that rank $\beta \geqq$ rank α. By Theorem 2.9(ii), α therefore belongs to the principal ideal of $\mathscr{T}_X$ generated by β. Hence $\alpha \in I$; and this shows that $I_\xi \subseteq I$. Hence $I = I_\xi$, as required.

The second kind of congruence on $\mathscr{T}_X$ is introduced and defined in the next theorem. Here, and throughout this section, ι denotes the identity congruence on $\mathscr{T}_X$. The identity element of $\mathscr{T}_X$ will be denoted by ι_X.

THEOREM 10.60. *Let n be a finite cardinal number such that $1 < n \leqq |X|$, let σ be a congruence on I_{n+1}/I_n, not the universal congruence, and let $\sigma^\dagger$ be the relation on $\mathscr{T}_X$ defined thus:*

$$\sigma^\dagger = \iota \cup [\sigma \cap (D_n \times D_n)] \cup [I_n \times I_n].$$

Then $\sigma^\dagger$ is a congruence on $\mathscr{T}_X$.

PROOF. It is quickly verified that $\sigma^\dagger$ is an equivalence relation and we need only show it is compatible.

Let $(\alpha, \beta) \in \sigma^\dagger$ and $\gamma \in \mathscr{T}_X$. If $\alpha = \beta$ or if $\alpha, \beta \in I_n$, then clearly $(\alpha\gamma, \beta\gamma) \in \sigma^\dagger$. The remaining possibility is that $(\alpha, \beta) \in \sigma \cap (D_n \times D_n)$. In this event, if $\gamma \in I_{n+1}$, $(\alpha\gamma, \beta\gamma) \in \sigma^\dagger$ because σ is a 0-restricted congruence on I_{n+1}/I_n. If $\gamma \in \mathscr{T}_X \backslash I_{n+1}$, we proceed as follows.

As noted in the paragraph preceding Theorem 10.58, the only permissible partition of the rectangle of $\mathscr{H}$-classes of nonzero elements of I_{n+1}/I_n is the identity partition. Hence $(\alpha, \beta) \in \sigma$ implies $(\alpha, \beta) \in \mathscr{H}$. Thus, in particular, by Lemma 10.56, $X\alpha = X\beta$. Hence $X\alpha\gamma = X\beta\gamma$; so that rank $(\alpha\gamma) =$ rank $(\beta\gamma)$. If both ranks are less than n, then $(\alpha\gamma, \beta\gamma) \in I_n \times I_n \subseteq \sigma^\dagger$. Otherwise, $X\alpha\gamma = X\beta\gamma = \{c_1, c_2, \cdots, c_n\}$, say. We know that $X\alpha = X\beta = \{a_1, a_2, \cdots, a_n\}$, say; and we may suppose that $a_i\gamma = c_i$, $i = 1, 2, \cdots, n$. Define δ by agreeing that $(X \backslash X\alpha)\delta = c_1$ and that $a_i\delta = c_i$, $i = 1, 2, \cdots, n$. Then $\delta \in I_{n+1}/I_n$ and so $(\alpha\delta, \beta\delta) \in \sigma$. But $\alpha\delta = \alpha\gamma$ and $\beta\delta = \beta\gamma$. Hence $(\alpha\gamma, \beta\gamma) \in \sigma \cap (D_n \times D_n) \subseteq \sigma^\dagger$.

This completes the proof that $\sigma^\dagger$ is right compatible. A similar argument shows that $\sigma^\dagger$ is left compatible and so is a congruence on $\mathscr{T}_X$.

We observe that this second kind of congruence is determined completely by the congruences on the semigroups I_{n+1}/I_n and that these latter congruences are specified in Theorem 10.58.

The third type of congruence on $\mathscr{T}_X$ is only relevant if X is infinite. If α and β are elements of $\mathscr{T}_X$, define

$$X_0 = X_0(\alpha, \beta) = \{x \in X \colon x\alpha \neq x\beta\}.$$

If $X_0 = \square$ (i.e. $\alpha = \beta$), define $\eta = 0$; if $X_0 \neq \square$, let $\eta = \max\{|X_0\alpha|, |X_0\beta|\}$. We call η the *difference rank* of the pair (α, β), and write $\eta = \mathrm{dr}(\alpha, \beta)$. If ξ is an infinite cardinal, or if $\xi = 1$, we define

$$\Delta_\xi = \{(\alpha, \beta) \in \mathscr{T}_X \times \mathscr{T}_X \colon \mathrm{dr}(\alpha, \beta) < \xi\}.$$

LEMMA 10.61. *Let ξ be an infinite cardinal. Then Δ_ξ is a congruence on $\mathscr{T}_X$.*

PROOF. Δ_ξ is clearly reflexive and symmetric. To show transitivity, consider (α, β) and (β, γ) both in Δ_ξ. Set $M = X_0(\alpha, \beta)$ and $N = X_0(\beta, \gamma)$. Then, by assumption, $|M\alpha|$, $|M\beta|$, $|N\beta|$ and $|N\gamma|$ are less than ξ. Set $Q = X_0(\alpha, \gamma)$. Then Q may be partitioned, thus:

$$\begin{aligned} Q &= \{x \in X: x\alpha \neq x\beta \quad \text{and} \quad x\alpha \neq x\gamma\} \\ &\quad \cup \{x \in X: x\alpha = x\beta \quad \text{and} \quad x\alpha \neq x\gamma\} \\ &= A \cup B, \end{aligned}$$

say, where $A \subseteq M$ and $B \subseteq N$. Moreover, $A\alpha \subseteq M\alpha$ and $B\alpha = B\beta \subseteq N\beta$. Hence

$$Q\alpha = A\alpha \cup B\alpha \subseteq M\alpha \cup N\beta;$$

whence

$$|Q\alpha| \leqq |M\alpha| + |N\beta| < \xi,$$

the latter inequality holding since ξ is infinite. Similarly, since

$$\begin{aligned} Q &= \{x \in X: x\beta \neq x\gamma \quad \text{and} \quad x\alpha \neq x\gamma\} \\ &\quad \cup \{x \in X: x\beta = x\gamma \quad \text{and} \quad x\alpha \neq x\gamma\} \\ &= C \cup D, \end{aligned}$$

say, where $C \subseteq N$ and $D \subseteq M$, we infer that $|Q\gamma| < \xi$. Thus $(\alpha, \gamma) \in \Delta_\xi$ and we have shown that Δ_ξ is an equivalence relation on $\mathscr{T}_X$.

Let $(\alpha, \beta) \in \Delta_\xi$ and $\gamma \in \mathscr{T}_X$. Set $M = X_0(\alpha, \beta)$, $R = X_0(\gamma\alpha, \gamma\beta)$ and $S = X_0(\alpha\gamma, \beta\gamma)$. Then clearly $R\gamma \subseteq M$. Hence $|R\gamma\alpha|$ and $|R\gamma\beta|$ are less than ξ, i.e. $(\gamma\alpha, \gamma\beta) \in \Delta_\xi$. Furthermore, $S \subseteq M$ and so $S\alpha\gamma \subseteq M\alpha\gamma$ and $S\beta\gamma \subseteq M\beta\gamma$. Since $|M\alpha\gamma| \leqq |M\alpha|$ and $|M\beta\gamma| \leqq |M\beta|$ therefore $|S\alpha\gamma|$ and $|S\beta\gamma|$ are less than ξ, i.e. $(\alpha\gamma, \beta\gamma) \in \Delta_\xi$. Thus we have shown that Δ_ξ is compatible and the proof of the lemma is complete.

Observe that, if X is finite, then Δ_ξ is the universal congruence on $\mathscr{T}_X$, for any infinite cardinal ξ. The congruences Δ_ξ are thus only of interest if X is infinite.

To avoid later interruption, we give at this point two technical lemmas which we shall need later.

LEMMA 10.62. (i) *If α and β are elements of $\mathscr{T}_X$ of unequal rank, and not both finite, then*

$$\text{dr}\,(\alpha, \beta) = \max\,\{\text{rank}\,\alpha, \text{rank}\,\beta\}.$$

(ii) *If ξ is an infinite cardinal, then*

$$I_\xi^* \subseteq \Delta_\xi \subseteq I_\xi^* \cup \mathscr{D}.$$

PROOF. (i) Let $\lambda = \text{rank}\,\alpha$, $\mu = \text{rank}\,\beta$, and suppose $\lambda > \mu$. Let $X\alpha = \{a_i: i \in I\}$, and let $R_i = \{x \in X: x\alpha = a_i\}$. For each i in the index set I, select an element r_i in R_i such that $r_i \in X_0 = X_0(\alpha, \beta)$ if $R_i \cap X_0 \neq \square$. Let $I_0 = \{i \in I: r_i \in X_0\}$ and $I_1 = I \backslash I_0$. Then $I_1 = \{i \in I: r_i\alpha = r_i\beta\}$. Since $a_i \in X\beta$ if $i \in I_1$, $|I_1| \leq |X\beta| = \mu < \lambda$. Since, by hypothesis, λ is infinite, $|I_0| = \lambda$. Hence

$$|X_0\alpha| = |\{a_i: i \in I_0\}| = |I_0| = \lambda,$$

and $\mathrm{dr}\,(\alpha, \beta) \geq \lambda = \max\{\mathrm{rank}\,\alpha, \mathrm{rank}\,\beta\}$. Since the opposite inequality is trivial, (i) is established.

(ii) The first inclusion is trivial. To show the second, let $(\alpha, \beta) \in \Delta_\xi \backslash \mathscr{D}$. By Theorem 2.9(iii), $\mathrm{rank}\,\alpha \neq \mathrm{rank}\,\beta$. If both ranks are finite, then $(\alpha, \beta) \in I_\xi^*$, since ξ is infinite. Otherwise, $\max\{\mathrm{rank}\,\alpha, \mathrm{rank}\,\beta\} < \xi$ by (i), whence $(\alpha, \beta) \in I_\xi^*$.

Lemma 10.63. (i) *Let η_1 and η_2 be infinite cardinals satisfying $\eta_1 \leq \eta_2$. Let α and β be elements of $\mathscr{T}_X$ such that*

$$\mathrm{rank}\,\alpha = \mathrm{rank}\,\beta = \eta_2, \qquad \mathrm{dr}\,(\alpha, \beta) = \xi \leq \eta_1.$$

Then there exists γ in $\mathscr{T}_X$ such that

$$\mathrm{rank}\,\alpha\gamma = \mathrm{rank}\,\beta\gamma = \eta_1, \qquad \mathrm{dr}\,(\alpha\gamma, \beta\gamma) = \xi.$$

(ii) *If ξ and η are cardinals such that η is infinite and $\xi \leq \eta \leq |X|$, there exist α and β in $\mathscr{T}_X$ such that*

$$\mathrm{rank}\,\alpha = \mathrm{rank}\,\beta = \eta, \qquad \mathrm{dr}\,(\alpha, \beta) = \xi.$$

Proof. (i) Let $X_0 = X_0(\alpha, \beta)$, and let $C = X_0\alpha \cup X_0\beta$. Clearly $X\alpha \backslash C = X\beta \backslash C$. Since the assertion to be proved is trivial for $\eta_1 = \eta_2$ (taking $\gamma = \iota_X$), we can assume $\eta_1 < \eta_2$. Then $|C| < \eta_2 = |X\alpha|$, so $|X\alpha \backslash C| = \eta_2$. Let γ be an element of $\mathscr{T}_X$ mapping $X\alpha \backslash C$ onto a set Y of cardinal η_1, and mapping the rest of X identically. Then

$$|X\alpha\gamma| = |Y \cup (C \cap X\alpha)| = |Y| = \eta_1,$$
$$|X\beta\gamma| = |Y \cup (C \cap X\beta)| = |Y| = \eta_1.$$

Thus the ranks of $\alpha\gamma$ and $\beta\gamma$ are both η_1, and it remains to show that their difference rank is ξ.

Clearly $X_0(\alpha\gamma, \beta\gamma) \subseteq X_0(\alpha, \beta) = X_0$. Conversely, if $x \in X_0$, then $x\alpha \neq x\beta$; and since γ is the identity on C, and both $x\alpha$ and $x\beta$ belong to C, it follows that $x\alpha\gamma \neq x\beta\gamma$. Hence $X_0(\alpha\gamma, \beta\gamma) = X_0$. Since $X_0\alpha\gamma = X_0\alpha$ and $X_0\beta\gamma = X_0\beta$, we conclude that $\mathrm{dr}\,(\alpha\gamma, \beta\gamma) = \xi$.

(ii) Let Y be a subset of X of cardinal ξ. Let δ be any permutation of X which leaves each element of $X \backslash Y$ fixed, and moves every element of Y. Then $\mathrm{dr}\,(\iota_X, \delta) = \xi$. By (i), there exists γ in $\mathscr{T}_X$ such that

$$\mathrm{rank}\,\gamma = \mathrm{rank}\,\delta\gamma = \eta, \qquad \mathrm{dr}\,(\gamma, \delta\gamma) = \xi.$$

The result follows, taking $\alpha = \gamma$ and $\beta = \delta\gamma$.

The theorem of Malcev which we are going to prove shows that any congruence on $\mathscr{T}_X$ is finitely determined by congruences of the three kinds we have now defined. The proof is long. We first establish some basic facts about congruences on $\mathscr{T}_X$.

For any x in X, denote by ζ_x the mapping of X onto x.

LEMMA 10.64. *Let ρ be a congruence on $\mathscr{T}_X$, not the identity congruence. Then all the elements of D_1 belong to a single ρ-class K_ρ, and K_ρ is an ideal of $\mathscr{T}_X$.*

PROOF. By hypothesis, there exist α, β in $\mathscr{T}_X$ such that $\alpha \neq \beta$ and $(\alpha, \beta) \in \rho$. Since $\alpha \neq \beta$, there exists c in X with $c\alpha \neq c\beta$. Let $a, b \in X$, and let γ be any transformation of X mapping $c\alpha$ onto a and $c\beta$ onto b. Then $\zeta_c\alpha\gamma = \zeta_a$ and $\zeta_c\beta\gamma = \zeta_b$. From $(\alpha, \beta) \in \rho$ we infer that $(\zeta_a, \zeta_b) \in \rho$.

This shows that all the elements of D_1 belong to a single ρ-class; call it K_ρ. Since D_1 is an ideal of $\mathscr{T}_X$, the same is true of K_ρ; for any congruence class containing an ideal is itself an ideal.

By Theorem 10.59, $K_\rho = I_\eta$ for some cardinal η. We call η the *primary cardinal* of ρ, and denote it by $\eta(\rho)$. We define $\eta(\iota) = 1$.

Thus $(\alpha, \beta) \in \rho$ if the ranks of α and β are both less than $\eta(\rho)$. The next theorem gives a partial converse: *if $(\alpha, \beta) \in \rho$, and* rank $\alpha \neq$ rank β, *then the ranks of α and β are both less than $\eta(\rho)$.*

THEOREM 10.65. *Let ρ be a congruence on $\mathscr{T}_X$, not the identity congruence, and let $\eta(\rho)$ be the primary cardinal of ρ. Then*

$$I^*_{\eta(\rho)} \subseteq \rho \subseteq I^*_{\eta(\rho)} \cup \mathscr{D}.$$

PROOF. That $I^*_{\eta(\rho)} \subseteq \rho$ follows from Lemma 10.64 and the definition of $\eta(\rho)$. Let $(\alpha, \beta) \in \rho$, and let $\eta =$ rank $\alpha >$ rank β. To establish $\rho \subseteq I^*_{\eta(\rho)} \cup \mathscr{D}$, it suffices to prove that $\eta < \eta(\rho)$. We break the proof into two cases, according to whether η is infinite or finite.

(i) *η is infinite.* Since $|X\alpha| > |X\beta|$, $|X\alpha \backslash X\beta| = |X\alpha|$. Let γ be any transformation of X which maps $X\beta$ onto a single element c, say, and which maps $X\alpha \backslash X\beta$ onto $X\alpha$. Then rank $(\alpha\gamma) =$ rank α and $\beta\gamma = \zeta_c$. Also $(\alpha, \beta) \in \rho$ implies $(\alpha\gamma, \zeta_c) \in \rho$. Hence $\alpha\gamma \in K_\rho = I_{\eta(\rho)}$. Since rank $(\alpha\gamma) = \eta$, this proves that $\eta < \eta(\rho)$.

(ii) *$\eta = r$, is finite.* Let $|X\beta| = s$, so that $s < r$. If $X\alpha \cap X\beta = \square$, let γ be the transformation of X which maps $X\beta$ onto the element c, say, and which maps $X \backslash X\beta$ identically. Then $\alpha\gamma = \alpha$ and $\beta\gamma = \zeta_c$; whence $(\alpha, \zeta_c) \in \rho$, $\alpha \in K_\rho = I_{\eta(\rho)}$, and again $\eta < \eta(\rho)$.

Otherwise, suppose that $C = X\alpha \cap X\beta = \{c_1, c_2, \cdots, c_t\}$, where $0 < t \leqq s < r$. Let γ_0 map $X \backslash X\alpha$ onto c_1 and leave all other elements of X fixed. Then $\alpha\gamma_0 = \alpha$ and $X\beta\gamma_0 = C$. For $i = 1, 2, \cdots, t$, let γ_i be the transformation mapping c_i onto c_1 and leaving all other elements of X fixed. Set $\alpha\gamma_0\gamma_1 \cdots \gamma_i = \alpha_i$ and $\beta\gamma_0\gamma_1 \cdots \gamma_i = \beta_i$. Then $(\alpha, \beta) \in \rho$ implies $(\alpha_i, \beta_i) \in \rho$ for $i = 0, 1, \cdots, t$.

Now β_0 is of rank t, α_0 is of rank r, and for $i = 1, 2, \cdots, t$, β_i is of rank $t + 1 - i$ and α_i is of rank $r + 1 - i$. Thus β_t is of rank 1 and so $(\alpha_t, \beta_t) \in \rho$ implies $\alpha_t \in K_\rho$. Since $r > t$, rank $\alpha_t = r + 1 - t > 1$. Hence $\eta(\rho) > 2$; whence, since β_{t-1} is of rank 2, $\beta_{t-1} \in K_\rho$. From $(\alpha_{t-1}, \beta_{t-1}) \in \rho$ it follows that $\alpha_{t-1} \in K_\rho$. In turn we deduce that $\beta_{t-2}, \alpha_{t-2}, \beta_{t-3}, \cdots, \alpha_1$ all belong to K_ρ. Since α_1 is of rank r, $r < \eta(\rho)$.

We remark here that if $\rho = \Delta_\xi$, then $\eta(\rho) = \xi$. This is immediate from Lemma 10.62(i) and the definition of $\eta(\rho)$. Part (ii) of that lemma is thus a consequence of Theorem 10.65.

The discussion now takes one of two alternative paths according to whether or not $\eta(\rho)$ is finite. When $\eta(\rho)$ is finite we can quickly determine the various possibilities open for ρ; and this we now proceed to do.

LEMMA 10.66. *Let ρ be a congruence on $\mathscr{T}_X$ and suppose that $\eta(\rho)$ is finite. Let α be an element of finite rank greater than or equal to $\eta(\rho)$. Let $(\alpha, \beta) \in \rho$. Then $(\alpha, \beta) \in \mathscr{H}$.*

PROOF. If $\rho = \iota$, then $\alpha = \beta$ and $(\alpha, \beta) \in \mathscr{H}$. Suppose that $\rho \neq \iota$. From the preceding theorem, rank β = rank α; hence $|X\alpha| = |X\beta| = r$, say. Since $\eta(\rho) > 1$, $r > 1$. Hence, if $X\alpha \neq X\beta$, we may choose $c \in X\beta \backslash X\alpha$ and define γ as a transformation of X which maps c into $X\beta \backslash \{c\}$ and which leaves all other elements of X fixed. Then $\alpha\gamma = \alpha$; whence $(\alpha, \beta\gamma) \in \rho$. But α is of rank r and $\beta\gamma$ is of rank $r - 1$, so by Theorem 10.65, $\eta(\rho) > r$. This conflicts with our hypotheses; hence $X\alpha = X\beta$.

Suppose that $\alpha \circ \alpha^{-1} \neq \beta \circ \beta^{-1}$; then there exists $(a, b) \in \alpha \circ \alpha^{-1} \backslash \beta \circ \beta^{-1}$. Choose B as a cross section of $\beta \circ \beta^{-1}$ and such that $a, b \in B$. Let γ map $X\beta = X\alpha$ onto B in such a fashion that each element of $X\beta$ is mapped onto that element of B which β maps onto it, and suppose that γ leaves all other elements of X fixed. Then $X\beta\gamma = B$ and, for x in B, $x\beta\gamma = x$. Hence $(\beta\gamma)^2 = \beta\gamma$ is an idempotent. On the other hand, while $X\alpha\gamma = B$, $X(\alpha\gamma)^2 \subset B$, since $a, b \in B$ and $a\alpha = b\alpha$. Thus rank $(\alpha\gamma)^2 <$ rank $(\alpha\gamma)$. Now $((\alpha\gamma)^2, (\beta\gamma)^2) = ((\alpha\gamma)^2, \beta\gamma) \in \rho$ because $(\alpha, \beta) \in \rho$; and since $(\alpha\gamma)^2$ and $\beta\gamma$ are of different rank, we conclude from Theorem 10.65, that $\eta(\rho) > \mathrm{r} = \max\{\text{rank } (\alpha\gamma)^2, \text{rank } (\beta\gamma)\}$. From this contradiction we infer that $\alpha \circ \alpha^{-1} = \beta \circ \beta^{-1}$.

The lemma now follows immediately from Lemmas 2.6 and 2.7.

LEMMA 10.67. *Let ρ be a congruence on $\mathscr{T}_X$ and suppose that $\eta(\rho)$ is finite. Let α be an element of finite rank $r \geqq \eta(\rho)$. If there exists β such that $(\alpha, \beta) \in \rho$ and $\alpha \neq \beta$, then $r = \eta(\rho)$.*

PROOF. By Lemma 10.66, $(\alpha, \beta) \in \mathscr{H}$; whence $\alpha \circ \alpha^{-1} = \beta \circ \beta^{-1}$ and $X\alpha = X\beta$. Let $M_1, M_2, \cdots, M_r$ be the equivalence classes of $\alpha \circ \alpha^{-1}$ and let $M_i\alpha = a_i$, say, $i = 1, 2, \cdots, r$. Then $M_i\beta = a_{i\sigma}$, $i = 1, 2, \cdots, r$, where σ is a permutation of $\{1, 2, \cdots, r\}$. Since $\alpha \neq \beta$ there exists p such that $p \neq p_\sigma$; and, since $r > 1$, there exists an integer k, say, with $k \neq p$ and $1 \leqq k \leqq r$.

Now define the transformation γ of X, thus:

$$\gamma: \begin{cases} M_k \cup M_p \to m_k, \\ \qquad M_j \to m_j, \end{cases}$$

if $j \neq k$, p, where $m_i \in M_i$, for $i \neq p$. Then

$$X\gamma\alpha = \{a_i: i \neq p\},$$
$$X\gamma\beta = \{a_{i\sigma}: i \neq p\}.$$

Since $p \neq p\sigma$, $X\gamma\alpha \neq X\gamma\beta$. Hence $(\gamma\alpha, \gamma\beta) \notin \mathscr{H}$ and so, by Lemma 10.66, the rank $r - 1$ of $\gamma\alpha$ is less than $\eta(\rho)$. It immediately follows that $r = \eta(\rho)$.

THEOREM 10.68. *Let ρ be a congruence on $\mathscr{T}_X$ and suppose that $\eta(\rho)$ is finite. Then if ρ is neither the identity congruence nor the universal congruence on $\mathscr{T}_X$, $\rho = \sigma^\dagger$, where σ is a congruence on $I_{n+1}|I_n$, where $n = \eta(\rho)$ and where $\sigma^\dagger$ is defined as in Theorem* 10.60.

PROOF. If $\eta(\rho) = 1$, ρ is the identity congruence and if $\eta(\rho) = |X|'$, ρ is the universal congruence. If $n = \eta(\rho)$ is neither 1 nor $|X|'$, I_n is a proper ideal of $\mathscr{T}_X$ equal to the ρ-class K_ρ. By Lemma 10.67, if α is of finite rank greater than n, then $(\beta, \alpha) \in \rho$ implies $\beta = \alpha$. Suppose that α is of infinite rank and that $(\alpha, \beta) \in \rho$. By Theorem 10.65, rank α = rank β. Suppose, if possible, that $\alpha \neq \beta$. Then there exists $c \in X$ such that $c\alpha \neq c\beta$ and we may define a transformation γ of X as follows. Set $c\alpha = a$ and $c\beta = b$ and let γ map $(X \backslash X\alpha) \cup \{a, b\}$ identically, and map $X\alpha \backslash \{a, b\}$ onto a finite set of more than n elements. Then $c\alpha\gamma = a$ and $c\beta\gamma = b$; whence $\alpha\gamma \neq \beta\gamma$. Moreover rank $(\alpha\gamma)$ is finite and greater than n and $(\alpha\gamma, \beta\gamma) \in \rho$. This conflicts with what we have already shown. Consequently, ρ reduces to the identity relation for all elements of rank greater than n.

The restriction of ρ to I_{n+1} is a congruence on I_{n+1} and, since I_n is a ρ-class, ρ induces a congruence σ, say, on I_{n+1}/I_n. It is clear now that ρ coincides with $\sigma^\dagger$.

The above theorem gives a complete description of the possible congruences ρ on $\mathscr{T}_X$ for which $\eta(\rho)$ is finite. In particular, if X is finite, then the theorem gives a description of all congruences on $\mathscr{T}_X$. Making use of the fact that the congruences on a finite symmetric group form a chain, the lattice of congruences on $\mathscr{T}_X$ is then easily seen to form a chain.

We now turn to the case where $\eta(\rho)$ is infinite. The key to the situation is contained in the following theorem.

THEOREM 10.69. *Let ρ be a congruence on $\mathscr{T}_X$. Suppose there are elements α and β of infinite rank η and difference rank ξ such that $(\alpha, \beta) \in \rho$. Then*

(i) *if ξ is infinite,*

$$(I_{\eta'} \times I_{\eta'}) \cap \Delta_{\xi'} \subseteq \rho;$$

(ii) *if ξ is finite and nonzero,*

$$(I_{\eta'} \times I_{\eta'}) \cap \Delta_{\aleph_0} \subseteq \rho.$$

We remark that (i) holds if ξ is finite, (ii) being a stronger statement than (i) in this case.

Before embarking on the lengthy proof of this theorem, we show how the second part of Malcev's theorem follows from it.

For each cardinal number λ satisfying $\eta(\rho) \leqq \lambda \leqq |X|$, define λ^* to be the smallest cardinal exceeding every cardinal ξ for which there exist α, β in $\mathscr{T}_X$ such that $(\alpha, \beta) \in \rho$,

$$\text{rank } \alpha = \text{rank } \beta = \lambda, \qquad \text{and} \qquad \text{dr } (\alpha, \beta) = \xi.$$

Lemma 10.70. *Let* λ *and* μ *be cardinal numbers in the interval* $[\eta(\rho), |X|]$. *Then* (i) $\lambda^* \leqq \eta(\rho)$, *and* (ii) $\lambda < \mu$ *implies* $\mu^* \leqq \lambda^*$.

Proof. (i) Suppose, by way of contradiction, that $\lambda^* > \eta(\rho)$. By definition of λ^*, there must exist a pair of elements α, β of $\mathscr{T}_X$ of rank λ such that $(\alpha, \beta) \in \rho$ and $\xi = \text{dr } (\alpha, \beta) \geqq \eta(\rho)$. By Theorem 10.69, ρ contains every pair in $I_{\lambda'} \times I_{\lambda'}$ of difference rank $\leqq \xi$. Since $\eta(\rho) \leqq \xi$, it follows that every pair of elements of $I_{\eta(\rho)'}$ belongs to ρ, contrary to the definition of $\eta(\rho)$.

(ii) Suppose, by way of contradiction, that $\lambda < \mu$ and $\lambda^* < \mu^*$. By definition of μ^*, there exist α and β in $\mathscr{T}_X$ of rank μ such that $(\alpha, \beta) \in \rho$ and $\lambda^* \leqq \xi = \text{dr } (\alpha, \beta) < \mu^*$. By part (i), $\mu^* \leqq \eta(\rho)$, and so $\xi < \eta(\rho) \leqq \lambda < \mu$. Applying Lemma 10.63(i) with $\eta_1 = \lambda$ and $\eta_2 = \mu$, there exists γ in $\mathscr{T}_X$ such that

$$\text{rank } \alpha\gamma = \text{rank } \beta\gamma = \lambda, \quad \text{and} \quad \text{dr } (\alpha\gamma, \beta\gamma) = \xi.$$

Clearly, $(\alpha\gamma, \beta\gamma) \in \rho$, which implies $\xi < \lambda^*$, contrary to $\lambda^* \leqq \xi$.

The foregoing lemma shows that $\lambda \to \lambda^*$ is a monotone non-increasing mapping of the interval $[\eta(\rho), |X|]$ into the interval $[1, \eta(\rho)]$. Since the cardinals are well-ordered, the range of this mapping must be finite: otherwise it would contain an infinite ascending sequence $\lambda_1^* < \lambda_2^* < \cdots$, which would imply the existence of an infinite descending sequence $\lambda_1 > \lambda_2 > \cdots$ of cardinals. Denote the range of the mapping $\lambda \to \lambda^*$ by $\{\xi_1, \xi_2, \cdots, \xi_k\}$, with $\xi_1 > \xi_2 > \cdots > \xi_k$. For each i $(= 1, \cdots, k)$, let η_i be the least cardinal such that $\eta_i^* = \xi_i$. It will be convenient to define $\eta_{k+1} = |X|'$, the least cardinal exceeding $|X|$. We then have

$$\xi_k < \xi_{k-1} < \cdots < \xi_2 < \xi_1 \leqq \eta(\rho) = \eta_1 < \eta_2 < \cdots < \eta_k < \eta_{k+1} = |X|',$$

and $\{\xi_k, \cdots, \xi_1, \eta_1, \cdots, \eta_k\}$ will be called the *sequence of cardinals of* ρ. We remark also that all the ξ_i are infinite, with the possible exception of ξ_k; and if ξ_k is finite, then $\xi_k = 1$. For, if $1 < \xi_i = r < \aleph_0$, then there exist two ρ-equivalent elements each of rank η_i and of nonzero difference rank $r - 1$. From Theorem 10.69(ii), it then follows that any two elements of rank η_i with finite difference rank are ρ-equivalent. This is contrary to the assumption that ξ_i is finite.

LEMMA 10.71. *For every* λ *in the interval* $[\eta(\rho), |X|]$,

$$\Delta_{\lambda^*} \cap (D_\lambda \times D_\lambda) \subseteq \rho.$$

Hence, if $\eta_i \leqq \eta < \eta_{i+1}$ $(1 \leqq i \leqq k)$, *then*

$$\Delta_{\xi_i} \cap (D_\eta \times D_\eta) \subseteq \rho.$$

PROOF. Let α and β be elements of $\mathscr{T}_X$ of rank λ such that $\xi = \mathrm{dr}\,(\alpha, \beta) < \lambda^*$. By definition of λ^*, there exist γ and δ in $\mathscr{T}_X$ of rank λ such that $(\gamma, \delta) \in \rho$ and $\mathrm{dr}\,(\gamma, \delta) \geq \xi$. By Theorem 10.69, $(\alpha, \beta) \in \rho$.

From the definition of η_i,

$$\eta_i \leqq \eta < \eta_{i+1} \quad \text{implies} \quad \eta^* = \xi_i, \quad (1 \leqq i \leqq k).$$

The second assertion of the lemma is now immediate from the first.

We come now to the second part of Malcev's theorem.

THEOREM 10.72. *Let* X *be an infinite set, and let* k *be a natural number. Let* ξ_i, η_i $(i = 1, 2, \cdots, k)$ *be* $2k$ *cardinal numbers satisfying*:

(i) $\xi_k < \xi_{k-1} < \cdots < \xi_1 \leqq \eta_1 < \eta_2 < \cdots < \eta_k \leqq |X|$;

(ii) *all the* ξ_i *and* η_i *are infinite except possibly* ξ_k, *and if* ξ_k *is finite then* $\xi_k = 1$.

Define a relation τ *on* $\mathscr{T}_X$ *by*

$$\tau = I^*_{\eta_1} \cup (\Delta_{\xi_1} \cap I^*_{\eta_2}) \cup \cdots \cup (\Delta_{\xi_{k-1}} \cap I^*_{\eta_k}) \cup \Delta_{\xi_k}.$$

Then τ *is a congruence on* $\mathscr{T}_X$, *and* (i) *is its sequence of cardinals.*

Conversely, if ρ *is a congruence on* $\mathscr{T}_X$, *not the universal congruence, such that* $\eta(\rho)$ *is infinite, and if* (i) *is its sequence of cardinals (with* $\eta_1 = \eta(\rho)$*), then* $\rho = \tau$, *as defined above.*

PROOF. For convenience, define $\xi_0 = |X|'$ as well as $\eta_{k+1} = |X|'$. Then Δ_{ξ_0} and $I^*_{\eta_{k+1}}$ are each the universal congruence on $\mathscr{T}_X$, and we may write

$$\tau = \bigcup_{i=0}^{k} (\Delta_{\xi_i} \cap I^*_{\eta_{i+1}}).$$

Clearly τ is reflexive and symmetric. To show transitivity, let $(\alpha, \beta) \in \tau$ and $(\beta, \gamma) \in \tau$. Then $(\alpha, \beta) \in \Delta_{\xi_i} \cap I^*_{\eta_{i+1}}$ and $(\beta, \gamma) \in \Delta_{\xi_j} \cap I^*_{\eta_{j+1}}$ for some i and j. We may assume that $\alpha \neq \beta$ and $\beta \neq \gamma$, and that $i \leqq j$.

If α and β have unequal rank, then ξ_i is not finite; for only ξ_1 can be finite, and when finite, $\xi_1 = 1$, and $\Delta_{\xi_1} = \iota$. Hence, if α and β are both of finite, but different rank, both ranks are less than ξ_i. If α and β have different rank and not both ranks are finite, then by Lemma 10.62(i), both ranks are less than ξ_i. Hence, in all cases, both ranks are less than η_1. Similarly, if β and γ have unequal rank, then their ranks are less than η_1. In this case we have $(\alpha, \gamma) \in I^*_{\eta_1} \subseteq \tau$. The same conclusion evidently holds if β and γ have equal rank, since rank $\beta < \eta_1$.

We may therefore assume that the ranks of α, β, γ are all the same, say η. Hence rank $\gamma =$ rank $\alpha < \eta_{i+1}$, so $(\alpha, \gamma) \in I^*_{\eta_{i+1}}$. From $i \leqq j$ we have $\xi_i \geq \xi_j$, and $\Delta_{\xi_j} \subseteq \Delta_{\xi_i}$. Hence $(\beta, \gamma) \in \Delta_{\xi_i}$. Since Δ_{ξ_i} is a congruence (Lemma 10.61), $(\alpha, \gamma) \in \Delta_{\xi_i}$. Hence $(\alpha, \gamma) \in \Delta_{\xi_i} \cap I^*_{\eta_{i+1}} \subseteq \tau$.

This shows that τ is an equivalence relation on $\mathscr{T}_X$. But τ is a union of compatible relations, namely the congruences $\Delta_{\xi_i} \cap I^*_{\eta_{i+1}}$, and so is itself compatible. Hence τ is a congruence on $\mathscr{T}_X$.

Since $I^*_{\eta_1} \subseteq \tau$, we know that $\eta(\tau) \geqq \eta_1$. To show that $\eta_1 = \eta(\tau)$, let $\alpha \in K_\tau$. Then $(\alpha, \zeta_a) \in \tau$ for some ζ_a in D_1 ($x\zeta_a = a$ for all x in X). We are to show that $\eta =$ rank $\alpha < \eta_1$. Assume the contrary, $\eta \geq \eta_1$. Then $(\alpha, \zeta_a) \in \Delta_{\xi_i} \cap I^*_{\eta_{i+1}}$ for some i $(1 \leqq i \leqq k)$, and we may assume that i is chosen so that $\eta_i \leqq \eta < \eta_{i+1}$. Let $X_0 = X_0(\alpha, \zeta_a)$. Then $|X_0\alpha| < \xi_i \leqq \eta_i \leqq \eta$, so that $|X\alpha \backslash X_0\alpha| = \eta$. But $y \in X\alpha \backslash X_0\alpha$ implies $y = x\alpha$ with $x \notin X_0$, so $y = x\alpha = x\zeta_a = a$, which would imply $\eta = 1$, contrary to $\eta \geq \eta_1$ and the hypothesis that η_1 is infinite.

We show next that if $\eta_i \leqq \eta < \eta_{i+1}$ $(1 \leqq i \leqq k)$, then $\eta^* = \xi_i$. Suppose $(\alpha, \beta) \in \tau$, and rank $\alpha =$ rank $\beta = \eta$. Then $(\alpha, \beta) \in I^*_{\eta_{i+1}}$ and, from the definition of τ, for some j, $0 \leqq j \leqq k$, $(\alpha, \beta) \in \Delta_{\xi_j} \cap I^*_{\eta_{j+1}}$. Since $(\alpha, \beta) \notin I^*_{\eta_i}$ and $\Delta_{\xi_i} \supseteq \Delta_{\xi_j}$ and $I^*_{\eta_{i+1}} \subseteq I^*_{\eta_{j+1}}$ if $i \leqq j$, it follows that $(\alpha, \beta) \in \Delta_{\xi_i} \cap I^*_{\eta_{i+1}}$. Thus dr $(\alpha, \beta) < \xi_i$. Let ξ be any cardinal less than ξ_i. By Lemma 10.63(ii), there exist γ and δ in $\mathscr{T}_X$ such that

$$\text{rank } \gamma = \text{rank } \delta = \eta, \qquad \text{dr } (\gamma, \delta) = \xi,$$

and so $(\gamma, \delta) \in \Delta_{\xi_i} \cap I^*_{\eta_i+1}$. Hence ξ_i is the least cardinal exceeding dr (α, β) for all (α, β) in $\tau \cap (D_\eta \times D_\eta)$; that is $\xi_i = \eta^*$.

The foregoing also shows that η_i is the least of all the cardinals η for which $\eta^* = \xi_i$, which concludes the proof that (i) is the sequence of cardinals of τ.

Turning to the converse, let ρ be a congruence on $\mathscr{T}_X$ with $\eta_1 = \eta(\rho)$ infinite, and assume that ρ is not the universal congruence, so that $\eta_1 \leqq |X|$. Let (i) be the sequence of cardinals of ρ, and again let $\eta_{k+1} = |X|'$. We are to show that $\rho = \tau$, as defined in the theorem.

We show first that $\rho \subseteq \tau$. Let $(\alpha, \beta) \in \rho$. By Theorem 10.65, either $(\alpha, \beta) \in I^*_{\eta_1} \subseteq \tau$, or rank $\alpha =$ rank $\beta = \eta$, say, where $\eta \geq \eta_1$. In the latter event, $\eta_i \leqq \eta < \eta_{i+1}$ for some i $(1 \leqq i \leqq k)$. By definition of η^*, dr $(\alpha, \beta) < \eta^*$. But $\eta^* = \xi_i$. Hence $(\alpha, \beta) \in \Delta_{\xi_i} \cap I^*_{\eta_{i+1}} \subseteq \tau$.

Conversely, let $(\alpha, \beta) \in \tau$. If $(\alpha, \beta) \in I^*_{\eta_1}$, then, since $\eta_1 = \eta(\rho)$, $(\alpha, \beta) \in \rho$. Otherwise $(\alpha, \beta) \in \Delta_{\xi_i} \cap I^*_{\eta_{i+1}}$ for some i $(1 \leqq i \leqq k)$. Now the ranks of α and β are not both finite, since $(\alpha, \beta) \notin I^*_{\eta_1}$ and η_1 is infinite. Assume they are not equal, say rank $\alpha >$ rank β. By Lemma 10.62(i), $\xi =$ dr $(\alpha, \beta) =$ rank α. Hence rank $\alpha < \xi_i \leq \eta_1$, which is impossible. Hence rank $\alpha =$ rank $\beta = \eta$, say. We can assume i chosen so that $\eta_i \leqq \eta < \eta_{i+1}$. But then $(\alpha, \beta) \in \Delta_\xi \cap (D_\eta \times D_\eta) \subseteq \rho$, by Lemma 10.71.

This concludes the proof that Theorem 10.72 follows from Theorem 10.69, and we now turn to the proof of the latter. Malcev's method was to reduce the result, stage by stage, until its proof depended on the determination of the normal subgroups of the infinite symmetric groups, by Schreier and Ulam [1933], for the countable case, and by Baer [1934] in general. The first stage of this reduction is followed in our proof of Theorem 10.69, which comes after Lemma 10.76 below. We have replaced the remaining stages by a different method, not depending upon the results of Schreier, Ulam and Baer, and we have preferred to do this by means of a sequence of preliminary lemmas, rather than by a reduction process. The reason for this replacement is that we never succeeded in evolving a satisfactory presentation of Malcev's method.

In the authors' original treatment, each of the two parts of Theorem 10.69 required a sequence of three tedious lemmas. The authors are indebted to R. S. Buckdale for providing a lemma which led to the present proof of Lemma 10.73. The discovery of this proof greatly shortened the proof of part (i) of Theorem 10.69. A corresponding simplification of the sequence (Lemmas 10.74, 10.75, and 10.76) pertaining to part (ii) of Theorem 10.69 has unfortunately not yet been found.

Lemma 10.73. *Let X be an infinite set, and let ρ be a congruence on $\mathscr{T}_X$. If ρ contains a pair (α, β) such that* dr $(\alpha, \beta) = \xi$, *say, where ξ is infinite, then $I_{\xi'} \times I_{\xi'} \subseteq \rho$ (that is, $\xi < \eta(\rho)$).*

Proof. Let $X_0 = X_0(\alpha, \beta)$. Then dr $(\alpha, \beta) = \max\{|X_0\alpha|, |X_0\beta|\} = \xi$. There will be no loss of generality in assuming that $|X_0\beta| = \xi$.

Let $\mathscr{F}$ be the family of all subsets Y of X such that $Y\alpha \cap Y\beta = \square$. $\mathscr{F}$ is nonempty; for it contains each set $\{y\}$, where $y \in X_0$. Let $\mathscr{C}$ be a chain contained in $\mathscr{F}$. Let $Z = \bigcup\{Y\colon Y \in \mathscr{C}\}$. Then, it is easily verified that $Z \in \mathscr{F}$; so applying Zorn's lemma we infer that $\mathscr{F}$ contains maximal elements.

Let M be a maximal element of $\mathscr{F}$. Suppose that $|M\alpha|$ or $|M\beta|$, say $|M\beta|$, is greater than or equal to ξ. Let γ be any element of $\mathscr{T}_X$ which maps X onto M and let δ be an element of $\mathscr{T}_X$ which maps $X\backslash M\alpha$ identically and which maps $M\alpha$ onto a, where $a \in X$. Then $\gamma\alpha\delta = \zeta_a$, the constant mapping of X onto a, and, since $M\alpha \cap M\beta = \square$, $X\gamma\beta\delta = M\beta\delta = M\beta$, so that rank $(\gamma\beta\delta) \geqq \xi$. Now $(\alpha, \beta) \in \rho$ implies $(\gamma\alpha\delta, \gamma\beta\delta) \in \rho$, i.e. $(\zeta_a, \gamma\beta) \in \rho$. Hence $\gamma\beta$, of rank $\geqq \xi$, belongs to K_ρ (Lemma 10.64); whence, $\xi < \eta(\rho)$.

If $|M\alpha|$ and $|M\beta|$ are both less than ξ, set $B = (M\alpha \cup M\beta)\beta^{-1} \cap X_0$ and $A = X_0\backslash B$. We shall show that $A \in \mathscr{F}$ and that $|A\beta| = \xi$. We may then argue for the set A exactly as for the set M in the preceding paragraph; whence it will follow again that $\xi < \eta(\rho)$, and the proof of the lemma will be complete.

Observe first that if $x \in X_0$ then either $x\alpha \in M\alpha \cup M\beta$ or $x\beta \in M\alpha \cup M\beta$; for since M is a maximal element of $\mathscr{F}$, if $x \notin M$, then $(M\alpha \cup x\alpha) \cap (M\beta \cup x\beta) \neq$

$\square$. Hence, since $x\alpha \neq x\beta$, either $x\alpha \in M\beta$ or $x\beta \in M\alpha$. Consequently, since, by the definition of A, $x \in A$ implies $x\beta \notin M\alpha \cup M\beta$, we have $A\alpha \subseteq M\alpha \cup M\beta$, and $A\alpha \cap A\beta = \square$. Thus $A \in \mathscr{F}$. Moreover, $B\beta = M\alpha \cup M\beta$ and so $|B\beta| = |M\alpha \cup M\beta| < \xi$, and $|X_0\beta| = |(A \cup B)\beta| = |A\beta \cup B\beta| = \xi$. Hence $|A\beta| = \xi$; and the proof of the lemma is complete.

The next sequence of lemmas leads to a proof of part (ii) of Theorem 10.69.

LEMMA 10.74. *Let X be a set of cardinal $\aleph_0$, let ρ be a congruence on $\mathscr{T}_X$, and let β be an element of $\mathscr{T}_X$ such that $(\iota_X, \beta) \in \rho$ and $\beta \neq \iota_X$, where ι_X is the identity element of $\mathscr{T}_X$. Assume further that β moves only a finite number of elements of X. Then ρ contains every pair (ι_X, β') such that β' is an element of $\mathscr{T}_X$ moving only a finite number of elements of X.*

PROOF. We may assume that β has infinite rank, since otherwise ρ is the universal congruence, by Theorem 10.65. Since β moves only a finite number of elements of X, the set F of fixed points of β is infinite. Since $\beta \neq \iota_X$, there exists x_0 in X such that $x_0\beta \neq x_0$.

Now let y_1 and y_2 be any two distinct elements of X. Let γ be an element of $\mathscr{T}_X$ mapping y_1 onto x_0, and mapping $X\backslash y_1$ in one-to-one fashion onto $F\backslash x_0\beta$. Define an element δ of $\mathscr{T}_X$ as follows: let δ act as the inverse of γ on $(F\backslash x_0\beta) \cup x_0$, and let δ map all other elements of X onto y_2. Thus

$$\begin{aligned} x\delta &= x\gamma^{-1} && \text{if } x \in F \quad \text{and} \quad x \neq x_0\beta, \\ &= y_1 && \text{if } x = x_0, \\ &= y_2 && \text{otherwise.} \end{aligned}$$

Then $\gamma\delta = \iota_X$, $y_1\gamma\beta\delta = x_0\beta\delta = y_2$; and, for x in $X\backslash y_1$, $x\gamma\beta\delta = x\gamma\delta = x$, since $x\gamma \in F$, and β is the identity on F. Hence ρ contains $(\iota_X, \gamma\beta\delta)$, where $\gamma\beta\delta$ maps y_1 onto y_2 and leaves all other elements of X fixed. Denote this transformation by

$$\begin{pmatrix} y_1 \\ y_2 \end{pmatrix}.$$

Since y_1 and y_2 were arbitrary distinct elements of X, ρ contains all (ι_X, τ) where

$$\tau = \begin{pmatrix} y_1 \\ y_2 \end{pmatrix}.$$

Now the set S of all α in $\mathscr{T}_X$ such that $(\iota_X, \alpha) \in \rho$ is a subsemigroup of $\mathscr{T}_X$, and we are to show that S contains every element of $\mathscr{T}_X$ which moves only a finite number of elements of X.

We note that if $\alpha \in S$, and if $\lambda\mu = \iota_X$, then $\lambda\alpha\mu \in S$; for $(\iota_X, \alpha) \in \rho$ implies $(\iota_X, \lambda\alpha\mu) = (\lambda\iota_X\mu, \lambda\alpha\mu) \in \rho$.

Let x_1, x_2 be two arbitrary, distinct elements of X. We show that S contains the transposition (x_1x_2). Let y_1, y_2, y_3, y_4 be four elements of X

distinct from each other and from x_1, x_2. Let the remaining elements of X be $z_1, z_2, z_3, \cdots$. Let

$$\lambda = \begin{pmatrix} x_1 & x_2 & y_1 & y_2 & y_3 & y_4 & z_1 & z_2 & z_3 & \cdots \\ y_1 & y_2 & z_1 & z_2 & z_3 & z_4 & z_5 & z_6 & z_7 & \cdots \end{pmatrix},$$

$$\mu = \begin{pmatrix} x_1 & x_2 & y_1 & y_2 & y_3 & y_4 & z_1 & z_2 & z_3 & z_4 & z_5 & z_6 & z_7 & \cdots \\ z_1 & z_2 & x_1 & x_2 & x_2 & x_1 & y_1 & y_2 & y_3 & y_4 & z_1 & z_2 & z_3 & \cdots \end{pmatrix}.$$

Then $\lambda\mu = \iota_X$ and

$$\lambda \begin{pmatrix} y_1 & y_2 \\ y_3 & y_4 \end{pmatrix} \mu = (x_1 x_2).$$

Since S contains

$$\begin{pmatrix} y_1 \\ y_3 \end{pmatrix} \begin{pmatrix} y_2 \\ y_4 \end{pmatrix} = \begin{pmatrix} y_1 & y_2 \\ y_3 & y_4 \end{pmatrix},$$

it contains $(x_1 x_2)$.

We proceed to prove that S contains every element α of $\mathscr{T}_X$ which moves exactly n elements of X (n finite) by induction on n. We have shown its truth for $n = 1$. Assume its truth for $n - 1$, and let

$$\alpha = \begin{pmatrix} x_1 & x_2 & \cdots & x_n \\ x_1' & x_2' & \cdots & x_n' \end{pmatrix}.$$

Here the x_i in the top row are all distinct, and α leaves fixed every other element of X.

Suppose that some element in the top row, say x_1, does not occur in the bottom row. Then

$$\alpha = \alpha' \begin{pmatrix} x_1 \\ x_1' \end{pmatrix} \quad \text{with} \quad \alpha' = \begin{pmatrix} x_2 & \cdots & x_n \\ x_2' & \cdots & x_n' \end{pmatrix}.$$

By the inductive hypothesis, $\alpha' \in S$. Since S contains

$$\begin{pmatrix} x_1 \\ x_1' \end{pmatrix},$$

it contains α.

Suppose that some element in the bottom row, say x_1', does not occur in the top row. Then from

$$\alpha = \begin{pmatrix} x_1 \\ x_1' \end{pmatrix} \begin{pmatrix} x_2 & \cdots & x_n \\ x_2' & \cdots & x_n' \end{pmatrix}$$

we again conclude by the hypothesis for induction that $\alpha \in S$.

Hence we may assume that α induces a permutation of the set $\{x_1, x_2, \cdots, x_n\}$. Since any such is a product of transpositions, and S contains all of these, we again conclude that $\alpha \in S$.

This concludes the proof of the lemma.

LEMMA 10.75. *Let* $|X| = \aleph_0$, *and let* ρ *be a congruence on* $\mathscr{T}_X$. *Suppose* *there is a pair of ρ-equivalent elements α, β, of finite nonzero difference rank, and each of rank $\aleph_0$. Then $\Delta_{\aleph_0} \subseteq \rho$.*

PROOF. By Theorem 10.68, since $\alpha \neq \beta$ and α and β are of infinite rank, $\eta(\rho) = \aleph_0$. Hence, to prove the lemma, it suffices to consider only elements of $\mathscr{T}_X$ of infinite rank.

Let α, β satisfy the hypotheses stated. Let $X\alpha = \{a_i: i \in I\}$ and let $R_i = a_i\alpha^{-1}$. Assume the notation chosen so that R_1 contains an element r_1 of $X_0 = X_0(\alpha, \beta)$. Choose r_i in R_i for each $i \in I$, with r_1 as already chosen. Let γ be a one-to-one mapping of X onto $\{r_i: i \in I\}$. Denote by x_i the element of X mapped onto r_i by γ. Thus $X = \{x_i: i \in I\}$ and $x_i\gamma = r_i$. Define δ by: $a_i\delta = x_i$ and, if $X \backslash X\alpha \neq \square$, $(X \backslash X\alpha)\delta = x_2$, where 2 denotes an element of I different from 1.

Then $x_i\gamma\alpha\delta = r_i\alpha\delta = a_i\delta = x_i$, so $\gamma\alpha\delta = \iota_X$. We proceed to show that $\gamma\beta\delta \neq \iota_X$. Now $x_1\gamma\beta\delta = r_1\beta\delta$. Since $r_1 \in X_0$, $r_1\beta \neq r_1\alpha = a_1$. If $r_1\beta \in X\alpha$, then $r_1\beta = a_i$ for some i in $I \backslash 1$. Then $r_1\beta\delta = a_i\delta = x_i \neq x_1$. If $r_1\beta \notin X\alpha$, then $r_1\beta\delta = x_2 \neq x_1$. In both cases, $x_1\gamma\beta\delta \neq x_1$, whence $\gamma\beta\delta \neq \iota_X$.

Since $\gamma\alpha\delta = \iota_X$, we have from $(\alpha, \beta) \in \rho$, $(\iota_X, \gamma\beta\delta) \in \rho$. If $\gamma\beta\delta$ moves an infinite number of elements of X, then we can apply Lemma 10.73 to the pair $(\iota_X, \gamma\beta\delta)$. We conclude that ρ is the universal congruence on $\mathscr{T}_X$, and the result of the lemma is evident. Hence we can assume that $\gamma\beta\delta$ moves only a finite number of elements of X. We can therefore apply Lemma 10.74 to the pair $(\iota_X, \gamma\beta\delta)$, and conclude that ρ contains all pairs (ι_X, λ), where λ is any element of $\mathscr{T}_X$ moving only a finite number of elements. Consequently, ρ contains all pairs (λ, μ), where λ and μ move only a finite number of elements of X.

Now let (α', β') be an arbitrary pair of elements of $\mathscr{T}_X$ of infinite rank and of finite difference rank. We proceed to show that $(\alpha', \beta') \in \rho$, which will complete the proof of the lemma.

Let $X_0 = X_0(\alpha', \beta')$. By hypothesis, $D = X_0\alpha' \cup X_0\beta'$ is finite, say $D = \{c_1, c_2, \cdots, c_n\}$, while $C = X\alpha' \cup X\beta'$ is infinite. Since $D \subseteq C$ we may write $C = \{c_1, \cdots, c_n, c_{n+1}, \cdots\}$.

Let $R_{ij} = \{x \in X: x\alpha' = c_i, x\beta' = c_j\}$. If $x \in R_{ij}$ with $i \neq j$, then $x \in X_0$, so $x\alpha'$ and $x\beta'$ belong to D, whence $i \leqq n$ and $j \leqq n$. Hence $R_{ij} = \square$ if $i \neq j$ and either $i > n$ or $j > n$.

From each nonempty R_{ij} $(i, j \leqq n)$, select a representative r_{ij}, and let Q be the set of all these r_{ij}. Let $Y = X \backslash (D \cup Q)$. Since D and Q are finite, Y is infinite. Let the elements of Y be denoted as follows: $Y = \{y_{n+1}, y_{n+2}, \cdots\}$ We now define elements γ', δ', λ', μ' of $\mathscr{T}_X$ as follows:

$$\begin{cases} R_{ij}\gamma' = r_{ij} & (i, j \leqq n,\ R_{ij} \neq \square), \\ R_{ii}\gamma' = y_i & (i > n). \end{cases}$$

$$\begin{cases} x\delta' = x & (x \in D \cup Q), \\ y_i\delta' = c_i & (i > n). \end{cases}$$

$$\begin{cases} r_{ij}\lambda' = c_i & (r_{ij} \in Q), \\ x\lambda' = x & (x \in X \backslash Q). \end{cases}$$

$$\begin{cases} r_{ij}\mu' = c_j & (r_{ij} \in Q), \\ x\mu' = x & (x \in X \backslash Q). \end{cases}$$

Then

$$\begin{aligned} R_{ij}\gamma'\lambda'\delta' &= r_{ij}\lambda'\delta' = c_i\delta' = c_i = R_{ij}\alpha' && (i, j \leqq n), \\ R_{ii}\gamma'\lambda'\delta' &= y_i\lambda'\delta' = y_i\delta' = c_i = R_{ii}\alpha' && (i > n); \\ R_{ij}\gamma'\mu'\delta' &= r_{ij}\mu'\delta' = c_j\delta' = c_j = R_{ij}\beta' && (i, j \leqq n), \\ R_{ii}\gamma'\mu'\delta' &= y_i\mu'\delta' = y_i\delta' = c_i = R_{ii}\beta' && (i > n). \end{aligned}$$

Hence $\gamma'\lambda'\delta' = \alpha'$ and $\gamma'\mu'\delta' = \beta'$. But λ' and μ' move only a finite number of elements of X, and hence $(\lambda', \mu') \in \rho$. Since ρ is a congruence on $\mathscr{T}_X$, $(\gamma'\lambda'\delta', \gamma'\mu'\delta') = (\alpha', \beta') \in \rho$; which completes the proof of the lemma.

In the sequel we shall use the notation

$$\alpha = \begin{pmatrix} P_i \\ a_i \end{pmatrix}$$

to mean that α is an element of $\mathscr{T}_X$ with range $X\alpha = \{a_i\}$, and that $P_i = a_i\alpha^{-1}$. In the interests of economy, we shall not specify the index set over which i ranges. The appropriate index sets will always be clear from the context. Furthermore, we shall not reserve to a subscript, such as i, a unique meaning. In any particular use, the meaning will be unique; but in varying context, the same index may indicate varying index sets.

Lemma 10.76. *Let X be an infinite set and let ρ be a congruence on $\mathscr{T}_X$. Suppose that ρ contains a pair (α, β) of finite nonzero difference rank and such that* rank α = rank $\beta = \aleph_0$. *Then*

$$\Delta_{\aleph_0} \cap (I_{\aleph_0'} \times I_{\aleph_0'}) \subseteq \rho.$$

Proof. By Theorem 10.68, $\eta(\rho)$ is not finite. So to prove the lemma it suffices to show that, if rank α' = rank $\beta' = \aleph_0$ and $(\alpha', \beta') \in \Delta_{\aleph_0'}$ then $(\alpha', \beta') \in \rho$.

Let α, β satisfy the hypotheses of the lemma. Since dr $(\alpha, \beta) \neq 0$, there is an element $x_0 \in X$ such that $x_0\alpha \neq x_0\beta$. Let $x_0\alpha = a$ and $x_0\beta = b$. Then $x_0 \in X_0(\alpha, \beta)$ and $a, b \in C = X_0\alpha \cup X_0\beta$. Let $C = A \cup B$, with $A \cap B = \square$, and $a \in A, b \in B$. By assumption C is finite; so $F = X\alpha \backslash C = X\beta \backslash C$ is countably infinite, and we may write $F = \{f_j : j = 1, 2, \cdots\}$.

Let δ be a transformation of X which maps A onto f_1, B onto f_2 and leaves all other elements of X fixed. Then $x_0\alpha\delta = a\delta = f_1$, and $x_0\beta\delta = b\delta = f_2$. Thus $\alpha\delta \neq \beta\delta$. Clearly dr $(\alpha\delta, \beta\delta) \leqq 2$, and we may write $\alpha\delta$ and $\beta\delta$ in the forms,

$$\alpha\delta = \begin{pmatrix} K_1 & K_2 & K_3 & \cdots \\ f_1 & f_2 & f_3 & \cdots \end{pmatrix},$$

$$\beta\delta = \begin{pmatrix} L_1 & L_2 & K_3 & \cdots \\ f_1 & f_2 & f_3 & \cdots \end{pmatrix}.$$

Since $x_0\alpha\delta = f_1$, $x_0 \in K_1$; and similarly $x_0 \in L_2$. It is possible that $K_2 = \square$ or $L_1 = \square$ (for example both are empty if $X_0 = \{x_0\}$). The sets K_j, for $j > 2$, are all nonempty.

For each $j \neq 2$, choose an element $k_j \in K_j$, with, in particular, $k_1 = x_0$. If $K_2 \neq \square$, choose $k_2 \in K_2$. Set

$$\gamma = \begin{pmatrix} K_j \\ k_j \end{pmatrix},$$

where the subscript $j = 2$ is omitted if $K_2 = \square$. Then $\gamma\alpha\delta = \alpha\delta$ and

$$\gamma\beta\delta = \begin{pmatrix} K_1 & K_2 & K_3 & \cdots \\ f_2 & f_n & f_3 & \cdots \end{pmatrix}$$

because $k_1 = x_0 \in L_2$. Here, if $K_2 \neq \square$, $n = 1$ or 2; if $K_2 = \square$, the column

$$\begin{pmatrix} K_2 \\ f_n \end{pmatrix}$$

may be omitted.

If $K_2 \neq \square$ define ϵ to be an element of $\mathcal{T}_X$ which maps f_j onto k_j for each j. If $K_2 = \square$, ϵ is taken to be an element of $\mathcal{T}_X$ which maps f_j onto k_j for $j \neq 2$ and which maps f_2 onto k_3. In the former case

$$\gamma\alpha\delta\epsilon = \begin{pmatrix} K_1 & K_2 & K_3 & \cdots \\ k_1 & k_2 & k_3 & \cdots \end{pmatrix},$$

$$\gamma\beta\delta\epsilon = \begin{pmatrix} K_1 & K_2 & K_3 & \cdots \\ k_2 & k_n & k_3 & \cdots \end{pmatrix};$$

while in the latter case

$$\gamma\alpha\delta\epsilon = \begin{pmatrix} K_1 & K_3 & K_4 & \cdots \\ k_1 & k_3 & k_4 & \cdots \end{pmatrix},$$

$$\gamma\beta\delta\epsilon = \begin{pmatrix} K_1 & K_3 & K_4 & \cdots \\ k_3 & k_3 & k_4 & \cdots \end{pmatrix}.$$

In both cases $\gamma\alpha\delta\epsilon$ and $\gamma\beta\delta\epsilon$ are ρ-equivalent and have nonzero finite difference rank.

Let $Y = \{k_j\}$ (with k_2 missing if $K_2 = \square$) and denote by $\mathcal{T}_Y^*$ the subset of $\mathcal{T}_X$ consisting of those elements of $\mathcal{T}_X$ which induce a mapping of $\{K_j\}$ (with K_2 missing if $K_2 = \square$) into Y. Then $\mathcal{T}_Y^*$ is naturally isomorphic to $\mathcal{T}_Y$ under the mapping

$$\begin{pmatrix} K_j \\ k_{j\sigma} \end{pmatrix} \to \begin{pmatrix} k_j \\ k_{j\sigma} \end{pmatrix},$$

where $\sigma: j \to j\sigma$ is an arbitrary transformation of the set $\{1, 2, 3, \cdots\}$, or of the set $\{1, 3, 4, \cdots\}$ in the case $K_2 = \square$.

Let ρ_Y be the congruence on $\mathscr{T}_Y$ corresponding, under this isomorphism, to the congruence $\rho \cap (\mathscr{T}_Y^* \times \mathscr{T}_Y^*)$ on $\mathscr{T}_Y^*$. The images of $\gamma\alpha\delta\epsilon$ and $\gamma\beta\delta\epsilon$ are two ρ_Y-equivalent elements. They are each of rank $|Y| = \aleph_0$ and have nonzero finite difference rank. By Lemma 10.75, therefore any two elements of $\mathscr{T}_Y$ of rank $\aleph_0$ and of finite difference rank are ρ_Y-equivalent; whence any two elements of $\mathscr{T}_Y^*$ of rank $\aleph_0$ and of finite difference rank are ρ-equivalent.

It will be convenient now to write $Y = \{y_j : j = 1, 2, \cdots\}$ and to denote the set $y_j\gamma^{-1}$ by Y_j. This avoids the necessity of dealing with the two cases $K_2 = \square$ and $K_2 \neq \square$, separately.

Now let α' and β' be any two elements of $\mathscr{T}_X$ of rank $\aleph_0$ and finite difference rank. Let

$$\alpha' = \begin{pmatrix} A_i \\ r_i \end{pmatrix}, \quad \beta' = \begin{pmatrix} B_i \\ s_i \end{pmatrix}.$$

The set of nonempty intersections $A_i \cap B_j$ is countably infinite, and so can be put into one-to-one correspondence with the set Y. Denote by y_{ij} the element y_n of Y corresponding to $A_i \cap B_j$, and let $Y_{ij} = Y_n$ if $y_{ij} = y_n$. Pick d_{ij} in $A_i \cap B_j$. Let

$$\theta = \begin{pmatrix} Y_{ij} \\ d_{ij} \end{pmatrix}, \quad \theta' = \begin{pmatrix} A_i \cap B_j \\ y_{ij} \end{pmatrix}.$$

Then

$$\theta'\theta = \begin{pmatrix} A_i \cap B_j \\ d_{ij} \end{pmatrix},$$

and, since $d_{ij}\alpha' = r_i$ and $d_{ij}\beta' = s_j$,

$$\theta'\theta\alpha' = \begin{pmatrix} A_i \cap B_j \\ r_i \end{pmatrix} = \alpha',$$

and

$$\theta'\theta\beta' = \begin{pmatrix} A_i \cap B_j \\ s_j \end{pmatrix} = \beta'.$$

Now

$$\theta\alpha' = \begin{pmatrix} Y_{ij} \\ r_i \end{pmatrix} \quad \text{and} \quad \theta\beta' = \begin{pmatrix} Y_{ij} \\ s_j \end{pmatrix}$$

have the same range as α' and β', respectively, and so have rank $\aleph_0$, and have finite difference rank, since in general

$$\text{dr}\,(\theta\alpha', \theta\beta') \leqq \text{dr}\,(\alpha', \beta').$$

We may assume that $|X| > \aleph_0$, since otherwise Lemma 10.76 reduces to Lemma 10.75. Hence there exists a permutation π, say, of X mapping $X\alpha' \cup X\beta'$ onto Y. Then $\theta\alpha'\pi$ and $\theta\beta'\pi$ belong to $\mathscr{T}_Y^*$, and have rank $\aleph_0$ and finite difference rank, whence $(\theta\alpha'\pi, \theta\beta'\pi) \in \rho$. Applying θ' on the left and π^{-1} on the right, we conclude that $(\alpha', \beta') \in \rho$.

This concludes the proof of the lemma.

We now turn to the proof of Theorem 10.69.

Proof of Theorem 10.69. By hypothesis, there are elements α and β of infinite rank η, of difference rank ξ, and such that $(\alpha, \beta) \in \rho$.

Let $X_0 = X_0(\alpha, \beta)$, and let $C = X_0\alpha \cup X_0\beta$. Then $X\alpha \backslash C = X\beta \backslash C = D$, say. Let $C = \{c_i\}$, $D = \{d_j\}$, $M_i = c_i\alpha^{-1}$, $N_i = c_i\beta^{-1}$, and $R_j = d_j\alpha^{-1} = d_j\beta^{-1}$. Then we may write, following the convention introduced before Lemma 10.76,

$$\alpha = \begin{pmatrix} M_i & R_j \\ c_i & d_j \end{pmatrix}, \tag{1}$$

$$\beta = \begin{pmatrix} N_i & R_j \\ c_i & d_j \end{pmatrix}. \tag{2}$$

In the present case we are extending this convention, in that some of the M_i and N_i may be empty, since, for example, a particular c_i need not be in $X\alpha$. However, for any given i, at least one of M_i and N_i is nonempty.

For each j, select an element r_j of R_j. Let γ be the element of $\mathscr{T}_X$ which maps d_j onto r_j, for each j, and leaves the rest of X unchanged. Then

$$\alpha\gamma = \begin{pmatrix} M_i & R_j \\ c_i & r_j \end{pmatrix}, \tag{3}$$

$$\beta\gamma = \begin{pmatrix} N_i & R_j \\ c_i & r_j \end{pmatrix}, \tag{4}$$

and these elements have rank η and difference rank ξ.

If C has any elements in common with $R = \bigcup R_j$, then we change our notation by adjoining to C those elements r_j such that $R_j \cap C \neq \square$. If ξ is infinite, then $|C|$ remains equal to ξ. If ξ is finite, then the new C remains finite. If C is finite we now make a further change: we adjoin $\aleph_0$ of the r_j to C in such a fashion that the set of the remaining r_j not adjoined to C still has cardinal η. If $\eta > \aleph_0$ then the latter condition is automatically satisfied. For the new set C, $|C| = \aleph_0$. With this notation an r_j transferred to C becomes one of the c_i and, correspondingly, its associated R_j becomes one of the M_i in (3) and one of the N_i in (4). If an r_j happens to lie in C, say $r_j = c_i$, then we replace M_i by $M_i \cup R_j$ and N_i by $N_i \cup R_j$.

We have $\bigcup M_i = \bigcup N_i$. Set $M = \bigcup M_i$ and, as before (but now with the possibly changed meaning), $R = \bigcup R_j$. Then $\alpha\gamma$ and $\beta\gamma$ belong to

$$\mathscr{T}_M^* = \{\phi \in \mathscr{T}_X : M\phi \subseteq M,\ R_j\phi = r_j\}.$$

If $\phi \in \mathscr{T}_M^*$, then $\phi|M$, the restriction of ϕ to M, belongs to $\mathscr{T}_M$ and $\phi \to \phi|M$ is an isomorphism of $\mathscr{T}_M^*$ onto $\mathscr{T}_M$. The restriction of ρ to $\mathscr{T}_M^*$ determines, under this isomorphism, a congruence ρ_M, say, on $\mathscr{T}_M$, such that $(\phi|M, \psi|M) \in \rho_M$ if and only if

$$(\phi, \psi) \in \rho \cap (\mathscr{T}^*_M \times \mathscr{T}^*_M).$$

Now $\alpha\gamma|M$ and $\beta\gamma|M$ are two ρ_M-equivalent elements of $\mathscr{T}_M$ which, if ξ is infinite, are each of rank ξ and have difference rank ξ; and if ξ is finite are of rank $\aleph_0$ and have finite difference rank. Thus (i) if ξ is infinite, by Lemma 10.73, any two elements of $\mathscr{T}_M$, each of rank $\leqq \xi$, are ρ_M-equivalent; and (ii) if ξ is finite, by Lemma 10.76, any two elements of $\mathscr{T}_M$, of finite difference rank and each of rank $\aleph_0$, are ρ_M-equivalent. We now argue case (i) and case (ii) separately.

(i) Since $|M| \geqq \xi$, there is an element of $\mathscr{T}_M$,

$$\mu = \begin{pmatrix} M'_i \\ m_i \end{pmatrix},$$

say, where $m_i \in M'_i$ for each i, and where $|\{m_i\}| = \xi$. This element is ρ_M-equivalent to

$$\begin{pmatrix} M \\ m_1 \end{pmatrix}.$$

Reverting from $\mathscr{T}_M$ to $\mathscr{T}^*_M$, it follows that

$$\begin{pmatrix} M'_i & R_j \\ m_i & r_j \end{pmatrix} \rho \begin{pmatrix} M & R_j \\ m_1 & r_j \end{pmatrix}. \tag{5}$$

If $|\{r_j\}| < \eta$ (which, since $\alpha\gamma$ is of rank η, can happen only if $\xi = \eta$), then one of these elements is of rank η and the other is of rank less than η. Conclusion (i) of Theorem 10.69 then follows immediately from Theorem 10.65. It remains to deal with the case when $|\{r_j\}| = \eta$.

Let $(\alpha', \beta') \in (I_{\eta'} \times I_{\eta'}) \cap \Delta_{\xi'}$. Following, for α' and β', the procedure adopted earlier to write α and β in the forms (1) and (2), we can write α' and β' in the forms

$$\alpha' = \begin{pmatrix} P_n & S_l \\ b_n & t_l \end{pmatrix},$$

$$\beta' = \begin{pmatrix} Q_n & S_l \\ b_n & t_l \end{pmatrix}.$$

For each n, at least one of P_n and Q_n is nonempty. Observe that our notation does not imply that α' and β' have the same rank.

Since ξ is infinite, $|\{b_n\}| \leqq \xi$. Since also $|\{t_l\}| \leqq \eta$, there exists an element δ, say, of $\mathscr{T}_X$, which induces one-to-one mappings

$$b_n \to m_{i_n}, \qquad t_l \to r_{j_l},$$

say, of $\{b_n\}$ into $\{m_i\}$ and of $\{t_l\}$ into $\{r_j\}$, respectively. Let ϵ be an element of $\mathscr{T}_X$ which induces the inverses of these mappings, i.e. which is such that

$$m_{i_n} \to b_n, \qquad r_{j_l} \to t_l$$

under ϵ.

Multiplying (5) on the left by

$$\alpha'\delta = \begin{pmatrix} P_n & S_l \\ m_{i_n} & r_{j_l} \end{pmatrix}$$

and letting $P = \bigcup P_n$, we obtain

$$\alpha'\delta \; \rho \begin{pmatrix} P & S_l \\ m_1 & r_{j_l} \end{pmatrix}.$$

Hence

$$\alpha' = \alpha'\delta\epsilon \; \rho \begin{pmatrix} P & S_l \\ m_1\epsilon & t_l \end{pmatrix}.$$

Similarly,

$$\beta' = \beta'\delta\epsilon \; \rho \begin{pmatrix} P & S_l \\ m_1\epsilon & t_l \end{pmatrix},$$

since $\bigcup Q_n = \bigcup P_n = P$. It follows that $(\alpha', \beta') \in \rho$; and this completes the proof of part (i) of Theorem 10.69.

(ii) *ξ is finite.* Let

$$(\alpha', \beta') \in (I_{\eta'} \times I_{\eta'}) \cap \Delta_{\aleph_0}.$$

Following an earlier procedure, we may write

$$\alpha' = \begin{pmatrix} P_n & S_l \\ b_n & t_l \end{pmatrix},$$

$$\beta' = \begin{pmatrix} Q_n & S_l \\ b_n & t_l \end{pmatrix},$$

where $\{b_n\} = \{b_1, b_2, \cdots, b_k\}$, where $|\{t_l\}| \leqq \eta$ and where $\operatorname{dr}(\alpha', \beta') \leqq k$. We may assume that $\operatorname{dr}(\alpha', \beta') \neq 0$. As before, P_n or Q_n may be empty but, for each n, one of P_n and Q_n is nonempty.

Now, from Lemma 10.76, as already remarked, we know that any two elements of $\mathscr{T}_M$, of finite difference rank and each of rank $\aleph_0$, are ρ_M-equivalent. Hence, since $|M| \geqq \aleph_0$, there exist ρ_M-equivalent elements

$$\begin{pmatrix} M_1' & M_2' & \cdots & M_{k+1}' & M_{k+1+i}' \\ m_1 & m_2 & \cdots & m_{k+1} & m_{k+1+i} \end{pmatrix},$$

$$\begin{pmatrix} M_1' & M_2' & \cdots & M_{k+1}' & M_{k+1+i}' \\ m_1 & m_1 & \cdots & m_1 & m_{k+1+i} \end{pmatrix}$$

in $\mathscr{T}_M$, with $m_t \in M_t'$, for all t, each of rank $\aleph_0$, and which, as exhibited, have difference rank k. Reverting from $\mathscr{T}_M$ to $\mathscr{T}_M^*$, therefore the elements

$$\begin{pmatrix} M_1' & \cdots & M_{k+1}' & M_{k+1+i}' & R_j \\ m_1 & \cdots & m_{k+1} & m_{k+1+i} & r_j \end{pmatrix},$$

$$\begin{pmatrix} M_1' & \cdots & M_{k+1}' & M_{k+1+i}' & R_j \\ m_1 & \cdots & m_1 & m_{k+1+i} & r_j \end{pmatrix}$$

are ρ-equivalent elements of $\mathscr{T}_X$. Recall that in the construction of M for the case of ξ finite it was arranged that $|\{r_j\}| = \eta$ (which by assumption is $\geqq \aleph_0$). Hence there exists an element δ, say, of $\mathscr{T}_X$, which induces one-to-one mappings

$$b_n \to m_n \qquad (n = 1, 2, \cdots, k),$$

$$t_l \to r_{j_l}$$

of $\{b_n\}$ into $\{m_n\}$ and of $\{t_l\}$ into $\{r_j\}$, respectively. Let ϵ be an element of $\mathscr{T}_X$ which induces the inverse mappings

$$m_n \to b_n \qquad (n = 1, 2, \cdots, k),$$

$$r_{j_l} \to t_l$$

of $\{m_n: n = 1, 2, \cdots, k\}$ into $\{b_n\}$ and of $\{r_{j_l}\}$ into $\{t_l\}$.

We may now proceed to complete the argument to show that $(\alpha', \beta') \in \rho$ precisely as the argument was completed for the proof of part (i).

This concludes the proof of Theorem 10.69, and hence of Theorem 10.72.

Malcev's theorems 10.68 and 10.72 together show that the lattice of congruences on $\mathscr{T}_X$ is generated by congruences of the type $\sigma^\dagger$ (in the notation of Theorem 10.68), Δ_ξ and I_η^*. We have already observed that, when X is finite, the lattice of congruences on $\mathscr{T}_X$ is a chain. In general, as we show in the ensuing corollary, the lattice of congruences on $\mathscr{T}_X$ is a sublattice of the lattice of binary relations on $\mathscr{T}_X$, i.e. of the lattice of all subsets of $\mathscr{T}_X \times \mathscr{T}_X$. So far as we know, this result has not appeared in the literature.

THEOREM 10.77. *The lattice of congruences on $\mathscr{T}_X$ is a sublattice of the lattice of all binary relations on $\mathscr{T}_X$. In particular, it is a distributive lattice.*

PROOF. In the lattice of congruences on any semigroup the meet operation is set-theoretic intersection. To prove the corollary it therefore suffices to show that the join operation in the lattice of congruences on $\mathscr{T}_X$ is set-theoretic union.

Let ρ, τ be two congruences on $\mathscr{T}_X$. Denote by $\vee$ the join operation in the lattice of congruences on $\mathscr{T}_X$ so that $\rho \vee \tau$ is the intersection of all congruences on $\mathscr{T}_X$ which contain each of ρ and τ. We have to show that $\rho \vee \tau = \rho \cup \tau$. The proof proceeds by listing cases.

(i) If one of ρ and τ is either the identity congruence or the universal congruence, then clearly $\rho \vee \tau = \rho \cup \tau$.

(ii) If $\eta(\rho) = n$ is finite, so that, by Theorem 10.68, $\rho = \sigma^{\dagger}$, where σ is a congruence on I_{n+1}/I_n, and if $\eta(\tau)$ is infinite, then $\rho \vee \tau = \tau = \rho \cup \tau$, since $I^*_{\eta(\tau)} \subseteq \tau$ and so $\rho \subseteq \tau$.

(iii) If $\eta(\rho) = n$ and $\eta(\tau) = m$ and both m and n are finite, then $\rho = \sigma^{\dagger}$, where σ is a congruence on I_{n+1}/I_n, and $\tau = \theta^{\dagger}$, where θ is a congruence on I_{m+1}/I_m. If $n < m$, then $\rho \subseteq \tau$ and so $\rho \vee \tau = \rho \cup \tau = \tau$. If $m = n$, then one of ρ, τ is contained in the other, by Theorems 10.68 and 10.49 since the normal subgroups of the symmetric group $\mathscr{G}_n$ form a chain.

(iv) If $\eta(\rho)$ and $\eta(\tau)$ are each infinite, then expressing ρ and τ in the form given by the Theorem 10.72, and applying the distributive law, we can express $\rho \cup \tau$ as the intersection of relations of the types I^*_{η}, Δ_{ξ} and $I^*_{\eta} \cup \Delta_{\xi}$, where η and ξ are each infinite. If we show that $I^*_{\eta} \cup \Delta_{\xi}$ is always a congruence on $\mathscr{T}_X$, it will therefore follow that $\rho \cup \tau$ is a congruence on $\mathscr{T}_X$; and this will show that $\rho \vee \tau = \rho \cup \tau$.

Consider therefore the relation $I^*_{\eta} \cup \Delta_{\xi}$. If $\eta \leqq \xi$, then any two elements of I_{η} have difference rank less than ξ, and so $I^*_{\eta} \subseteq \Delta_{\xi}$. Thus $I^*_{\eta} \cup \Delta_{\xi} = \Delta_{\xi}$. If $\xi < \eta$, then the result is already contained in Theorem 10.72, for $k = 1$.

CHAPTER 11

REPRESENTATION BY TRANSFORMATIONS OF A SET

In Chapter 5 we presented the theory of representations of a semigroup by linear transformations of a vector space. Here we discuss representations by transformations of a set without structure. This is an extension of the classical theory of representations of groups by permutations of a set. For this theory, we recommend the book of H. Zassenhaus [1937], especially since our approach is patterned on his.

Among the first to consider extensions of this theory from groups to semigroups were A. K. Suschkewitsch [1922, 1926] (see Appendix), R. R. Stoll [1944], V. V. Vagner [1956], and E. S. Lyapin [1960b]. The material in this chapter is drawn chiefly from the work of E. J. Tully [1960], B. M. Šaĭn [1961, 1962], and H.-J. Hoehnke [1963, 1966].

A closely related matter is the representation of semigroups by partial transformations. The theory of representations of inverse semigroups by one-to-one partial transformations was initiated by V. V. Vagner [1952] and G. B. Preston [1954c], and greatly extended by B. M. Šaĭn [1962]. A detailed account of the latter was given in Chapter 7. In the present chapter (§11.4) we give an account of Šaĭn's theory [1961] of representations of an arbitrary semigroup by one-to-one partial transformations. By a device due to Vagner [1956], it is possible to bring the theory of partial transformations under that of total (i.e., ordinary) transformations. For the most part, we adopt the latter viewpoint in this chapter, although it is not always natural or convenient to do this. It should be borne in mind that any result on representations by total transformations having a given set of fixed points can be translated into one on representations by partial transformations by removing one or more of the fixed points.

Another related matter is the representation of a semigroup by monomial matrices over a group with zero, which was treated in Chapter 3. In §11.8 we indicate the connexion. As with the corresponding theory for groups (Zassenhaus, loc. cit., Chapter V, §1), any irreducible representation ϕ of a semigroup S by transformations of a set is equivalent to a representation of S by row-monomial matrices over a group with zero G^0, where G is any group which is anti-isomorphic to some subgroup of the group of operator automorphisms of the operand associated with ϕ.

The notion of "operand" is the obvious analogue of that of "representation space" (Chapter 5), and there is the same intimate connexion between an operand and its associated representation. Any term, like "irreducible", that is defined for one is applied without comment to the other.

11.1 Basic definitions

By a *representation of a semigroup S by transformations of a set M* we mean a homomorphism ϕ of S into $\mathscr{T}_M$, where $\mathscr{T}_M$ is the full transformation semigroup on M (§1.3). In other words, to each element a of S there corresponds a transformation $a\phi$ of M such that

$$(ab)\phi = (a\phi)(b\phi) \qquad (\text{all } a, b \in S).$$

By a *right operand* (or *right S-system*) M_S *over a semigroup* S we mean a set M together with a mapping $(x, a) \rightarrow xa$ of $M \times S$ into M satisfying

$$x(ab) = (xa)b \qquad (\text{all } x \in M;\ a, b \in S).$$

Dually, by a *left operand* ${}_SM$ we mean a set M acted on by S from the left:

$$(ab)x = a(bx) \qquad (\text{all } x \in M;\ a, b \in S).$$

Since we shall deal mostly with right operands, we shall use the term *operand*, simply, to mean "right operand".

Given an operand M_S over S, we obtain a representation ϕ of S by transformations of M by defining

$$x(a\phi) = xa \qquad (\text{all } x \in M,\ a \in S).$$

Conversely, given a representation ϕ of M, the same equation serves to define xa for all $x \in M$ and $a \in S$, whereby M is turned into an operand M_S over S. We call ϕ *the representation associated with the operand M_S, and vice-versa.*

If ${}_SM$ is a left operand over S, and if we define $\psi: S \rightarrow \mathscr{T}_M$ by

$$x(a\psi) = ax \qquad (\text{all } x \in M,\ a \in S),$$

then

$$(ab)\psi = (b\psi)(a\psi) \quad (\text{all } a, b \in S).$$

We call ψ the *anti-representation* of S associated with ${}_SM$. Conversely, if ψ is an anti-representation of S by transformations of M, then $ax = x(a\psi)$ defines M as a left operand ${}_SM$ over S.

By the *kernel* of a representation ϕ of S we mean the congruence $\kappa = \phi \circ \phi^{-1}$; in other words,

$$a\,\kappa\, b \text{ if and only if } a\phi = b\phi \quad (a, b \in S).$$

The representation ϕ is *true* or *faithful* if $\kappa = \iota_S$, the identity relation on S. Any representation ϕ clearly induces a faithful representation ϕ' of the factor semigroup S/κ (§1.5), defined by

$$(a\kappa)\phi' = a\phi \quad (a \in S).$$

The image $S\phi$ of S under ϕ is a subsemigroup of $\mathscr{T}_M$ isomorphic with S/κ; in fact ϕ' is an isomorphism of S/κ onto $S\phi$.

By an *operator homomorphism* (or *S-homomorphism*) $\theta: M_S \to M'_S$ of one operand M_S over S into another M'_S, we mean a mapping $\theta: M \to M'$ such that

$$(xa)\theta = (x\theta)a \quad (\text{all } x \in M,\ a \in S).$$

If θ is one-to-one and onto, we call it an *operator isomorphism* (or *S-isomorphism*), and then the operands M_S and M'_S, and their associated representations ϕ and ϕ' of S, are said to be *equivalent*.

An *operator endomorphism* (or *S-endomorphism*) of an operand M_S is just an operator homomorphism of M_S into itself; it is an *operator automorphism* (or *S-automorphism*) if it is one-to-one and onto. The set $\mathscr{E}(M_S)$ of all operator endomorphisms of M_S is a subsemigroup of $\mathscr{T}_M$; it is in fact the *centralizer semigroup of $S\phi$ in $\mathscr{T}_M$*, the set of all elements of $\mathscr{T}_M$ which commute with each element of $S\phi$. The set $\mathscr{A}(M_S)$ of all operator automorphisms of M_S is the group of units (§1.7) of $\mathscr{E}(M_S)$. If T is any subsemigroup of the dual of $\mathscr{E}(M_S)$, we can set up M_S as a (T, S)-bioperand ${}_TM_S$ by writing θx for $x\theta$ ($x \in M$, $\theta \in T$). We shall do this in §11.8 with T contained in the dual of $\mathscr{A}(M_S)$. (See next page for the definition of "bioperand".)

An equivalence relation σ on an operand M_S will be called an *operator equivalence*, or *S-equivalence*, or *congruence*, if $x\,\sigma\,y$ ($x, y \in M$) implies $xa\,\sigma\,ya$ for all $a \in S$. The *factor operand* M_S/σ consists of all the σ-classes $x\sigma$ of M with the product of $x\sigma$ by any element a of S defined by

$$(x\sigma)a = (xa)\sigma \quad (x \in M,\ a \in S).$$

This definition is independent of the choice of the element x of the σ-class $x\sigma$; for if $x\sigma = y\sigma$, then $x\,\sigma\,y$, whence $xa\,\sigma\,ya$, and so $(xa)\sigma = (ya)\sigma$.

If ρ and σ are operator equivalences on M_S such that $\rho \supseteq \sigma$, then we can define a relation $\rho' = \rho/\sigma$ on $M'_S = M_S/\sigma$ as follows:

$$\rho' = \{(x\sigma, y\sigma) \in M'_S \times M'_S\colon (x, y) \in \rho\}.$$

It is easily seen that ρ' is an operator equivalence on M'_S. Conversely, if ρ' is an operator equivalence on M'_S, and we define ρ on M_S by

$$\rho = \{(x, y) \in M_S \times M_S\colon (x\sigma, y\sigma) \in \rho'\},$$

then ρ is an operator equivalence on M_S, $\rho \supseteq \sigma$ and $\rho' = \rho/\sigma$.

The following lemma may be compared with Lemma 9.52(i).

LEMMA 11.1. *If σ is an operator equivalence on an operand M_S over a semigroup S, then the mapping $\rho \to \rho/\sigma$ is a lattice isomorphism of the lattice of all operator equivalences ρ on M containing σ, and the lattice of all operator equivalences on the factor operand M_S/σ. We have*

$$(x\sigma, y\sigma) \in \rho/\sigma \text{ if and only if } (x, y) \in \rho, \quad (x, y \in M).$$

A number of results in §1.5 can be carried over almost verbatim to the present (more general) situation. For purposes of reference, we state a combined analogue of Theorem 1.5 (Main Homomorphism Theorem) and Theorem 1.6 (Induced Homomorphism Theorem).

THEOREM 11.2. *Let θ be an operator homomorphism of an operand M_S over a semigroup S onto another operand M'_S over S. Then $\theta \circ \theta^{-1}$ is an operator equivalence on M_S. If σ is any operator equivalence on M_S such that $\sigma \subseteq \theta \circ \theta^{-1}$ then $(x\sigma)\theta' = x\theta$ $(x \in M)$ defines an operator homomorphism θ' of M_S/σ onto M'_S such that $\sigma^\natural\theta' = \theta$. If $\sigma = \theta \circ \theta^{-1}$, then θ' is an operator isomorphism.*

A subset $N \neq \square$ of an operand M_S is called *invariant*, or a *suboperand*, or an *S-subsystem*, if $NS \subseteq N$; in other words, if $x \in N$ and $a \in S$ imply $xa \in N$. N is of course also an operand N_S over S. If N consists of a single element z, then $za = z$ for all $a \in S$; such an element z we call an *invariant* or *fixed* element of M. (Cf. §6.1.)

An invariant subset N of M_S determines an operator equivalence ν on M as follows: $x \nu y$ $(x, y \in M)$ if $x = y$ or if both x and y belong to N. We write M_S/N for M_S/ν, and call it the *Rees factor operand* of M by N, by analogy with §1.5. Clearly M_S/N has the invariant element N.

By a *decomposition* of an operand M_S we mean a partition of M into a set $\{N_i: i \in I\}$ of mutually disjoint invariant subsets N_i. If no such partition is possible with $|I| > 1$, then M_S is said to be *indecomposable*. Conversely, if $\{N_{iS}: i \in I\}$ is a set of mutually disjoint operands over S, then their union $M = \bigcup_i N_i$ can be made into an operand over S in exactly one way so that each N_{iS} is a suboperand of M_S.

An operand M_S is called *transitive* (cf. §6.1) if, for each pair of elements x, y of M, there exists a in S such that $xa = y$. Clearly M_S is transitive if and only if M contains no proper invariant subset. A transitive operand is evidently indecomposable.

An operand M_S is called *unital* if S has an identity element 1, and if $x1 = x$ for every x in M. The units of S (§1.7) are then represented by permutations of M.

Let S and T be two semigroups, not necessarily disjoint. By an (S, T)-*bioperand* ${}_SM_T$ we mean a set M which is both a left operand ${}_SM$ over S and a right operand M_T over T, and satisfies the condition

$$(sx)t = s(xt), \quad (\text{all } x \text{ in } M, s \text{ in } S, t \text{ in } T).$$

We then write sxt for the common value of $(sx)t$ and $s(xt)$. A subset N of M is called *left* [*right*] *invariant* if N is invariant when M is regarded as a left operand ${}_SM$ [right operand M_T]; and N is called *invariant* if it is both left and right invariant. An equivalence relation σ on M is called a *left* [*right*] *operator equivalence*, or a *left* [*right*] *congruence*, on M if it is an operator equivalence on M regarded as a left operand ${}_SM$ [right operand M_T]; and σ is called an *operator equivalence*, or *congruence*, on M if it is both a left and a right congruence on M. If ${}_SM'_T$ is a second bioperand, then a mapping $\theta\colon M \to M'$ is called an *operator homomorphism* if $(sx)\theta = s(x\theta)$ for all s in S and x in M, and $(xt)\theta = x(t\theta)$ for all t in T and x in M. A bioperand ${}_SM_T$ is called *transitive* if, given x and y in M, there exist s in S and t in T such that $sxt = y$. A bioperand ${}_SM_T$ is called *unital* if both S and T have identity elements, 1_S and 1_T, say, and if $1_Sx = x = x1_T$, for all x in M.

An (S, T)-bioperand ${}_SM_T$ determines a right operand M_P over the direct product $P = S^* \times T$ of the dual S^* of S with T, as follows:

$$x(s, t) = sxt, \quad (\text{all } x \text{ in } M,\ s \text{ in } S,\ t \text{ in } T).$$

(By the *(left-right) dual* S^* of a semigroup S we mean the semigroup $S(\circ)$, the elements of which are the same as those of S, and in which the binary operation $\circ$ is defined by $a \circ b = ba$ (all a, b in S).) We call M_P the right operand *associated* with ${}_SM_T$. If S and T have identity elements, then there is a one-to-one correspondence between unital (S, T)-bioperands and unital right operands over $P = S^* \times T$ (Exercise 4 below). In this important case, the concepts "invariant subset", "operator equivalence", "operator homomorphism", and "transitive", as defined in the previous paragraph for an (S, T)-bioperand, are easily seen to be equivalent to the corresponding concepts for the associated right operand. For this reason, we shall not state the analogues of Lemma 11.1 and Theorem 11.2 for bioperands.

In the right regular representation (§1.3) of a semigroup S, S is itself the operand S_S. Similarly, in the left regular anti-representation, S is the left operand ${}_SS$. The notions defined above for arbitrary operands M_S then specialize to S_S [${}_SS$] as follows:

operator equivalence $\to$ right [left] congruence,
invariant subset $\to$ right [left] ideal,
transitive operand $\to$ right [left] simple semigroup.

We may also regard S as an (S, S)-bioperand ${}_SS_S$. The three notions above then specialize respectively to "congruence", "ideal", and "simple semigroup".

By a *centered operand* M_S over a semigroup S we mean an operand containing a unique invariant element 0_M. If M_S is centered, and S has a zero element 0_S, then $x0_S = 0_M$ for every $x \in M$. For $x0_S$ is clearly an invariant element of M.

If M_S is a centered operand over a semigroup S with zero, then every invariant subset N of M_S contains 0_M, and N_S is also a centered operand. By a *0-decomposition* of M_S we mean an expression $M = \bigcup\{N_i\colon i \in I\}$ of M as a union of invariant subsets N_i which are 0-disjoint in the sense that $N_i \cap N_j = 0_M$ for $i \neq j$ in I. We call M_S *0-indecomposable* if no non-trivial 0-decomposition of M_S exists. We say that M_S is *0-transitive* if, for each pair x, y of elements of M such that $x \neq 0_M$, there exists a in S such that $xa = y$.

An operand M_S is called *fixed* (or *trivial*) if $xa = x$ for all x in M and all a in S, and we say that *S acts identically* (or *trivially*) on M. A centered operand is called *null* if $xa = 0_M$ for all x in M and all a in S.

By a *partial transformation* of a set M we mean a binary relation α on M such that, for each x in M, there is at most one element y of M such that $x \, \alpha \, y$. If α and β are partial transformations of M, so is their composition $\alpha \circ \beta$ (§1.4), and hence the set $\mathscr{PT}_M$ of all partial transformations of M is a subsemigroup of the semigroup $\mathscr{B}_M$ of all binary relations on M. The *domain* $D(\alpha)$ of α is the set of all x in M such that $x \, \alpha \, y$ for some y in M; we write $y = x\alpha$. We may then think of α as a mapping of $D(\alpha)$ into M, or onto its *range* $M\alpha = D(\alpha)\alpha$. We may also denote $D(\alpha)$ by $M\alpha^{-1}$. $D(\alpha \circ \beta)$ consists of all those x in $D(\alpha)$ such that $x\alpha \in D(\beta)$, and then $x(\alpha \circ \beta) = (x\alpha)\beta$. α is called *one-to-one* if $\alpha^{-1} \in \mathscr{PT}_M$; we may then think of α as a one-to-one mapping of $D(\alpha)$ onto $M\alpha$, and α^{-1} is the inverse mapping.

The theory of partial transformations of a set M can be included in the theory of total (i.e., ordinary) transformations of a set $M^0 = M \cup 0_M$, consisting of M and one additional element 0_M, by the following device due to V. V. Vagner [1956]. For α in $\mathscr{PT}_M$, define the transformation α^0 of M^0 by

$$x\alpha^0 = \begin{cases} x\alpha & \text{if } x \in D(\alpha), \\ 0_M & \text{if } x \in M^0 \backslash D(\alpha). \end{cases}$$

Then α^0 belongs to the subsemigroup ${}^0\mathscr{T}_M$ of $\mathscr{T}_{M^0}$ consisting of all those transformations of M^0 leaving 0_M fixed. Conversely, if $\beta \in {}^0\mathscr{T}_M$, then its restriction to M,

$$\beta | M = \beta \cap (M \times M),$$

is a partial transformation of M. The domain of $\beta|M$ is the set of all x in M for which $x\beta \neq 0_M$. Clearly the mappings $\alpha \to \alpha^0$ and $\beta \to \beta|M$ are mutually inverse isomorphisms of $\mathscr{PT}_M$ onto ${}^0\mathscr{T}_M$ and vice-versa.

By a *representation of a semigroup S by partial transformations of a set M* we mean a homomorphism ϕ of S into $\mathscr{PT}_M$. If we define, when $x(a\phi)$ is defined,

$$xa = x(a\phi) \quad (x \in M, a \in S)$$

then M becomes a *partial operand over* S in the sense that $x(ab)$ is defined ($x \in M$; $a, b \in S$) if and only if both xa and $(xa)b$ are defined, and then

$$x(ab) = (xa)b.$$

Conversely, if M is a partial operand over S, then the equation $x(a\phi) = xa$ serves to define, for each a in S, a partial transformation $a\phi$ of M such that the mapping ϕ of S into $\mathscr{PT}_M$ is a homomorphism.

The theory of representations of a semigroup S by partial transformations may be included in that of representations of S by ordinary transformations, by means of the device of Vagner mentioned above. This is equivalent to embedding the partial operand M in the ordinary operand $M^0 = M \cup 0_M$ in which we define xa to be 0_M if xa is not defined in M $(x \in M, a \in S)$, and $0_M a = 0_M$ for all a in S. Conversely, if M^0 is an operand over S with a fixed element 0_M (not necessarily the only one), then $M = M^0 \backslash 0_M$ is a partial operand over S.

We shall deal in this chapter almost exclusively with centered operands M_S over a semigroup S with zero. For semigroups S with zero, there is no loss in generality in doing this, since every operand over S decomposes uniquely into centered ones (Exercise 2 below). If S is a semigroup without zero, then we can adjoin one (§1.1); let us denote it by 0_S, and let $S^0 = S \cup 0_S$. If M_S is a centered operand over S, then we can make M into a centered operand M_{S^0} over S^0 by defining $x0_S = 0_M$ for all $x \in M$. If M_S is any operand over S, let $M^0 = M \cup 0_M$ and define $x0_S = 0_M a = 0_M 0_S = 0_M$ for all $x \in M$, $a \in S$. M^0 becomes thereby a centered operand $M^0{}_{S^0}$ over S^0 which differs only trivially from M_S. We note in particular that $M^0{}_{S^0}$ is 0-transitive if and only if M is transitive, and that there is a one-to-one correspondence between 0-decompositions of $M^0{}_{S^0}$ and decompositions of M_S.

Exercises for §11.1

1. There is a one-to-one correspondence between decompositions of an operand M_S over a semigroup S and operator equivalences σ on M with the property that $x \,\sigma\, xa$ for all x in M and a in S.

2. Every operand M_S over a semigroup S with zero 0_S decomposes uniquely into centered operands. The operator equivalence σ on M which gives this decomposition can be defined by $\sigma = \{(x, y) \in M \times M : x0_S = y0_S\}$.

An operand M_S over a semigroup S is called an *inflation* of an operand N_S if N is a proper invariant subset of M with the following property. With each element of x of N there is associated a subset N_x of M such that (i) $N_x \cap N = \{x\}$, (ii) $M = \bigcup \{N_x : x \in N\}$, and (iii) $xa = ya$ for each $x \in N$, each $y \in N_x$, and each $a \in S$. It is clear that, for given N_S, we can construct all inflations M_S of N_S in a trivial way: simply pick some set N_x for each x in N so that no two intersect, and such that $x \in N_x$; let M be their union, and let the action of S on M be defined by (iii).

The following exercise is taken from Tully [1960]. The result is also given by Lyapin [1960a, p. 32].

3. (a) Let M_S be an operand over a semigroup S. Let $\sigma = \{(x, y) \in M \times M : xa = ya \text{ for all } a \in S\}$. Choose a representative $r(A)$ in each σ-class A such that $r(A) \in MS$ if $A \cap MS \neq \square$, and arbitrarily otherwise. Let P be the set of all these representatives $r(A)$, and let $N = P \cup MS$. For each x in N, define N_x to be $\{x\} \cup (x\sigma \backslash MS)$ if $x \in P$, and $\{x\}$ if $x \notin P$. Then either $N = M$ or M_S is an inflation of N_S, and N_S is not an inflation of any proper suboperand.

(b) An operand M_S is an inflation of a proper suboperand if and only if, in the notation of part (a), either

(i) there exists $x \in M \backslash MS$ and $y \in MS$ such that $x \, \sigma \, y$, or

(ii) there exists a σ-class A, with $A \cap MS = \square$, containing more than one element.

4. If S and T are semigroups with identity elements, and $P = S^* \times T$, where S^* is the dual of S, then every unital right operand M_P over P is the associated right operand of a unital bioperand ${}_S M_T$.

5. Let S be a subsemigroup of the full transformation semigroup $\mathcal{T}_M$ on a set M. Regard M as a (faithful) operand M_S over S. For each operator equivalence λ on M, we define

$$\lambda^{\dagger} = \{(\alpha, \beta) \in S \times S \colon \alpha^{-1} \circ \beta \subseteq \lambda\}.$$

Then $\lambda^{\dagger}$ is a congruence on S; in fact, it is the kernel of the representation of S corresponding to the factor operand M/λ. If ρ is a congruence on S, we define $\rho^{\downarrow}$ to be the intersection of all operator equivalences λ on M such that $\rho \subseteq \lambda^{\dagger}$. Then

(i) $\lambda \subseteq \mu$ implies $\lambda^{\dagger} \subseteq \mu^{\dagger}$. (ii) $\rho \subseteq \sigma$ implies $\rho^{\downarrow} \subseteq \sigma^{\downarrow}$.

(iii) $\lambda^{\dagger\downarrow} \subseteq \lambda$. (iv) $\rho \subseteq \rho^{\downarrow\dagger}$.

(v) $\rho^{\downarrow} \subseteq \lambda$ if and only if $\rho \subseteq \lambda^{\dagger}$.

Call $\lambda[\rho]$ *closed* if $\lambda^{\dagger\downarrow} = \lambda\,[\rho^{\downarrow\dagger} = \rho]$. Then $(^{\dagger})$ is an isomorphism of the lattice of all closed operator equivalences on M onto the lattice of all closed congruences on S, and $(^{\downarrow})$ is its inverse. (Thurston [1952].)

11.2 Decomposition of an operand; fully reducible operands and semigroups

In this section we give the analogue for semigroups of the classical theorem that every representation of a group by permutations of a set decomposes into transitive ones, and we find the class of semigroups for which this classical theorem holds (when suitably expressed). As remarked at the end of the last section, we deal only with centered operands over a semigroup with zero.

Let M_S be a centered operand over a semigroup S with zero 0_S. If N is any invariant subset of M, we shall write N^- for $N \backslash 0_M$.

We define a binary relation τ on M^- as follows:

$$\tau = \{(x, y) \in M^- \times M^- \colon x = y \text{ or } xa = y \text{ for some } a \in S\},$$

and call τ the *transitivity relation* on M^-. By the *connectedness relation* τ^* on M^- we mean the smallest equivalence relation on M^- containing τ. Clearly $y \, \tau^* z$ $(y, z \in M^-)$ if and only if $y = z$ or there exists a finite sequence of elements $x_1, x_2, \cdots, x_n$ of M^- such that $x_1 = y$, $x_n = z$, and $(x_i, x_{i+1}) \in \tau \cup \tau^{-1}$ for $i = 1, 2, \cdots, n-1$.

For each x in M^- and each a in S, either $xa = 0_M$ or $xa\,\tau^*x$. It follows that $x\tau^* \cup 0_M$ is an invariant subset of M, and hence that

$$(1) \qquad M = \bigcup \{x\tau^* \cup 0_M : x \in M^-\}$$

is a 0-decomposition of the operand M into invariant subsets. The proof of the next theorem will show that this is the unique 0-decomposition of M into indecomposable suboperands. The unbracketed version of the theorem (which follows from the bracketed version by the adjunction of 0_M, as described at the end of the last section) was first proved by R. R. Stoll [1944] for finite operands.

Theorem 11.3. *Every [centered] operand over a semigroup [with zero] is uniquely [0-]decomposable into [0-]indecomposable suboperands.*

Proof. Let us say that a subset N of a centered operand M_S over a semigroup S with zero is *coinvariant* if both N and $M\backslash N^-$ are invariant subsets. This is equivalent to saying that N is a term in some 0-decomposition of M into invariant subsets, and hence each $x\tau^* \cup 0_M$ ($x \in M^-$) is coinvariant. We proceed to show that if N is coinvariant and $y \in N^-$, then $y\tau^* \cup 0_M \subseteq N$.

Suppose the contrary. Then there would exist z in $M\backslash N^-$ such that $y\,\tau^*z$. Clearly $y \neq z$, and hence there exists a finite sequence of elements $x_1, x_2, \cdots, x_n$ of M^- such that $x_1 = y$, $x_n = z$, and $(x_i, x_{i+1}) \in \tau \cup \tau^{-1}$ for $i = 1, 2, \cdots, n-1$. Each x_i belongs either to N^- or to $M^-\backslash N^-$, and, since $x_1 \in N^-$, and $x_n \in M^-\backslash N^-$, there must exist i such that $x_i \in N^-$ and $x_{i+1} \in M^-\backslash N^-$.

But $(x_i, x_{i+1}) \in \tau$ would imply $x_{i+1} \in N^-$ since N is invariant and $x_{i+1} \neq 0_M$. Similarly, $(x_i, x_{i+1}) \in \tau^{-1}$ would imply $x_i \in M^-\backslash N^-$, and hence we arrive at a contradiction.

The foregoing shows that if N is a coinvariant subset of M then

$$(2) \qquad N = \bigcup \{y\tau^* \cup 0_M : y \in N^-\}.$$

Hence either N reduces to just one set $y\tau^* \cup 0_M$, or else the operand N_S is 0-decomposable. Thus (1) is a 0-decomposition of M into 0-indecomposable suboperands, and it is unique because there are no other 0-indecomposable suboperands of M beyond those appearing in (1).

The expression (2) shows also that any 0-decomposition of M can be obtained by lumping terms of (1) together.

A [centered] operand M_S is called *fully* (or *completely*) [0-] *reducible* if it is [0-]decomposable into [0-]transitive suboperands. This is clearly equivalent to saying that the [0-]indecomposable constituents of M_S are [0-]transitive. A semigroup S is called *right fully* [0-] *reducible* if its right regular operand S_S is fully [0-] reducible. (Cf. Vagner [1957] and Šaĭn [1963].)

The following analogue of the Main Representation Theorem for Semisimple Algebras (§5.1) is substantially due to H.-J. Hoehnke [1965]. Observe that condition (A) of the theorem implies that every operand M_S over S satisfying $MS = M$ is fully [0-] reducible. But the latter condition would not be strong enough to imply full [0-] reducibility of S, since it would hold vacuously for any nilpotent semigroup. Condition (C) of the theorem is equivalent to saying that S coincides with its right socle Σ_r (§6.3), and that S contains no degenerate 0-minimal right ideals (§6.1).

THEOREM 11.4. *For any semigroup S [with zero], the following assertions are equivalent*:

(A) *If M_S is any [centered] operand over S, then $(MS)_S$ is fully* [0-] *reducible.*

(B) *S is right fully* [0-] *reducible.*

(C) *S is the union of its [non-degenerate* 0-] *minimal right ideals.*

If any one of these three equivalent conditions on S is satisfied, then every [0-] *transitive operand over S is an operator homomorphic image of some* [0-] *minimal right ideal of S.*

PROOF. We prove only the bracketed statement; the unbracketed follows by adjunction of 0_M (see end of §11.1).

Assuming (A), let M_S be the extended right regular operand $(S^1)_S$ of S (§1.3). Then $(MS)_S$ is the right regular operand S_S of S. Hence (A) implies (B).

(B) implies that S is the union of its 0-minimal right ideals. No 0-minimal right ideal R of S can be degenerate; for this would mean $R = \{0, a\}$ with $aS = 0$, and R_S would not be 0-transitive. Thus (B) implies (C).

Assume (C), and let M_S be any centered operand over S. Let $\{R_i : i \in I\}$ be the 0-minimal right ideals of S. By (C), S is their union, and no R_i is degenerate. If $x \in M \backslash 0_M$, then either $xR_i = 0_M$ or xR_i is a 0-transitive invariant subset of M. For if $xa \in xR_i \backslash 0_M$ $(a \in R_i)$, then $aS = R_i$, since R_i is non-degenerate, and hence $(xa)S = xR_i$. Since

$$N = \bigcup\{y\tau^* \cup 0_M : y \in N^-\}.$$

we obtain, after removing superfluous terms, a 0-decomposition of MS into 0-transitive invariant subsets. This establishes (A).

To prove the final assertion of the theorem, let S satisfy (C), and let M_S be a 0-transitive operand over S. Then $M = xS$ for some $x \in M$. By (C), there must exist a 0-minimal right ideal R of S such that $xR \neq 0_M$, and hence $xR = M$. The mapping $r \to xr$ is clearly an operator homomorphism of R_S onto M_S.

EXERCISES FOR §11.2

1. Let S be a left zero semigroup. Let M be a set, and let F be any subset of M. For each a in S, let $a\phi$ be any projection of M onto F. Then ϕ is a representation of S, and every representation of S is so obtained.

2. Let $S = \{a, b\}$ be the left zero semigroup of order two. Let M be the set of non-negative integers. Define

$$xa = \begin{cases} x - 1 & \text{if } x \text{ is odd,} \\ x & \text{if } x \text{ is even;} \end{cases} \qquad xb = \begin{cases} x + 1 & \text{if } x \text{ is odd,} \\ x & \text{if } x \text{ is even.} \end{cases}$$

The resulting operand M_S is indecomposable, and does not arise by inflation from any proper suboperand (Exercise 3(b) for §11.1). (Tully [1960].)

3. The extended right regular representation (§1.3) of any semigroup is indecomposable.

4. A semigroup S with identity [and zero] has the property that every [centered] unital operand over S is fully reducible if and only if S is a group [with zero].

5. Every suboperand, and every operator homomorphic image, of a fully reducible operand is fully reducible. (Hoehnke [1965].)

6. Let ϕ be a representation of a semigroup S by partial transformations of a set M, and let M_S be the corresponding partial operand over S. Let $M_S^0 = M_S \cup 0_M$ as given in §11.1. We say that ϕ (and M_S) are fully reducible if M_S^0 is fully reducible. This is the case if and only if the transitivity relation τ (on $(M^0)^- = M$) is symmetric. (Vagner [1957].)

7. Let $\{\phi_i : i \in I\}$ be any set of representations of a semigroup S by partial transformations. Then there exists a semigroup T with zero 0 containing S as a subsemigroup such that, for each i in I, there exists a right ideal R_i of T such that ϕ_i is equivalent to the representation of S associated with the partial operand $R_i \backslash 0$ over S.

Such a semigroup T may be constructed as follows. Let M_i be a partial operand over S associated with ϕ_i. Choose the M_i so that they are disjoint from S and from each other. Let 0 be a symbol not in S or any M_i. Let $T = \bigcup\{M_i : i \in I\} \cup S \cup 0$, and define a product in T as follows (wherein $a, b \in S$; $x_i \in M_i$): (i) ab as given; (ii) $x_i a$ as given, if it is defined, and otherwise $x_i a = 0$; (iii) all other products $= 0$. (Lyapin [1960b].)

11.3 Strictly cyclic operands and modular right congruences

An operand M_S over a semigroup S is called [*strictly*] *cyclic* if there exists x in M such that $M = xS \cup x$ [$M = xS$]. Such an element x is called a [*strict*] *generator* of M_S.

Every transitive or 0-transitive operand M_S is strictly cyclic, with any element of $M \backslash 0_M$ a strict generator. If M_S is strictly cyclic, then any generator is a strict generator. If M_S is cyclic, but not strictly cyclic, say $M = x \cup xS$ with x not in xS, then x is the only generator of M.

For any cyclic operand M_S, the set $I(M)$ of nongenerators of M is either empty or an invariant subset of M. The following lemma is evident.

Lemma 11.5. *Let M_S be a cyclic operand over a semigroup S. If M_S is* *not strictly cyclic, say* $M = x \cup xS$ *with* x *not in* xS, *then* $I(M) = xS$, *and* $M/I(M)$ *is the null operand of order two. Assume* M_S *is strictly cyclic. If* $I(M) = \square$, *then* M_S *is transitive. If* $I(M) \neq \square$, *then* $M/I(M)$ *is* 0-*transitive.*

If M_S is any operand over S, and $x \in M$, then $C_x = x \cup xS$ is a cyclic suboperand of M. Thus, as pointed out by Stoll [1944], *every operand is a union of cyclic ones.* (See also Exercise 2 below.) Defining the *principal factors* of M_S to be the operands $C_x/I(C_x)$, agreeing that $C_x/\square = C_x$, the above lemma tells us that every principal factor is transitive, 0-transitive, or null of order two. A case of importance is $M_S = S_S$. Here $C_a = R(a) = a \cup aS$ $(a \in S)$, and $C_a \backslash I(C_a)$ is the $\mathscr{R}$-class R_a containing a. We shall return to this in §11.8.

A right congruence ρ on a semigroup S is called *modular* if there exists e in S such that $ea \, \rho \, a$ for all a in S; such an element e is called a *left identity of* S *modulo* ρ. We recall that the factor operand S_S/ρ, which we shall abbreviate to S/ρ, consists of the set of ρ-classes $a\rho$ of S $(a \in S)$, with multiplication of $a\rho$ by an element b of S defined by $(a\rho)b = (ab)\rho$. Of course, if S has a left identity, then every right congruence on S is modular.

The following theorem, due to Tully [1960, 1961], is the natural generalization to semigroups of the classical theorem that every transitive representation of a group G by permutations is equivalent to that obtained by letting G act on the set of right cosets $\{Ha : a \in G\}$ of some subgroup H of G. It was found independently by Šaĭn [1961].

THEOREM 11.6. *Let* M_S *be a strictly cyclic operand over a semigroup* S, *with strict generator* x. *Let*

$$\rho = \{(a, b) \in S \times S : xa = xb\}.$$

Then ρ *is a modular right congruence on* S, *and the operand* S/ρ *is equivalent to* M_S. *If* $xe = x$, *then* e *is a left identity of* S *modulo* ρ.

Conversely, if ρ *is a modular right congruence on* S, *then* S/ρ *is strictly cyclic. If* e *is a left identity of* S *modulo* ρ, *then* $e\rho$ *is a strict generator of* S/ρ. *The set of left identities of* S *modulo* ρ *is a left unitary subsemigroup of* S *saturated by* ρ (*i.e., it is a union of* ρ-*classes*).

PROOF. It is clear that ρ is a right congruence on S. Since $x \in M = xS$, there exists e in S such that $xe = x$. If $xe = x$, then $xea = xa$ for all a in S, whence $ea \, \rho \, a$ for all a in S; thus e is a left identity of $S \bmod \rho$, and ρ is modular.

Define $\theta : S \to M$ by $a\theta = xa$. This is an operator homomorphism of S_S onto M_S with kernel $\theta \circ \theta^{-1} = \rho$. By Theorem 11.2, $\theta' : S/\rho \to M$ defined by $(a\rho)\theta' = a\theta$ $(a \in S)$ is an operator isomorphism. Hence S/ρ and M_S are equivalent.

Conversely, let ρ be a modular right congruence on S, and let e be a left identity of $S \bmod \rho$. Then, for any a in S, $(e\rho)a = (ea)\rho = a\rho$. Hence $e\rho$ is a strict generator of S/ρ, and S/ρ is strictly cyclic. Let U be the set of left identities of $S \bmod \rho$. If e and f belong to U, then $efa \,\rho\, fa \,\rho\, a$ for all a in S, so $ef \in U$. If $e \in U$, $s \in S$, and $es \in U$, then, for all a in S, $sa \,\rho\, esa \,\rho\, a$, so $s \in U$; thus U is left unitary. If $e \in U$, $s \in S$, and $e \,\rho\, s$, then $ea \,\rho\, sa$ for all a in S, and hence $sa \,\rho\, ea \,\rho\, a$; thus $s \in U$, showing that U is saturated by ρ.

Let ρ be a right congruence on a semigroup S. An element c of S will be called a *right unit of S modulo ρ* if, for each element a of S, there exists a corresponding s in S such that $cs \,\rho\, a$. In particular, any left identity of $S \bmod \rho$ is a right unit of $S \bmod \rho$. Clearly c is a right unit of $S \bmod \rho$ if and only if $c\rho$ is a strict generator of S/ρ.

We now have the following result.

COROLLARY 11.7. *All transitive representations of a semigroup S are obtained from operands of the form S/ρ, where ρ is a modular right congruence on S such that every element of S is a right unit of S modulo ρ. All 0-transitive representations of S are obtained from operands of the form S/ρ, where ρ is a modular right congruence on S such that one of its congruence classes W is a right ideal of S, and every element of $S \backslash W$ is a right unit of S modulo ρ.*

If ρ is a modular right congruence on S, then the kernel $\phi \circ \phi^{-1}$ of the representation ϕ of S associated with the operand S/ρ is just the relation ρL introduced in §10.1:

$$\rho L = \{(a, b) \in S \times S \colon sa \,\rho\, sb \quad \text{for all } s \in S^1\}.$$

For $(s\rho)(a\phi) = (sa)\rho$, and hence $a\phi = b\phi$ if and only if $(sa)\rho = (sb)\rho$ for all $s \in S$. Taking for s any left identity of $S \bmod \rho$, we conclude $a\rho = b\rho$, so that S may be replaced by S^1. By Lemma 10.3 (or directly) we see that ρL is the largest congruence (also the largest left congruence) on S contained in ρ. In particular, *S/ρ is faithful if and only if ρ contains no [left] congruence other than ι.*

We call ρL the *kernel* of ρ. When S is a group, and ρ is right congruence modulo a subgroup H of S, then ρL is congruence modulo the largest normal subgroup of S contained in H (= the intersection of all the conjugates of H).

Having in mind the problem of finding a complete set of inequivalent strictly cyclic (in particular, transitive and 0-transitive) representations of a semigroup S, Theorem 11.6 suggests the question: when do two modular right congruences on S yield equivalent operands? Theorem 11.10 below is a step in this direction. The remainder of this section is new.

If ρ is a right congruence on a semigroup S, and if $c \in S$, then

$$\rho^c = \{(s, t) \in S \times S \colon (cs, ct) \in \rho\}$$

is also a right congruence on S, which we call the *translate of ρ by c*. If c is a right unit of $S \bmod \rho$, then we call ρ^c a *conjugate* of ρ. We remark that $(\rho^c)^d = \rho^{cd}$ $(c, d \in S)$, and that $c \,\rho\, d$ implies $\rho^c = \rho^d$.

LEMMA 11.8. *Let ρ be a modular right congruence on a semigroup S, and* *let c be a right unit of S modulo ρ. Then ρ^c is also a modular right congruence on S, and the mapping $\theta: S/\rho^c \to S/\rho$ defined by $(s\rho^c)\theta = (cs)\rho$ is an operator isomorphism.*

Conjugacy is an equivalence relation on the set of modular right congruences on S.

Proof. Since c is a right unit of $S \bmod \rho$, we can solve $cf \, \rho \, c$ for f in S. Then, for any a in S, $cfa \, \rho \, ca$, hence $fa \, \rho^c \, a$. Hence f is a left identity of $S \bmod \rho^c$, and so ρ^c is modular.

To see that θ is single-valued and one-to-one, we observe that $s\rho^c = t\rho^c$ if and only if $(cs)\rho = (ct)\rho$, since the former is equivalent to $s \, \rho^c \, t$ and the latter to $cs \, \rho \, ct$. For given a in S, we can solve $cs \, \rho \, a$ for s in S, so that $(s\rho^c)\theta = a\rho$; hence θ is onto. If $a, s \in S$, then

$$((s\rho^c)a)\theta = ((sa)\rho^c)\theta = (csa)\rho = ((cs)\rho)a = ((s\rho^c)\theta)a,$$

and hence θ is an operator isomorphism.

Let ρ be a modular right congruence on S, and let e be a left identity of $S \bmod \rho$. Then $ea \, \rho \, a$ for all a in S, so $a \, \rho \, b$ if and only if $ea \, \rho \, eb$. Hence $\rho^e = \rho$. Since e is a right unit of $S \bmod \rho$, this shows that ρ is conjugate to itself.

Let ρ^c be a conjugate of ρ, and $(\rho^c)^d$ a conjugate of ρ^c, where $c[d]$ is a right unit of $S \bmod \rho[\rho^c]$. Since $(\rho^c)^d = \rho^{cd}$, transitivity will follow when we show that cd is also a right unit of $S \bmod \rho$. Let $a \in S$. Solve $ct \, \rho \, a$ for $t \in S$, and then solve $ds \, \rho^c \, t$ for $s \in S$. Then $cds \, \rho \, ct \, \rho \, a$, so that s is a solution of $(cd)s \, \rho \, a$.

Finally, to show symmetry, let ρ^c be a conjugate of ρ. Solve $cd \, \rho \, e$, where e is a left identity of $S \bmod \rho$. Then $(\rho^c)^d = \rho^e = \rho$. To complete the proof that ρ is a conjugate of ρ^c, we must show that d is a right unit of $S \bmod \rho^c$. Let $a \in S$, and let $s = ca$. Then $cds \, \rho \, es \, \rho \, s = ca$, so that $ds \, \rho^c \, a$.

Lemma 11.9. *If ρ and σ are modular right congruences on S such that $S/\rho \cong S/\sigma$, then ρ and σ are conjugate.*

Proof. Let θ be an operator isomorphism of S/σ onto S/ρ. Let e be a left identity of $S \bmod \sigma$. Let $c \in (e\sigma)\theta$, so that $(e\sigma)\theta = c\rho$. Then, for any s in S,

$$(s\sigma)\theta = ((es)\sigma)\theta = ((e\sigma)s)\theta = ((e\sigma)\theta)s = (c\rho)s = (cs)\rho.$$

Since θ is one-to-one, $s\sigma = t\sigma$ $(s, t \in S)$ if and only if $(cs) \, \rho \, (ct)$, which shows that $\sigma = \rho^c$. Since θ is onto, given a in S, there exists s in S such that $(s\sigma)\theta = a\rho$, hence $cs \, \rho \, a$. Hence c is a right unit of $S \bmod \rho$, and $\sigma = \rho^c$ is a conjugate of ρ.

From Theorem 11.6 and Lemmas 11.8 and 11.9, the following is immediate.

Theorem 11.10. *Every strictly cyclic operand over S is equivalent to S/ρ for some modular right congruence ρ on S. Two such are equivalent if and only if the associated right congruences are conjugate.*

Exercises for §11.3

1. Let S be the semigroup $\{a, b, 0\}$ with $a^2 = a$, $ba = b$, and all other products defined to be 0. Let ρ $[\sigma]$ be the right congruence on S defined by the partition of S into $A \cup B$ $[C \cup Z]$, where

$$A = \{a\}, \quad B = \{b, 0\}, \quad C = \{a, b\}, \quad Z = \{0\}.$$

Then S/ρ and S/σ are equivalent operands, although ρ is modular and σ is not.

2. Let M_S be a partial operand over a semigroup S, and let τ be the transitivity relation on M (§11.2). A subset N of M is called *dense* if $N\tau = M$. The *weight* of M_S is defined to be the smallest cardinal number w such that there exists a dense subset of M of cardinal w.

(a) M_S is cyclic if and only if its weight is 1.

(b) Any partial operand of weight w is the union of w suboperands each of weight 1. (Vagner [1956].)

3. A strictly cyclic operand over a completely simple semigroup is transitive. (Stoll [1944].)

11.4 Representations by one-to-one partial transformations

By a representation of a semigroup S by one-to-one partial transformations of a set M we mean a homomorphism ϕ of S into the symmetric inverse semigroup $\mathscr{I}_M$ of all one-to-one partial transformations of M (§1.9). In §§7.2 and 7.3 we gave an account of B. M. Šaĭn's theory [1962] of such representations of an inverse semigroup S. We now present his theory [1961] for an arbitrary semigroup S. Since every inverse semigroup T can be embedded in $\mathscr{I}_T$, by Theorem 1.20, a semigroup S can be embedded in an inverse semigroup if and only if it admits a faithful representation by one-to-one partial transformations of some set M.

In §11.1 we discussed Vagner's device for converting a partial transformation α of a set M into a total transformation α^0 of the set $M^0 = M \cup 0_M$ leaving 0_M fixed. We see that α will be one-to-one if and only if α^0 has the property

$$x\alpha^0 = y\alpha^0 \neq 0_M \ (x, y \in M^0) \text{ implies } x = y.$$

A transformation α^0 of M^0 leaving 0_M fixed and having this property will be called *partially one-to-one*. The set of all such transformations is a subsemigroup ${}^0\mathscr{I}_M$ of ${}^0\mathscr{T}_M$ which is isomorphic with $\mathscr{I}_M$ under the correspondence $\alpha \leftrightarrow \alpha^0$.

Thus there is no essential difference between the theory of representations of a semigroup S by one-to-one partial transformations of a set M and that of representations of S by partially one-to-one transformations of M^0. We adopt the latter point of view in this section in the development, but we may express the results in either or both forms.

A partial operand M_S over a semigroup S is called *cancellative* if $xa = ya$

$(x, y \in M; a \in S)$ implies $x = y$. Clearly the representation ϕ of S associated with M is by one-to-one partial transformations of M if and only if M_S is cancellative. By Theorem 11.3, every partial operand M_S over S decomposes uniquely into indecomposable ones, and clearly M_S is cancellative if and only if this is true for each of its indecomposable constituents. We may therefore restrict our attention to indecomposable, cancellative, partial operands M_S over S.

We note first that an indecomposable, cancellative, partial operand M_S over S with $|M| > 1$ cannot contain a fixed element. For suppose z were a fixed element of M; that is, $za = z$ for all a in S. If $z = xb$ for some $x \in M$ and $b \in S$, then $zb = xb$, whence $x = z$. Thus $M = \{z\} \cup (M \backslash z)$ would be a decomposition of M. Consequently, if we adjoin an *origin* 0_M to M, then 0_M is the only fixed element in the set $M^0 = M \cup 0_M$, so that M_S^0 is a centered operand. Of course, M_S^0 is also 0-indecomposable.

A centered operand $M_S^0 = M_S \cup 0_M$ is called *0-cancellative* if $xa = ya \neq 0_M$ $(x, y \in M; a \in S)$ implies $x = y$. Clearly M_S^0 is 0-cancellative if and only if the partial operand $M_S = M_S^0 \backslash 0_M$ is cancellative.

A right congruence ρ on a semigroup S will be called *centered* if exactly one ρ-class W_ρ is a right ideal of S. Clearly this is the case if and only if the operand S/ρ is centered, with origin W_ρ. We say that a centered right congruence ρ is *right 0-cancellative* if

$$ac \, \rho \, bc \notin W_\rho \ (a, b, c \in S) \text{ implies } a \, \rho \, b.$$

Clearly ρ is right 0-cancellative if and only if S/ρ is 0-cancellative.

A right congruence ρ on a semigroup S will be called *right cancellative* if $ac \, \rho \, bc$ $(a, b, c \in S)$ implies $a \, \rho \, b$. This is of course equivalent to the assertion that the operand S/ρ is cancellative. For the sake of uniformity, we agree to consider a right congruence ρ on S as centered if no ρ-class is a right ideal of S. We then define $W_\rho = \square$, and regard W_ρ as an adjoined origin to S/ρ. With this convention, such a right congruence ρ is right cancellative if and only if it is right 0-cancellative. But as in Corollary 11.11 below, we may prefer to give separate expression to this case, for the sake of emphasis.

In the second half of Corollary 11.7, it is clear that W is the only ρ-class which is a right ideal, and hence that ρ is centered, with origin $W_\rho = W$. The following is an immediate consequence of Corollary 11.7.

COROLLARY 11.11. *All 0-transitive representations of a semigroup S by partially one-to-one transformations (hence all transitive representations of S by one-to-one partial transformations) are obtained from operands of the form S/ρ, where ρ is a centered, right 0-cancellative, modular right congruence on S, such that every element of $S \backslash W_\rho$ is a right unit of S modulo ρ.*

All transitive representations of S by one-to-one (total) transformations are obtained from operands of the form S/ρ, when ρ is a right cancellative, modular right congruence on S, such that every element of S is a right unit of S modulo ρ.

Thanks to Dubreil [1941], we have a method for constructing all such right congruences ρ on a semigroup S. Let H be a strong, unitary subsemigroup of S (§10.2). Let $\mathscr{R}_H$ be Dubreil's principal right congruence on S, and let W_H be the right residue of H (§10.2). By Theorem 10.22, $\rho = \mathscr{R}_H$ has all the properties stated for ρ in Corollary 11.11, with $W_\rho = W_H$, and every such ρ is an $\mathscr{R}_H$ for some strong unitary subsemigroup H of S. Moreover, H is the set of all left identities of S modulo $\mathscr{R}_H$, and is a single $\mathscr{R}_H$-class. In particular, the correspondence $H \leftrightarrow \mathscr{R}_H$ is one-to-one. Combining this with Corollary 11.11, we obtain the following definitive result.

THEOREM 11.12. *Let H be a strong unitary subsemigroup of a semigroup S. Let $\mathscr{R}_H$ be Dubreil's principal right congruence, and let W_H be the right residue of H. If $W_H = \square$, then the operand $S/\mathscr{R}_H$ is transitive and cancellative. If $W_H \neq \square$, then $S/\mathscr{R}_H$ is 0-transitive and 0-cancellative, with origin W_H. Conversely, every [0-] transitive, [0-] cancellative operand over S is equivalent to $S/\mathscr{R}_H$ for some strong unitary subsemigroup H of S.*

Thus we get all transitive representations of a semigroup S by one-to-one transformations of a set from the strong unitary subsemigroups H of S with $W_H = \square$; and all the 0-transitive representations of S by partially one-to-one transformations, hence all the transitive representations of S by one-to-one partial transformations, from the strong subsemigroups H of S with $W_H \neq \square$. The natural question arises: when do two strong unitary subsemigroups of S give rise to equivalent representations?

For the special case of an inverse semigroup S this problem was solved by Theorem 7.27. A subsemigroup of an inverse semigroup is unitary if and only if it is a closed inverse subsemigroup (Exercise 3 for §7.3) and any such subsemigroup is necessarily strong (Exercise 4 for §7.2). We now give the generalization to an arbitrary semigroup S. The result (Theorem 11.13) is new.

Two strong unitary subsemigroups H and K of a semigroup S are called *conjugate* if there exist elements c, d of S such that

$$cd \in H, \quad dc \in K, \quad cKd \subseteq H, \quad dHc \subseteq K. \tag{1}$$

If H and K are closed inverse subsemigroups, i.e., strong unitary subsemigroups, of an inverse semigroup S, then (1) implies that $cKc^{-1} \subseteq H$ and $c^{-1}Hc \subseteq K$, so that, by Theorem 7.27, H and K are conjugate in our earlier sense. For if $k \in K$, then $(ckc^{-1})(cd) = (ckd)(cd)^{-1}(cd) \in H$, whence $ckc^{-1} \in H$; hence $cKc^{-1} \subseteq H$, and, similarly, $c^{-1}Hc \subseteq K$. The reverse implication is immediate.

THEOREM 11.13. *Let H and K be strong unitary subsemigroups of a semigroup S. Then the operands $S/\mathscr{R}_H$ and $S/\mathscr{R}_K$ are equivalent if and only if H and K are conjugate.*

PROOF. By Theorem 11.10, we need only show that $\mathscr{R}_H$ and $\mathscr{R}_K$ are conjugate if and only if H and K are conjugate.

Assume first that H and K are conjugate. Then there exist elements c, d of S such that $cd \in H$, $dc \in K$, and (1) holds. We shall show that $\mathscr{R}_K = \mathscr{R}^c_H$. Since $cd \in H$, and every element of H is a left identity of $S \bmod \mathscr{R}_H$, c is a right unit of $S \bmod \mathscr{R}_H$, and hence it will follow that $\mathscr{R}_K$ is conjugate to $\mathscr{R}_H$.

Let $a\mathscr{R}^c_H b$ $(a, b \in S)$. This means that $ca\mathscr{R}_H cb$. Let $as \in K$ $(s \in S)$. Then $(ca)(sd) \in cKd \subseteq H$ by (1), and $ca\mathscr{R}_H cb$ implies $(cb)(sd) \in H$. Hence $(dc)(bs)(dc) \in dHc \subseteq K$. Since $dc \in K$, and K is unitary, we conclude that $bs \in K$. Similarly, we can show that $bs \in K$ implies $as \in K$, and we conclude that $a\mathscr{R}_K b$.

Let $a\mathscr{R}_K b$ $(a, b \in S)$. We proceed to show that $ca\mathscr{R}_H cb$, that is, $a\mathscr{R}^c_H b$, which will then establish $\mathscr{R}_K = \mathscr{R}^c_H$. Let $(ca)s \in H$ $(s \in S)$. Then $(dc)(asc) \in dHc \subseteq K$, and so $asc \in K$, since K is unitary. From $a\mathscr{R}_K b$ we conclude that $bsc \in K$, and hence $(cbs)(cd) \in cKd \subseteq H$. Since $cd \in H$ and H is unitary, we conclude that $(cb)s \in H$, Similarly we can prove that $(cb)s \in H$ implies $(ca)s \in H$, and hence $ca\mathscr{R}_H cb$.

Conversely, assume that $\mathscr{R}_H$ and $\mathscr{R}_K$ are conjugate. Then there exists a right unit c of $S \bmod \mathscr{R}_H$ such that $\mathscr{R}_K = \mathscr{R}^c_H$. By the definition of right unit, there exists d in S such that $cd \in H$.

We note first that $ca\mathscr{R}_H c$ implies $a \in K$. For, let k be any element of K. Then $cak\mathscr{R}_H ck$, whence $ak\mathscr{R}^c_H k$, or $ak\mathscr{R}_K k$. But this implies $ak \in K$, and hence $a \in K$.

Now $cd \in H$, and every element of H is a left identity of $S \bmod \mathscr{R}_H$. Hence $cdc\mathscr{R}_H c$, which implies $dc \in K$ from the foregoing remark. Let $h \in H$. Then $cdh \in H^2 \subseteq H$. Hence, as before, $cdhc\mathscr{R}_H c$, and we conclude that $dhc \in K$. This shows that $dHc \subseteq K$.

Before showing that $cKd \subseteq H$, we note first that $dbc \in K$ implies $b \in H$. For, since K is strong, $dbc \in K$ and $dc \in K$ imply $db\mathscr{R}_K d$. But this implies $cdb\mathscr{R}_H cd$. Since $cd \in H$, and H is unitary, we conclude that $b \in H$.

Now let $k \in K$, and let $b = ckd$. Then $dbc = (dc)k(dc) \in K$, and we conclude that $b \in H$ from the foregoing remark. Thus $cKd \subseteq H$, and we have shown that H and K are conjugate, thereby establishing the theorem.

Let M_S be a 0-cancellative, centered operand over a semigroup S with identity 1. For each x in M, $(x1)1 = x1$, and so $x1$ is either x or 0_M. Let

$$M_0 = \{x \in M : x1 = 0_M\} \qquad M_1 = \{x \in M : x1 = x\}.$$

Then $M = M_0 \cup M_1$ is a 0-decomposition of M into a null operand M_0 and a unital operand M_1.

If S does not have an identity, then any unital, 0-cancellative, centered operand over S^1 is a 0-cancellative, centered operand over S. Conversely, if M_S is a 0-cancellative, centered operand over S, and we define $x1 = x$ for all x in M, then M_S becomes a unital, 0-cancellative, centered operand over S^1.

These considerations show that there will be no loss of generality if we confine our attention to unital, 0-cancellative, centered operands over a semigroup with identity, and this we shall do for the rest of this section. We shall also regard a unital cancellative operand M_S as a unital, 0-cancellative, centered operand, since $M_S^0 = M_S \cup 0_M$ is such.

If H is any strong subset of a semigroup S with identity (not necessarily a unitary subsemigroup of S), then $S/\mathscr{R}_H$ is a unital, 0-cancellative operand over S. Let us agree that if $W_H = \Box$, then W_H is adjoined as origin to $S/\mathscr{R}_H$. Let ϕ_H be the representation of S associated with the operand $S/\mathscr{R}_H$. The kernel $\mathscr{P}_H$ of ϕ_H is Croisot's *principal bilateral equivalence* (§10.4) defined by H (cf. Lemma 7.23):

$$\mathscr{P}_H = \{(a, b) \in S \times S \colon sat \in H \text{ if and only if } sbt \in H \ (s, t \in S)\}. \tag{2}$$

For, as defined above in §11.3,

$$\mathscr{P}_H = \mathscr{R}_H L = \{(a, b) \in S \times S \colon (sa, sb) \in \mathscr{R}_H \text{ for all } s \text{ in } S^1\},$$

and here $S^1 = S$ by assumption. Probably the first one to observe this fact was Schützenberger [1955/6]; with applications to coding theory in mind, he called it "syntactic equivalence".

Let $\{H_\delta \colon \delta \in \Delta\}$ be the set of all strong subsets of S. Let $M_\delta = S/\mathscr{R}_{H_\delta}$, and let us regard M_{δ_1} and M_{δ_2} as disjoint if $\delta_1 \neq \delta_2$. (We could, for example, take the elements of M_δ to be pairs (X, δ) with X in $S/\mathscr{R}_{H_\delta}$.) The union M of all the M_δ is an operand over S in an obvious way, and the set W of all the origins $W_\delta = W_{H_\delta}$ $(\delta \in \Delta)$ is an invariant subset of M. The Rees factor $\hat{M} = M/W$ is then a unital, 0-cancellative, centered operand over S.

Let $\hat{\phi}$ be the representation of S associated with the operand $\hat{M}_S$, and let $\hat{\kappa}$ be its kernel. Since $\hat{M}_S$ is 0-decomposable into operands equivalent to the M_δ (in fact identical with the M_δ except that the origin W_δ of M_δ has been replaced by that of $\hat{M}$), it is clear that

$$\hat{\kappa} = \bigcap \{\mathscr{P}_{H_\delta} \colon \delta \in \Delta\}.$$

Now the intersection of any set of strong subsets of S is, if not empty, also a strong subset of S. For each a in S, let $\hat{a}$ be the intersection of all the strong subsets of S (such as S itself) containing a. Let $\hat{\epsilon}$ be the equivalence on S defined by

$$\hat{\epsilon} = \{(a, b) \in S \times S \colon \hat{a} = \hat{b}\}. \tag{3}$$

We proceed to show that $\hat{\epsilon}$ is a congruence on S. First we note that if H is strong and $c \in S$, then

$$Hc^{[-1]} = \{x \in S \colon xc \in H\}$$

is also strong, by Corollary 10.13.

Now suppose that $a \hat{\epsilon} b$, and let $c \in S$. To show that $ac \hat{\epsilon} bc$ it suffices, by symmetry, to show that if H is any strong subset of S containing ac, then H

contains bc also. But $ac \in H$ implies $a \in Hc^{[-1]}$. Since $a \hat{\epsilon} b$, and $Hc^{[-1]}$ is strong, $b \in Hc^{[-1]}$, and so $bc \in H$. Similarly, we can show that $ca \hat{\epsilon} cb$.

We come now to a fundamental result due to Šaĭn [1961].

THEOREM 11.14. *If S is any semigroup with identity, and $\hat{\epsilon}$ is the congruence on S defined by (3), then the factor semigroup $S/\hat{\epsilon}$ admits a faithful representation by partially one-to-one (or one-to-one partial) transformations. If ϕ is any such representation (not necessarily faithful) of S, then the kernel of ϕ contains $\hat{\epsilon}$, and so ϕ induces a representation ψ of $S/\hat{\epsilon}$ such that $\phi = \hat{\epsilon}^{\natural}\psi$.*

PROOF. We shall prove the first part by showing that $\hat{\epsilon}$ coincides with the kernel $\hat{\kappa}$ of the representation $\hat{\phi}$ of S constructed above from the set $\{H_\delta : \delta \in \Delta\}$ of all strong subsets of S. For, by Theorem 11.2, $\hat{\phi}$ induces a faithful representation $\hat{\psi}$ of $S/\hat{\kappa}$ such that $\hat{\phi} = \hat{\kappa}^{\natural}\hat{\psi}$.

We prove first that $\hat{\kappa} \subseteq \hat{\epsilon}$. Let $a \hat{\kappa} b$ $(a, b \in S)$. Then $a \mathscr{P}_{H_\delta} b$ for every $\delta \in \Delta$, hence $sat \in H_\delta$ if and only if $sbt \in H_\delta$. Putting $s = t = 1$, we conclude that $a \in H_\delta$ if and only if $b \in H_\delta$ (all $\delta \in \Delta$), which is equivalent to $a \hat{\epsilon} b$.

The converse inclusion $\hat{\epsilon} \subseteq \hat{\kappa}$ is a special case of the second part of the theorem, that $\hat{\epsilon} \subseteq \phi \circ \phi^{-1}$ for any representation ϕ of S by partially one-to-one transformations, and so it suffices to prove the latter.

Let M_S be the 0-cancellative, centered operand over S associated with ϕ. For each pair of elements x, y of $M \backslash 0_M$ such that $(x, y) \in \tau$, the transitivity relation for ϕ, let

$$H_{x,y} = \{a \in S : xa = y\}.$$

Then $H_{x,y}$ is strong. For it is not empty, and if $a_1b_1, a_1b_2, a_2b_1 \in H_{x,y}$, then $xa_1b_1 = y = xa_2b_1$. Since $y \neq 0_M$ and M_S is 0-cancellative, this yields $xa_2 = xa_1$. Hence $xa_2b_2 = xa_1b_2 = y$, and $a_2b_2 \in H_{x,y}$.

Now let $a \hat{\epsilon} b$, and let $x \in M$. If $xa \neq 0_M$, then $x \neq 0_M$ also, and $a \in H_{x,xa}$. Since $H_{x,xa}$ is strong, and $a \hat{\epsilon} b$, we conclude that $b \in H_{x,xa}$, hence $xb = xa$. Similarly, if $xb \neq 0_M$, then $xa = xb$. The only remaining possibility is $xa = 0_M = xb$, and hence $xa = xb$ in all cases. Thus $a\phi = b\phi$, or $(a, b) \in \phi \circ \phi^{-1}$, which proves that $\hat{\epsilon} \subseteq \phi \circ \phi^{-1}$. The final assertion of the theorem is immediate from Theorem 11.2.

COROLLARY 11.15. *A semigroup S with identity can be embedded in an inverse semigroup if and only if, for each pair of distinct elements of S, there exists a strong subset of S containing one member of the pair but not the other.*

PROOF. As pointed out at the beginning of this section, S can be embedded in an inverse semigroup if and only if it admits a faithful representation by one-to-one partial transformations of a set. By Theorem 11.14, this is so if and only if $\hat{\epsilon}$ is the identity relation on S, and to say that $(a, b) \notin \hat{\epsilon}$ is to say that there is a strong subset H of S such that $a \in H$ and $b \notin H$, or else $a \notin H$ and $b \in H$.

Exercises for §11.4

1. Let M_S be a 0-cancellative, 0-transitive, centered operand over a semigroup S. For each x in $M_S \backslash 0_M$, let $H_x = \{a \in S: xa = x\}$.

   (a) H_x is a strong unitary subsemigroup of S, and $M_S \cong S/\mathscr{R}_{H_x}$, where $\mathscr{R}_{H_x}$ is the Dubreil principal right congruence defined by H_x.

   (b) If K is a strong unitary subsemigroup of S, then K is conjugate to H_x if and only if $K = H_y$ for some y in $M \backslash 0_M$.

2. Let H be a strong unitary subsemigroup of a semigroup S, and let W_H be its right residue. Then every idempotent element of S belongs to $H \cup W_H$.

11.5 Irreducible and transitive operands and semigroups

Let FM denote the set of fixed (= invariant) elements of an operand M_S over a semigroup S. Following Hoehnke [1966], we shall say that M_S is *irreducible* if $MS \nsubseteq FM$ and the only invariant subset of M of cardinal greater than one is M itself. We apply the same term, of course, to the representation of S associated with M_S. In Lemma 11.16 below, we show that M_S is irreducible if and only if it is either transitive or 0-transitive (§11.1).

Following Tully [1960, 1961], we call a semigroup S *right* [0-] *transitive* if it has a faithful [0-] transitive operand (or representation). Hoehnke calls a semigroup S "primitive" if it has a faithful irreducible representation; hence, by Lemma 11.16, S is primitive in his sense if and only if it is either right transitive or right 0-transitive in Tully's sense. We shall call such a semigroup *right irreducible*. S is called *left* [0-] *transitive* if it has a faithful [0-] transitive left operand (or anti-representation). S is left [0-] transitive if and only if its dual S^* is right [0-] transitive.

A number of examples of semigroups which are transitive on one side but not the other, and of others which are transitive on both sides, are given in the exercises. The remaining results of this section, all due to Tully [1960, 1961], describe a number of classes of transitive semigroups which are not so easily handled. Of especial interest is the fact (Theorem 11.20) that, with one exception, any free product is transitive, and hence (Corollary 11.21) any free semigroup on more than one generator is transitive.

Lemma 11.16. (A) *Let M_S be an operand over a semigroup S such that the only invariant subset of M of cardinal greater than one is M itself. Then either M_S is irreducible, or else $|M| = 2$ and M_S is either null or fixed.*

(B) *An operand M_S over a semigroup S is irreducible if and only if it is either transitive or 0-transitive.*

Proof. (A) If $FM = \square$, then M_S is irreducible by definition. Hence we may assume $FM \neq \square$. Since FM is an invariant subset of M, we must have either $FM = M$ or $|FM| = 1$. If $FM = M$, then M is fixed; and since every subset of M is invariant, we must have $|M| = 2$. If $|FM| = 1$, let $FM = \{0_M\}$. If M_S is not irreducible, then $MS \subseteq FM$, that is, $MS = 0_M$. Thus M_S is null, and since every subset of M containing 0_M is invariant, we must again have $|M| = 2$.

(B) Let M_S be irreducible. The possibility $FM = M$ is excluded by the condition $MS \nsubseteq FM$, and hence $FM = \square$ or $|FM| = 1$. Assume the latter, and again let $FM = \{0_M\}$. For any x in $M \backslash 0_M$, xS is an invariant subset of M, and hence either $|xS| = 1$ or $xS = M$. That M is 0-transitive will follow when we show that $|xS| = 1$ is impossible. If $xS = \{y\}$, then $y \in FM$, hence $y = 0_M$. Then $\{x, 0_M\}$ is an invariant subset of M of cardinal greater than one, and hence is equal to M. But then $MS = \{0_M\}$, contrary to $MS \nsubseteq FM$.

If $FM = \square$, we can show by a similar proof that M_S is transitive.

Conversely, let M_S be 0-transitive. If N is an invariant subset of M with $|N| > 1$, then N contains an element x of $M \backslash 0_M$. Then $M = xS \subseteq N$, so $N = M$. Clearly $FM = \{0_M\}$, and if $x \in M \backslash 0_M$ then $xa = x$ for some a in S, so that $MS \nsubseteq FM$. Hence M_S is irreducible.

Similarly, if M_S is transitive, then $FM = \square$, and M itself is the only invariant subset of M, so that M_S is irreducible.

THEOREM 11.17. *Let S be a semigroup containing a non-degenerate* [0-] *minimal right ideal R. Then S is right* [0-] *transitive if and only if $ra = rb$ $(a, b \in S)$ for all r in R implies $a = b$.*

PROOF. We prove the bracketed statement; the proof of the unbracketed one is similar.

If the stated condition holds, then R_S is clearly a faithful operand over S. If $r \in R \backslash 0$, then $R = rS$, by Lemma 6.1, and hence R_S is 0-transitive.

Assume conversely that M_S is a faithful, 0-transitive operand over S. Since M_S is faithful, $MR \neq 0_M$, and hence there exists x in $M \backslash 0_M$ such that $xR \neq 0_M$. Since xR is an invariant subset $\neq 0_M$ of M, and M_S is irreducible by Lemma 11.16, we conclude that $xR = M$. The mapping $r \rightarrow xr$ is an operator homomorphism of R_S onto M_S. Let a and b be elements of S such that $ra = rb$ for all $r \in R$. If $y \in M$, then $y = xr$ for some r in R, and hence $ya = xra = xrb = yb$. Since M_S is faithful, we conclude $a = b$.

The next corollary has connexions with Exercise 2 for §3.6. The relationship between 0-transitive representations and monomial representations will be taken up in §11.8.

COROLLARY 11.18. *A regular Rees matrix semigroup $S = \mathscr{M}^0(G; I, \Lambda; P)$ is right 0-transitive if and only if no two different columns of the sandwich matrix P are proportional.*

PROOF. Let $i \in I$, and let R_i be the set of all elements $(c; i, \lambda)$ of S with c in G and λ in Λ. By Lemma 3.2, $R_i^0 = R_i \cup 0$ is a 0-minimal right ideal of S.

Let $(a; j, \mu)$ and $(b; k, \nu)$ be non-zero elements of S. The condition

$$(c; i, \lambda)(a; j, \mu) = (c; i, \lambda)(b; k, \nu)$$

for all $(c; i, \lambda)$ in R_i^0 becomes

$$(cp_{\lambda j}a; i, \mu) = (cp_{\lambda k}b; i, \nu)$$

for all c in G and λ in Λ. This in turn is equivalent to $\mu = \nu$ and

$$p_{\lambda j}a = p_{\lambda k}b \quad (\text{all } \lambda \in \Lambda),$$

which is what we mean by saying that the jth and kth columns of P are proportional ($a, b \in G\backslash 0$).

Hence the condition of Theorem 11.17 is equivalent in this case to the assertion that no two different columns of P are proportional.

We now have the following result (also proved by Šaĭn [1963]).

LEMMA 11.19. *A semigroup S is right transitive if and only if S^1 is right transitive.*

PROOF. Since $S = S^1$ if S has an identity, we may assume that S has no identity.

Let S be right transitive. Then there exists a faithful transitive operand M_S over S. We make M into an operand M_{S^1} over S^1 by defining $x1 = x$ for all x in M. Clearly M_{S^1} is transitive. To show that it is faithful, assume that $x1 = xa$ for all x in M and some element a in S. Then, for every s in S, $xas = x1s = xs$ and $xsa = xs1 = xs$. Since M_S is faithful, we conclude that $as = s = sa$ for all s in S, contrary to the assumption that S has no identity.

Conversely, let S^1 be right transitive, and let M_{S^1} be a faithful transitive operand over S^1. M_{S^1} is unital, for if $x \in M$ then $xa = x$ for some a in S^1, and so $x1 = xa1 = xa = x$. M is also an operand M_S over S, and is clearly faithful. If $x, y \in M$, there exists a in S^1 such that $xa = y$. If $x \neq y$, then $a \in S$, since $x1 = x$. If $|M| = 1$, then clearly M_S is transitive. Assume $|M| > 1$, and let $x, y \in M$ with $x \neq y$. Then $xa = y$ and $yb = x$ with $a, b \in S$, and hence $xab = x$. This shows that $xS = M$ for any x in M, and hence M_S is transitive.

THEOREM 11.20. *Let P be the free product of two semigroups S and T. Then P is transitive on both sides, except when $|S| = |T| = 1$.*

PROOF. We may assume that $|S| > 1$. (For the excluded case, see Exercise 6 below.) By symmetry, we need only show that P is right transitive. We shall prove this for P^1, and it follows for P by Lemma 11.19.

We proceed to construct what, in analogy with Croisot's procedure of §9.5, might be termed a set of canonical forms for a right congruence on P^1. As in the application of Theorem 9.54 in §9.5, we begin by means of an auxiliary transformation E.

Let a be a fixed element of S.

Each element of P is uniquely expressible as a product of elements taken alternately from S and T. Let $p \in P$. Then pE is defined as follows:

(1) if p begins with an element t of T, cancel t in p;
(2) if p begins with st ($s \in S \backslash a$, $t \in T$), cancel st;
(3) if p begins with a factor of the form

$$(at_1)(at_2)\cdots(at_n)(s_1t_1')(s_2t_2')\cdots(s_nt_n')$$

where $t_i, t_i' \in T$, $s_i \in S$ ($i = 1, \cdots, n$), and $s_1 \neq a$, then cancel this factor.

(4) If no such cancellation in p is possible, we define $pE = p$.

The final alternative $pE = E$ occurs if and only if $p = 1$, or $p \in S$, or

$$p = (at_1)(at_2)\cdots(at_n)q$$

where q is an element of P^1 of length less than $2n$. We call these *canonical forms* under E.

E may reduce p to 1; for example, if $p = st$. We agree also that $1E = 1$. We remark that E is a transformation of P^1. Since either $pE = p$ or the length of pE is less than that of p, there must exist a positive integer k such that $\overline{p} = pE^k$ is a canonical form. We call $\overline{p}$ the *canonical form of* p.

We observe that

$$\overline{\overline{p}q} = \overline{pq} \quad \text{(all } p, q \text{ in } P^1).$$

This is trivial if $\overline{p} = p$. Otherwise, it follows by induction on the least integer k such that $\overline{p} = pE^k$, from the fact that if $pE \neq p$ then $(pq)E = (pE)q$. Consequently, the equivalence relation σ on P^1 defined by $p \,\sigma\, q$ ($p, q \in P^1$) if $\overline{p} = \overline{q}$ is a right congruence. For if $p \,\sigma\, q$, and $r \in P^1$, then

$$\overline{pr} = \overline{\overline{p}r} = \overline{\overline{q}r} = \overline{qr},$$

hence $pr \,\sigma\, qr$. We proceed to show that the operand P^1/σ is transitive and faithful.

The transitivity will follow when we show that, for given p in P^1, there exists q in P^1 such that $pq \,\sigma\, 1$, since then $(p\sigma)(qr) = r\sigma$ for arbitrary r in P^1. It suffices to take $q = (bt)^n$ with b in $S \backslash a$, t in T, and n equal to the length of $\overline{p}$.

To show that P^1/σ is faithful, we must show that if $p \neq q$ in P^1, then there exists r in P^1 such that $(rp, rq) \notin \sigma$. It is easily seen, from the nature of P, that either $ap \neq aq$ or $tp \neq tq$. Thus if we take

$$r = \begin{cases} (at)^n & \text{if } ap = aq \quad \text{(and so } tp \neq tq), \\ (at)^n a & \text{if } ap \neq aq, \end{cases}$$

then $rp \neq rq$. If we take n greater than the lengths of p and q, then $\overline{rp} = rp$ and $\overline{rq} = rq$, whence $(rp, rq) \notin \sigma$.

Corollary 11.21. *A free semigroup on more than one generator is transitive on both sides.*

Theorem 11.22. *Any Baer-Levi semigroup is transitive on both sides.*

Proof. Let S be a Baer-Levi semigroup of type (p, q). By definition

(§8.1), S consists of all one-to-one transformations α of a set M of infinite cardinal p such that $|M \backslash M\alpha| = q$. Clearly, since S is right simple, the identity mapping on S is a transitive representation of S, so that all we need to prove is that S is left transitive.

Well-order M by means of the first ordinal having cardinal p. Then each subset N of M is well-ordered by restriction to N of the ordering on M. If $|N| = p$, the resulting ordinal precedes or equals that of M, and since it has cardinal p, it must be equal to that of M. Hence there is a (unique) order-preserving, one-to-one mapping π_N of N onto M. For each α in S, it follows that $\alpha\pi_{M\alpha}$ is a permutation of M.

For α, β in S, define $\alpha\,\sigma\,\beta$ to mean that $\alpha\pi_{M\alpha} = \beta\pi_{M\beta}$. It is clear that σ is an equivalence relation on S. We show next that $\alpha\,\sigma\,\beta$ if and only if

$$(1) \qquad x\alpha < y\alpha \quad \text{if and only if } x\beta < y\beta \quad (x, y \in M).$$

It is immediate from this that $\alpha\,\sigma\,\beta$ implies $\gamma\alpha\,\sigma\,\gamma\beta$ for all γ in S, whence σ is a left congruence on S.

Since π_N is order-preserving, (1) is equivalent to

$$(2) \qquad x\alpha\pi_{M\alpha} < y\alpha\pi_{M\alpha} \quad \text{if and only if} \quad x\beta\pi_{M\beta} < y\beta\pi_{M\beta}.$$

If $\alpha\,\sigma\,\beta$, this is trivial, and so (1) holds. But (2) implies $\alpha\,\sigma\,\beta$. For if we set $\phi = \alpha\pi_{M\alpha}$ and $\psi = \beta\pi_{M\beta}$, and if $x < y$, then $(x\phi^{-1})\phi < (y\phi^{-1})\phi$, so that (2) implies that $x\phi^{-1}\psi < y\phi^{-1}\psi$. Hence $\phi^{-1}\psi$ is an order-preserving permutation of the well-ordered set M. Since the only such is the identity, we conclude that $\phi = \psi$, and hence $\alpha\,\sigma\,\beta$.

To show that the left operand ${}_S(S/\sigma)$ is transitive, let $\alpha, \beta \in S$. We are to show that there exists γ in S such that $\gamma\alpha\,\sigma\,\beta$. Choose any subset P of $M\alpha$ such that $|P| = p$ and $|M\alpha \backslash P| = q$, and let $\delta = \beta\pi_{M\beta}\pi_P^{-1}$. Since δ is a one-to-one mapping of M onto P, and

$$|M \backslash P| = |M \backslash M\alpha| + |M\alpha \backslash P| = q + q = q,$$

it follows that $\delta \in S$. For x in M, $x\delta \in P \subseteq M\alpha$, and there is a unique y in M such that $x\delta = y\alpha$; define $x\gamma = y$. Then $x\delta = y\alpha = x\gamma\alpha$ for all x in M, so $\delta = \gamma\alpha$. Since δ is one-to-one, so is γ. Moreover,

$$M\gamma = \{y \in M : y\alpha \in M\delta\} = P\alpha^{-1},$$

$$|M \backslash M\gamma| = |M \backslash P\alpha^{-1}| = |M\alpha \backslash P| = q,$$

whence $\gamma \in S$. That $\delta\,\sigma\,\beta$ is clear from $P = M\delta$ and the definition of δ.

To show that S/σ is faithful, let α and β be distinct elements of S. We are to show that there exists γ in S such that $\alpha(\gamma\sigma) \neq \beta(\gamma\sigma)$, that is, $(\alpha\gamma, \beta\gamma) \notin \sigma$.

Suppose first that $(\alpha, \beta) \notin \sigma$. Then, using (1), there exist x and y in M such that $x\alpha < y\alpha$ and $x\beta > y\beta$. There evidently exists an element γ in S leaving fixed each of the (finitely many) elements $x\alpha$, $y\alpha$, $x\beta$, $y\beta$. Then $x\alpha\gamma < y\alpha\gamma$ and $x\beta\gamma > y\beta\gamma$, so that $(\alpha\gamma, \beta\gamma) \notin \sigma$, by (1).

We may therefore assume $(\alpha, \beta) \in \sigma$. Since $\alpha \neq \beta$, there exists x in M such that $x\alpha \neq x\beta$, and we may assume $x\alpha < x\beta$. Choose $y \neq x$ in M, and let

$$u_1 = x\alpha, \quad v_1 = y\alpha, \quad u_2 = x\beta, \quad v_2 = y\beta.$$

Then $u_1 < u_2$. Since α is one-to-one, $u_1 \neq v_1$. If $u_1 < v_1$ then $u_2 < v_2$, and conversely, since $\alpha \,\sigma\, \beta$. Hence there are two cases:

$$\text{(i) } u_1 < v_1,\, u_1 < u_2 < v_2; \qquad \text{(ii) } u_2 > v_2,\, u_2 > u_1 > v_1.$$

In case (i), let γ be an element of S which transposes u_2 and v_2, leaves u_1 fixed, and leaves v_1 fixed unless $v_1 = u_2$ or $v_1 = v_2$. Clearly such an element γ exists in S. Then $v_1\gamma \in \{v_1, u_2, v_2\}$, so that $u_1\gamma < v_1\gamma$. But $u_2\gamma > v_2\gamma$, showing that $(\alpha\gamma, \beta\gamma) \notin \sigma$. We arrive at the same conclusion in case (ii) by choosing γ in S so that γ transposes u_1 and v_1, leaves u_2 fixed, and leaves v_2 fixed unless $v_2 = u_1$ or $v_2 = v_1$.

Exercises for §11.5

1. A right transitive semigroup S of order greater than one cannot have a left zero element.

2. The full transformation semigroup $\mathscr{T}_M$ on a set M with $|M| > 1$ is right transitive but not left transitive. (Hoehnke [1966].)

3. (a) A right transitive semigroup is left reductive (§1.3).

(b) A right simple, left reductive semigroup is right transitive; in particular, a right group is right transitive.

(c) If a right group is left transitive, it is a group. (Tully [1960].)

4. A commutative semigroup S is [0-] transitive if and only if it is a group [with zero]. (Tully [1960] and Šaĭn [1963].)

5. The bicyclic semigroup is transitive on both sides. (Tully [1960, 1961].)

6. Let $S = \{e\}$ and $T = \{f\}$ be one-element semigroups, and let P be their free product. Any transitive operand over P must be finite, and so cannot be faithful; thus P is not a transitive semigroup (cf. Theorem 11.20).

7. A 0-bisimple inverse semigroup is 0-transitive on both sides. (See Exercise 5(e) for §7.3.)

11.6 Various radicals of a semigroup

The notions of the $\mathscr{C}$-radical of a semigroup and the derived type $\mathscr{C}'$ of a type $\mathscr{C}$ of semigroups, and the first three theorems of this section, are due to E. J. Tully (unpublished). We then turn to the case $\mathscr{C} = \mathscr{I}$, the class of right irreducible semigroups (§11.5), and give an interesting connexion, due to H. Seidel [1965], between the $\mathscr{I}$-radical and the nilradical of a semigroup.

Apart from the exposition we have given of Tully's unpublished theory, our only claim to novelty in this section is Theorem 11.25. This theorem involves the notion of maximal homomorphic image of type $\mathscr{C}$, and we need a definition of this concept somewhat more restrictive than that given in Proposition 1.7 of §1.5.

By a *type* of semigroups we mean a class $\mathscr{C}$ of semigroups (in general not a set) such that (i) if $S \in \mathscr{C}$ and S is isomorphic to S', then $S' \in \mathscr{C}$, and (ii) any one-element semigroup belongs to $\mathscr{C}$.

If S is any semigroup, and $\mathscr{C}$ is any type of semigroups, then a semigroup S^* is called a *maximal homomorphic image of S of type $\mathscr{C}$* if $S^* \in \mathscr{C}$ and if there exists a homomorphism η of S onto S^* with the *factorization property*: if ϕ is a homomorphism of S onto a semigroup T of type $\mathscr{C}$, then there exists a homomorphism θ of S^* onto T such that $\eta\theta = \phi$.

The factorization property of η was omitted in the earlier definition. In the strengthened form, it is easy to show that if a maximal homomorphic image of S of type $\mathscr{C}$ exists, then it is unique to within isomomorphism. This is not true for the weaker version, since it is possible for two non-isomorphic semigroups (in fact groups) to be homomorphic images of each other.

Proposition 1.7 remains true when "maximal homomorphic image of type $\mathscr{C}$" is construed in its new meaning. Assume that the intersection ρ of all congruences σ on S having type $\mathscr{C}$ (i.e., $S/\sigma \in \mathscr{C}$) also has type $\mathscr{C}$. Then S/ρ is a maximal homomorphic image of S of type $\mathscr{C}$ in the strong sense, with η in the definition taken to be the natural homomorphism $\rho^\natural$ of S onto S/ρ. For let ϕ be a homomorphism of S onto a semigroup T of type $\mathscr{C}$. Then $\sigma = \phi \circ \phi^{-1}$ is a congruence on S of type $\mathscr{C}$, since $S/\sigma \cong T$ and $T \in \mathscr{C}$, so that $S/\sigma \in \mathscr{C}$ by condition (i) on $\mathscr{C}$. Hence $\rho \subseteq \sigma$ by definition of ρ. Since $\rho = \rho^\natural \circ \rho^{\natural-1}$, it follows from Theorem 1.6 that there exists a (unique) homomorphism θ of S/ρ onto T such that $\rho^\natural\theta = \phi$.

In this connexion, we do not know if Exercise 16 for §2.7 is correct as it stands, in the new sense. It is correct if we assume that H_e is a "homomorphic retract" of S; that is, that there exists a homomorphism of S onto H_e leaving the elements of H_e fixed. (See Appendix.)

Let S be a semigroup, and let $\mathscr{C}$ be a type. Let $\mathscr{C}_S$ be the set of all congruences on S of type $\mathscr{C}$. $\mathscr{C}_S \neq \square$ since it contains the universal congruence ω_S on S, by condition (ii) of the definition of "type". The *$\mathscr{C}$-radical* of S is defined by

$$\mathscr{C}\text{-rad}\, S = \bigcap \{\sigma \colon \sigma \in \mathscr{C}_S\}.$$

It is clearly a congruence on S. (It was denoted by ρ in the above proof of the new Proposition 1.7.) The *derived type* $\mathscr{C}'$ of $\mathscr{C}$ is the class of all semigroups S such that $\mathscr{C}\text{-rad}\, S = \iota_S$, the equality relation on S. If $S \in \mathscr{C}'$, we say that S is *$\mathscr{C}$-radical-free*. Clearly $\mathscr{C} \subseteq \mathscr{C}'$; for if $S \in \mathscr{C}$ then $\iota_S \in \mathscr{C}_S$.

THEOREM 11.23. *If S is any semigroup, and $\mathscr{C}$ is any type of semigroups, then $S/\mathscr{C}$-rad S is of type $\mathscr{C}'$.*

PROOF. Let $\rho = \mathscr{C}\text{-rad}\, S$, and let $S' = S/\rho$. We are to show that $\mathscr{C}\text{-rad}\, S' = \iota_{S'}$.

If σ is a congruence on S containing ρ, then σ induces a congruence $\sigma' = \sigma/\rho$ on S' defined as follows:

$$(a\rho)\,\sigma'\,(b\rho) \quad \text{if} \quad a\,\sigma\, b \quad (a, b \in S).$$

The mapping $\lambda\colon \sigma \to \sigma'$ is a lattice isomorphism of the lattice of all congruences on S containing ρ onto the lattice of all congruences on S'. Moreover,

$$S'/\sigma' = (S/\rho)/(\sigma/\rho) \cong S/\sigma.$$

Consequently, $\sigma' \in \mathscr{C}_{S'}$ if and only if $\sigma \in \mathscr{C}_S$. Since λ is intersection-preserving,

$$\rho\lambda = (\bigcap\{\sigma\colon \sigma \in \mathscr{C}_S\})\lambda = \bigcap\{\sigma'\colon \sigma' \in \mathscr{C}_{S'}\} = \mathscr{C}\text{-rad}\, S'.$$

But $\rho\lambda = \rho/\rho = \iota_{S'}$, whence $\mathscr{C}\text{-rad}\, S' = \iota_{S'}$, and $S' \in \mathscr{C}'$.

THEOREM 11.24. *For any type $\mathscr{C}$ of semigroups, $\mathscr{C}'' = \mathscr{C}'$. For any semigroup S, $\mathscr{C}'\text{-rad}\, S = \mathscr{C}\text{-rad}\, S$.*

PROOF. We prove the second assertion first.

Let $\rho \in \mathscr{C}'_S$, and let $S' = S/\rho$. Then $S' \in \mathscr{C}'$, and so

$$\bigcap\{\sigma'\colon \sigma' \in \mathscr{C}_{S'}\} = \iota_{S'}.$$

Let $\lambda\colon \sigma \to \sigma' = \sigma/\rho$ be as in the proof of Theorem 11.23. Then

$$(\bigcap\{\sigma\colon \sigma \in \mathscr{C}_S,\ \sigma \supseteq \rho\})\lambda = \bigcap\{\sigma'\colon \sigma' \in \mathscr{C}_{S'}\} = \iota_{S'}.$$

Since $\rho\lambda = \rho/\rho = \iota_{S'}$, and λ is one-to-one, we conclude that

$$\bigcap\{\sigma\colon \sigma \in \mathscr{C}_S,\ \sigma \supseteq \rho\} = \rho.$$

In other words, every congruence ρ on S of type $\mathscr{C}'$ is the intersection of all congruences σ of type $\mathscr{C}$ containing ρ.

We conclude that

$$\mathscr{C}'\text{-rad}\, S = \bigcap\{\rho\colon \rho \in \mathscr{C}'_S\} = \bigcap\{\rho\colon \rho \in \mathscr{C}_S\} = \mathscr{C}\text{-rad}\, S.$$

From this, the first assertion of the theorem follows. For if $S \in \mathscr{C}''$, then $\iota_S = \mathscr{C}'\text{-rad}\, S = \mathscr{C}\text{-rad}\, S$, whence $S \in \mathscr{C}'$; and the converse is trivial.

THEOREM 11.25. (A) *A semigroup S admits a maximal homomorphic image S^* of type $\mathscr{C}$ if and only if $S/\mathscr{C}$-rad S has type $\mathscr{C}$, and then we may take $S^* = S/\mathscr{C}$-rad S.*

(B) *Every semigroup S has a maximal homomorphic image of type $\mathscr{C}'$, namely $S/\mathscr{C}$-rad S.*

(C) *A type $\mathscr{C}$ has the property that every semigroup S has a maximal homomorphic image of type $\mathscr{C}$ if and only if $\mathscr{C}' = \mathscr{C}$.*

PROOF. (A) The "if" part is just the new version of Proposition 1.7, proved above. Conversely, let η be a homomorphism of S onto S^* with the factorization property. Then $\rho = \eta \circ \eta^{-1}$ has type $\mathscr{C}$, and it will follow that $\rho = \mathscr{C}\text{-rad}\, S$ when we show that if σ is any congruence on S of type $\mathscr{C}$ then $\rho \subseteq \sigma$.

Since $\sigma^\natural$ is a homomorphism of S onto the semigroup S/σ of type $\mathscr{C}$, it follows from the factorization property of η that there exists a homomorphism θ of S^* onto S/σ such that $\eta\theta = \sigma^\natural$. If a and b are elements of S such that $a\,\rho\,b$, then $a\eta = b\eta$ (since $\rho = \eta \circ \eta^{-1}$), and hence $a\sigma^\natural = a\eta\theta = b\eta\theta = b\sigma^\natural$. But this means that $a\,\sigma\,b$, proving that $\rho \subseteq \sigma$.

(B) $S/\mathscr{C}\text{-rad}\, S$ is of type $\mathscr{C}'$ by Theorem 11.23. It is the same as $S/\mathscr{C}'\text{-rad}\, S$ by Theorem 11.24, and the assertion then follows from (A) with $\mathscr{C}$ replaced by $\mathscr{C}'$.

(C) The "if" part is immediate from (B). Conversely, assume that every semigroup has a maximal homomorphic image of type $\mathscr{C}$. We need only prove that $\mathscr{C}' \subseteq \mathscr{C}$, the converse inclusion being trivial. Let $S \in \mathscr{C}'$. From (A) and our hypothesis we have that $S/\mathscr{C}\text{-rad}\, S$ belongs to $\mathscr{C}$. But $\mathscr{C}\text{-rad}\, S = \iota_S$ by definition of $\mathscr{C}'$, and hence $S = S/\iota_S \in \mathscr{C}$.

For an alternative characterization of the types $\mathscr{C}$ of semigroups for which $\mathscr{C} = \mathscr{C}'$, see Exercise 2(b) below.

We turn now to the class $\mathscr{I}$ of right irreducible semigroups, called "primitive" by Hoehnke, namely those admitting a faithful irreducible representation. We examined various subclasses of $\mathscr{I}$ in §11.5. We shall write $\text{rad}\, S$ for $\mathscr{I}\text{-rad}\, S$. We begin with a description of $\text{rad}\, S$ due to H. Seidel [1965].

For any element a of a semigroup S, let

$$\mu(a) = \{(b, c) \in S \times S \colon a^m b = a^n c \text{ for some nonnegative integers } m, n\}.$$

In this, a^0 is to be interpreted as the element 1 of S^1. Clearly $\mu(a)$ is a right congruence on S; in fact, it is modular, since a is a left identity of $S \bmod \rho$ (§11.3).

THEOREM 11.26. *For any semigroup S,*

$$\text{rad}\, S = \{(a, b) \in S \times S \colon (a, b) \in \mu(as) \cap \mu(bs) \text{ for all } s \text{ in } S^1\}.$$

PROOF. Let $(a, b) \notin \text{rad}\, S$. Then there exists an irreducible operand M_S over S and an element x of M such that $xa \neq xb$. Not both xa and xb can be 0_M (if such exists), and we may assume $xa \neq 0_M$. Then $xaS = M$. Let s be an element of S such that $xas = x$. We proceed to show that $(a, b) \notin \mu(as)$. For otherwise, $(as)^m a = (as)^n b$ for some $m, n \geq 0$, and we conclude that

$$xa = x(as)^m a = x(as)^n b = xb.$$

Conversely, assume $(a, b) \notin \mu(as) \cap \mu(bs)$ for some s in S^1. We may then assume that $(a, b) \notin \mu(as)$. Let $\sigma = \mu(as)$. By Theorem 11.6, $M_S = S/\sigma$ is a strictly cyclic operand over S with strict generator $(as)\sigma$, and such that

$a\sigma \neq b\sigma$. By Lemma 11.5, $M/I(M) = M'_S$ is transitive or 0-transitive, and hence irreducible by Lemma 11.16. Since as is a left identity for $S \bmod \sigma$, we have

$$((as)\sigma)a = a\sigma \neq b\sigma = ((as)\sigma)b.$$

Hence $a\phi \neq b\phi$, where ϕ is the representation of S associated with M'_S. Since $S/\phi \circ \phi^{-1}$ is isomorphic with the irreducible semigroup $S\phi$ of transformations of M', $\phi \circ \phi^{-1}$ is a congruence on S of type $\mathscr{I}$. Since $(a, b) \notin \phi \circ \phi^{-1}$, and $\operatorname{rad} S$ is the intersection of all congruences on S of type $\mathscr{I}$, it follows that $(a, b) \notin \operatorname{rad} S$.

For a semigroup S with zero element 0, Hoehnke [1963] defines $\operatorname{rad}^0 S$ to be the $(\operatorname{rad} S)$-class containing 0. We recall (Exercise 13 for §6.6) that the *nilradical* $N(S)$ of S is defined to be the union of all the nilideals of S, and an ideal of S is *nil* if every element of it is nilpotent. The following is due to Seidel [1965].

COROLLARY 11.27. $\operatorname{rad}^0 S = N(S)$.

PROOF. Let $a \in N(S)$. Let M_S be any irreducible operand over S. By Lemma 11.16, M_S is transitive or 0-transitive, and it must be the latter since S has a zero element. If $Ma \neq 0_M$, then there exists x in M such that $xa \neq 0_M$, and so $xaS = M$. Hence $xas = x$ for some s in S. But $as \in N(S)$, and so $(as)^n = 0$ for some n, whence $x = x(as)^n = x0 = 0_M$. But this contradicts $xa \neq 0_M$, and we conclude that $Ma = 0_M$. Since this holds for every irreducible operand M_S over S, $(a, 0) \in \operatorname{rad} S$; that is, $a \in \operatorname{rad}^0 S$.

Conversely, let $a \in \operatorname{rad}^0 S$; i.e., $(a, 0) \in \operatorname{rad} S$. Evidently $\mu(0) = \omega_S$, and we conclude from Theorem 11.26 that $(a, 0) \in \mu(as)$ for all s in S^1. But this means that there exist non-negative integers m and n such that

$$(as)^m a = (as)^n 0.$$

Hence $(as)^{m+1} = 0$, and a belongs to the nilideal $aS^1 \subseteq N(S)$.

EXERCISES FOR §11.6

1. Let $\mathscr{G}$ be the class of groups, and let $\mathscr{G}'$ be the derived class. Any infinite cyclic semigroup belongs to $\mathscr{G}'$. Hence $\mathscr{G} \neq \mathscr{G}'$. (See Appendix.)

2. (a) Let $\mathscr{C}$ be any type of semigroups, and let $\mathscr{C}'$ be the derived type. Then a semigroup belongs to $\mathscr{C}'$ if and only if it is isomorphic to a subdirect product of semigroups of type $\mathscr{C}$. (By a theorem of Birkhoff, *Lattice theory*, [1948, p. 92], there is a one-to-one correspondence between the representations of a semigroup S as a subdirect product of semigroups S_i $(i \in I)$, and sets $\{\theta_i : i \in I\}$ of congruence relations on S such that $\bigcap \{\theta_i : i \in I\} = \iota_S$. Here $S_i \cong S/\theta_i$.)

(b) Let $\mathscr{C}$ be a type of semigroups. Then $\mathscr{C}$ is such that any semigroup has a maximal homomorphic image of type $\mathscr{C}$ if and only if $\mathscr{C}$ is closed under the formation of subdirect products, i.e. if and only if any subdirect product of semigroups of type $\mathscr{C}$ is also of type $\mathscr{C}$.

11.7 The normalizer of a right congruence ρ and the endomorphisms of S/ρ

Let M_S be an operand over a semigroup S, and let ϕ be the associated representation of S by transformations of M. The set $\mathscr{E}(M_S)$ of operator endomorphisms of M_S is a semigroup under iteration; in fact it is the centralizer semigroup of $S\phi$ within $\mathscr{T}_M$. The one theorem of this brief section, found by Tully [1960] and Hoehnke [1966], tells us how to construct $\mathscr{E}(M_S)$ when M_S is strictly cyclic. By Theorem 11.6, this is equivalent to the construction of $\mathscr{E}(S/\rho)$, where ρ is a modular right congruence on S.

We define the *normalizer* $\mathscr{N}(\rho)$ of a right congruence ρ on S as follows:

$$\mathscr{N}(\rho) = \{a \in S\colon s\,\rho\,t\ (s, t \in S) \text{ implies } as\,\rho\,at\}.$$

Clearly $\mathscr{N}(\rho)$ is a subsemigroup of S. Moreover, it is a union of ρ-classes. For if $a \in \mathscr{N}(\rho)$ and $a\,\rho\,b$, then $as\,\rho\,bs$ and $at\,\rho\,bt$ for any $s, t \in S$; hence if $s\,\rho\,t$, we have $as\,\rho\,at$, and we conclude $bs\,\rho\,bt$ from the transitivity of ρ. It is evident that $\rho' = \rho \cap (\mathscr{N}(\rho) \times \mathscr{N}(\rho))$ is a (two-sided) congruence on $\mathscr{N}(\rho)$. The factor semigroup $\mathscr{N}(\rho)/\rho'$ consists of all the ρ-classes $a\rho$ of S with a in $\mathscr{N}(\rho)$, and we shall denote it by $\mathscr{N}(\rho)/\rho$.

For each element a of $\mathscr{N}(\rho)$ we define a transformation λ_a of S/ρ by

$$(s\rho)\lambda_a = (as)\rho \quad (\text{all } s \in S).$$

λ_a is single-valued since $a \in \mathscr{N}(\rho)$. Moreover, it is an operator endomorphism of S/ρ. For, if $s, t \in S$,

$$((s\rho)t)\lambda_a = ((st)\rho)\lambda_a = (ast)\rho = ((as)\rho)t = ((s\rho)\lambda_a)t.$$

Clearly $\lambda_a\lambda_b = \lambda_{ba}$ for all a, b in $\mathscr{N}(\rho)$. If $a\,\rho\,b$ $(a, b \in \mathscr{N}(\rho))$, then $as\,\rho\,bs$ for all s in S; hence

$$(s\rho)\lambda_a = (as)\rho = (bs)\rho = (s\rho)\lambda_b,$$

and $\lambda_a = \lambda_b$. We conclude that $\lambda\colon a\rho \to \lambda_a$ is an anti-homomorphism of $\mathscr{N}(\rho)/\rho$ into the semigroup $\mathscr{E}(S/\rho)$ of operator endomorphisms of S/ρ.

Now assume that ρ is modular, and let e be a left identity of $S \bmod \rho$. Then $e \in \mathscr{N}(\rho)$. For if $s\,\rho\,t$, then $es\,\rho\,s\,\rho\,t\,\rho\,et$. If $a \in \mathscr{N}(\rho)$, then $(e\rho)(a\rho) = (ea)\rho = a\rho$, so that $e\rho$ is a left identity of $\mathscr{N}(\rho)/\rho$. Let

$$\mathscr{N}_e(\rho) = \{a \in \mathscr{N}(\rho)\colon ae\,\rho\,a\}.$$

Clearly $\mathscr{N}_e(\rho)$ is a left ideal of $\mathscr{N}(\rho)$, and is a union of ρ-classes. $\mathscr{N}_e(\rho)/\rho$ is the principal left ideal of $\mathscr{N}(\rho)/\rho$ generated by the idempotent $e\rho$.

Theorem 11.28. *Let ρ be a modular right congruence on a semigroup S, and let e be a left identity of S modulo ρ. Let $\mathscr{N}(\rho)$ be the normalizer of ρ, and let $\mathscr{N}_e(\rho)$ be the left ideal of $\mathscr{N}(\rho)$ defined above. Let $\lambda\colon a\rho \to \lambda_a$ be the anti-homomorphism of $\mathscr{N}(\rho)/\rho$ into the semigroup $\mathscr{E}(S/\rho)$ of operator endomorphisms of S/ρ defined above. Then the restriction of λ to $\mathscr{N}_e(\rho)/\rho$ is an anti-isomorphism of $\mathscr{N}_e(\rho)/\rho$ onto $\mathscr{E}(S/\rho)$.*

PROOF. Let $\mu \in \mathscr{E}(S/\rho)$, and let $(e\rho)\mu = c\rho$. For any s in S,

$$(1) \qquad (s\rho)\mu = ((es)\rho)\mu = ((e\rho)s)\mu = ((e\rho)\mu)s = (c\rho)s = (cs)\rho.$$

If $s\,\rho\,t$, this implies

$$(cs)\rho = (s\rho)\mu = (t\rho)\mu = (ct)\rho,$$

or $cs\,\rho\,ct$, whence $c \in \mathscr{N}(\rho)$. Hence, by definition of λ_c, (1) becomes $(s\rho)\mu = (s\rho)\lambda_c$, whence $\mu = \lambda_c$. For $s = e$, (1) and the definition of c give $c\rho = (e\rho)\mu = (ce)\rho$. Hence $ce\,\rho\,c$, and $c \in \mathscr{N}_e(\rho)$. Thus every element μ of $\mathscr{E}(S/\rho)$ has the form λ_c with c in $\mathscr{N}_e(\rho)$, showing that the restriction of λ to $\mathscr{N}_e(\rho)/\rho$ is onto $\mathscr{E}(S/\rho)$.

To show that λ, so restricted, is one-to-one, let a and b be elements of $\mathscr{N}_e(\rho)$ such that $\lambda_a = \lambda_b$. Then

$$a\rho = (ae)\rho = (e\rho)\lambda_a = (e\rho)\lambda_b = (be)\rho = b\rho,$$

so that a and b determine the same element of $\mathscr{N}_e(\rho)/\rho$.

EXERCISE FOR §11.7

1. The choice of the left identity e of S mod ρ in Theorem 11.28 is immaterial in the following sense. If f is any other, and if $a \in \mathscr{N}_e(\rho)$, then $af \in \mathscr{N}_f(\rho)$ and $\lambda_{af} = \lambda_a$.

11.8 REPRESENTATIONS BY MONOMIAL MATRICES

We recall (§3.1) that a matrix A over a group with zero G^0 is called *row-monomial* if each row of A contains at most one non-zero element of G^0. Any representation ϕ of a semigroup S by transformations of a set M can be regarded as a representation of S by $M \times M$ row-monomial matrices over the two-element group with zero $E^0 = \{0, 1\}$; for each element a of S, let $N(a) = (n_{xy}(a))$ be defined as follows ($x, y \in M$):

$$(1) \qquad n_{xy}(a) = \begin{cases} 1, & \text{if } x(a\phi) = y, \\ 0, & \text{otherwise.} \end{cases}$$

(Note Exercise 4 for §3.5.) One easily verifies that $N(ab) = N(a)N(b)$ for all a, b in S.

We say "can be regarded as", since $N(a)$ is simply a kind of incidence matrix that describes the action of $a\phi$ on M. But we can also formulate the relationship as an equivalence. Let V be the set of monomial M-vectors over E^0. For each x in M, let $x\theta$ be the element of V having 1 in the x-position and zeros elsewhere. Define V as an operand V_S over S by $(x\theta)a = (x\theta)N(a)$. Then θ is an operator isomorphism of M_S onto V_S.

In some cases, especially if the given representation ϕ is irreducible, we can convert ϕ into a representation of S by row-monomial matrices over a non-trivial group with zero, and this will be our concern in the present section. This is well-known for representations of groups by permutations; see, for example, Chapter V, §1, of Zassenhaus [1937]. It was first extended to semigroups by Tully [1960], and independently by Hoehnke [1963].

As an important special case, we shall show that, for the representation of a semigroup S induced in a principal right ideal of S by the right regular representation of S, this leads to the Schützenberger representation (§3.5). This connexion was first pointed out by Tully [1960].

The results of this section can be dualized in an obvious way to convert anti-representations by transformations to anti-representations by column-monomial matrices. *We shall treat explicitly only* 0*-transitive representations.* The results for transitive ones can easily be inferred; we remark that they lead to representations by *strictly* row-monomial matrices, those having exactly one non-zero entry in each row.

Let M_S be a 0-transitive operand over a semigroup S, and let 0_M be its origin. Let $\mathscr{A} = \mathscr{A}(M_S)$ be the group of operator automorphisms of M_S; it is, of course, the group of units of the semigroup $\mathscr{E}(M_S)$ discussed in the last section. Let G be any subgroup of the left-right dual $\mathscr{A}^*$ of $\mathscr{A}$. Then we can set up M as a (G, S)-bioperand by defining $gx = x\gamma$, where $x \in M$, $g \in G$, and γ is the element of $\mathscr{A}$ corresponding to the element g of $\mathscr{A}^*$. We note that ${}_GM$ is faithful and unital with respect to G, and that $M\backslash 0_M$ is invariant under G.

By the left-right dual of Theorem 11.3, $M\backslash 0_M$ decomposes into the union of disjoint, transitive, left operands N_i $(i \in I)$ over G. Moreover, each N_i is *simply transitive*, i.e. if u and v are any elements of N_i, there exists a unique element g of G such that $gu = v$. Since N_i is transitive, we need only show that $gu = hu$ $(g, h \in G)$ implies $g = h$. Let $x \in M$. Since $u \neq 0_M$, and M_S is 0-transitive, there exists a in S such that $ua = x$. Hence $gx = gua = hua = hx$. Since this holds for every x in M, and ${}_GM$ is faithful, we conclude that $g = h$.

From each N_i select an element u_i. Since each N_i is simply transitive over G, every element of $M\backslash 0_M$ is uniquely expressible in the form gu_i with $i \in I$ and $g \in G$.

Let V be the set of monomial I-vectors over G^0. An element v of V is a mapping of I into G^0 which has a non-zero value for at most one element of I. If $iv = g \in G$, $jv = 0$ for all $j \neq i$, we write $v = (g)_i$. The zero vector will be denoted variously by 0_V or $(0)_i$ for any i in I. Define $\theta\colon M \to V$ by $0_M\theta = (0)_i$, $(gu_i)\theta = (g)_i$. Clearly this is a one-to-one mapping of M onto V. We make V into an operand V_S over S equivalent to M_S by defining $(x\theta)a = (xa)\theta$ for every x in M, a in S.

For each element a of S we define an $I \times I$ matrix $F(a) = (f_{ij}(a))$ over G^0 as follows:

$$f_{ij}(a) = \begin{cases} h, & \text{if } u_i a = hu_j, \\ 0, & \text{if } u_i a \notin N_j \ (= Gu_j). \end{cases} \tag{2}$$

Since either $u_i a = 0$ or $u_i a \in N_j$ for precisely one j in I, $F(a)$ is row-monomial. We proceed to show that $va = vF(a)$ for every v in V, where $vF(a)$ means the usual matrix product.

Now $v = x\theta$ for some x in M. If $x = 0_M$, both va and $vF(a)$ are 0_V, so we may assume $x \neq 0_M$. Then $x = gu_i$ for some $g \in G$, $i \in I$, and so $v = x\theta = (gu_i)\theta = (g)_i$. Assume first that $u_i a \neq 0_M$. Then $u_i a = hu_k$ for some $h \in G$, $k \in I$, and (2) gives

$$f_{ij}(a) = \begin{cases} h, & \text{if } j = k, \\ 0, & \text{if } j \neq k. \end{cases}$$

Hence the ith row $(1)_i F(a)$ of $F(a)$ is the vector $(h)_k$ of V. We thus have

$$va = (x\theta)a = (xa)\theta = (gu_i a)\theta = (ghu_k)\theta = (gh)_k,$$

$$vF(a) = (g)_i F(a) = g(1)_i F(a) = g(h)_k = (gh)_k.$$

If $u_i a = 0_M$, then $va = (gu_i a)\theta = 0_M\theta = 0_V$. Also, by (2), the ith row $(1)_i F(a)$ of $F(a)$ is 0_V, and so $vF(a) = g0_V = 0_V$.

If $a, b \in S$ and $v \in V$, we have

$$vF(ab) = v(ab) = (va)b = (vF(a))F(b) = v(F(a)F(b)),$$

and, since this holds for every v in V,

$$F(ab) = F(a)F(b).$$

Thus $a \to F(a)$ is a representation of S by $I \times I$ row-monomial matrices over G^0.

THEOREM 11.29. *Let M_S be a 0-transitive operand over a semigroup S. Let G be any subgroup of the left-right dual of the group of operator automorphisms of M_S, and set up M as a bioperand ${}_G M_S$. Then there exists a subset $\{u_i : i \in I\}$ of $M \backslash 0_M$ such that every element of $M \backslash 0_M$ is uniquely expressible in the form gu_i with $g \in G$, $i \in I$. For each element a of S, define an $I \times I$ matrix $F(a)$ over G^0 by (2). Then $F(a)$ is row-monomial, and $F(ab) = F(a)F(b)$ for all a, b in S.*

This representation is equivalent to the original one in the following sense. Let V be the set of monomial I-vectors $(g)_i$ over G^0 regarded as a bioperand ${}_G V_\Sigma$, where Σ is the semigroup of all $I \times I$ row-monomial matrices over G^0. Turn V into an operand over S by defining $va = vF(a)$ for every $v \in V$ and $a \in S$. Then the mapping θ defined by $(gu_i)\theta = (g)_i$, $0_M\theta = 0_V$, is an operator isomorphism of M_S onto V_S.

We proceed to tie in the foregoing with the (right-hand) Schützenberger representation (1) of §3.5.

Let R be an $\mathscr{R}$-class of S, and let $a \in R$. We shall exclude the trivial case $|R| = 1$; hence $a \in aS$, since otherwise $R = \{a\}$. aS is thus a strictly cyclic operand over S. By Lemma 11.5, the factor operand $R^0 = aS/I(aS)$ is 0-transitive if $I(aS) \neq \square$. Denoting the origin of R^0 by 0_R, we have $R^0 =$

$R \cup 0_R$. If $I(aS) = \square$, then R itself is transitive. For the sake of uniformity, let us adjoin an origin 0_R to R in this case.

Let H be an $\mathscr{H}$-class of S contained in R. In §2.4 we defined

$$T'(H) = \{u \in S^1 \colon uH \subseteq H\}.$$

We let λ_u denote the inner left translation $s \to us$ of S^1, and we let $\gamma'_u = \lambda_u | H$.

Let $u \in T'(H)$ and let $h \in H$. Then $uh\mathscr{H}h$, and hence $uh\mathscr{L}h$. By the dual of Green's Lemma 2.2, $\lambda_u | R_h$ is a one-to-one, $\mathscr{L}$-class preserving, mapping of R_h onto R_{uh}. Here $R_h = R_{uh} = R$, and so $\gamma''_u = \lambda_u | R$ is an $\mathscr{H}$-class preserving permutation of R. Hence $\Gamma'' = \{\gamma''_u \colon u \in T'\}$ is a group. Since Γ'' acts transitively on H, which could be any $\mathscr{H}$-class of S contained in R, it is clear that the $\mathscr{H}$-classes of S contained in R are the transitivity classes of R with respect to Γ''. Clearly $\gamma''_u \to \gamma'_u$ $(= \gamma''_u | H)$ is an isomorphism of Γ'' onto the dual Schützenberger group $\Gamma'(H)$, as defined in §2.4.

Since, for each u in T', λ_u commutes with each transformation of R^0 induced by an inner right translation of S, the same is true of each γ''_u. It follows that Γ'' is a subgroup of the group $\mathscr{A}(R^0)$ of operator automorphisms. (If R is regular, then $\Gamma'' = \mathscr{A}(R^0)$—Exercise 2 below.) We take for G the left-right dual of Γ''.

By Theorem 2.24, G is isomorphic with the (right) Schützenberger group $\Gamma(H)$. In order to reconcile (2) above with the formula (1) of §3.5 for the (right) Schützenberger representation $M_D(s) = (m_{\lambda\mu}(s))$ of S, we must set up a specific isomorphism between G and Γ; or, what amounts to the same, a specific anti-isomorphism between Γ'' and Γ. What we do is suggested by the proof of Theorem 2.24. Select and fix an element h_1 of H. If $\gamma'' \in \Gamma''$, then $h_1\gamma'' \in H$, and so $h_1\gamma'' = h_1\gamma$ for a unique γ in Γ. Setting $\gamma = \gamma''\phi$, it is clear that ϕ is an anti-isomorphism of Γ'' onto Γ.

Let the $\mathscr{H}$-classes of R be $\{H_\lambda \colon \lambda \in \Lambda\}$; and for each $\lambda \in \Lambda$, select an element h_λ in H_λ. In doing this, let $H = H_1$, and let h_1 be the element already selected in H. Then the representation F defined by (2) takes the form

$$f_{\lambda\mu}(a) = \begin{cases} \gamma''(u)\phi, & \text{if } h_\lambda a = uh_\mu \quad (u \in T'), \\ 0, & \text{if } h_\lambda a \notin H_\mu. \end{cases} \tag{3}$$

In other words, $f_{\lambda\mu}(a)$ is 0 unless $h_\lambda a \in H_\mu$; in the latter case, there exists u in T' such that $h_\lambda a = uh_\mu$, and the corresponding element $\gamma''(u)$ of Γ'' is uniquely determined, since $u_1h_\mu = u_2h_\mu$ implies $\gamma''(u_1) = \gamma''(u_2)$; we then take $f_{\lambda\mu}(a)$ to be the corresponding element $\gamma''(u)\phi$ of Γ, which is serving for G in the present situation.

As in §3.5, there exist elements q_λ and q'_λ in S such that $h_\lambda = h_1q_\lambda$ and $h_1 = h_\lambda q'_\lambda$. The relation $h_\lambda a = uh_\mu$ can now be rewritten as follows:

$$h_1q_\lambda a = uh_1q_\mu,$$

or

$$h_1 q_\lambda a q'_\mu = u h_1 q_\mu q'_\mu = u h_1 = h_1 \gamma''(u).$$

Hence $\gamma''(u)\phi = \gamma(q_\lambda a q'_\mu)$, and (3) becomes

$$f_{\lambda\mu}(a) = \begin{cases} \gamma(q_\lambda a q'_\mu), & \text{if } h_\lambda a \in H_\mu, \\ 0, & \text{otherwise.} \end{cases} \tag{4}$$

Noting that $h_\lambda a \in H_\mu$ implies $H_\lambda a = H_\mu$, we see that (4) is the same as (1) of §3.5.

Exercises for §11.8

1. In the representation $a \to F(a)$ defined by (2), the rows of F(a) can be filled arbitrarily, in the sense that, for given $i, j \in I$ and $g \in G$, there exists a in S such that $f_{ij}(a) = g$. (Tully [1960].)
2. Let R be a regular $\mathscr{R}$-class of a semigroup S. Then the left Schützenberger group of R (extended to operate on all of R, as in the text) coincides with the group $\mathscr{A}(R^0)$ of operator automorphisms of R^0_S. (It is not known if this is true for irregular R.)

11.9 Other types of representations

The purpose of this concluding section is to call attention to certain types of operands (or representations) which are, apart from operands of cardinal two, more special than irreducible ones. We do not go deeply into their theory, which is still in embryonic form, but refer the reader for further details to Tully [1960, 1961] and Hoehnke [1963, 1966].

An operand M_S over a semigroup S is called *primitive* (Tully) if the only operator equivalences on M are the equality relation ι_M and the universal relation ω_M. This condition is equivalent to the following: if θ is an operator homomorphism of M_S onto some operand M'_S over S, then either $|M'| = 1$ or else θ is an operator isomorphism. Hence the endomorphism semigroup $\mathscr{E}(M_S)$ of a primitive operand M_S is a group or a group with zero.

Let FM denote the set of fixed elements of an operand M_S. M_S is called *totally irreducible* (Hoehnke) if it is primitive and if $MS \nsubseteq FM$. Clearly an operand is primitive if and only if it is either totally irreducible, or of cardinal two or less and is either fixed or null. A primitive operand can not have a proper invariant subset of cardinal greater than one; hence, by Lemma 11.16, it is either irreducible or else of cardinal two or less and null or fixed. Hence a totally irreducible operand is irreducible.

We call a right congruence ρ on a semigroup S *maximal* if $\rho \subseteq \sigma \subseteq \omega_S$ (σ a right congruence on S) implies $\sigma = \rho$ or $\sigma = \omega_S$. (The possibility $\rho = \omega_S$ is allowed.) If ρ and σ are right congruences on S, $\rho \subseteq \sigma$, and ρ is modular, then σ is modular; consequently there is no ambiguity in the term "maximal modular right congruence". The following is due both to Tully [1960] and Hoehnke [1966].

LEMMA 11.30. (i) *If an operand M_S over a semigroup S is totally irreducible, then there exists a maximal modular congruence ρ on S such that M_S is equivalent to S/ρ.*

(ii) *Let ρ be a right congruence on a semigroup S. Then S/ρ is primitive if and only if ρ is maximal. S/ρ is totally irreducible if and only if ρ is maximal and satisfies the following two conditions:* (C1) *there exists at most one ρ-class which is a right ideal of S, and* (C2) *if R is a ρ-class which is a right ideal of S, then $S^2 \nsubseteq R$.*

PROOF. (i) If M_S is totally irreducible, then it is irreducible, as remarked above, and by Lemma 11.16, it is strictly cyclic. By Theorem 11.6, there exists a modular right congruence ρ on S such that M_S is equivalent to S/ρ. If ρ were not maximal, then there would exist a right congruence σ on S such that $\rho \subset \sigma \subset \omega_S$. By Lemma 11.1, there would exist an operator equivalence on S/ρ which is neither the equality relation nor the universal relation; and since S/ρ is equivalent to M_S, this conflicts with our hypotheses.

(ii) Let ρ be a right congruence on S. Since, by Lemma 11.1, there is a lattice isomorphism between the lattice of operator equivalences on S/ρ and that of all operator equivalences (i.e. right congruences) on S containing ρ, it is clear that S/ρ is primitive if and only if ρ is maximal. Assuming that S/ρ is primitive, (C1) asserts that S/ρ is not fixed, unless it has only one element and (C2) asserts that S/ρ is not null and that, if it is fixed, then S/ρ has more than one element.

Hoehnke calls a semigroup *totally primitive* if it admits a faithful, totally irreducible representation, and investigates the $\mathscr{C}$-radical of a semigroup S, for $\mathscr{C}$ the class of totally primitive semigroups [1966]. A group is totally primitive if and only if it is primitive in the classical sense; this is the case if and only if it contains a maximal subgroup H such that the intersection of all the conjugates of H is the identity subgroup.

Tully [1960] calls an operand M_S over a semigroup S *disjunctive* if the identity relation ι_M on M is the only operator equivalence on M having a one-element equivalence class. He proves the following.

LEMMA 11.31. (i) *If M_S is a disjunctive operand over a semigroup S, then M contains no proper invariant subset of cardinal greater than one, and so is one of the following: transitive, 0-transitive, null of cardinal two or less, fixed of cardinal two or less.*

(ii) *Every primitive operand is disjunctive. Every disjunctive operand containing an invariant element is primitive.*

PROOF. (i) Suppose N is a proper invariant subset of M, and $|N| > 1$. If ν is the corresponding Rees equivalence on M, then $\nu \neq \iota_M$ since $|N| > 1$, while every element of $M\backslash N$ ($\neq \square$) constitutes a one-element ν-class. Then M_S is not disjunctive. The second assertion of (i) is immediate from Lemma 11.16.

(ii) It is immediate from the definitions that every primitive operand is disjunctive. Let M_S be a disjunctive operand containing an invariant element z. Let σ be any operator equivalence on M. Then $(z\sigma)a = (za)\sigma = z\sigma$, for all a in S. Hence $N = z\sigma$ is an invariant subset of M. By (i), $N = M$ or $|N| = 1$. If $N = M$, then $\sigma = \omega_M$. If $|N| = 1$, then $\sigma = \iota_M$, since M_S is disjunctive. Hence ω_M and ι_M are the only operator equivalences on M, which means that M_S is primitive.

The following lemma, due to Tully [1960], gives us a way of telling when an operand is disjunctive.

LEMMA 11.32. *An operand M_S over a semigroup S is disjunctive if and only if it satisfies the following condition: if x, y, z are any three distinct elements of M, then there exists an element a of S such that either $xa = z$ and $ya \neq z$, or else $xa \neq z$ and $ya = z$.*

If $|M|$ is two or less, this condition is vacuously satisfied and so, as is otherwise clear, M is disjunctive.

PROOF. For each element z of M, define a relation τ_z on M as follows:

$$\tau_z = \{(z, z)\} \cup \{(x, y) \in (M\backslash z) \times (M\backslash z) \colon xs = z \text{ if and only if } ys = z \ (s \in S)\}.$$

This is analogous to Dubreil's principal right regular equivalence defined by the set $\{z\}$. It is clear that τ_z is an operator equivalence on M having the equivalence class $\{z\}$ of cardinal one, and $\tau_z \neq \omega_M$ except when $|M| = 1$.

It is readily seen that the condition of the lemma is equivalent to the assertion that $\tau_z = \iota_M$ for every $z \in M$. Clearly this is so if M_S is disjunctive. Assume that M_S is not disjunctive. Then there exists an operator equivalence σ on M having a one-element σ-class $\{z\}$, and such that $\sigma \neq \iota_M$. Now $\sigma \subseteq \tau_z$. For if $(x, y) \in \sigma$ then $(xs, ys) \in \sigma$ for every s in S, so that $xs = z$ [or $x = z$] if and only if $ys = z$ [or $y = z$]. Hence $\tau_z \neq \iota_M$, and the condition of the lemma is not satisfied.

An operand M_S is called *doubly transitive* if, for any elements u, v, x, y of M with $u \neq v$, there exists a in S such that $ua = x$ and $va = y$.

Any doubly transitive operand is primitive. For suppose σ is an operator equivalence on M, and $\sigma \neq \iota_M$. Then there exist $u \neq v$ in M such that $u \,\sigma\, v$. Let $x, y \in M$. Then there exists a in S such that $ua = x$ and $va = y$. But $u \,\sigma\, v$ implies $ua \,\sigma\, va$. Hence $x \,\sigma\, y$, showing that $\sigma = \omega_M$.

Let M_S be an operand without an invariant element. Then we have the following implications: doubly transitive *implies* primitive *implies* disjunctive *implies* transitive. Examples are given in the exercises to show that none of these implications can be reversed.

EXERCISES FOR §11.9

1. Let G be a cyclic group of prime order. Then G_G is a primitive operand over G which is not doubly transitive.

2. Let G be an abelian group which is not of prime order. Then G_G is a disjunctive operand over G which is not primitive.

3. Let S consist of the constant transformations of a set M with $|M| > 2$. Then the operand M_S, defined in the obvious way, is transitive but not disjunctive.

CHAPTER 12

EMBEDDING A SEMIGROUP IN A GROUP

That a right reversible, cancellative semigroup can be embedded in a group was shown in Theorem 1.23. Here we consider the general problem of embedding a semigroup in a group. A. I. Malcev first showed that a cancellative semigroup was not necessarily embeddable in a group [1937]. We refer to his counter-example in §12.6. Necessary and sufficient conditions for the embeddability of a semigroup in a group were then given by Malcev in his paper [1939]. The conditions of Malcev are each of the form that a certain finite set of equations can hold in a semigroup only if another related equation also holds. Malcev's necessary and sufficient conditions are countably infinite in number and no finite subset of them will suffice to ensure embeddability of a semigroup. We give an account of this work in §12.6 and §12.8. A similar set of conditions involving equational implications was provided by J. Lambek [1951]. §12.5 presents Lambek's results and §12.7 establishes the relation between the Lambek conditions and the Malcev conditions.

A feature of Lambek's proofs is the use of quotients of elements of a semigroup which are a generalization of those used in the special case of right reversible semigroups (Theorem 1.23). In §12.4 we introduce what we call the quotient condition, which a semigroup necessarily satisfies if it can be embedded in a group, and show that a "group of quotients" can always be constructed from a semigroup satisfying the quotient condition. A semigroup is embeddable in a group if and only if it is embedded, by a natural mapping, in its group of quotients.

§§12.1 and 12.2 introduce the concept of a free group on a semigroup which is fundamental to our treatment. As a corollary to the discussion it is shown that a semigroup can be embedded in a group if and only if every finitely generated subsemigroup can be so embedded. From the fact that a semigroup can be embedded in a group if and only if it can be embedded in the free group on it, we derive in §12.3 the necessary and sufficient conditions for embeddability of V. Pták [1949].

12.1 The free group on a semigroup

Theorem 9.4 showed that if M is a set and $\mu\colon M \rightarrow S$ is a one-to-one mapping of M into S, then S is a free semigroup on $M\mu$ if and only if, for any semigroup T and mapping $\nu\colon M \rightarrow T$, there exists a homomorphism $\phi\colon S \rightarrow T$

such that $\mu\phi = \nu$. We shall call such a free semigroup the *free semigroup* (S, μ) and, by abuse of language, call it a free semigroup on M.

An analogue to Theorem 9.4 holds for free groups. Let M be a set and let $\nu: M \to G$ be a one-to-one mapping of M into G. Then (G, ν) *is a free group on* M (strictly, on $M\nu$) if and only if, for any group H and mapping $\delta: M \to H$, there exists a homomorphism $\theta: G \to H$, say, such that $\nu\theta = \delta$. Using the definition of a free group on a set given in §1.12 this characterization follows by a proof similar to that of Theorem 9.4 for semigroups.

If S and S' are semigroups and μ and μ' are mappings of M into S and S', respectively, then (S, μ) and (S', μ') are said to be *equivalent* if there exists an isomorphism σ, say, of S onto S' such that $\mu\sigma = \mu'$ and (consequently) $\mu'\sigma^{-1} = \mu$. If (S, μ) is a free semigroup on M and (S', μ') is equivalent to (S, μ), then we easily see that (S', μ') is also a free semigroup on M. Similarly, if (G, ν) is a free group on M and (G', ν') is equivalent to (G, ν), then (G', ν') is a free group on M. The converse holds in each case and we have

LEMMA 12.1. (a) *Let* (S, μ) *be a free semigroup on the set* M. *Then* (S', μ') *is a free semigroup on* M, *where* $\mu': M \to S'$, *if and only if* (S', μ') *is equivalent to* (S, μ).

(b) *Let* (G, ν) *be a free group on the set* M. *Then* (G', ν') *is a free group on* M, *where* $\nu': M \to G'$, *if and only if*, (G', ν') *is equivalent to* (G, ν).

PROOF. Both (a) and (b) are proved in a similar fashion. We prove (a). The sufficiency of the condition has already been remarked on. Suppose then that (S, μ) and (S', μ') are free semigroups on M. Then, by Theorem 9.4, there exist homomorphisms $\phi: S \to S'$ and $\phi': S' \to S$, say, such that $\mu\phi = \mu'$ and $\mu'\phi' = \mu$. Hence $\mu\phi\phi' = \mu$ and $\mu'\phi'\phi = \mu'$. Thus $\phi\phi'$ is the identical mapping on $M\mu$ and $\phi'\phi$ is the identical mapping on $M\mu'$. Consequently, since $M\mu$ is a set of generators for S, $\phi\phi'$ must be the identical mapping on the whole of S; similarly, $\phi'\phi$ is the identical mapping on the whole of S'. Hence ϕ is one-to-one and onto S' (with inverse ϕ') and so ϕ is an isomorphism of S onto S'. From $\mu\phi = \mu'$ we now have that (S, μ) and (S', μ') are equivalent. This shows the necessity of the condition and completes the proof of the lemma.

Let A be a subset of the group G. The intersection of all subgroups of G which contain A is a subgroup, which we shall denote by $[A]$, said to be the subgroup of G *generated by* A. The set A is said to be a set of *group generators* of $[A]$. The reason for the term "group generators" is to enable a distinction to be drawn between the semigroup $\langle A \rangle$ generated by A and the group $[A]$ generated by A. $\langle A \rangle$ is the intersection of all subsemigroups of G which contain A and, as before, we continue to say that A is a set of generators of $\langle A \rangle$. In general $\langle A \rangle$ is not equal to $[A]$. For example, if (G, ν) is a free group on M, then $[M\nu] = G$ whereas $(\langle M\nu \rangle, \nu)$ is a free semigroup on M.

The pair (H, η) will be said to be a *group on the semigroup* S, or, simply, an S-*group*, if H is a group and η is a homomorphism of S into H such that $S\eta$

is a set of group generators of H. The pair (G, γ) will be called a *free group on the semigroup* S, or a *free* S*-group*, if (G, γ) is an S-group and if, further, for any S-group (H, η) there exists a homomorphism θ of G into H such that $\gamma\theta = \eta$ (see diagram below).

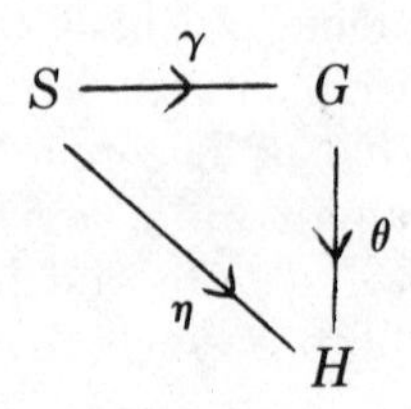

If α and β are two homomorphisms of the semigroup [group] S into the semigroup T then $\alpha = \beta$ if α and β agree on the elements of any set of [group] generators of S. Consequently, since by hypothesis $S\gamma$ is a set of group generators of G, the homomorphism θ of G into H is uniquely determined by the condition $\gamma\theta = \eta$.

As for free semigroups and free groups, we shall say that two S-groups (H, η) and (H', η') are *equivalent* if there exists an isomorphism τ, say, of H onto H' such that $\eta\tau = \eta'$ and (consequently) $\eta'\tau^{-1} = \eta$. Using a proof closely following that of Lemma 12.1 we have

Lemma 12.2. *Let (G, γ) be a free group on the semigroup S. Then the S-group (G', γ') is a free group on the semigroup S if and only if it is equivalent to (G, γ).*

We now show, in the following construction, that, given any semigroup S, there exists a free group on S. The construction given is more complicated than would be necessary merely to provide an existence theorem. The generality of the construction will prove convenient in the sequel.

Construction 12.3. *Let S be a semigroup and let Γ be a set of generators of S. Let M be any set of the same cardinal as Γ and let σ be a one-to-one mapping of M onto Γ. Let (T, τ) be a free semigroup on M and let (F, ϕ) be a free*

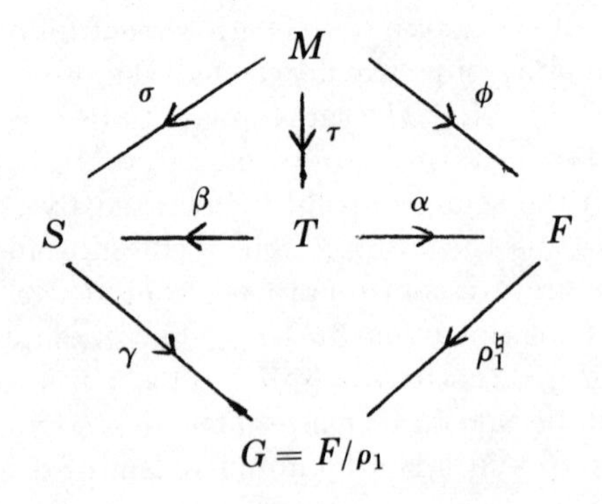

group on M. Let α be the homomorphism of T into F such that $\tau\alpha = \phi$ and let β be the homomorphism of T into S such that $\tau\beta = \sigma$.

Let ρ be the relation $\rho = \alpha^{-1} \circ \beta \circ \beta^{-1} \circ \alpha$ on F and let ρ_1 be the congruence on F generated by ρ. Let $G = F/\rho_1$ and let γ be the homomorphism of S into G such that $\sigma\gamma = \phi\rho_1^\natural$. Then (G, γ) is a free group on S.

JUSTIFICATION. The homomorphisms α and β of T into F and of T into S, respectively, we know to exist by Theorem 9.4. Moreover α and β are the only such mappings with the asserted properties (for $\langle M\sigma\rangle = S$ and $[M\phi] = F$). Thus ρ (and so also ρ_1) is a uniquely defined relation on the group F. Further the equation $\sigma\gamma = \phi\rho_1^\natural$ defines the mapping γ unambiguously on the subset $M\sigma$ of S.

Let us first show that if γ is to be a homomorphism of S then γ can be extended from $M\sigma$ to S in only one way. Let $s \in S$. Then, since, by hypothesis, $M\sigma$ generates S, $s = m_1\sigma m_2\sigma \cdots m_n\sigma$ for some m_i in M, and so we must have $s\gamma = m_1\sigma\gamma m_2\sigma\gamma \cdots m_n\sigma\gamma = m_1\phi\rho_1^\natural \cdots m_n\phi\rho_1^\natural$. On the other hand, the latter expression serves to define $s\gamma$ uniquely. For suppose that also $s = m_1'\sigma \cdots m_p'\sigma$ for m_j' in M. Let $u = m_1\tau m_2\tau \cdots m_n\tau$ and $u' = m_1'\tau \cdots m_p'\tau$. Then, since $\tau\beta = \sigma$, $u\beta = s = u'\beta$. Hence $(u\alpha, u'\alpha) \in \rho$ $(= \alpha^{-1} \circ \beta \circ \beta^{-1} \circ \alpha)$. Thus $u\alpha\rho_1^\natural = u'\alpha\rho_1^\natural$. But, using the fact that $\tau\alpha = \phi$, this is the same as $m_1\phi\rho_1^\natural \cdots m_n\phi\rho_1^\natural = m_1'\phi\rho_1^\natural \cdots m_p'\phi\rho_1^\natural$ which is the same as $m_1\sigma\gamma m_2\sigma\gamma \cdots m_n\sigma\gamma = m_1'\sigma\gamma \cdots m_p'\sigma\gamma$.

It is now almost immediate that γ is in fact a homomorphism. For if $s = m_1\sigma \cdots m_n\sigma$ and $t = m_1'\sigma \cdots m_p'\sigma$ then $st = m_1\sigma \cdots m_n\sigma m_1'\sigma \cdots m_p'\sigma$. Hence $(st)\gamma = m_1\sigma\gamma \cdots m_n\sigma\gamma m_1'\sigma\gamma \cdots m_p'\sigma\gamma = (s\gamma)(t\gamma)$.

Now $M\phi$ is a set of group generators of F and hence $M\phi\rho_1^\natural$ is a set of group generators of $G = F/\rho_1$. Hence, since $M\phi\rho_1^\natural = M\sigma\gamma \subseteq S\gamma$, it follows that $S\gamma$ is a set of group generators of G. Thus we have shown that (G, γ) is an S-group.

Let (H, η) be any S-group. Then $\sigma\eta$ is a mapping of M into H. Since (F, ϕ) is a free group on M there exists a homomorphism λ of F into H such that $\sigma\eta = \phi\lambda$. We will show that $\rho \subseteq \lambda \circ \lambda^{-1}$. Let $(w, w') \in \rho$ so that there exist u, u' in T such that $u\alpha = w$, $u'\alpha = w'$ and $u\beta = u'\beta$. Let $u = m_1\tau \cdots m_n\tau$ and $u' = m_1'\tau \cdots m_p'\tau$ $(m_i, m_j' \in M)$. Then

$$\begin{aligned} w\lambda &= (m_1\phi \cdots m_n\phi)\lambda \\ &= m_1\phi\lambda \cdots m_n\phi\lambda \\ &= m_1\sigma\eta \cdots m_n\sigma\eta \\ &= (m_1\sigma \cdots m_n\sigma)\eta \\ &= (u\beta)\eta; \end{aligned}$$

and, similarly, $w'\lambda = (u'\beta)\eta$.

Hence, since $u\beta = u'\beta$, we have $w\lambda = w'\lambda$, i.e. $(w, w') \in \lambda \circ \lambda^{-1}$. Thus $\rho \subseteq \lambda \circ \lambda^{-1}$ and so also $\rho_1 \subseteq \lambda \circ \lambda^{-1}$.

It now follows from Lemma 1.6 that there exists a homomorphism θ, say, of G into H such that $\rho_1^\natural\theta = \lambda$. We then have that $\sigma(\gamma\theta) = (\sigma\gamma)\theta = \phi(\rho_1^\natural\theta) = \phi\lambda = \sigma\eta$. Thus η agrees with $\gamma\theta$ on the elements of $M\sigma$. Since $M\sigma$ generates S it follows that $\eta = \gamma\theta$.

The construction of θ satisfying this equation completes the justification of the assertions made in the construction and shows that (G, γ) is a free group on S.

12.2 The general problem of embedding a semigroup in a group

The fundamental importance of the free group on a semigroup for the problem of embedding a semigroup in a group is made clear in the next theorem.

Theorem 12.4. *Let (G, γ) be a free group on the semigroup S. Then S can be embedded in a group if and only if γ embeds S in G.*

Proof. We need only prove necessity. Suppose that S can be embedded in a group so that there exists $\eta: S \to H$, an isomorphism of S into the group H. We may clearly suppose that $S\eta$ is a set of group generators of the group H. Thus (H, η) is an S-group and so, since (G, γ) is a free S-group, there exists a homomorphism $\theta: G \to H$ such that $\gamma\theta = \eta$. Since η is one-to-one, it immediately follows that γ is also one-to-one. Thus γ embeds S in G.

Let us say that a group H is a *receiving group* for a semigroup S if there exists an isomorphism ϕ of S into H such that $S\phi$ generates H. The proof of Theorem 12.4 shows that if S can be embedded in a group, then the free group G on S is the largest receiving group for S, in the sense that any other receiving group H for S is a homomorphic image of G. Exercise 1 below shows that minimal receiving groups for S also exist, and Exercise 2 shows that two minimal receiving groups H_1 and H_2 for S may exist such that no isomorphism of H_1 onto H_2 over S exists.

Whether or not a semigroup can be embedded in a group depends only on its finitely generated subsemigroups. We use Construction 12.3 to deduce this result from Theorem 12.4. We need a preliminary lemma.

Lemma 12.5. (a) *Let (S, μ) be a free semigroup on M. Let M' be a non-empty subset of M. Set $\mu' = \mu|M'$ and $S' = \langle M'\mu'\rangle$. Then (S', μ') is a free semigroup on M'.*

(b) *Let (G, γ) be a free group on M. Let M' be a non-empty subset of M. Set $\gamma' = \gamma|M'$ and $G' = [M'\gamma']$. Then (G', γ') is a free group on M'.*

Proof. We prove (b); the proof of (a) is similar. Let $\delta': M' \to H$, where H is a group. Let $\delta: M \to H$ be any mapping such that $\delta|M' = \delta'$. Since (G, γ) is free, there exists a homomorphism θ, say, $\theta: G \to H$ such that $\gamma\theta = \delta$. Set $\theta' = \theta|G'$. Let $m' \in M'$. Then $m'\gamma = m'\gamma' \in G'$; and so $m'\gamma\theta = m'\gamma'\theta'$.

Also $m'\delta = m'\delta'$. Hence $m'\gamma'\theta' = m'\delta'$. Consequently, $\gamma'\theta' = \delta'$; which shows that (G', γ') is a free group on M'.

THEOREM 12.6. *A semigroup is embeddable in a group if and only if every finitely generated subsemigroup is so embeddable.*

PROOF. We need only prove the sufficiency of the condition. Assume therefore that every finitely generated subsemigroup of S can be embedded in a group and suppose that S cannot be so embedded. In particular it then follows that if (G, γ) is any free group on S then γ is not one-to-one.

Choose (G, γ) as in Construction 12.3. As $M\sigma$, the set of generators of S, we take S itself. Since γ is not one-to-one there exist m, m' in M such that $m\sigma\gamma = m'\sigma\gamma$ but $m\sigma \neq m'\sigma$. Since $\sigma\gamma = \phi\rho_1^\natural$, $m\sigma\gamma = m'\sigma\gamma$ implies that $(m\phi, m'\phi) \in \rho_1$. Since ρ_1 is the congruence generated by ρ it follows that there exist elements $w_0 = m\phi, w_1, \cdots, w_n = m'\phi$ in F such that $w_{i-1} \rightarrow w_i$ $(i = 1, 2, \cdots, n)$ is an elementary ρ-transition. Thus $w_{i-1} = a_i y_i b_i$ and $w_i = a_i x_i b_i$, where $(y_i, x_i) \in \rho \cup \rho^{-1} \cup \iota_F$ and $a_i, b_i \in F$, for $i = 1, 2, \cdots, n$.

Since $M\phi$ is a set of group generators of F, each of the elements a_i, b_i, x_i, y_i is a finite product of elements of $M\phi \cup (M\phi)^{-1}$. Choose a fixed (but arbitrary) expression for each of the a_i, b_i, x_i, y_i as such a product and let N be the set of all elements m of M such that either $m\phi$ or $(m\phi)^{-1}$ occurs as a factor of one of the a_i, b_i, x_i, y_i (expressed as a product of elements of $M\phi \cup (M\phi)^{-1}$ in the fixed manner we have selected). Then N is a finite set.

Let S' be the subsemigroup of S generated by $N\sigma$ and let $M' = S'\sigma^{-1}$. Let F' be the subgroup of F generated by $M'\phi$ and let $\phi' = \phi|M'$. Then (F', ϕ') is a free group on M' (Lemma 12.5(b)). Let T' be the subsemigroup of T generated by $M'\tau$ and let $\tau' = \tau|M'$. Then (T', τ') is a free semigroup on M' (Lemma 12.5(a)). Further if $\alpha' = \alpha|T'$ and $\beta' = \beta|T'$ then $\tau'\alpha' = \phi'$ and $\tau'\beta' = \sigma'$.

Now let $\rho' = \rho \cap (F' \times F') = (\alpha')^{-1} \circ \beta' \circ (\beta')^{-1} \circ \alpha'$. Then, by the Construction 12.3, $(F'/\rho_1', \gamma')$, where ρ_1' is the congruence on F' generated by ρ' and $\gamma' = \gamma|S'$, is a free group on S'. Our hypothesis implies, since S' is finitely generated by the finite set $N\sigma$, that S' can be embedded in a group. We now apply Theorem 12.4 to deduce that γ' is an isomorphism. This is a contradiction. For F' was so constructed that the transition from $m\phi = w_0$ to $m'\phi = w_n$ by means of the elementary ρ-transitions $w_{i-1} \rightarrow w_i$ $(i = 1, 2, \cdots, n)$ in fact takes place entirely within F'. Hence $(m\phi, m'\phi) \in \rho_1'$ and so $m\phi'\rho_1'^\natural = m'\phi'\rho_1'^\natural$, i.e. $m\sigma'\gamma' = m'\sigma'\gamma'$. But $m\sigma' = m\sigma$ and $m'\sigma' = m'\sigma$ are, by hypothesis, distinct elements of S and so also of S'. Thus γ' is in fact not one-to-one. This completes the proof of the theorem.

EXERCISES FOR §12.2

1. Let S be a semigroup embeddable in a group and let (G, ϕ) be a free group on S. Let P denote the set of congruences ρ on G with the property

that $\phi\rho^\natural$ embeds S in G/ρ. Then every element of P is contained in a maximal element of P.

Let ρ_1, ρ_2 be two distinct maximal elements of P. Then G/ρ_1 is not isomorphic over S to G/ρ_2, i.e. there does not exist an isomorphism μ of G/ρ_1 onto G/ρ_2 such that $s\phi\rho_1^\natural\mu = s\phi\rho_2^\natural$ for all $s \in S$. (Malcev [1939].)

2. Let S be a free semigroup on the two-element set $\{a, b\}$. Let (G, ϕ) be a free group on S. Let ρ_1 be the congruence on G generated by the relation $\{((a\phi)(b\phi)^{-1}, (b\phi)(a\phi)^{-1})\}$. Let ρ_2 be the congruence on G generated by $\{((a\phi)(b\phi)^{-1}(a\phi)(b\phi)^{-1}, (b\phi)(a\phi)^{-1})\}$. Then, in the notation of Exercise 1, $\rho_1, \rho_2 \in P$ and G/ρ_1 is not isomorphic over S to G/ρ_2. (Malcev [1939].)

3. Let S be a semigroup and let ρ_i, $i \in I$, be congruences on S such that each S/ρ_i is a group. Set $\rho = \bigcap\{\rho_i : i \in I\}$. Then S/ρ can be embedded in a group, in fact in the direct product of the groups S/ρ_i. Indeed, S/ρ is a subdirect product of the groups S/ρ_i (see Exercise 2 for §11.6).

12.3 Pták's conditions for embeddability

In this section we consider some conditions (necessary and sufficient) for embedding a semigroup in a group due to V. Pták [1949]. We find it more convenient to work initially with congruences instead of normal subgroups. We derive Pták's basic result in the form that Pták gave it as a Corollary 12.8 to Theorem 12.7. The corollary and theorem state equivalent results, one phrased in terms of subgroups, the other phrased in terms of congruences. From the corollary we then derive an interesting sufficient condition for embeddability also obtained by Pták.

THEOREM 12.7. *The semigroup S can be embedded in a group if and only if*

$$\beta \circ \beta^{-1} = \alpha \circ \rho_1 \circ \alpha^{-1}.$$

[*For notation see Construction* 12.3.]

PROOF. In the proof we continue to use without comment the notation of Construction 12.3.

Suppose that S is embeddable in a group. Then, by Theorem 12.4, γ is one-to-one. Hence $\gamma \circ \gamma^{-1} = \iota_S$. Thus $\beta \circ \beta^{-1} = \beta \circ \gamma \circ \gamma^{-1} \circ \beta^{-1} = (\beta\gamma) \circ (\beta\gamma)^{-1}$. But $\sigma\gamma = \phi\rho_1^\natural$, and so, since $\tau\beta = \sigma$ and $\tau\alpha = \phi$, it follows that $\tau\beta\gamma = \tau\alpha\rho_1^\natural$. Thus $\beta\gamma$ and $\alpha\rho_1^\natural$ agree on the set of generators $M\tau$ of T; hence $\beta\gamma = \alpha\rho_1^\natural$. Consequently, $\beta \circ \beta^{-1} = (\beta\gamma) \circ (\beta\gamma)^{-1} = (\alpha\rho_1^\natural) \circ (\alpha\rho_1^\natural)^{-1} = \alpha \circ \rho_1 \circ \alpha^{-1}$. Hence we have shown that, if S is embeddable in a group, then $\beta \circ \beta^{-1} = \alpha \circ \rho_1 \circ \alpha^{-1}$.

Conversely, assume that $\beta \circ \beta^{-1} = \alpha \circ \rho_1 \circ \alpha^{-1}$. Then

$$\beta \circ \beta^{-1} = \alpha \circ \rho_1^\natural \circ (\rho_1^\natural)^{-1} \circ \alpha^{-1} = \alpha\rho_1^\natural \circ (\alpha\rho_1^\natural)^{-1}$$
$$= (\beta\gamma) \circ (\beta\gamma)^{-1}.$$

It follows immediately that γ is one-to-one. For let $(s, s') \in \gamma \circ \gamma^{-1}$. Since β is onto S there exist t, t' in T such that $t\beta = s$, $t'\beta = s'$. Then (t, t'), which by construction belongs to $\beta \circ (\gamma \circ \gamma^{-1}) \circ \beta^{-1}$, belongs also to $\beta \circ \beta^{-1}$. Thus $s = s'$; which shows that γ is one-to-one.

This completes the proof of the theorem.

COROLLARY 12.8. *Let A denote the subset of F*

$$A = \{ab^{-1};\, a, b \in T\alpha \quad \textit{and} \quad (a, b) \in \rho\}$$

and let A^* be the normal subgroup of F generated by A. Then S is embeddable in a group, if and only if, for a, b in $T\alpha$, $ab^{-1} \in A^*$ implies $ab^{-1} \in A$.

[*For notation again see Construction* 12.3.]

PROOF. The corollary is merely a restatement of the theorem. For A^* is the congruence class modulo ρ_1 containing the identity. Hence $ab^{-1} \in A^*$ if and only if $(a, b) \in \rho_1$. Consequently, for a, b in $T\alpha$, the implication "$ab^{-1} \in A^*$ implies $ab^{-1} \in A$" holds if and only if $\rho_1 \cap (T\alpha \times T\alpha) = \rho$, i.e. if and only if $\rho_1 \cap (T\alpha \times T\alpha) = \alpha^{-1} \circ \beta \circ \beta^{-1} \circ \alpha$. But $\alpha \circ [\rho_1 \cap (T\alpha \times T\alpha)] \circ \alpha^{-1} = \alpha \circ \rho_1 \circ \alpha^{-1}$ and $\alpha \circ \alpha^{-1} = \iota_T$. Hence the required implication holds if and only if $\beta \circ \beta^{-1} = \alpha \circ \rho_1 \circ \alpha^{-1}$. In view of the theorem the corollary now follows.

The sufficient condition for embeddability [Pták, loc. cit.] that we are now going to derive follows from the fact that if S is a cancellative semigroup then it is always true (again in the notation of Construction 12.3) that $\beta \circ \beta^{-1} = \alpha \circ R(\rho) \circ \alpha^{-1}$, where $R(\rho)$ here denotes the right congruence on F generated by ρ. It will be more convenient now to work in terms of subgroups and we restate this result in terms of subgroups in the following lemma.

A comment first about free groups. The elements of F, except for its identity element, are products of the elements of $M\phi \cup (M\phi)^{-1}$; and an element of F is said to be in *reduced form* as such a product if the product contains no adjacent pair $(m\phi)(m\phi)^{-1}$ where $m \in M$. Each element of F, other than the identity, has a unique such reduced form (see, for example, M. Hall [1959], chapter 7).

LEMMA 12.9. *Let S be a cancellative semigroup. Then for a, b in $T\alpha$, $ab^{-1} \in [A]$ if and only if $ab^{-1} \in A$.* [*For notation see Construction* 12.3 *and Corollary* 12.8.]

PROOF. It is clear from the definition of ρ that $\rho = \rho^{-1}$. Thus $ab^{-1} \in A$ if and only if $ba^{-1} = (ab^{-1})^{-1} \in A$. Hence $A = A^{-1}$, where A^{-1} denotes the set of inverses of elements of A. It follows that

$$[A] = \bigcup \{A^n\colon n \text{ a positive integer}\},$$

where, as usual, A^n denotes the set of all products of n elements of A. We prove, by induction on n, that for a, b in $T\alpha$, $ab^{-1} \in A^n$ implies that $ab^{-1} \in A$.

We deal first with the cases $n = 1$ and $n = 2$. For $n = 1$ the assertion is trivially true. Suppose then that a, b, c, d, f, g are elements of $T\alpha$, that $cd^{-1} \in A$, $fg^{-1} \in A$ and that $ab^{-1} = cd^{-1}fg^{-1}$. Since S is cancellative and

$\alpha \circ \alpha^{-1} = \iota_T$ we easily have that ρ is cancellative when restricted to $T\alpha$. Hence we can assume that ab^{-1}, cd^{-1} and fg^{-1} are each in reduced form. With this assumption the equality $ab^{-1} = cd^{-1}fg^{-1}$ can only hold if either (1) $f = dr$ or (2) $d = fr$, where, in each case, r belongs to $T\alpha$ or alternatively is the identity of F. In case (1) $cd^{-1}fg^{-1} = crg^{-1}$. If $r \in T\alpha$ then crg^{-1} is clearly in reduced form and we must have $a = cr$ and $b = g$. Hence $a = cr\,\rho\, dr = f \rho g = b$, so that $a \rho b$, i.e. $ab^{-1} \in A$. If $f = d$ then $(c, d) \in \rho$ and $(d, g) \in \rho$; whence, by the transitivity of ρ on $T\alpha$, $(c, g) \in \rho$. Hence $ab^{-1} = cg^{-1} \in A$. Case (2) is dealt with similarly; and this completes the proof of our assertion for $n = 2$.

We now proceed by induction for $n > 2$ and assume that for a, b in $T\alpha$, $ab^{-1} \in A^j$ implies $ab^{-1} \in A$ for all j such that $1 \leqq j \leqq n - 1$.

Let $a_i b_i^{-1} \in A$, $i = 1, 2, \cdots, n$ and suppose that

$$ab^{-1} = a_1 b_1^{-1} a_2 b_2^{-1} \cdots a_n b_n^{-1}.$$

We may again assume that each of the $a_i b_i^{-1}$ and also ab^{-1} is in its reduced form. Let k be the greatest integer such that

$$a_1 b_1^{-1} a_2 b_2^{-1} \cdots a_{k-1} b_{k-1}^{-1} a_k \in T\alpha.$$

Since $a_1 \in T\alpha$ such an integer k exists. We deal with three cases.

(a) $k = n$. Let $g = a_1 b_1^{-1} \cdots a_{n-1} b_{n-1}^{-1}$; then $g a_n = c$, say, is in $T\alpha$. Hence $g = c a_n^{-1}$ and by the inductive assumption it follows that $(c, a_n) \in \rho$. Since $ab^{-1} = c a_n^{-1} a_n b_n^{-1}$, by our inductive hypothesis, we have $ab^{-1} \in A$.

(b) $1 < k < n$. As for case (a), setting $a_1 b_1^{-1} \cdots a_{k-1} b_{k-1}^{-1} a_k = c$, $c \in T\alpha$ and $c b_k^{-1} \in A$ by the inductive hypothesis (since $k < n$). Hence $ab^{-1} = c b_k^{-1} a_{k+1} b_{k+1}^{-1} \cdots a_n b_n^{-1} \in A^{n-k+1} \subseteq A$, again by the inductive hypothesis, since $n - k + 1 < n$.

(c) $k = 1$. Since ab^{-1} and each $a_i b_i^{-1}$ are in reduced form and since, for each $k > 1$, $a_1 b_1^{-1} \cdots b_{k-1}^{-1} a_k \notin T\alpha$ it follows that $a = a_1$. To see this consider firstly $a_1 b_1^{-1} a_2$. This element does not lie in $T\alpha$. Consequently, $b_1^{-1} a_2$, when reduced, must be the form $c_1^{-1} d_2$ where $c_1 \in T\alpha$. Hence, since $a_2 b_2^{-1}$ is in reduced form so also is $d_2 b_2^{-1}$, and it follows that $a_1 c_1^{-1} d_2 b_2^{-1}$ is the reduced form of $a_1 b_1^{-1} a_2 b_2^{-1}$. A similar argument applies for each k and we finally have that $a_1 b_1^{-1} a_2 b_2^{-1} \cdots a_n b_n^{-1}$ has reduced form $a_1 f_1^{-1} \cdots$ where $f_1 \in T\alpha$. Thus, as already asserted, $a = a_1$.

It now follows that

$$b_1 b^{-1} = a_2 b_2^{-1} \cdots a_n b_n^{-1}$$

a product of $n - 1$ elements of A. By the inductive hypothesis for $j = n - 1$ we have $b_1 b^{-1} \in A$. Hence finally, $ab^{-1} = ab_1^{-1} b_1 b^{-1}$, a product of two elements of A, and so $ab^{-1} \in A$.

This completes the proof of the lemma.

As an immediate corollary to the lemma and to Corollary 12.8 we have Pták's theorem. Again we use the notation of Construction 12.3 and Corollary 12.8.

THEOREM 12.10. *Let S be a cancellative semigroup. Then S is embeddable in a group if the subgroup of F generated by A is a normal subgroup of F.*

Pták, in the paper cited, uses this result to provide a further proof of Ore's Theorem (Theorem 1.23) by showing that if S also is right reversible (viz. $Sa \cap Sb \neq \square$ for all a, b, in S) then the subgroup of F generated by A is also normal in F.

12.4 THE CONSTRUCTION OF QUOTIENTS

If a semigroup S can be embedded in a group then it can be embedded in a group for which S is a set of group generators. Every element of such a group will be a finite product of elements of S and of inverses (in the group) of elements of S. We consider in this section a condition, the *quotient condition*, which a semigroup necessarily satisfies if it can be embedded in a group. For a semigroup S satisfying the quotient condition we construct a *group of right* [*left*] *quotients* and show that, with the appropriate associated mapping, this group forms a free group on S. Another proof of Ore's Theorem (Theorem 1.23), which follows more closely Ore's original proof for rings, is given. In the next section we apply the construction of a group of right quotients to obtain Lambek's necessary and sufficient conditions for embedding a semigroup in a group.

A semigroup S will be said to satisfy the *quotient condition* if for any a, b, c, d, x, y, u, v in S the three equations

$$\left.\begin{aligned} xa &= yb \\ xc &= yd \\ ua &= vb \end{aligned}\right\} \quad \text{———} \quad (Z)$$

together imply the equation

$$uc = vd \quad \text{———} \quad (z).$$

Note that the quotient condition is a symmetrical condition. For the implication that $cu = dv$ follows from the equations $ax = by$, $cx = dy$ and $au = bv$ is, except for a renaming of the symbols involved, identical with the implication (Z) implies (z).

The quotient condition was called condition Z by Malcev [1937].

Suppose that S can be embedded in a group G (by an inclusion mapping). Suppose that the equations (Z) hold in S. Then in G we have

$$\begin{aligned} uc &= ux^{-1}(xc) = ux^{-1}(yd) = ux^{-1}(yb)b^{-1}d \\ &= ux^{-1}(xa)b^{-1}d = (ua)b^{-1}d = (vb)b^{-1}d \\ &= vd. \end{aligned}$$

Thus equation (z) also holds. We have proved

LEMMA 12.11. *A semigroup can be embedded in a group only if it satisfies the quotient condition.*

This result was used by Malcev in [1937] to settle the question, which was then open, of whether any cancellative semigroup could be embedded in a group. Malcev's counterexample is the special case $n = 1$ of the semigroups S_n constructed in §12.8.

When the quotient condition is satisfied quotients may be constructed in a satisfactory way. Let S be a cancellative semigroup satisfying the quotient condition. For a, b in S such that $Sa \cap Sb \neq \square$ we define the *right quotient*

$$a/b = \{(x, y): xa = yb\}.$$

If $Sa \cap Sb = \square$ then a/b is not defined.

Similarly we define the *left quotient*

$$a\backslash b = \{(x, y): ax = by\}$$

when $aS \cap bS \neq \square$.

LEMMA 12.12. *Let S be a cancellative semigroup satisfying the quotient condition. Let $a, b, c, d \in S$. Then*

(i) *$a/b = c/d$ if and only if $a/b \cap c/d \neq \square$;*
(ii) *$a\backslash b = c\backslash d$ if and only if $a\backslash b \cap c\backslash d \neq \square$.*

PROOF. (i) If $a/b = c/d$ then both quotients are defined and so $a/b \cap c/d = a/b \neq \square$.

Conversely, let $(x, y) \in a/b \cap c/d$. Then $xa = yb$ and $xc = yd$. Suppose that $(u, v) \in a/b$ so that $ua = vb$. Then the quotient condition immediately gives $uc = vd$, i.e. $(u, v) \in c/d$. Hence $a/b \subseteq c/d$. Similarly, $c/d \subseteq a/b$. Thus $a/b = c/d$.

(ii) The proof of (ii) now follows when we take into account the remark made above about the symmetry of the quotient condition.

CONSTRUCTION 12.13. *Let S be a cancellative semigroup satisfying the quotient condition. Let Q be the set of right quotients of pairs of elements of S:*

$$Q = \{a/b: a, b \in S, Sa \cap Sb \neq \square\}.$$

Let κ be the one-to-one mapping $a \to (ax)/x$ of S into Q. Let (T, τ) be a free semigroup on Q (where, without loss of generality, we suppose that τ is an inclusion mapping and so write a/b for $(a/b)\tau$). Let ρ be the relation on T defined thus:

$$\rho = \{(xy, z): x = a/b, y = b/c, z = a/c \text{ for some } a/b, b/c, a/c \text{ in } Q\}.$$

Let ρ_1 be the congruence on T generated by ρ. Let $G = T/\rho_1$ and let γ be the mapping $\gamma = \kappa\rho_1{}^\natural$ $(= \kappa\tau\rho_1{}^\natural)$ of S into G. Then (G, γ) is a free group on S.

The pair (G, γ) will be called *the group of right quotients of S*.

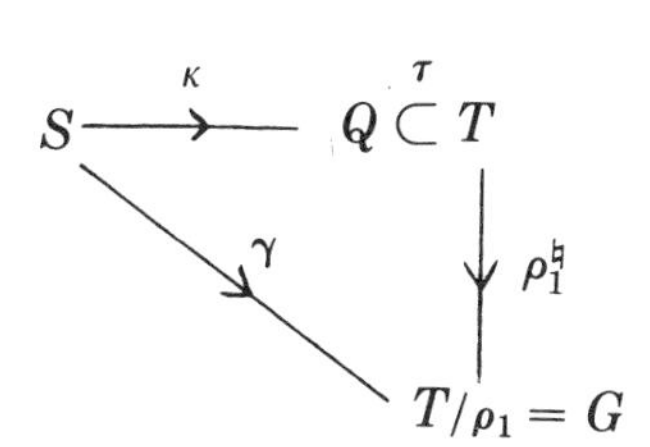

Justification. The verification is routine. The first assertion to be verified is that κ is a one-to-one mapping of S into Q. This follows firstly because $(ax)/x$ is always defined and is independent of x in S since $(a, a^2) \in (ax)/x$ for any x (Lemma 12.12). Further $(ax)/x = (by)/y$ implies $aby = a^2y$ since then $(a, a^2) \in (by)/y$. Hence, by cancellation, $b = a$.

We will now show that T/ρ_1 is a group. For any a in S, a/a is clearly defined for it contains (x, x) for any x in S. Hence, from Lemma 12.12, we also have that $a/a = b/b$ for any a, b in S. We denote a/a by E. For any quotient a/b in Q it is clear that $(a/b \cdot E, a/b) = (a/b \cdot b/b, a/b)$ belongs to ρ. Hence E is a right identity element for T/ρ_1.

Now observe that a/b exists if and only if b/a exists; for $(x, y) \in a/b$ if and only if $(y, x) \in b/a$. Let $t = (a_1/b_1) \cdots (a_n/b_n)$ be any element of T. Then $t^* = (b_n/a_n) \cdots (b_1/a_1)$ is also an element of T. Further $(tt^*, E) \in \rho_1$. For it is clear, for any a/b in Q, that $(a/b \cdot b/a, E) \in \rho$. Hence by a sequence of $2n - 1$ elementary ρ-transitions tt^* can be reduced to E. Thus every element of T/ρ_1 has a right inverse with respect to E. This proves that $G = T/\rho_1$ is a group.

For any a, b in S we may write $a\kappa = (abx)/(bx)$, $b\kappa = (bx)/x$ and $(ab)\kappa = (abx)/x$. Observing further that $((abx)/(bx) \cdot (bx)/x, (abx)/x) \in \rho$ we deduce that $a\kappa\rho_1^\natural \cdot b\kappa\rho_1^\natural = (ab)\kappa\rho_1^\natural$, i.e. $(a\gamma)(b\gamma) = (ab)\gamma$. Thus γ is a homomorphism of S into G.

It is easy to see that for a/b in Q, $(a/b)\rho_1^\natural = (a\gamma)(b\gamma)^{-1}$. Hence it follows that $S\gamma$ is a set of group generators for G. This shows that (G, γ) is a group on the semigroup S. It remains to prove that (G, γ) is a free S-group.

Let (H, η) be any S-group. Construct the mapping $\chi: Q \to H$ as follows: $(a/b)\chi = (a\eta)(b\eta)^{-1}$. Then it is easily verified that χ is well-defined. Since T is a free semigroup on the set Q there exists a homomorphism θ, say, of T into H such that $\tau\theta = \chi$. We then easily verify that ρ, and so also ρ_1, is contained in $\theta \circ \theta^{-1}$. By the Induced Homomorphism Theorem (Theorem 1.6) it now follows that there exists a homomorphism ψ of $G = T/\rho_1$ into H such that $\rho_1^\natural\psi = \theta$.

We now have $\gamma\psi = \eta$. For, firstly, $\kappa\chi = \eta$ since $a\kappa\chi = ((ax)/x)\chi = (ax)\eta(x\eta)^{-1} = a\eta$. Whence we have $a\eta = a\kappa\chi = a\kappa\tau\theta = a(\kappa\tau\rho_1^\natural)\psi = a\gamma\psi$. Thus $\psi: G \to H$ satisfies the equation $\gamma\psi = \eta$. This shows that (G, γ) is a free group on S and completes the justification of the construction.

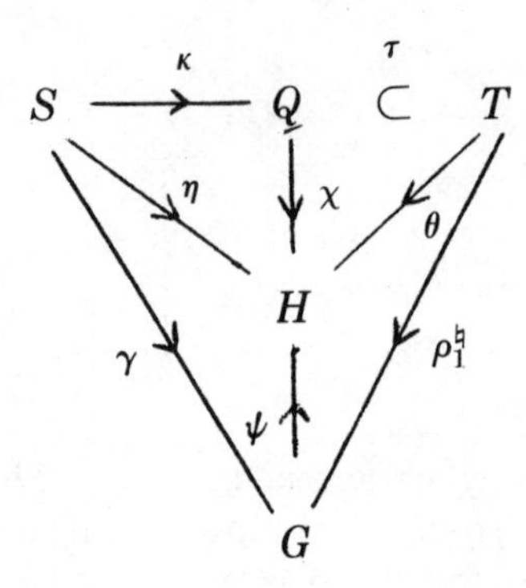

As an immediate corollary to this construction, using Theorem 12.4, we have

LEMMA 12.14. *Let S be a cancellative semigroup satisfying the quotient condition. Then S can be embedded in a group if and only if it can be embedded in its group of right quotients.*

We now consider again Ore's condition for embeddability of a semigroup. Recall that the semigroup S is said to be *left reversible* if $aS \cap bS \neq \Box$ for any a, b in S.

LEMMA 12.15. *A left reversible cancellative semigroup satisfies the quotient condition.*

PROOF. Let S be a left reversible cancellative semigroup and suppose that for a, b, c, d, x, y, u, v in S we have

$$\left.\begin{aligned} xa &= yb \\ xc &= yd \\ ua &= vb \end{aligned}\right\} \quad \text{———} \quad (Z).$$

To show that the quotient condition is satisfied we must show that the equations (Z) imply that

$$uc = vd \quad \text{———} \quad (z).$$

Since S is left reversible there exist p and q in S such that $ap = cq$. Hence $xap = xcq$, whence, by (Z), $ybp = ydq$. Cancelling y gives $bp = dq$. Thus $ucq = uap = vbp = vdq$, i.e. $ucq = vdq$. Cancelling q gives the required equation, $uc = vd$.

Because of this lemma we may apply Construction 12.13 to a left reversible cancellative semigroup. This we now do and show that such a semigroup is embedded by Construction 12.13 in its group of right quotients. We use the notation of Construction 12.13 without further comment. We proceed by means of a series of observations. S will denote a cancellative left reversible semigroup.

(a) *Let* $b_1, a_2, b_2, a_3, b_3, \cdots, b_{m-1}, a_m$ *belong to* S. *Then there exist* $x_1, x_2, \cdots, x_m$ *in* S *such that* $b_i x_i = a_{i+1}x_{i+1}$, $i = 1, 2, \cdots, m-1$.

This follows because the left reversibility of S implies that we can select in turn u_1, y_2; u_2, y_3; $\cdots$; u_{m-1}, y_m such that $b_1u_1 = a_2y_2$; $b_2y_2u_2 = a_3y_3$; $b_3y_3u_3 = a_4y_4$; $\cdots$; $b_{m-1}y_{m-1}u_{m-1} = a_my_m$. Setting $x_1 = u_1u_2 \cdots u_{m-1}$, $x_2 = y_2u_2 \cdots u_{m-1}$, $\cdots$, $x_{m-1} = y_{m-1}u_{m-1}$, $x_m = y_m$ gives a set of x_i with the required properties.

(b) *If* $a/b,\ b/c \in Q$ *then* $(a/b \cdot b/c,\ (ax)/x \cdot x/(cx)) \in \rho_1$.

We have $((ax)/x \cdot x/(bx),\ a/b) \in \rho$ and $((bx)/x \cdot x/(cx),\ b/c) \in \rho$. Hence

$$\begin{aligned} a/b \cdot b/c &\rightarrow (ax)/x \cdot x/(bx) \cdot b/c \\ &\rightarrow (ax)/x \cdot x/(bx) \cdot (bx)/x \cdot x/(cx) \\ &\rightarrow (ax)/x \cdot x/x \cdot x/(cx) \\ &\rightarrow (ax)/x \cdot x/(cx) \end{aligned}$$

is a sequence of elementary ρ-transitions.

Note that if a/b and b/c exist it does not necessarily follow that a/c exists.

(c) *If* $a/b, b/c, c/\cdots/e, e/f$ *exist then* $a/b \cdot b/c \cdot c/ \cdot \cdots \cdot /e \cdot e/f \rho_1 (ax)/x \cdot x/(fx)$.

This is merely the extension of (b) to an arbitrary number of factors.

(d) *Any element of* T *is* ρ_1*-equivalent to an element* $(ax)/x \cdot x/(bx)$ *for some* a, b *in* S.

Let $a_1/b_1 \cdot a_2/b_2 \cdots a_m/b_m$ be an arbitrary element of T. By (a) there exist $x_1, x_2, \cdots, x_m$ in S such that $b_ix_i = a_{i+1}x_{i+1}$, $i = 1, 2, \cdots, m-1$. Hence $a_1/b_1 \cdot a_2/b_2 \cdots a_m/b_m = a/c_1 \cdot c_1/c_2 \cdots c_{m-1}/b$ where $a_1x_1 = a$, $b_1x_1 = a_2x_2 = c_1$, $b_2x_2 = a_3x_3 = c_2$, $\cdots$, $b_mx_m = b$. Applying (c) we have the required result.

(e) *Let* $a_1/a_2 \cdot a_2/a_3 \cdot \cdots a_{m-1}/a_m = t$ *be an element of* T. *Let* $t \rightarrow t'$ *be an elementary* ρ*-transition. Then* $t' = (a_1x)/b_2 \cdot b_2/b_3 \cdot \cdots \cdot b_k/(a_mx)$ *for some* $b_2, b_3, \cdots, b_k$ *in* S.

There are two possibilities. Firstly, the ρ-transition may replace $a_{i-1}/a_i \cdot a_i/a_{i+1}$ by a/b. If this occurs then $a_{i-1}/a_i = a/c$ and $a_i/a_{i+1} = c/b$ for some c in S. Since S is left reversible there exist x, y in S such that $a_{i-1}x = ay$. Then also $a_ix = cy$ and $a_{i+1}x = by$. For $(a_{i-1}x)/(a_ix) = a_{i-1}/a_i = a/c = (ay)/(cy)$. Hence if $(u, v) \in (a_{i-1}x)/(a_ix)$ so that $ua_{i-1}x = va_ix$, then also $uay = vcy$. Since $a_{i-1}x = ay$, it follows that $va_ix = vcy$ and cancelling v we get $a_ix = cy$. Similarly, $a_{i+1}x = by$. Hence

$$t' = (a_1x)/(a_2x) \cdots (a_{i-2}x)/(a_{i-1}x) \cdot (a_{i-1}x)/(a_{i+1}x) \cdot (a_{i+1}x)/(a_{i+2}x) \cdots (a_{m-1}x)/(a_mx)$$

which is of the required form.

Alternatively the ρ-transition may replace a_i/a_{i+1} by $a/b \cdot b/c$. It then follows that $a_i/a_{i+1} = a/c$. Choose x, y so that $a_ix = ay$. It then follows, as in the first case, that $a_{i+1}x = cy$ whence we again obtain t' in the required form.

Now consider any quotient a/b in T. If t is ρ_1-equivalent to a/b, then it follows from remark (e), since t can be obtained from a/b by a sequence of elementary ρ-transitions, that $t = (ax)/b_1 \cdot b_1/b_2 \cdots b_k/(bx)$ for some x, b_i in S. Hence, if t is itself a quotient, then $t = (ax)/(bx)$ for some x in S. But $(ax)/(bx) = a/b$. Thus the only quotient ρ_1-equivalent to a/b is a/b itself. It immediately follows, in particular, that $\gamma = \kappa\rho_1^\natural$ is a one-to-one mapping, i.e. that γ embeds S in G.

Further it follows from remark (d) above that any element of G can be written in the form $a\gamma(b\gamma)^{-1}$ for some a, b in S.

Suppose conversely that $\gamma: S \to H$ embeds S in a group H and that every element of H can be written in the form $a\gamma(b\gamma)^{-1}$ for a, b in S. Then, in particular, given a, b in S, there must exist x, y in S such that $(a\gamma)^{-1}b\gamma = x\gamma(y\gamma)^{-1}$, i.e. such that $(by)\gamma = (ax)\gamma$. Since γ is one-to-one, therefore $by = ax$. Thus, given any a, b in S, $aS \cap bS \neq \square$, i.e. S is left reversible.

We have thus completed our alternative proof of Ore's theorem (Theorem 1.23) and Dubreil's theorem (Theorem 1.24): *A cancellative semigroup S can be embedded in a group of right quotients* of S if and only if it is left reversible.*

Exercises for §12.4

1. Let R be a left reversible, cancellative semigroup with identity element and let ϕ be an endomorphism of R onto a proper left unitary subsemigroup $R\phi$ of R. Let Z denote the additive semigroup of nonnegative integers. Let $S = Z \times R$, and define a product in S by $(m, a)(n, b) = (m + n, a\phi^n b)$, where $m, n \in Z$, $a, b \in R$ and where ϕ^0 is interpreted as the identity mapping. Then S is a left reversible, left cancellative semigroup with identity element and S is not right reversible. If ϕ is also one-to-one then S is cancellative.

2. Let S be a semigroup containing cancellable elements. The product of any two cancellable elements is also cancellable and hence the subset C, say, of cancellable elements of S forms a subsemigroup of S. Let M be a subsemigroup of C. Then a semigroup G is said to be a *right quotient semigroup of S by M*, if (1) $S \subseteq G$, (2) G contains an identity element 1, (3) if $m \in M$ then there exists $g \in G$ such that $mg = gm = 1$, and (4) if $g \in G$ then there exists $m \in M$ such that $gm \in S$.

S will be said to be *left M-reversible* if $sM \cap mS \neq \square$, for all $m \in M$, $s \in S$.

Let S be left M-reversible. Then:

(i) If, for $s_1, s_2 \in S$ and $m_1, m_2 \in M$, there exist $x, y \in S$ such that $s_1x = s_2y$ and $m_1x = m_2y \in M$, then, for any $x', y' \in S$, $m_1x' = m_2y'$ implies $s_1x' = s_2y'$.

(ii) Define the relation ρ on $S \times M$ by agreeing that $(s_1, m_1)\,\rho\,(s_2, m_2)$ if and only if there exist $x, y \in S$ such that $s_1x = s_2y$ and $m_1x = m_2y \in M$. Then ρ is an equivalence on $S \times M$.

(iii) The equation $(s_1, m_1)\rho \cdot (s_2, m_2)\rho = (s_1s_3, m_2m_3)\rho$, where $s_2m_3 = m_1s_3$, is a well-defined product on $(S \times M)/\rho$ relative to which $(S \times M)/\rho$ is a semigroup.

* The term "group of right quotients" is here used in its earlier sense, Vol. 1, p. 36.

(iv) The mapping $s \to (sm, m)\rho$ embeds S in the semigroup $(S \times M)/\rho$ and, identifying S with its image under this mapping, $(S \times M)/\rho$ becomes a right quotient semigroup of S by M.

(v) A right quotient semigroup of S by M is uniquely determined to within isomorphism.

Conversely, suppose that a right quotient semigroup of S by M exists. Then S is left M-reversible. (Murata [1950]; cf. Asano [1949] and Schiefer-decker [1955].)

3. Let S be a left cancellative semigroup which is left M-reversible (in the terminology of the preceding exercise). Then the right quotient semigroup of S by M is left cancellative. If S is also right cancellative then the right quotient semigroup of S by M is cancellative.

4. An element m of a semigroup S is said to be a *left reversible* element of S if $aS \cap mS \neq \square$ for all $a \in S$. Let N denote the set of all left reversible elements of S. Then, if non-empty, N is a subsemigroup of S which is left consistent (i.e. $xy \in N$ implies $x \in N$). If S is left cancellative then N is consistent. S is said to be *left quasi-reversible* if N is non-empty and if, for $a, b \in S$, $aS \cap bS \neq \square$ implies that either $aN \cap bS \neq \square$ or $aS \cap bN \neq \square$. If S is left cancellative and left quasi-reversible then S is left N-reversible, in the sense of Exercise 2. (Doss [1948].)

5. Let w, w' be elements of a free semigroup F. Then the *law* $w = w'$ is said to be *satisfied* in the semigroup S, if $w\phi = w'\phi$ for all homomorphisms $\phi\colon F \to S$. If w and w' are distinct elements of F, then the law $w = w'$ is said to be a *non-trivial* law.

Let S be a cancellative semigroup satisfying a non-trivial law. Then S is left reversible and so can be embedded in a group (Malcev [1953]).

12.5 Lambek's polyhedral conditions for embeddability

In this section is presented a geometrical interpretation of the conditions for embedding a semigroup in a group due to J. Lambek [1951]. We follow Lambek's treatment. The conditions for embeddability are stated in terms of Eulerian polyhedra, i.e. polyhedra for which $V + F = E + 2$, where V is the number of vertices, F is the number of faces, and E is the number of edges, which are homeomorphic to the 2-sphere in Euclidean 3-space. Throughout this section by *polyhedron*, simply, will be meant such a polyhedron which is homeomorphic to the 2-sphere. It will suffice for our purposes to treat such polyhedra informally and to illustrate the argument by drawing pictures. For a purely algebraic treatment, see Bush [1963].

Each edge of a polyhedron is an edge to two faces. We will say that each edge has two *sides*, one on each face. Each edge has two vertices, one at each end. At each vertex of an edge the edge has two *angles*, one on each side of the edge. For example, in the diagram below, the edge drawn with two vertices has the symbols x and y attached to its two sides, the symbols a

and b attached to its angles at one vertex and the symbols c and d attached to its angles at the other vertex. The angles lettered a and c are on the side lettered x and the angles b and d are on the side lettered y.

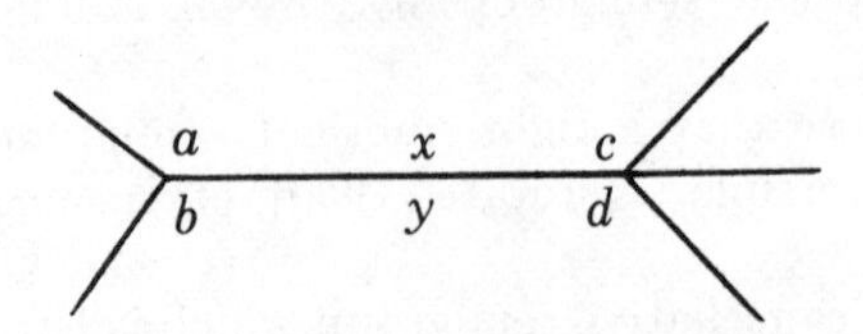

Each edge will be said to consist of two *half-edges*, one corresponding to each vertex. Each half-edge has two sides, the sides of the edge to which it belongs, and two angles, one on each side, at its vertex. In the following diagram we have lettered the sides of an edge and the angles of a half-edge of this edge. When so lettered the equation $xa = yb$ will be called the *half-edge equation* corresponding to this half-edge.

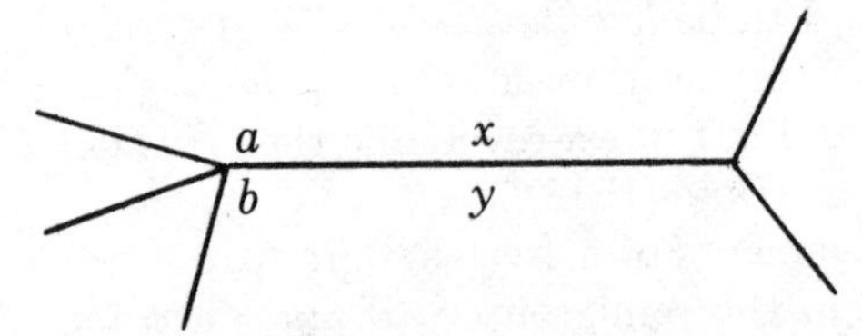

Diagram 1

Now let S be any semigroup. Then S will be said to satisfy the *polyhedral condition* if, for any polyhedron, when all the sides and angles of the polyhedron are lettered by elements of S, then any half-edge equation is a consequence of all the other half-edge equations.

We can now state Lambek's theorem.

THEOREM 12.16 (Lambek). *A semigroup can be embedded in a group if and only if* (1) *it is cancellative and* (2) *it satisfies the polyhedral condition.*

The rest of this section is devoted to the proof of this theorem. We first prove the necessity of the conditions. Clearly, that S is cancellative, is necessary. We prove the necessity of the polyhedral condition by induction on the size of the polyhedron.

Let P be any polyhedron lettered (i.e. whose sides and angles are lettered) by elements of S. Triangulate P by taking a point, the *center*, in the interior of each face and for each face, drawing a line from its center to each vertex of that face and from an interior point, the *mid-point*, of each edge of that face to its center. Then orient this triangulation by attaching the positive directions (1) from center to vertex (2) from mid-point of edge to center and (3) from mid-point to vertex.

The given lettering of the polyhedron P by elements of S determines, as follows, a lettering of the triangulated polyhedron. Suppose that a half-edge of P is lettered as in Diagram 1. Then we attach the letter x to the line from the mid-point of the edge to the center on the same side as x and attach the letter a to the line from this center to the vertex of the half-edge. To the half-edge itself we attach the letter p, where $xa = p$. If the half-edge equation holds we can proceed similarly for y and b and since then $yb = xa = p$ only one letter is attached to the half-edge. The resulting lettering is illustrated in Diagram 2.

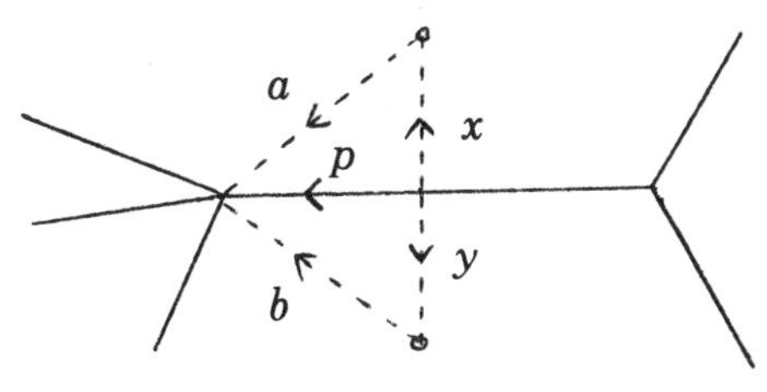

Diagram 2

Suppose now that all the half-edge equations, except possibly one, are known to hold in S. Then all the edges of the triangulation can be consistently lettered by a single letter by the above procedure with the possible exception of a single half-edge, the half-edge whose equation is not assumed to be valid. If the half-edge in question is that shown lettered in Diagram 1, then the procedure for lettering the triangulated polyhedron has attached letters a, x, b, y to the edges of the triangulation, as in Diagram 2, but no letter has been attached to the half-edge of P lettered by p in Diagram 2.

We wish to deduce from the assumption that S can be embedded in a group that $xa = yb$. Suppose that $yb = p$ and attach this p to the remaining unlettered half-edge. Then it will suffice to show that $xa = p$.

Suppose that S is embedded, by inclusion, in a group G with identity 1. Each triangle, except possibly one, of the triangulation corresponds to an equation such as $yb = p$. This equation can also be written in any one of the six forms: $ybp^{-1} = 1$, $bp^{-1}y = 1$, $p^{-1}yb = 1$, $y^{-1}pb^{-1} = 1$, $b^{-1}y^{-1}p = 1$ and $pb^{-1}y^{-1} = 1$. These six equations correspond to the six ways of traversing the edges of the triangle, taking into account the orientation, starting at any edge and going round the triangle in either direction. More precisely we get the left-hand side of each of these equations as a special case of the following procedure.

Consider any connected path C consisting entirely of the edges of the triangulation. Let $E_1, E_2, \cdots, E_n$ be the successive edges of the path in the order in which they are traversed. Let x_i be the letter attached to the edge E_i. Then this path, together with the given direction $E_1, E_2, \cdots, E_n$ of traversing it, determines the product $C\alpha = x_1^{\epsilon_1}x_2^{\epsilon_2} \cdots x_n^{\epsilon_n}$ where ϵ_i is $+1$ if E_i is traversed in the positive direction and ϵ_i is -1 if E_i is traversed in the negative direction.

We have seen that a simple closed path which bounds a single triangle, for which the triangle equation holds, determines a product equal to 1. Proceed now by induction. Suppose that this is true for all simple closed paths of edges of the triangulation which form a boundary for a simply connected set of less than k triangles for each of which the triangle equation holds. Let C be a simple closed path which is the boundary of a simply connected set of k triangles for each of which the triangle equation holds. Then there exists a connected path B whose edges are selected from the edges of the k triangles bounded by C and such that the end points of B divide C into two connected paths C_1 and C_2 such that (1) the path B followed by the path C_1 is a simple closed path for which our inductive hypothesis holds and (2) the path B^{-1} (B traversed in the opposite direction) followed by the path C_2 is a simple closed path for which our inductive hypothesis holds.

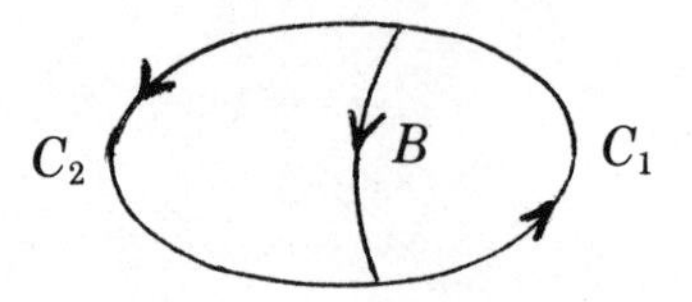

Hence we have $(BC_1)\alpha = (B\alpha)(C_1\alpha) = 1$ and $(B^{-1}C_2)\alpha = (B^{-1}\alpha)(C_2\alpha) = 1$. Whence it follows that $C\alpha = (C_1\alpha)(C_2\alpha) = (B\alpha)^{-1}(B^{-1}\alpha)^{-1} = 1$ since clearly $(B\alpha)(B^{-1}\alpha) = (BB^{-1})\alpha = 1$.

It thus follows by induction that $C\alpha = 1$ for any simple closed path C which bounds any simply connected set of triangles for each of which the triangle equation holds.

Now let C be the path bounding all the triangles of the triangulated polyhedron except the triangle whose equation we have to deduce. Then $C\alpha = 1$. But the path C is merely a path along the sides of the excluded triangle. Hence $C\alpha = 1$ implies $xa = p$.

This completes the proof of the necessity of the polyhedral condition.

We now prove sufficiency. Let S be a cancellative semigroup satisfying the polyhedral condition. Then S satisfies the quotient condition (§12.4). For consider the polyhedron lettered in Diagram 3 below.

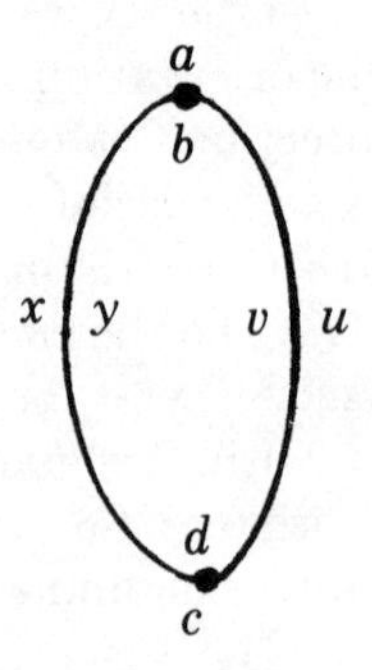

Diagram 3

Suppose that the equations $xa = yb$, $xc = yd$ and $ua = vb$ hold in S. Then from Diagram 3, using the polyhedral condition, we have $uc = vd$. Thus S satisfies the quotient condition. To show that S can be embedded in a group we can therefore proceed by showing that the mapping γ of Construction 12.13 is an isomorphism. It will suffice to show that two elements of Q (we are now using the terminology of Construction 12.13) are ρ_1-equivalent only if they are equal. This follows directly from the polyhedral condition by constructing the appropriate polyhedra. The whole procedure we are about to describe is exemplified below by a special case which is illustrated in Diagram 5.

Let a/b, c/d belong to Q and suppose that a/b is ρ_1-equivalent to c/d. Then there exist w_0 $(= a/b)$, $w_1, w_2, \cdots, w_n$ $(= c/d)$, elements of T, such that $w_{i-1} \rightarrow w_i$ is an elementary ρ-transition for $i = 1, 2, \cdots, n$. Let w_i be a word of length n_i (see §9.1) in the elements of Q $(= Q_T)$ for $i = 0, 1, 2, \cdots, n$. Then, relative to ordinary Cartesian axes, each word w_i determines a set of n_i integral lattice points: (j, i), $j = 1, 2, \cdots, n_i$. We use the set of all these lattice points (j, i), $j = 1, 2, \cdots, n_i$, $i = 0, 1, 2, \cdots, n$ to determine a polyhedron. Suppose firstly that (j, i) corresponds to the jth factor (in Q, counting from the left) of the word w_i. The lattice point (j, i) will then determine a vertex of the polyhedron if either (a) the jth factor of w_i resulted from replacing the jth and $(j + 1)$th factors of w_{i-1} in the ρ-transition $w_{i-1} \rightarrow w_i$ or (b) the jth factor of w_i is replaced by the jth and $(j + 1)$th factors of w_{i+1} in the ρ-transition $w_i \rightarrow w_{i+1}$.

We make the assumption that if the jth factor of w_i results from a contraction of the jth and $(j + 1)$th factors of w_{i-1}, then the transition $w_i \rightarrow w_{i+1}$ does not replace this jth factor by two factors in w_{i+1}. This assumption is possible for if such a sequence $w_{i-1} \rightarrow w_i \rightarrow w_{i+1}$ did occur then it could be replaced by the sequence $w_{i-1} \rightarrow w_i \rightarrow w_i \rightarrow w_{i+1}$, obtained by a repetition of w_i. In such a circumstance there will be no ambiguity if we refer to $w_i \rightarrow w_i$ as a ρ-transition.

We then construct the edges of the polyhedron by drawing lines as follows. If a factor remains unchanged in the transition $w_{i-1} \rightarrow w_i$ the corresponding lattice points are joined by a straight line. If two adjacent factors of w_{i-1} are replaced in w_i by a single factor, then the two corresponding lattice points are joined by straight lines to the corresponding vertex. If a single factor is replaced by two adjacent factors then the corresponding vertex is joined by straight lines to the two corresponding lattice points. Finally we join the lattice points $(1, 0)$ and $(1, n)$ by any simple curve not intersecting, except at its end points, any of the other edges.

The polyhedron so constructed (note the assumption about repetitions we made) has three edges, and so three angles, at each vertex. We now letter these angles as follows. Suppose that p/q in w_{i-1} is replaced by $p/r \cdot r/q$ in w_i. Then the angles at the vertex determined by p/q are lettered p, r, q in a clockwise fashion, the letter r being attached to the angle between the two edges going upwards from the vertex (see Diagram 4).

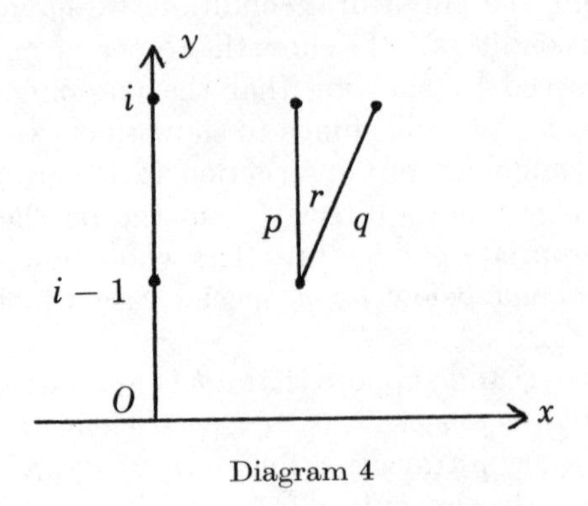

Diagram 4

Similarly, if $p/r \cdot r/q$ in w_{i-1} is replaced by p/q in w_i then the angles of the vertex determined by p/q are lettered in a counter-clockwise fashion by p, r, q, the letter r being attached to the angle between the two edges going downward from the vertex.

The whole procedure, for a special case, is illustrated in Diagram 5 in which the vertices are the encircled lattice points.

This diagram shows the polyhedron determined by the sequence of ρ-transitions

$$
\begin{aligned}
w_0 = a/b \to w_1 &= a/z \cdot z/b = x/y \cdot z/b \\
\to w_2 &= x/v \cdot v/y \cdot z/b = x/v \cdot s/t \cdot t/u \\
\to w_3 &= x/v \cdot s/u = x/v \cdot q/r \\
\to w_4 &= x/v \cdot q/p \cdot p/r = l/m \cdot m/n \cdot p/r \\
\to w_5 &= l/n \cdot p/r = c/h \cdot h/d \\
\to w_6 &= c/d.
\end{aligned}
$$

Note the repetition of w_3 necessary to carry out the construction. Equality denotes equality in T, the free semigroup on Q.

Consider any edge in the polyhedron except that connecting the vertices $(1, 0)$ and $(1, n)$. The choice of lettering of angles in such that the quotient

(letter attached to angle on side 1)/(letter attached to angle on side 2)

is the same for each half-edge of the edge. Thus there exists a pair of elements in S that may be attached to the sides of the edge so that the half-edge equations then hold for that edge. Attach in this way elements of S to the sides of each edge of the polyhedron except the edge joining $(1, 0)$ and $(1, n)$. This latter edge has the quotient a/b as the quotient of the letters attached to its angles at $(1, 0)$ and similarly c/d at $(1, n)$. There exist x, y in S such that $xa = yb$. Attach x, y to the sides of this edge so that the half-edge

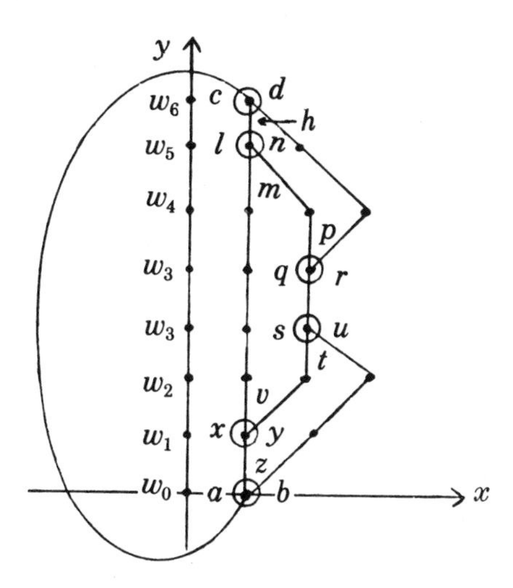

Diagram 5

equation (with vertex at (1, 0)) holds. By the polyhedral condition it then follows that $xc = yd$. Thus $a/b \cap c/d \neq \square$ and so, finally, $a/b = c/d$.

This completes the proof of Lambek's theorem.

12.6 Malcev's conditions for embeddability

In this section we give an account of what was historically the first adequate discussion of the problem of embedding a semigroup in a group. In [1939], A. Malcev established necessary and sufficient conditions for embedding a semigroup in a group in the form of an infinite set of implications similar to, but different from, those of Lambek, discussed in the last section.

A Lambek or a Malcev system consists of an even number q of equations of the form $x_iy_j = x_ky_l$ in $2q$ variables $x_1, \cdots, x_q; y_1, \cdots, y_q$ (the x's always on the left and the y's on the right) such that each variable occurs exactly twice. A given cancellative semigroup S can be embedded in a group if and only if, for each Lambek system, whenever the x's and y's are replaced by elements of S such that $q - 1$ equations of the system are satisfied, then the remaining equation is also satisfied. We shall see that the same assertion holds for Malcev systems, except that a definite one of the q equations is singled out to be a consequence of the remaining $q - 1$. Actually we shall refer to the $q - 1$ equations as the "Malcev system," and to the consequent equation as the "locked" equation of the system.

In §12.7 it is shown that a Lambek system is a Malcev system if and only if it arises from a two-vertex polyhedron. In particular, the quotient condition is a Malcev system. It would be interesting to find a general description of those systems, of the kind described above, which serve to determine when a semigroup is embeddable in a group.

Malcev [1940] also showed that no finite subset of his conditions would suffice for embeddability. The last section of this chapter, §12.8, will be devoted to this result. Throughout this section and §12.8, we follow Malcev's account.

We shall first describe how to form a Malcev system. We begin by introducing $2k + 2p$ symbols

$$L_1, \cdots, L_k; \quad L_1^*, \cdots, L_k^*;$$
$$R_1, \cdots, R_p; \quad R_1^*, \cdots, R_p^*.$$

A *Malcev sequence* (of length $q = 2k + 2p$) is a sequence of these q symbols, each occurring just once, such that

(i) the L's and R's occur in their natural order;

(ii) L_i^* $(1 \leqq i \leqq k)$ always occurs after L_i, and if L_j occurs between L_i and L_i^*, then L_j^* also occurs between them; and similarly for R_i^*.

For example (with $k = 3$, $p = 2$), the following is a Malcev sequence:

$$L_1 \, L_2 \, R_1 \, L_2^* \, R_2 \, L_3 \, R_2^* \, L_3^* \, L_1^* \, R_1^*.$$

It is easily seen, though we shall not need it in the sequel, that there is a one-to-one correspondence between Malcev sequences and sequences of integers $\{i_1, \cdots, i_q\}$ such that

(i) each i_r is $+1$, -1, $+2$, or -2;

(ii) for each n such that $1 \leqq n \leqq q$,

$$\sum_{r=1}^{n} \{i_r : i_r = \pm 1\} \geqq 0, \qquad \sum_{i=1}^{n} \{i_r : i_r = \pm 2\} \geqq 0,$$

with equality holding for $n = q$. We simply replace each L_i by $+1$, each L_i^* by -1, each R_i by $+2$, and each R_i^* by -2. Thus, for the above example, we get $\{1, 1, 2, -1, 2, 1, -2, -1, -1, -2\}$. It is easy to see, though clumsy to state, how to reverse the procedure.

We now describe how to get the Malcev system $\sigma(I)$ of equations corresponding to a given Malcev sequence I. If I has length q, then the $q - 1$ adjacent pairs of symbols in I determine the $q - 1$ equations in $\sigma(I)$ by means of the following table:

TABLE 1

L_i	L_i^*	R_i	R_i^*
$d_i a_i$	$c_i b_i$	$A_i D_i$	$B_i C_i$
$c_i a_i$	$d_i b_i$	$A_i C_i$	$B_i D_i$

We have here eight sequences of variables

$$a_i,\ b_i,\ c_i,\ d_i,\ A_i,\ B_i,\ C_i,\ D_i \quad (i = 1, 2, \cdots).$$

If, for example, the adjacent pair is $L_i^* R_j$, the equation is $c_i b_i = A_j C_j$. The first [second] member of the pair gives the left [right] member of the equation, and this is read off the first [second] row in the table, using the same subscript as on the symbol being read. The locked equation of the system is obtained in exactly the same way, using the last and the first member of I.

For example, the Malcev sequence

$$L_1\ L_2\ R_1\ L_2^*\ R_2\ L_3\ R_2^*\ L_3^*\ L_1^*\ R_1^*$$

leads to the system

$$\begin{aligned} d_1a_1 &= c_2a_2 \\ d_2a_2 &= A_1C_1 \\ A_1D_1 &= d_2b_2 \\ c_2b_2 &= A_2C_2 \\ A_2D_2 &= c_3a_3 \\ d_3a_3 &= B_2D_2 \\ B_2C_2 &= d_3b_3 \\ c_3b_3 &= d_1b_1 \\ c_1b_1 &= B_1D_1, \end{aligned}$$

the equation locked to this system being

$$B_1C_1 = c_1a_1.$$

We can now state Malcev's theorem.

Theorem 12.17 (Malcev). *Let S be a semigroup with identity. Then S can be embedded in a group if and only if, for each Malcev sequence I, the system of equations $\sigma(I)$ holds in S only if its locked equation also holds in S.*

To prove Malcev's theorem we must first give a further construction of a free group on a semigroup. To avoid some slight complications we restrict the discussion, as in the statement of the theorem, to semigroups with an identity.

Construction 12.18. *Let S be a semigroup with identity and let Γ be a set of generators of S. Let M be any set of the same cardinal as Γ. Let M^L and M^R be two disjoint sets also disjoint from M of the same cardinal as M. Let σ be a one-to-one mapping of M onto Γ and let*

$$m \to m^L \qquad (m \in M)$$

and

$$m \to m^R \qquad (m \in M)$$

be one-to-one mappings of M onto M^L and of M onto M^R, respectively.

Let μ be the inclusion mapping μ: $M \subset M \cup M^L \cup M^R$. Let (K, κ) be a free semigroup with identity on M. Let (T, τ) be a free semigroup with identity on $M \cup M^L \cup M^R$. Let α: $K \to T$ be a homomorphism such that $\kappa\alpha = \mu\tau$ and let β: $K \to S$ be a homomorphism such that $\kappa\beta = \sigma$.

Now let $\pi_1 = \alpha^{-1} \circ \beta \circ \beta^{-1} \circ \alpha$ and let $\pi_2 = \{(w, 1)\} \cup \{(1, w)\}$ where w ranges over all words in T of the form mm^R or of the form $m^L m$ $(m \in M = M\mu = M\mu\tau)$ and where 1 is the identity of T. Finally, let $\rho = \pi_1 \cup \pi_2$ and let ρ_1 be the congruence on T generated by ρ. Define the homomorphism γ: $S \to T/\rho_1$ by the equation $\sigma\gamma = \mu\tau\rho_1^\natural$. Then, setting $G = T/\rho_1$, (G, γ) is a free group on the semigroup S.

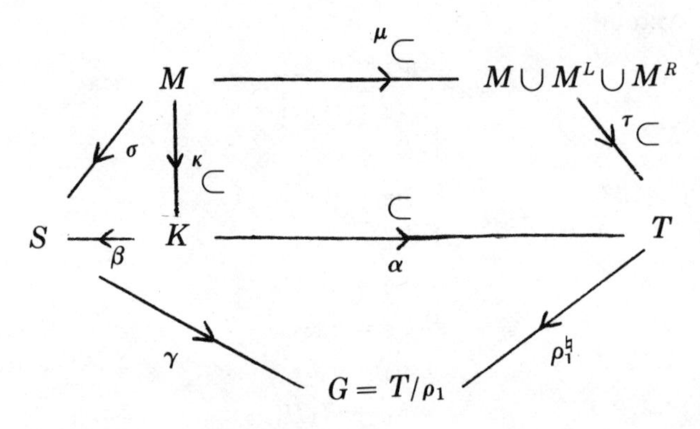

The verification of the assertions made in the construction is a routine matter similar to that of the previous constructions of this chapter. We take (without loss of generality) μ, τ, κ, α as inclusion mappings. The elements of K are then words in the elements of M and the elements of T are words in the elements of $M \cup M^L \cup M^R$ and the identity of K and of T may be taken as the empty word.

Since (G, γ) is a free group on S, S can be embedded in a group if and only if γ is a one-to-one mapping (Theorem 12.4). Note that, since $\kappa\beta = \sigma$ and $\kappa\alpha = \mu\tau$ and also $M\kappa$ is a set of generators for K, the equation $\sigma\gamma = \mu\tau\rho_1^\natural$ is equivalent to the equation $\beta\gamma = \alpha\rho_1^\natural$. Thus $\gamma \circ \gamma^{-1} = \beta^{-1} \circ \alpha \circ \rho_1 \circ \alpha^{-1} \circ \beta$. We then readily deduce that γ is a one-to-one mapping, i.e. that $\gamma \circ \gamma^{-1}$ is the identical relation on S, if and only if ρ_1 restricted to $K\alpha$ coincides with π_1. In other words, S can be embedded in a group if and only if, for w, w' in K $(= K\alpha)$ $(w, w') \in \rho_1$ implies $(w, w') \in \pi_1$, i.e. implies that $w\beta = w'\beta$ is an equation which holds in S.

We proceed to show that: for w, w' in K, $(w, w') \in \rho_1$ if and only if a certain set of equations holds in S.

Suppose then that $w, w' \in K$ and that $(w, w') \in \rho_1$. Then there exists a sequence X, say, of elements of T, $X = (w_0, w_1, \cdots, w_n)$, such that $w = w_0$, $w' = w_n$ and $w_i \to w_{i+1}$ is an elementary ρ-transition for $i = 0, 1, \cdots, n-1$.

Any other sequence of elements of T which derives w' from w by a series of elementary ρ-transitions will be said to be *equivalent* to X. Such sequences will be called *ρ-chains* (*in* T) from w to w'.

A ρ-chain X from w to w' will be called *normal* if w and w' belong to K and if no ρ-transition $w_i \to w_{i+1}$ of X takes place either to the right (in the obvious sense) of any element m^R or to the left of any element m^L in w_i ($m\mu_T = m \in M$). The following lemma is fundamental.

LEMMA 12.19. *Any ρ-chain in T from an element of K to an element of K is equivalent to a normal ρ-chain.*

PROOF. Let $X = (w_0, w_1, \cdots, w_n)$ be a ρ-chain such that $w_0, w_n \in K$. Thus w_0 and w_n are words in M, not involving elements of M^L and M^R. In the course of the chain X, elements of M^L and M^R will be introduced, and eventually suppressed. We shall denote the elements of M^L introduced in X by $d_1^L, d_2^L, \cdots$ in the order in which they are introduced, and those of M^R by $D_1^R, D_2^R, \cdots$, where $d_1, d_2, \cdots, D_1, D_2, \cdots$ belong to M. The various d_s^L (and D_t^R) need not be distinct, but this notation will enable us to distinguish between the different occurrences of the same element of M^L (and M^R). Thus the *symbol* d_s^L appears at a certain moment in X, lives a while, and then disappears never to appear again.

Suppose that k of the corresponding transitions T_i: $w_i \to w_{i+1}$ are π_2-transitions which introduce elements of M^L and that p of the T_i are π_2-transitions which introduce elements of M^R.

Suppose that $j - 1$ of the occurrences of the elements of M^L, $1 \leqq j \leqq k + 1$, which are introduced, are such that no ρ-transformation takes place to their left in any of the transitions T_i. Then, if $j < k + 1$, there is a further element of M^L, d_s^L say, not satisfying this condition. Hence for some i, l,

$$\begin{aligned} w_i = ab \to w_{i+1} &= ad_s^L d_s b \to \cdots \to w_{l-1} \\ &= a_1 d_s^L d_s b_1 \to w_l = a_1 b_1, \end{aligned}$$

where a, b, a_1, b_1 are elements of T.

Since d_s^L can take part in no ρ-transition except those made explicit in the above sequence, i.e. T_i and T_{l-1}, the ρ-transitions which transform a to a_1 are independent of those which transform $d_s b$ to $d_s b_1$. Hence we may rearrange the order of the ρ-transitions (keeping those from a to a_1, in the same order as before and those from $d_s b$ to $d_s b_1$ in the same order as before) to obtain the sequence

$$\begin{aligned} w_i = ab \to \cdots \to a_1 b \to a_1 d_s^L d_s b \cdots \\ \to a_1 d_s^L d_s b_1 \to a_1 b_1 = w_l. \end{aligned}$$

Inserting the new sequence between w_i and w_l we obtain a new sequence equivalent to X. Moreover in the new sequence there are j occurrences of elements of M^L such that no ρ-transition takes place to their left. To see this in detail let us number the stages in the above transformations thus:

$$\begin{aligned} &\text{stage 1:} && ab \to ad_s^L d_s b \\ &\text{stage 2:} && ad_s^L d_s b \to \cdots \to a_1 d_s^L d_s b_1 \\ &\text{stage 3:} && a_1 d_s^L d_s b_1 \to a_1 b_1 \end{aligned}$$

for the first ρ-chain; and

$$\begin{aligned} &\text{stage 1}': && ab \to \cdots \to a_1 b \\ &\text{stage 2}': && a_1 b \to a_1 d_s^L d_s b \\ &\text{stage 3}': && a_1 d_s^L d_s b \to \cdots \to a_1 d_s^L d_s b_1 \\ &\text{stage 4}': && a_1 d_s^L d_s b_1 \to a_1 b_1 \end{aligned}$$

for the new ρ-chain.

Now it is clear that the second ρ-chain has been so constructed that the element d_s^L occurring above is such that no ρ-transition takes place to its left. We will show that this property is preserved for the other $j - 1$ occurrences of elements of M^L which, by hypothesis, had this property in the first chain. For let d_u^L be one of these elements. Since no ρ-transition can take place to its left in any of the stages 1, 2, and 3, therefore it is immediate that no ρ-transition takes place to its left in either stage 1′ (from a to a_1) or stage 3′ (from $d_s b$ to $d_s b_1$). Further, since the transformation of stage 1 takes place to the left of b, d_u^L is not a factor of b; hence the transformation of stage 2′ does not take place to its left. Finally, since the transformation of stage 3 takes place to the left of b_1, it follows similarly that the transformation of stage 4′ does not take place to the left of d_u^L.

We can now, if necessary, repeat this process until we obtain a chain equivalent to X such that no ρ-transition takes place to the left of any of the k elements of M^L involved.

Similarly we could have started with the elements of M^R involved and obtained a ρ-chain with the required properties for these elements. Suppose then that we do this first and obtain a ρ-chain equivalent to X in which no ρ-transitions take place to the right of any of the elements of M^R used in the transitions. Now repeat the previous argument for the symbols of M^L involved. We will show that the replacement of stages 1–3 by stages 1′–4′ does not introduce transformations to the right of any element D_t^R.

For let D_t^R be an element of M^R occurring as a factor of a word in the chain. Since the ρ-transition of stage 1 takes place to the right of a, therefore D_t^R is not a factor of a. Similarly, because of stage 3, D_t^R is not a factor of a_1. Hence if D_t^R is introduced to the left of d_s^L at stage 2 it must also be removed at stage 2. Hence it must be introduced and removed in stage 1′ and no ρ-transition takes place to its right in stage 1′. Consequently, also, the ρ-transition of stage 2′ does not take place to its right. If D_t^R is involved in the transformations from $d_s b$ to $d_s b_1$ then any ρ-transitions taking place to its right must take place to its right at stage 2. There are no such transformations at stage 2 and hence there are also none at stage 3′. Finally, since D_t^R is not a factor of a_1, the transition of stage 4′ does not take place to its right.

This proves our assertion. The process of modification of X thus terminates in a normal chain from w_0 to w_n. This completes the proof of the lemma.

If w is any word in a normal chain, it is clear that every factor of w in M^L precedes every factor of w in M^R. By the *center* of w we mean the portion of w lying between the last factor in M^L and the first factor in M^R. For example, if w is a word with no factors from $M^L \cup M^R$, then w is its own center.

Let $X = (w_0, w_1, \cdots, w_n)$ be a normal chain. Then X is said to be *proper* if

(1) n is odd, $n = 2q + 1$, say;
(2) $T_{2j}: w_{2j} \to w_{2j+1}$, $j = 0, 1, \cdots, q$, is a π_1-transition;
(3) $T_{2j+1}: w_{2j+1} \to w_{2j+2}$, $j = 0, 1, \cdots, q - 1$, is a π_2-transition.

We can now strengthen the previous lemma and prove

LEMMA 12.20. *Any normal ρ-chain is equivalent to a proper normal ρ-chain.*

PROOF. The proof of the lemma follows from two observations. Firstly, since the pair $(1, 1)$ belongs to π_1, this trivial π_1-transition may be inserted between any two consecutive π_2-transitions. Secondly, since π_1 is a congruence on K and the centers of all words in a normal chain belong to K, and further a sequence of consecutive π_1-transitions is effectively a sequence of π_1-transitions on the centers of the words involved, therefore such a sequence can be combined to form a single π_1-transition. Both types of change considered replace a normal chain by a normal chain. It is clear that by a sequence of changes of these types we can replace a normal chain by an equivalent proper normal chain.

Now let $X = (w_0, w_1, \cdots, w_{2q+1})$ be a proper normal chain. Then X determines a Malcev sequence $I = (X_1, X_2, \cdots, X_q)$, of length q, as follows: X_j is L_i, L_i^*, R_i, or R_i^* according as T_{2j-1} inserts the pair $d_i^L d_i$, removes the pair $d_i^L d_i$, inserts the pair $D_i D_i^R$, or removes the pair $D_i D_i^R$, respectively. (As before, we denote the elements of $M^L[M^R]$ which are inserted by the T_{2j-1} by $d_1^L, d_2^L, \cdots, d_k^L$, $[D_1^R, D_2^R, \cdots, D_p^R]$ in the order of occurrence.) It is clear that the sequence I so determined satisfies the defining conditions (i) and (ii) for a Malcev sequence.

Suppose that T_{2j-1} inserts the pair $d_i^L d_i$ into w_{2j-1}. Since X is normal there must exist c_i, a_i in K such that $c_i a_i$ is the center of w_{2j-1} and $w_{2j} = \cdots c_i d_i^L d_i a_i \cdots$. The center of w_{2j} is then $d_i a_i$. Note also, again since X is normal, that the word c_i will remain to the left of d_i^L unaffected by further transitions until d_i^L is removed. Similarly, if T_{2j-1} inserts the pair $D_iD_i^R$ into w_{2j-1}, then we agree to write the center of w_{2j-1} as A_iC_i where $w_{2j} = \cdots A_iD_iD_i^RC_i \cdots$. The center of w_{2j} is then A_iD_i and C_i remains unaffected by any transition until D_i^R is removed. If T_{2j-1} removes the pair $d_i^Ld_i$ from w_{2j-1} then we agree to write the center of w_{2j-1} as d_ib_i. This is possible because, since X is normal, $d_i^Ld_i$ can only be removed from w_{2j-1} if the first factor of its center is d_i. Taking into account the above remark about c_i, we can say further that $w_{2j-1} = \cdots c_id_i^Ld_ib_i \cdots$ and that c_ib_i is the center of w_{2j}. Similarly, if T_{2j-1} removes the pair $D_iD_i^R$ from w_{2j-1}, we may write the center of w_{2j-1} as B_iD_i. The center of w_{2j} then becomes B_iC_i.

The above process determines a form of writing the centers of each element of X except w_0 and w_{2q+1}. We may summarize it in tabular form as follows

TABLE 2

T_{2j-1}	Center of w_{2j-1}	Center of w_{2j}
Inserts $d_i^Ld_i$	c_ia_i	d_ia_i
Removes $d_i^Ld_i$	d_ib_i	c_ib_i
Inserts $D_iD_i^R$	A_iC_i	A_iD_i
Removes $D_iD_i^R$	B_iD_i	B_iC_i

It is now clear how the chain X determines the Malcev system of equations $\sigma(I)$ and its locked equation. For the above table is merely, with appropriate interpretation, Table 1 we met earlier in this section. The equations of $\sigma(I)$ determine, and are determined by, the transformations on centers caused by the transitions T_{2j} $(j = 1, 2, \cdots, q-1)$. For example, suppose that T_{2j-1} inserts $d_i^Ld_i$ and that T_{2j+1} removes $D_kD_k^R$. Then the center of w_{2j} is d_ia_i and that of w_{2j+1} is B_kD_k. Thus T_{2j} has the effect of replacing d_ia_i by B_kD_k. The pair T_{2j-1}, T_{2j+1} determines the pair $L_iR_k^*$ in the Malcev sequence I and, from Table 1, the corresponding equation in $\sigma(I)$ is $d_ia_i = B_kD_k$. This is the same equation that is effectively determined by T_{2j}. The equation $d_ia_i = B_kD_k$ does not in general hold in T. However, since π_1 is a congruence on $K(\subseteq T)$ we do have $d_ia_i\ \pi_1 B_kD_k$ and so $(d_i\beta)(a_i\beta) = (B_k\beta)(D_k\beta)$ is an equation which holds in S. The locked equation of $\sigma(I)$ is similarly determined by the centers of w_{2q} and w_1. Since $(w_{2q}, w_{2q+1}) \in \pi_1$ and $(w_0, w_1) \in \pi_1$, therefore $(w_0, w_{2q+1}) \in \pi_1$ if and only if $(w_{2q}, w_1) \in \pi_1$, i.e. if and only if $w_{2q}\beta = w_1\beta$. As an example, we give the proper normal chain which leads to the Malcev system given as an example above. The centers of the words are underscored. Bracketed words give the equations of the system.

$$w_0$$

$$\underline{c_1a_1}$$

$$\left\{\begin{array}{l} c_1d_1^L\underline{d_1a_1} \\ c_1d_1^L\underline{c_2a_2} \end{array}\right.$$

$$\left\{\begin{array}{l} c_1d_1^Lc_2d_2^L\underline{d_2a_2} \\ c_1d_1^Lc_2d_2^L\underline{A_1C_1} \end{array}\right.$$

$$\left\{\begin{array}{l} c_1d_1^Lc_2d_2^L\underline{A_1D_1}D_1^RC_1 \\ c_1d_1^Lc_2d_2^L\underline{d_2b_2}D_1^RC_1 \end{array}\right.$$

$$\left\{\begin{array}{l} c_1d_1^L\underline{c_2b_2}D_1^RC_1 \\ c_1d_1^L\underline{A_2C_2}D_1^RC_1 \end{array}\right.$$

$$\left\{\begin{array}{l} c_1d_1^L\underline{A_2D_2}D_2^RC_2D_1^RC_1 \\ c_1d_1^L\underline{c_3a_3}D_2^RC_2D_1^RC_1 \end{array}\right.$$

$$\left\{\begin{array}{l} c_1d_1^Lc_3d_3^L\underline{d_3a_3}D_2^RC_2D_1^RC_1 \\ c_1d_1^Lc_3d_3^L\underline{B_2D_2}D_2^RC_2D_1^RC_1 \end{array}\right.$$

$$\left\{\begin{array}{l} c_1d_1^Lc_3d_3^L\underline{B_2C_2}D_1^RC_1 \\ c_1d_1^Lc_3d_3^L\underline{d_3b_3}D_1^RC_1 \end{array}\right.$$

$$\left\{\begin{array}{l} c_1d_1^L\underline{c_3b_3}D_1^RC_1 \\ c_1d_1^L\underline{d_1b_1}D_1^RC_1 \end{array}\right.$$

$$\left\{\begin{array}{l} \underline{c_1b_1}D_1^RC_1 \\ \underline{B_1D_1}D_1^RC_1 \end{array}\right.$$

$$\underline{B_1C_1}$$

$$w_{21}$$

We have now shown that any proper normal chain X determines (1) a Malcev sequence I ($= I_X$) (2) a set of elements of π_1 which determine the equations $\sigma(I)$ and (3) an element of π_1 which determines the locked equation of $\sigma(I)$. Suppose conversely that we are given a Malcev sequence I together with the set of equations $\sigma(I)$ and that we then substitute for the variables, a_i, b_i, c_i, d_i, A_i, B_i, C_i, D_i elements of K, selecting the d_i and D_i from M, in such a way that the two sides of each equation of $\sigma(I)$ then form the components of a pair of π_1. It is then easy to see that we may take the elements of π_1, so determined, as determining the successive centers of the elements of a proper normal chain which, in turn, by the process we have just described, will again determine I and the set of elements of π_1 which were derived from $\sigma(I)$. The proper normal chain determined is unique to within the adjunction of π_1-equivalent elements at each end.

We are now in a position to complete the proof of Theorem 12.17. Suppose firstly that S can be embedded in a group. Let I be any Malcev sequence and suppose that the symbols $a_i, b_i, \cdots, D_i$ are given values in S such that the equations $\sigma(I)$ all hold in S. Now choose the set of generators Γ of S so that all the d_i, D_j belong to Γ, i.e. so that there are elements of M which map onto the d_i and D_j under σ and so also under β. With this choice of M we may now select values for the symbols $a_i, b_i, \cdots, D_i$ in K such that the values we have given these symbols in S are, respectively, their images under β, and such that the d_i, D_j are given values in M. Then with this choice the left and right sides of each equation of $\sigma(I)$ determine an element of π_1. These elements of π_1 then determine, by the process we have discussed above, a proper normal chain X from w to w', say. We have $w, w' \in K$ and $(w, w') \in \rho_1$. Hence, since S is embeddable in a group, we also have $(w, w') \in \pi_1$. Taking into account the remarks made above about the construction of the locked equation from X, this is easily seen to imply, after mapping by β, that the locked equation of $\sigma(I)$ holds in S.

Conversely, assume that whenever, for any Malcev sequence I, all the equations of $\sigma(I)$ hold in S, then the locked equation of $\sigma(I)$ also holds in S. Let $w, w' \in K$ and let $(w, w') \in \rho_1$. Then we have seen that there exists a proper normal chain X, say, of ρ-transitions transforming w to w', which determines a Malcev sequence I_X and a set of elements of π_1 that, on mapping by β, determine a set of equations $\sigma(I_X)$ which hold in S. Our hypothesis is then that the locked equation of $\sigma(I_X)$ also holds in S. But this is equivalent to $(w_1, w_1') \in \pi_1$, where $w \rightarrow w_1$ is the first transition and $w_1' \rightarrow w'$ is the last transition of X. These two transitions are both π_1-transitions. Consequently, $(w, w') \in \pi_1$; whence it follows immediately that S can be embedded in a group.

This completes the proof of Malcev's theorem.

We remark, finally, that the above proof of sufficiency implies a stronger result than was stated in the theorem. The set M may be chosen so that $M\sigma = \Gamma$ is an arbitrary set of generators of S. It is then sufficient for embeddability of S in a group that when the $\sigma(I)$ hold in S with the d_i, D_j in Γ then the locked equation of $\sigma(I)$ also holds in S. We make use of this remark in §12.8.

Exercises for §12.6

1. Let I be a Malcev sequence. If XY is an adjacent pair of symbols of I, XY determines, using Table 1, an equation from $\sigma(I)$. We call the side of this equation read off under X in Table 1, the left-hand side of this equation. Form the equation (α), say, obtained by equating the product of all the left-

hand sides of the equations of $\sigma(I)$ to the product of all the right-hand sides (the equations may be taken in any specified order).

Let S be a cancellative, commutative semigroup. Suppose that the equations $\sigma(I)$ hold in S. Then (α) also holds in S and, by cancellation from the equation (α), it follows that the locked equation of $\sigma(I)$ holds in S. Consequently, S may be embedded in a group.

2. Let S be a cancellative left quasi-reversible semigroup (see Exercise 4 for §12.4). Then S is embeddable in a group. (Doss [1948].)

3. A semigroup with identity is cancellative if and only if, for every Malcev sequence I of length two, the system $\sigma(I)$ can hold in S only if its locked equation holds in S.

12.7 Comparison of Malcev and Lambek systems

In this section we show that, in a sense which is made precise in the following new theorem, Malcev systems which, together with their locked equations, form Lambek systems are precisely those Lambek systems which arise from 2-vertex polyhedra. Bush [1963] also considers the relation between Malcev and Lambek systems. We continue to use the notation of the preceding section.

To be precise, by a Lambek system we mean the system of half-edge equations which arises from lettering all the sides and angles of an Eulerian polyhedron in the manner described in §12.5. Moreover we will assume, for the purpose of comparison with Malcev systems, that no letter is used twice in lettering the polyhedron. This corresponds to our convention for Malcev systems whereby, in deriving a Malcev sequence from a chain of elementary transitions, we gave a different name to each distinct occurrence of a letter.

Theorem 12.21. *A Malcev system of equations $\sigma(I)$, together with its locked equation, forms a Lambek system of equations if and only if the sequence I has the form*

$$L_1 \cdots L_n R_1 L_n^* \cdots L_1^* L_{n+1} \cdots L_{n+m} R_1^* L_{n+m}^* \cdots L_{n+1}^*,$$

where $n + m > 0$. In this event the Lambek system arises from a two vertex polyhedron with $n + m + 1$ edges.

Conversely, let P be a polyhedron with two vertices and let $\mathscr{S}(P)$ be the Lambek system of equations determined by P. Select any equation $\bar{P}$, say, from $\mathscr{S}(P)$. Then there exists a Malcev system I, say, with $\bar{P}$ as its locked equation and such that $\mathscr{S}(P)\backslash\bar{P} = \sigma(I)$.

We first need a couple of lemmas. Also we use the following terminology. The equation read off Table 1 (§12.6) corresponding to a pair XY of symbols will be said to be *initiated* by X and *terminated* by Y. Further, if I is a Malcev sequence we will assume that its associated equations $\sigma(I)$, together with their locked equation, are ordered in the order that they are read off from Table 1 if we commence with the first symbol of I, go through the symbols of I successively, going full cycle and terminating with the first symbol of I, so that the last equation is the locked equation of $\sigma(I)$.

Two half-edge equations determined by a single edge of a polyhedron have the form $xa = yb$ and $xc = yd$, where x, y, a, b, c, d are, because of our convention, all distinct. These two equations have precisely two letters in common, viz. x and y, and both x and y occur as left factors in these equations. We shall say that the two equations have *two left factors in common.* It is a basic fact about a Lambek system that it is a set of pairs of equations, the two equations of each pair having two left factors in common. This fact is the basis of our proof of the theorem.

LEMMA 12.22. *If two successive equations of a Malcev system have two left factors in common, they correspond to a triple in the Malcev sequence of one of the following kinds;* $L_iR_jL_i^*$, $L_iR_j^*L_i^*$, $L_i^*R_j^*L_i$, *or* $L_i^*R_jL_i$ *(the latter two occurring only when the whole sequence is of the form* $L_i \cdots L_i^*R_j^*$, *or* $R_jL_i \cdots L_i^*$, *respectively).*

PROOF. The letters a_i, b_i, C_i and D_i occur only as right factors in a Malcev system, so we can confine our attention to the letters A_i, B_i, c_i, d_i. The letter c_i occurs only in the equation terminated by L_i and in the equation initiated by L_i^*. Hence, if it is a common left factor of two successive equations the corresponding triple in the Malcev sequence must be of the form $L_i^*XL_i$ or $L_i^*X^*L_i$. From Table 1, we quickly verify that $X = R_j$ is the only choice which gives a second common left factor. However, in a Malcev sequence L_i^* must occur after L_i. Also a Malcev sequence must commence with an unstarred symbol and terminate with a starred symbol. Hence these possibilities occur only when the whole sequence has one of the forms: $R_jL_i \cdots L_i^*$, or $L_i \cdots L_i^*R_j^*$.

The letter d_i occurs only in the equation initiated by L_i and in the equation terminated by L_i^*. A similar examination of the possibilities shows that two common left factors arise solely from the triples $L_iR_jL_i^*$ and $L_iR_j^*L_i^*$. The other letters A_i and B_i come under the cases we have already dealt with; for $A_i[B_i]$ is a common left factor solely when the central symbol of the triple is $R_i[R_i^*]$.

LEMMA 12.23. *If two non-successive equations in a Malcev system have two left factors in common, the corresponding pairs of adjacent symbols in the Malcev sequence must be one of the following kinds:* L_iL_{i+1} *and* $L_{i+1}^*L_i^*$; *or* $L_i^*L_j$ *and* $L_j^*L_i$ *(the latter occurring only when the whole sequence is of the form* $L_i \cdots L_i^*L_j \cdots L_j^*$*).*

PROOF. The letters A_i and B_i can occur as common left factors only for successive equations. Hence only the letters c_i and d_j can occur as common left factors of two non-successive equations. The letter c_i occurs in the equation initiated by L_i^* and in the equation terminated by L_i; and the letter d_i occurs in the equation initiated by L_i and in the equation terminated by L_i^*. Hence, the only possibilities for the pairs are (i) $L_i^* L_j$ and $L_j^* L_i$ and (ii) $L_j^* L_i^*$ and $L_i L_j$. From the defining conditions of a Malcev sequence it follows that, in case (ii), $j = i + 1$. For case (i), again from the definition of a Malcev sequence, we cannot have both pairs $L_i^* L_j$ and $L_j^* L_i$ in this order in the main sequence, and so one of them, say $L_j^* L_i$, must correspond to the locked equation of the system. But this implies that L_i is the first member, and L_j^* the last member, of the sequence.

This completes the proof of the lemma.

We now turn to the proof of the first part of the theorem. Suppose first, if possible, that no symbol R_j occurred in the Malcev sequence I, so that I consisted entirely of L_i's and corresponding L_i^*'s. If L_j were the last such symbol in the sequence I, then, from the defining conditions of a Malcev sequence, it follows that L_j^* is the next symbol of I. The pair $L_j L_j^*$ determines the equation $d_j a_j = d_j b_j$. This cannot be an equation of a Lambek system, bearing in mind our agreement to letter all sides and angles with distinct letters. Consequently, R_1 is a symbol in the sequence I.

From Lemmas 12.22 and 12.23 it follows that R_1 occurs in the sequence I in either of the two ways: $\cdots L_n R_1 L_n^* \cdots$, or $R_1 L_1 \cdots L_1^*$. In the former event L_n can be preceded only by L_j's and L_i^*'s; and L_i^*'s are ruled out by the argument just given. Hence, in this event, taking into account Lemma 12.22 again, the sequence I commences thus:

$$L_1 \cdots L_n R_1 L_n^* \cdots L_1^* \cdots.$$

From Lemma 12.22, the sequence may now terminate in R_1^* (the case $m = 0$, of the theorem) or, alternatively, since $L_1^* R_2$ is not allowed by our lemmas, the sequence continues $\cdots L_{n+1} \cdots$. Applying the lemmas again now quickly shows that the sequence I must have the form given in the theorem. A similar argument applies to the remaining case ($n = 0$, of the theorem).

We must now characterize our Lambek systems as those arising from 2-vertex polyhedra; and this we do by referring to pictures of the appropriately lettered polyhedra for the three cases $n = 0$, $m = 0$, and $mn \neq 0$. It will be clear from our discussion that the converse half of the theorem also holds. In each case $\bar{P}$ will denote an arbitrarily selected equation of the Lambek system $\mathscr{S}(P)$ determined by the polyhedron P; and in each illustrating diagram we take for the lower vertex the vertex at which the half-edge determining $\bar{P}$ terminates.

Case $n = 0$. The right side of the edge whose half-edge determines $\bar{P}$ is labelled A_1 and its left side is labelled c_1. The edge next on the right, or if there are none on the right, the left-most edge in the diagram, is labelled on its left side B_1 and on its right side d_m. There remain $m - 1$ left sides of edges to label; successively, starting from the edge which determined $\bar{P}$ and moving left-wards to the left-most edge and then starting at the right of the diagram and again moving left-wards these left sides are labelled $c_2 \cdots, c_m$, respectively. The right side of the edge whose left side is labelled c_i is then labelled d_{i-1}, for $i = 2, 3, \cdots, m$.

The angles are now labelled as follows. For the lower vertex, starting at the angle on the left side of the half-edge determining $\bar{P}$, proceeding counterclockwise, the angles are labelled, in turn, $b_1, \cdots, b_m, C_1$. At the upper vertex, starting at the left side of the edge whose lower half-edge determines $\bar{P}$, proceeding clockwise, the angles are labelled in turn, $a_1, a_2, \cdots, a_m, D_1$. The diagram illustrates one possibility for $m = 4$.

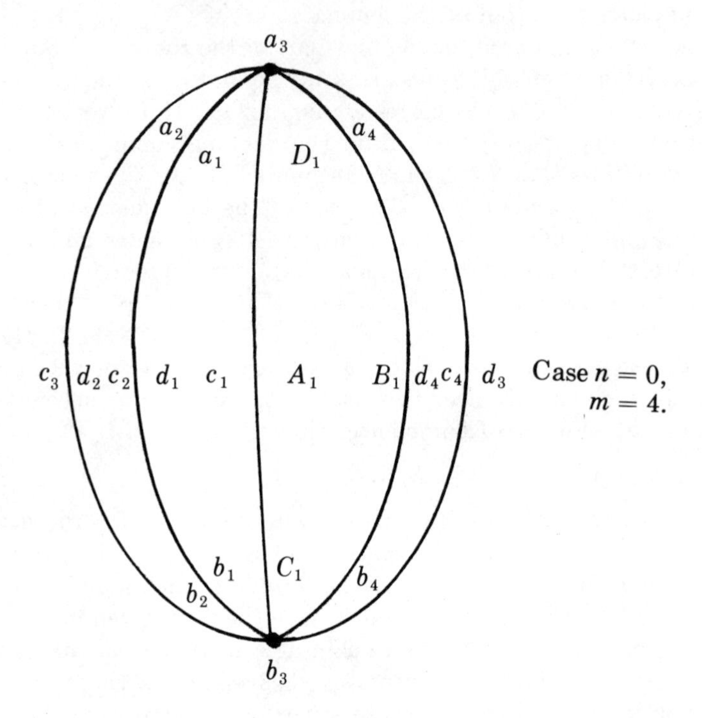

The Lambek system of equations is then seen to be the Malcev system of equations (including the locked equation $c_1b_1 = A_1C_1$) of the Malcev sequence:

$$R_1L_1 \cdots L_mR_1^*L_m^* \cdots L_1^*.$$

Case $m = 0$. To obtain the diagram and its labelling for this case we merely take the diagram for the previous case of $n = 0$, interchange A_1 with B_1 and C_1 with D_1, replace m by n and turn the diagram upside down. The locked equation selected is here $B_1C_1 = c_1a_1$.

Case $mn \neq 0$. We need here the appropriate combination of the two pictures just described. Begin by labelling the right side of the edge, whose half-edge determines $\bar{P}$, by c_1 and its left side by c_{n+1}. There remain $n + m$ edges to label. Starting from the edge already labelled and moving left-wards to the left-most edge and then starting at the right of the diagram and moving left-wards again, label the left sides of the edges, successively, with $c_{n+2}, \cdots c_{n+m}, B_1, d_n, d_{n-1}, \cdots, d_1$. Taking the edges in the same order, label their right sides, successively, with $d_{n+1}, \cdots d_{n+m}, A_1, c_n, c_{n-1}, \cdots, c_2$.

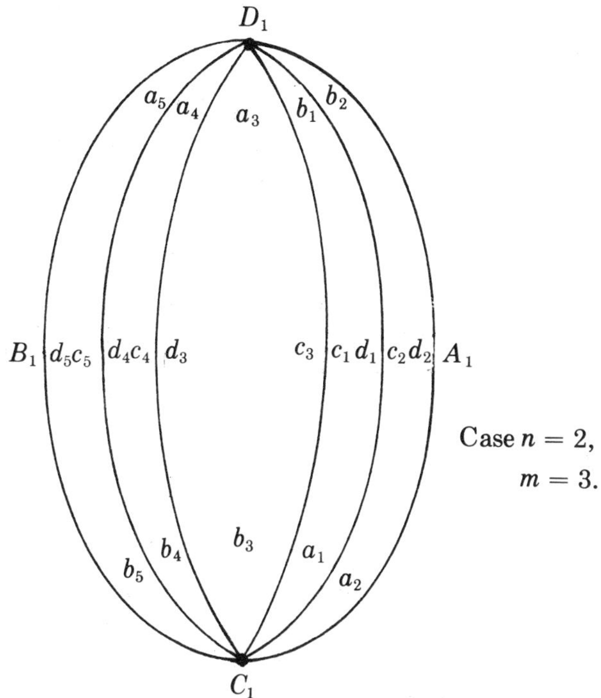

Case $n = 2$,
$m = 3$.

Now label the angles as follows. For the lower vertex, starting at the angle on the side labelled B_1, proceeding clockwise, the angles are labelled $C_1, b_{n+m}, \cdots, b_{n+1}, a_1, \cdots, a_n$. At the upper vertex, starting at the angle on the side labelled A_1 and proceeding counter clockwise, the angles are labelled $D_1, a_{n+m}, \cdots, a_{n+1}, b_1, \cdots, b_n$. The diagram shown illustrates the case $n = 2$, $m = 3$.

The Lambek system of equations is then seen to be the Malcev system of equations (including the locked equation $c_1a_1 = c_{n+1}b_{n+1}$) of the Malcev sequence

$$L_1 \cdots L_n R_1 L_n^* \cdots L_1^* L_{n+1} \cdots L_{n+m} R_1^* L_{n+m}^* \cdots L_{n+1}^*,$$

for which $n > 0$ and $m > 0$.

The proof of the theorem is complete.

12.8 Finite sets of equational implications

The set of all Malcev sequences is countable. Consequently, the set of necessary and sufficient conditions for embeddability of a semigroup into a group, given in Theorem 12.17 is countable. In this section we prove the further result of Malcev [1940] that no finite subset of these conditions will suffice to ensure embeddability.

In fact we shall prove a more general result also due to Malcev [1940]. Let us call an *equational implication* a condition that one finite set of equations should imply that another finite set of equations should hold. More precisely each equation will be an equation which may be written formally as $w = w'$, where w and w' are words of a free semigroup $\mathscr{F}$, say. We emphasize that we are, in this use of the word "equation," indulging in a convenient abuse of language: when we speak of the "equation" $w = w'$, we do not thereby imply that w and w' are equal in $\mathscr{F}$. Let $w_i = w_i'$, $i = 1, 2, \cdots, n$ be one finite set of equations and let $v_j = v_j'$, $j = 1, 2, \cdots, m$ be another finite set. The set of equations $w_i = w_i'$ will be said to *imply* the equations $v_j = v_j'$ in the semigroup S, if for all homomorphisms $\phi\colon \mathscr{F} \to S$, $w_i\phi = w_i'\phi$, $i = 1, 2, \cdots, n$, only if $v_j\phi = v_j'\phi$ for $j = 1, 2, \cdots, m$.

We shall show that no finite set of equational implications suffices to ensure that a semigroup be embeddable in a group. In addition to the Malcev system implications, this theorem also covers the Lambek system implications. Thus, no finite set of polyhedra is such that the set of equational implications it serves to determine suffices to ensure embeddability of a semigroup into a group.

The following lemma is fundamental. The notation used is that of Construction 12.18. We continue to regard α as an inclusion mapping, and will suppress α, and other inclusion mappings where there is no resulting ambiguity. Thus, for example, we will write $\pi_1 = \beta \circ \beta^{-1}$.

Lemma 12.24. *Let $M = P \cup Q$ where P and Q are disjoint sets. Suppose that $\beta \circ \beta^{-1}$ is generated* (*as a congruence on K*), *by a set of pairs of the form $(pq, p'q')$, where $p, p' \in P$ and $q, q' \in Q$. If w is an element of K, denote by $l(w)$ the length of w regarded as a word in the elements of M. Let $(w, w') \in \rho_1 \cap (K \times K)$.*

Then

$$w = a_1 a_2 \cdots a_n$$

and

$$w' = a_1' a_2' \cdots a_n'$$

where, for $i = 1, 2, \cdots, n$,

(i) $a_i, a_i' \in K$;

(ii) $(a_i, a_i') \in \rho_1$;

(iii) $l(a_i) = l(a_i')$;

(iv) *either* (a) $l(a_i) = 1$, *when a_i and a_i' both belong either to P or to Q; or* (b) $l(a_i) = 2$, *when $a_i = pq$, $a_i' = p'q'$ for some p, p' in P and q, q' in Q.*

Proof. By Lemma 12.20, there exist proper normal chains from w to w'. From the set of all such proper normal chains choose a chain X which contains a minimal number of π_2-transitions involving elements of $P^L \cup P^R$. Here $P^L = \{m^L: m^L \in M^L, m \in P\}$, $P^R = \{m^R: m^R \in M^R, m \in P\}$.

We need some definitions. Let π be the set of pairs $(pq, p'q')$ which, by assumption, generate the congruence π_1 on K. Any π_1-transition is thus equivalent to a finite sequence of π-transitions. In any π-transition, and hence in any π_1-transition, an element of $P[Q]$ is either left unchanged or is replaced by an element of $P[Q]$. Suppose, for example, that $p[q]$ is replaced by $p'[q']$ which in turn is replaced by $p''[q'']$, and so on. Then $p, p', p'', \cdots$ $[q, q', q'', \cdots]$ will be called *consequents* of $p[q]$. We do not include among the consequents of p the consequents of any other occurrence of p.

Consider now a π_2-transition of X which introduces an element of $P^L \cup P^R$, introducing the pair pp^R, say. Now p^R is removed at some later stage by deleting a pair pp^R. We will say that p^R is *monogamous* if the element p with which it is removed is a consequent of the element p with which it was introduced. If p^R is not monogamous, it will be said to be *polygamous*. Similarly we define monogamous and polygamous for elements of P^L.

We will now show that all the elements of $P^L \cup P^R$ introduced by transitions of X are monogamous.

Suppose the contrary, and let $p^L \in P^L$, say, be the last element of $P^L \cup P^R$ introduced which is polygamous. Then p^L cannot be removed from the chain X before the consequents of the element p with which it was inserted. For, firstly, p^L cannot be removed with an element which lies to its left (because it is an element of M^L) and, secondly, while the consequents of p are still present, it cannot be removed with any element lying to the right of any such consequent. Thus, if removed before these consequents are removed, it must be removed with an element introduced after p^L was introduced. But this is easily seen to be contrary to our hypothesis that p^L is the last polygamous element to be introduced in X.

Hence, some consequent, p', say, of p must be removed from X before p^L. Further p' must be removed with an element of P^R; for any element of P^L with which p' could be removed would have to be introduced after p^L and, by hypothesis, such elements are monogamous. For the same reason it follows that the element of P^R with which p' is removed must have been introduced before the element p^L. Thus X contains a subchain (ξ) of the following form:

$$(\xi) \qquad \begin{aligned} &UN \overset{(1)}{\rightarrow} Up'p'^RN \rightarrow \overset{(2)}{\cdots} \rightarrow KAp'^RN \overset{(3)}{\rightarrow} Kp^LpAp'^RN \rightarrow \\ &\rightarrow \overset{(4)}{\cdots} Kp^LBp'p'^RN \overset{(5)}{\rightarrow} Kp^LBN \rightarrow \overset{(6)}{\cdots} \rightarrow Kp^LpV \overset{(7)}{\rightarrow} KV. \end{aligned}$$

Here p'^R, introduced at stage (1), is removed at stage (5) with the consequent p' of p. We have introduced the convention that capital letters denote any elements of T, reserving lower case letters for elements of $M \cup M^L \cup M^R$.

Stage (4) consists of a sequence of transitions which change pA into Bp' where p' is a consequent of p. Because the elements of P occur only as left factors in the elements of the pairs in π it follows that neither p nor any of its consequents can be involved in any transition affecting elements lying to their left. Thus the transitions of stage (4) can be split up into two independent sequences, a sequence transforming pA to p' and a sequence transforming the empty word to B. Taking into account the normality of X, it follows that the transformation of pA to p' must take place by means of a sequence (η) of the following form:

$$(\eta) \qquad \begin{aligned} &pA \to \overset{(\eta_1)}{\cdots} \to pA_0 = pqA_1 \overset{(\eta_2)}{\to} p_1q_1A_1 \to \overset{(\eta_3)}{\cdots} \\ &\to p_1q_2A_2 \overset{(\eta_4)}{\to} p_2q_3A_2 \to \cdots \to p_{r-1}q_{2r-2}A_r \to \\ &\overset{(\eta_{2r})}{\to} p_rq_{2r-1}A_r \to \overset{(\eta_{2r+1})}{\cdots} \to p_r = p'. \end{aligned}$$

Here $\eta_2, \eta_4, \cdots, \eta_{2r}$ are each π-transitions while η_{2i+1} is a sequence of transitions not involving the p_i (putting $p = p_0$).

Now let (ζ) denote the following sequence of transitions transforming p into $p'C$, say:

$$(\zeta) \qquad \begin{aligned} &p \to pqq^R \to p_1q_1q^R \to p_1q_2q_2^Rq_1q^R \to p_2q_3q_2^Rq_1q^R \to \cdots \\ &\to p_rq_{2r-1}q_{2r-2}^Rq_{2r-3} \cdots q_1q^R = p'C. \end{aligned}$$

The chains (η) and (ζ) may be used to replace (ξ) by an equivalent sequence (χ) of transitions thus:

$$(\chi) \qquad \begin{aligned} &UN \overset{(4')}{\to} UBN \to \overset{(6)}{\cdots} \to UpV \to \overset{(\zeta)}{\cdots} \to Up'CV \to \\ &\overset{(2)}{\cdots} \to KACV \to \overset{(\eta_1)}{\cdots} \to KA_0CV = KqA_1CV \to \\ &\to Kqq_1^Lq_1A_1CV \to \overset{(\eta_3)}{\cdots} \to Kqq_1^Lq_2A_2CV \to \\ &\to Kqq_1^Lq_2q_3^Lq_3A_2CV \to \cdots \to \\ &\to \cdots \to Kqq_1^Lq_2q_3^L \cdots q_{2r-1}^Lq_{2r-1}A_rCV \to \overset{(\eta_{2r+1})}{\cdots} \\ &\to Kqq_1^Lq_2q_3^L \cdots q_{2r-1}^LCV = \\ &= Kqq_1^Lq_2q_3^L \cdots q_{2r-1}^Lq_{2r-1}q_{2r-2}^R \cdots q_1q^RV \to \cdots \to \\ &\to KV. \end{aligned}$$

Here (4′) denotes the sequence of transitions contained in the sequence (4) which transforms the empty word to B. The sequence (χ) is certainly normal, or rather the sequence Y, say, obtained from X by replacing (ξ) therein by (χ) is normal, and if not proper normal is trivially made so.

In the sequence (χ) no π_2-transitions involving the elements of $P^L \cup P^R$ occur which do not occur in X. In fact there are two fewer π_2-transitions in (χ) than (ξ) which involve elements of $P^L \cup P^R$, for the elements p'^R and p^L introduced in (ξ) are not introduced in (χ). Thus Y is equivalent to X and contains fewer π_2-transitions involving elements of $P^L \cup P^R$ than X. This is contrary to the choice of X. A similar discussion shows that a contradiction results if the last polygamous element of $P^L \cup P^R$ introduced into X belongs to P^R. It follows therefore that every element of $P^L \cup P^R$ introduced into X is monogamous.

We can now quickly complete the proof of the lemma. Since in the chain X from w to w' every element of $P^L \cup P^R$ introduced is monogamous, it follows that the consequents of any elements of P which are in w cannot be removed in the transformation to w'. Thus every element of P in w has a consequent in w'. The argument also shows, or we may argue from symmetry, that each element of P in w' is a consequent of an element of P in w. Similarly it can be shown that the factors from Q in w' are each consequents of the factors from Q in w and that each such element in w has a consequent in w'.

Now an element of $P[Q]$ and its consequents cannot be involved in any transition affecting elements lying to their left [right] (except in the case of the removal of the element or its consequents). This follows from the assumed form of π. Thus in the transformation from w to w' if w splits into two words, $w = w_1 w_2$, say, such that w_1 ends in q and/or w_2 begins with p, ($p \in P$, $q \in Q$), then w_1 and w_2 are effectively transformed independently to their corresponding subwords in w'. Applying this remark to w_1 and w_2, and so on, we see finally that w splits into subwords, each transformed independently, of three forms (a) an element of P (b) an element of Q and (c) a word pq where $p \in P$ and $q \in Q$. These observations complete the proof of the lemma.

We now construct an infinite sequence of semigroups S_n, $n = 1, 2, \cdots$, none of which, as we shall show, is embeddable in a group, but with the remarkable property that any finite set of equational implications, each of which holds in all groups, is satisfied in S_n for all sufficiently large n. The construction given and the result is that of Malcev [1940].

Let I_n be the Malcev sequence

$$I_n = L_1 R_1 \cdots R_n L_1^* R_n^* \cdots R_1^*.$$

Let π^n be the set of pairs read off Table 1 (§12.6) which determine the Malcev system $\sigma(I_n)$ and which, for later use, we display in tabular form below.

$$\text{Group I} \begin{cases} (d_1 a_1,\ A_1 C_1) \\ (A_1 D_1,\ A_2 C_2) \\ (A_2 D_2,\ A_3 C_3) \\ \quad \cdot \quad \cdot \quad \cdot \\ (A_{n-1} D_{n-1},\ A_n C_n) \\ (A_n D_n,\ d_1 b_1) \end{cases}$$

$$\text{Group II} \begin{cases} (c_1 b_1,\ B_n D_n) \\ (B_n C_n,\ B_{n-1} D_{n-1}) \\ \quad \cdot \quad \cdot \quad \cdot \\ (B_3 C_3,\ B_2 D_2) \\ (B_2 C_2,\ B_1 D_1) \end{cases}$$

The relation "locked" to π^n is then (B_1C_1, c_1a_1). Set

$$P_n = \{A_1, A_2, \cdots, A_n, B_1, B_2, \cdots, B_n, c_1, d_1\},$$

$$Q_n = \{a_1, b_1, C_1, C_2, \cdots, C_n, D_1, D_2, \cdots, D_n\}.$$

Then, all the elements of π^n are of the form $(pq, p'q')$, where $p, p' \in P_n$ and $q, q' \in Q_n$.

Now observe that π $(= \pi^n)$, P $(= P_n)$ and Q $(= Q_n)$ have the following properties:

(1) Each element of $P[Q]$ is a factor of at least one and at most two of the words occurring as elements of the pairs in π.

(2) The number of elements of $P[Q]$ that occur only once as such factors is at most two.

Let π' be a non-empty subset of π. Let $P'[Q']$ be the subset of $P[Q]$ which consists of those elements of $P[Q]$ which occur as factors in the words of the pairs forming π'.

LEMMA 12.25. *π', P' and Q' have properties* (1) *and* (2) *if and only if either $\pi = \pi'$ or π' consists of a single element of π.*

PROOF. The sufficiency of the conditions is clear. We prove necessity by examining the possibilities open for subsets of π.

Suppose that $\pi' \neq \pi$, $\pi' \neq \square$, and that π', P' and Q' have properties (1) and (2). It will suffice to show that π' is a one-element set.

We note that the members of π in Group I connect half the members of P in a single cycle:

$$d_1—A_1—A_2— \cdots —A_n—d_1,$$

while the members of π in Group II connect the remaining members of P in a single chain:

$$c_1—B_n—B_{n-1}— \cdots —B_1.$$

(This would also be a cycle if we included the locked relation.)

Consider the elements of π' falling in Group II, ordered as listed (from top to bottom). The first and last will each contain an unmated member of P'. Any break in the sequence would evidently cause further unmated members of P', and since (by hypothesis on π') there cannot be more than two, we conclude that the set of elements of π' in Group II must either be empty or an unbroken sequence.

Case (i). The set of elements of π' in Group II is empty. We note that all the members of Q occur in Group I, each exactly once. Since π' is contained in Group I, and only two unmated members of Q' can occur in π', we conclude that $|\pi'| = 1$.

Case (ii). The set of elements of π' in Group II consists of a non-empty unbroken sequence. There are exactly two unmated members of P', arising from the first and last term of the sequence. (These terms coincide if the sequence has length 1.) Hence no unmated members of P' can occur in Group I. This requires that the set of elements of π' occurring in Group I must be either empty or all of Group I. In the former event, since all elements of Q appearing in Group II occur only once in Group II, and now π' is contained in Group II, π' can contain only one member. In the latter event, when Group I is contained in π', we note that a_1 and C_1 are unmated elements of Q', and so all the rest must be mated. This clearly requires that π' contains all of Group II, and so $\pi' = \pi$, which is contrary to hypothesis.

Before proceeding, we need the analogue for groups of Lemma 9.11. We wish to show that if a group is given by generators and relations and one further generator together with a relation which expresses this generator in terms of the other generators, then the new group is isomorphic to the original group. Effectively new names have merely been given to some elements of the group.

In the present context it will be convenient to continue to use part of the notation of Construction 12.18. Observe that T/π_2^*, denoting by π_2^* the congruence on T generated by π_2, is a free group on M. Let ζ be a binary relation on K and let ξ be the congruence on T generated by $\zeta \cup \pi_2$. Then T/ξ we take as the original group given by generators and relations. Let n denote a further generator and let n^L, n^R denote two further symbols. Denote by T' the free semigroup on $M \cup M^L \cup M^R \cup \{n, n^L, n^R\}$ and containing this set, so that $T \subseteq T'$. Let K' denote the free semigroup on $M \cup \{n\}$, containing this set, so that $K \subseteq K'$. For a further relation we take (w_1, w_2), say, where $w_1 \in K'$ and contains n precisely once as a factor and where $w_2 \in K$. Let χ denote the congruence on T' generated by

$$\zeta \cup \pi_2 \cup \{(1, n^L n), (1, nn^R)\} \cup \{(w_1, w_2)\}.$$

Lemma 12.26. *The group T/ξ is isomorphic to the group T'/χ under the homomorphism $x\xi \to x\chi$, $x \in T$.*

Proof. We need three applications of Lemma 9.11. By assumption $w_1 = unv$, say, where $u, v \in K$. Introduce the notation that, if $w = m_1 m_2 \cdots m_k$, where $m_i \in M$, then $w^L = m_k^L \cdots m_2^L m_1^L$ and $w^R = m_k^R \cdots m_2^R m_1^R$. Then, a sequence of π_2-transitions followed by a (w_1, w_2)-transition, transforms n into $u^L w_2 v^R$. Conversely, observing that mm^L and $m^R m$, for $m \in M$, can each be derived from the empty word by a sequence of π_2-transitions, it follows that w_2 can be derived from w_1 by a $(n, u^L w_2 v^R)$-transition followed by a sequence of π_2-transitions. It follows that the congruence generated by $\xi \cup \{(w_1, w_2)\}$ coincides with that generated by $\xi \cup \{(n, u^L w_2 v^R)\}$.

Now $u^L w_2 v^R$ is a word in T so we are in a position to apply Lemma 9.11. Indeed, denote by T_1 the subsemigroup of T' generated by $T \cup \{n\}$, and by ξ_1 the congruence on T_1 generated by $\xi \cup \{(n, u^L w_2 v^R)\}$. Then, by Lemma 9.11, $x\xi \to x\xi_1$ $(x \in T)$ is an isomorphism of T/ξ onto T_1/ξ_1.

Denote by T_2 the subsemigroup of T' generated by $T_1 \cup \{n^L\}$ and by ξ_2 the congruence on T_2 generated by $\xi_1 \cup \{(n^L, vw_2^L u)\}$. Then, by Lemma 9.11, $x\xi_1 \to x\xi_2$ $(x \in T_1)$ is an isomorphism of T_1/ξ_1 onto T_2/ξ_2.

Finally, denote by ξ_3 the congruence on T' generated by $\xi_2 \cup \{(n^R, n^L)\}$. Then, as before $x\xi_2 \to x\xi_3$ $(x \in T_2)$ is an isomorphism of T_2/ξ_2 onto T'/ξ_3.

Since all the isomorphisms are onto, it follows that $x\xi \to x\xi_3$ $(x \in T)$ is an isomorphism of T/ξ onto T'/ξ_3. Since it is easily verified that $\xi_3 = \chi$, this completes the proof of the lemma.

Revert now to the notation of Lemma 12.24 and to that established immediately prior to Lemma 12.25. Let η be a non-empty subset of π $(= \pi^n)$. Select from η a pair $(p_1q_1, p_1'q_1')$, say, such that one of p_1, q_1, p_1', q_1', say, p_1, does not occur as a factor of the elements of any remaining pairs in η. Let T_1 be the subsemigroup of T generated by $(M \cup M^L \cup M^R)\backslash\{p_1, p_1^L, p_1^R\}$. Let ξ be the congruence on T_1 generated by

$$(\eta\backslash\{(p_1q_1, p_1'q_1')\}) \cup (\pi_2\backslash\{(1, p_1p_1^R), (p_1p_1^R, 1), (1, p_1^Lp_1), (p_1^Lp_1, 1)\}).$$

Then, by the preceding lemma, $x\xi \to x\chi$ $(x \in T_1)$ is an isomorphic mapping of T_1/ξ onto T/χ, where χ is the congruence on T generated by $\eta \cup \pi_2$. We infer

COROLLARY 12.27. *Let $w, w' \in T_1$. Then $(w, w') \in \chi$ if and only if $(w, w') \in \xi^*$, where ξ^* is the congruence on T generated by ξ.*

PROOF. Since $\xi \subseteq \chi$, $\xi^* \subseteq \chi$. By Lemma 9.9, $\chi \cap (T_1 \times T_1) = \xi$. Hence $\xi^* \cap (T_1 \times T_1) = \xi$. Consequently, $(w, w') \in \xi^*$ if and only if $(w, w') \in \xi$; whence, since $x\xi \to x\chi$ $(x \in T_1)$ is a one-to-one mapping, the corollary follows.

The following two lemmas now enable us to prove our theorem.

LEMMA 12.28. *Let π' be a nonempty subset of π, with $\pi' \neq \pi$. Let ρ_1' be the congruence on T generated by $\pi' \cup \pi_2$. Let $p, p' \in P$ and $q, q' \in Q$. Then $(pq, p'q') \in \rho_1'$ if and only if either $p = p'$ and $q = q'$, or $(pq, p'q') \in \pi'$.*

PROOF. Let π'' be a subset of π' obtained by constructing $\pi'\backslash\pi''$ as follows. Suppose that a subset σ, possibly empty, of $\pi'\backslash\pi''$, has already been selected. Then as a further element of $\pi'\backslash\pi''$ we take any element (w, w') in $\pi'\backslash\sigma$ such that either w or w' has a factor from $P \cup Q$ which does not occur again as a factor of any of the words which form the other elements of $\{\pi'\backslash\sigma\} \cup \{(pq, p'q')\}$. The construction is completed when no such further element can be found. π'' is then a proper subset of π. Further, if $P''[Q'']$ denotes the subset of $P[Q]$ involved in forming the words of the elements of π'', then the construction ensures that either π'' is empty or π'', P'', Q'' have properties (1) and (2) of the Lemma 12.25. Hence, by this lemma, either π'' is empty or π'' is a one-element set.

Now Corollary 12.27 was stated so that it applied to each step of the above construction of π''. Applying it at each stage, we conclude finally, that $(pq, p'q') \in \rho_1'$ if and only if $(pq, p'q') \in \rho_1''$, where ρ_1'' is the congruence on T generated by $\pi_2 \cup \pi''$.

The conclusion of the present lemma is evident if $\pi'' = \square$, for then T/ρ_1'' is a free group on $P \cup Q$. It remains to prove the lemma for the case when π' is a one-element set, $\pi' = \{(p_1q_1, p_1'q_1')\}$, say.

Note, firstly, that we can assume that $\{p_1, q_1, p_1', q_1'\} \subseteq \{p, q, p', q'\}$, for otherwise we can, by the above construction, replace π' by $\pi'' = \square$. Since π' is a subset of π we know that $p_1 \neq p_1'$ and $q_1 \neq q_1'$. Hence we must have $\{p_1, p_1'\} = \{p, p'\}$ and $\{q_1, q_1'\} = \{q, q'\}$. There is clearly no loss of generality in assuming that $p_1 = p$ and $p_1' = p'$. If $q_1 = q$, then the conclusion of the lemma holds. Suppose then that $q_1 = q'$, so that also $q_1' = q$. We shall show that this is not possible.

We have to show that if $\pi' = \{(pq', p'q)\}$ then $(pq, p'q')$ cannot belong to ρ_1'. Let ϕ be any homomorphism of the semigroup T onto the cyclic group $\langle a \rangle$ of order five which maps p, q, p' and q' onto a, a^3, a^2 and a^4, respectively, and which is such that for m in M, $m^L\phi = m^R\phi = (m\phi)^{-1}$. Then $\pi' \cup \pi_2 \subseteq \phi \circ \phi^{-1}$, and so, also $\rho_1' \subseteq \phi \circ \phi^{-1}$. But $(pq)\phi = aa^3 = a^4$ and $(p'q')\phi = a^2a^4 = a$. Hence $(pq, p'q') \notin \phi \circ \phi^{-1}$ and, *a fortiori*, $(pq, p'q') \notin \rho_1'$.

This completes the proof of the lemma.

Lemma 12.29. *If $p, p' \in P$ $[q, q' \in Q]$ then $(p, p') \in \rho_1$ $[(q, q') \in \rho_1]$, if and only if $p = p'$ $(q = q']$.*

Proof. Let ϕ be the homomorphism of T into the cyclic group $\langle a \rangle$ determined by the following mapping of the generators $M \cup M^L \cup M^R$ of T.

$$\begin{array}{ll} a_1 \to a^2 & b_1 \to a^{5n+2} \\ c_1 \to a & d_1 \to a^0 \\ A_i \to a^{2i} & B_i \to a^{2i+1} \\ C_i \to a^{3(i-1)} & D_i \to a^{3i+2} \\ m^L\phi = m^R\phi = (m\phi)^{-1} \quad (m \in M). & \end{array}$$

It is easily verified that $\pi \cup \pi_2 \subseteq \phi \circ \phi^{-1}$. Hence $\rho_1 \subseteq \phi \circ \phi^{-1}$. Now if the order of $\langle a \rangle$ is greater than or equal to $5n + 2$ then the images under ϕ of any two distinct elements of $P[Q]$ are distinct. The assertion of the lemma follows immediately.

We now define the sequence of semigroups S_n as the semigroups K_n/π_1^n, $n = 1, 2, \cdots$, where K_n $(= K)$ is the free semigroup on $M_n = P_n \cup Q_n$ $(M_n = M, P_n = P, Q_n = Q)$ and π_1^n $(= \pi_1)$ is the congruence on K_n generated by π^n $(= \pi)$.

Consider an equational implication that the set of equations $w_i = w_i'$, $i = 1, 2, \cdots, k$ imply the single equation $w = w'$, say. And suppose that this implication holds in any group. Let n be a natural number at least as great as the total length of all the words w_i, w_i' $(i = 1, 2, \cdots, k)$. In particular, these words may be regarded as elements of K_n. We shall show that if $\phi: K_n \to S_n$ is any homomorphism such that $w_i\phi = w_i'\phi$ $(i = 1, 2, \cdots, k)$ then $w\phi = w'\phi$.

We begin by defining a normal form for the words of K. The elements of

π, given in displayed form earlier, have a first component and a second component. The normal form of any word is obtained by replacing any two-element subword which is a second component of an element of π by the first component of that element of π. This can clearly be done in a unique fashion. Since the normal form of a word is obtained by applying a sequence of π-transitions, a word and its normal form are π_1-equivalent. We show that each π_1-class contains precisely one word in normal form.

To see this, observe firstly that, if u is a word, then the two-element subwords of u which are components of elements of π form a unique set of non-overlapping subwords of u; for an element of $P \cup Q$ cannot both terminate and begin components of the elements of π. A π-transition applied to u merely replaces one of these subwords by the other component of the element of π to which it belongs. It follows immediately that if u is in normal form, then the only word in normal form to which u is π_1-equivalent is u itself.

Let us suppose that

$$w_i = x_1x_2 \cdots x_l, \qquad w'_i = x'_1x'_2 \cdots x'_m$$

are the expressions of w_i and w'_i in the generators of the free semigroup K. Let $u_j[u'_j]$ be the (unique) word in normal form contained in the π_1-class $x_j\phi[x'_j\phi]$. Let

$$w_i\psi = u_1u_2 \cdots u_l, \qquad w'_i\psi = u'_1u'_2 \cdots u'_m.$$

Then $w_i\phi = w'_i\phi$ implies $(w_i\psi, w'_i\psi) \in \pi_1$. But this means that the two words $w_i\psi$ and $w'_i\psi$ of K have the same normal form. Since the $u_j[u'_j]$ are in normal form, the reduction of $w_i\psi[w'_i\psi]$ to normal form requires at most $l - 1$ $[m - 1]$ π-transitions. Hence $w_i\psi$ can be transformed to $w'_i\psi$ by less than $l + m$ π-transitions. The total number of π-transitions required to transform $w_i\psi$ to $w'_i\psi$, for $i = 1, 2, \cdots, k$, can therefore, by the choice of n, be taken to be fewer than n. Hence there exists a proper subset π' of π such that only π'-transitions are needed to derive $w'_i\psi$ from $w_i\psi$ for $i = 1, 2, \cdots, k$. Then $(w_i\psi, w'_i\psi) \in \pi'_1$, $i = 1, 2, \cdots, k$, where, π'_1 denotes the congruence on K generated by π'.

Consider now the congruence ρ'_1 on T generated by $\pi'_1 \cup \pi_2$. Since each $(w_i\psi, w'_i\psi) \in \rho'_1$ and T/ρ'_1 is a group, it follows, by hypothesis, that $(w\psi, w'\psi) \in \rho'_1$. From Lemma 12.24, it then follows that

$$w\psi = a_1a_2 \cdots a_s \quad \text{and} \quad w'\psi = a'_1a'_2 \cdots a'_s,$$

say, where (i) $a_j, a'_j \in K$, (ii) $(a_j, a'_j) \in \rho'_1$, (iii) $l(a_j) = l(a'_j)$ and (iv) either (a) $l(a_j) = 1$, when a_j and a'_j both belong either to P or to Q, or (b) $l(a_j) = 2$, when $a_j = pq$ and $a'_j = p'q'$ for some $p, p' \in P$ and $q, q' \in Q$. Since $\pi' \subseteq \pi$, but $\pi' \neq \pi$, from Lemma 12.29, in case (iv) (a), it follows that $a_j = a'_j$ and from Lemma 12.28, in case (iv) (b), it follows that $(a_j, a'_j) \in \pi'$. Consequently, $(w\psi, w'\psi) \in \pi'_1$ and, *a fortiori*, $(w\psi, w'\psi) \in \pi_1$. This completes the proof that the equations $w_i = w'_i$, $i = 1, 2, \cdots, k$ imply the equation $w = w'$ in S_n for all sufficiently large n.

Clearly we can extend our argument to deal with any finite number of such implications.

Observe finally that none of the semigroups S_n can be embedded in a group. For if S_n were embeddable in a group then the locked equation of π would have to hold in S_n. However the word B_1C_1, the first component of the locked relation of π, admits of no π-transition. Hence it is π_1-equivalent solely to itself. Thus we have proved

THEOREM 12.30. *The sequence of semigroups S_n, $n = 1, 2, \cdots$, is a sequence of semigroups none of which can be embedded in a group but such that, given any finite set of equational implications, each of which holds in a group, the implications all hold in S_n, for all sufficiently large n.*

On recalling the definition of the semigroups S_n, the following corollary is immediate. A similar statement can be made about Lambek systems, since Bush [1963] has shown that every Malcev system is a consequence of some Lambek system (as well as vice-versa).

COROLLARY 12.31. *Any equational implication which holds in any group (in particular, one which holds in any embeddable semigroup) is a consequence of some Malcev system; in fact, of some $\sigma(I_n)$, where I_n is the Malcev sequence $L_1R_1 \cdots R_nL_1^*R_n^* \cdots R_1^*$.*

BIBLIOGRAPHY

References to Mathematical Reviews give the review number, where there is one, and otherwise give the page on which the review occurs.

ANDERSEN, O.

1952 *Ein Bericht über die Struktur abstrakter Halbgruppen.* Thesis (Staatsexamensarbeit), Hamburg, 1952.

ASANO, K.

1949 *Über die Quotientenbildung von Schiefringen.* J. Math. Soc. Japan 1 (1949) 73–78 (MR 11, 154).

BACHMANN, H.

1955 *Transfinite Zahlen.* Ergebnisse der Mathematik, Heft 1 (N.F.), Springer, Berlin, 1955, vii + 204 pp. (MR 17, 134).

BAER, R.

1934 *Die Kompositionsreihe der Gruppe aller eineindeutigen Abbildungen einer unendlichen Menge auf sich.* Studia Math. 5(1934) 15–17.

BAER, R. AND LEVI, F.

1932 *Vollständige irreduzibele Systeme von Gruppenaxiomen.* (Beiträge zur Algebra No. 18.) Sitzber. Heidelberger Akad. Wiss., Abh. 2(1932) 1–12.

BIRKHOFF, G.

1948 *Lattice theory.* Amer. Math. Soc. Colloq. Publ., Vol. 25 (revised ed.), New York, 1948, xiii + 283 pp. (MR 10, 673).

BOURNE, S. G.

1949 *Ideal theory in a commutative semigroup.* Dissertation, Johns Hopkins University, Baltimore, 1949.

BRUCK, R. H.

1958 *A survey of binary systems.* Ergebnisse der Math. Heft 20, Springer, Berlin, 1958, 185 pp. (MR 20, 76).

BUSH, G. C.

1963 *The embedding theorems of Malcev and Lambek.* Canad. J. Math. 15(1963) 49–58 (MR 26, 1385).

CHAUDHURI, N. P.

1959 *Sur les complexes unitaires dans un demi-groupe.* C. R. Acad. Sci. Paris 248(1959) 1750–1752 (MR 21, 1353).

CLARK, W. E.

1965 *Remarks on the kernel of a matrix semigroup.* Czechoslovak Math. J. 15(1965) 305–310 (MR 31, 1311).

CLIFFORD, A. H.

1948 *Semigroups containing minimal ideals.* Amer. J. Math. 70(1948) 521–526 (MR 10, 12).

1949 *Semigroups without nilpotent ideals.* Amer. J. Math. 71(1949) 833–844 (MR 11, 327).

1953 *A class of d-simple semigroups.* Amer. J. Math. 75(1953) 547–556 (MR 15, 98).

1963 *Note on a double coset decomposition of semigroups due to Štefan Schwarz.* Mat.-Fyz. Časopis Sloven. Akad. Vied 13(1963) 55–57 (MR 28, 147).

COHN, P. M.

1956a *Embeddings in semigroups with one-sided division.* J. London Math. Soc. 31(1956) 169–181 (MR 18, 14).

1956b *Embeddings in sesquilateral division semigroups.* J. London Math. Soc. 31(1956) 181–191 (MR 18, 14).

1965 *Universal Algebra.* Harper and Row, New York (1965) (MR 31, 224).

Croisot, R.

1952 *Propriétés des complexes forts et symétriques des demi-groupes.* Bull. Soc. Math. France 80(1952) 217–223 (MR 14, 842).

1953 *Demi-groupes inversifs et demi-groupes réunions de demi-groupes simples.* Ann. Sci. École Norm. Sup. (3) 70(1953) 361–379 (MR 15, 680).

1954a *Demi-groupes simples inversifs à gauche.* C. R. Acad. Sci. Paris 239(1954) 845–847 (MR 16, 215).

1954b *Automorphismes intérieures d'un semi-groupe.* Bull. Soc. Math. France 82(1954) 161–194 (MR 16, 215).

1957 *Équivalences principales bilatères définies dans un demi-groupe.* J. Math. Pures Appl. (9) 36(1957) 373–417 (MR 19, 1037).

Dieudonné, J.

1942 *Sur le socle d'un anneau et les anneaux simples infinis.* Bull. Soc. Math. France 70(1942) 46–75 (MR 6, 144).

Doss, R.

1948 *Sur l'immersion d'un semi-groupe dans un groupe.* Bull. Sci. Math. (2) 72(1948) 139–150 (MR 10, 591).

Dubreil, P.

1941 *Contribution à la théorie des demi-groupes.* Mém. Acad. Sci. Inst. France (2) 63, no. 3(1941), 52 pp. (MR 8, 15).

1954 *Algèbre.* Tome I. *Équivalences, opérations, groupes, anneaux, corps.* 2 ème ed. Cahiers Scientifiques, Fasc. XX, Gauthier-Villars, Paris, 1954, 467 pp. (MR 16, 328).

Dubreil, P. and Dubreil-Jacotin, M.-L.

1939 *Théorie algébrique des relations d'équivalence.* J. Math. (9) 18(1939) 63–95.

1940 *Équivalences et opérations.* Ann. Univ. Lyon. Sect. A(3) 3(1940) 7–23 (MR 8, 254).

Dubreil-Jacotin, M.-L.

1947 *Sur l'immersion d'un semi-groupe dans un groupe.* C. R. Acad. Sci. Paris 225(1947) 787–788 (MR 9, 174).

Evans, T.

1952 *Embedding theorems for multiplicative systems and projective geometries.* Proc. Amer. Math. Soc. 3(1952) 614–620 (MR 14, 347).

Franklin, S. P. and Lindsay, J. W.

1960/61 *Straddles on semigroups.* Math. Mag. 34 (1960/61) 269–270 (MR 23, A1741)

Fulp, R. O.

1967 *On extending semigroup characters.* Proc. Edinburgh Math. Soc. 15(1967) 199–202 (MR 35, 4319).

Gluskin, L. M. (ГЛУСКИН, Л. М.)

1956 *Вполне простые полугруппы.* (*Completely simple semigroups.*) Uč. Zap. Kharkov. Ped. 18(1956) 41–55.

1959a *Полугруппы и кольца линейных преобразований.* (*Semigroups and rings of linear transformations.*) Doklady Akad. Nauk SSSR 127(1959) 1151–1154 (MR 26, 1382).

1959b *Полугруппы и кольца эндоморфизмов линейных пространств.* (*Semigroups and rings of endomorphisms of linear spaces.*) Izv. Akad. Nauk SSSR Ser. Mat. 23(1959) 841–870 (MR 26, 1383). Amer. Math. Soc. Translations (2) 45(1965) 105–139.

GOLDIE, A. W.
1950 *The Jordan-Hölder Theorem for general abstract algebras.* Proc. London Math. Soc. (2) 52(1950) 107–131 (MR 12, 238).
GREEN, J. A.
1951 *On the structure of semigroups.* Ann. of Math. (2) 54(1951) 163–172 (MR 13, 100).
HALL, M.
1959 *The theory of groups.* The Macmillan Co., New York, 1959. xiii + 434 pp. (MR 22, 1996).
HOEHNKE, H.-J.
1962 *Zur Theorie der Gruppoide,* I. Math. Nachr. 24 (1962) 137–168 (MR 28, 4092).
1963 *Zur Strukturtheorie der Halbgruppen.* Math. Nachr. 26(1963) 1–13 (MR 28, 4051).
1966 *Structure of semigroups.* Canad. J. Math. 18(1966) 449–491 (MR 33, 5762).
HOWIE, J. M.
1962 *Embedding theorems with amalgamation for semigroups.* Proc. London Math. Soc. (3) 12(1962) 511–534 (MR 25, 2139).
1963a *An embedding theorem with amalgamation for cancellative semigroups.* Proc. Glasgow Math. Assoc. 6(1963) 19–26 (MR 27, 221).
1963b *Subsemigroups of amalgamated free products of semigroups.* Proc. London Math. Soc. (3) 13(1963) 672–686 (MR 27, 3724).
1963c *Embedding theorems for semigroups.* Quart. J. Math. Oxford Ser. (2) 14(1963) 254–258 (MR 27, 3725).
1964a *Subsemigroups of amalgamated free products of semigroups,* II. Proc. London Math. Soc. (3) 14(1964) 537–544 (MR 28, 5128).
1964b *The maximum idempotent-separating congruence on an inverse semigroup.* Proc. Edinburgh Math. Soc. (2) 14(1964/65) 71–79 (MR 29, 1275).
1964c *The embedding of semigroup amalgams.* Quart. J. Math. Oxford Ser. (2) 15(1964) 55–68 (MR 28, 3107).
KIMURA, N.
1957 *On semigroups.* Doctoral Thesis, The Tulane University of Louisiana, 1957.
KOCH, R. J. AND WALLACE, A. D.
1957 *Stability in semigroups.* Duke Math. J. 24(1957) 193–195 (MR 18, 907).
LAJOS, S.
1961 *Generalized ideals in semigroups.* Acta. Sci. Math. Szeged 22(1961) 217–222 (MR 25, 136).
LAMBEK, J.
1951 *The immersibility of a semigroup into a group.* Canad. J. Math. 3(1951) 34–43 (MR 12, 481).
LANDAU, E.
1927 *Vorlesungen über Zahlentheorie,* Vol. 3. Hirzel, Leipzig (1927).
LEFEBVRE, P.
1962 *Sur certaines conditions minimales en théorie des demi-groupes.* Ann. Mat. Pura Appl. 59(1962) 77–163 (MR 27, 222).
LEVI, F. W.
1944 *On semigroups.* Bull. Calcutta Math. Soc. 36(1944) 141–146 (MR 6, 202).
1946 *On semigroups,* II. Bull. Calcutta Math. Soc. 38(1946). 123–124 (MR 8, 368).
LIBER, A. E. (ЛИБЕР, A. E.)
1954 *К теории обобщенных групп.* (*On the theory of generalized groups.*) Doklady Akad. Nauk SSSR (N.S.) 97(1954) 25–28 (MR 16, 9).
LUH, J.
1960 *On the concepts of radical of semigroup having kernel.* Portugal. Math. 19(1960) 189–198 (MR 24, A1332).

Lyapin, E. S. (ЛЯПИН, Е. С.)

1950 *Нормальные комплексы ассоциативных систем.* (*Normal complexes of associative systems.*) Izv. Akad. Nauk SSSR Ser. Mat. 14(1950) 179–192 (MR 11, 575).

1960a *Полугруппы.* (*Semigroups.*) Gosudarstv. Izdat. Fiz.-Mat. Lit., Moscow, 1960, 592 pp. (MR 22, 11054). English translation, American Mathematical Society, Providence, R.I., 1963.

1960b *О представлениях полугрупп частичными преобразованиями.* (*Representations of semigroups by partial mappings.*) Mat. Sbornik (N.S.) 52(1960) 589–596 (MR 22, 12155).

MacLane, S.

1965 *Categorical algebra.* Bull. Amer. Math. Soc. 71 (1965) 40–106 (MR 30, 2053).

Malcev, A. I. (МАЛЬЦЕВ, А. И.)

1937 *On the immersion of an algebraic ring into a field.* Math. Ann. 113(1937) 686–691.

1939 *О включении ассоциативных систем в группы.* (*On the immersion of associative systems in groups.*) Mat. Sbornik (N.S.) 6(1939) 331–336 (MR 2, 7).

1940 *О включении ассоциативных систем в группы.* (*On the immersion of associative systems in groups.*) Mat Sbornik (N.S.) 8(1940) 251–264 (MR, 2, 128).

1952 *Симметрические группоиды.* (*Symmetric groupoids.*) Mat. Sbornik (N.S.) 31(1952) 136–151 (MR 14, 349).

1953 *Нилпотентные полугруппы.* (*Nilpotent semigroups.*) Ivanov. Gos. Ped. Inst. Uč. Zap. Fiz.-Mat. Nauki 4(1953) 107–111 (MR 17, 825).

McAlister, D. B.

1967 *Characters on commutative semigroups.* Quart. J. Math. Oxford 19(1968) 141–157 (MR 37, 4183).

1955 *On semigroup algebras.* Proc. Cambridge Philos. Soc. 51(1955) 1–15 (MR 16, 561).

1957 *Semigroups satisfying minimal conditions.* Proc. Glasgow Math. Assoc. 3(1957) 145–152 (MR 22, 9543).

1961 *A class of irreducible matrix representations of an arbitrary inverse semigroup.* Proc. Glasgow Math. Assoc. 5(1961) 41–48 (MR 27, 3723).

1964a *Brandt congruences on inverse semigroups.* Proc. London Math. Soc. (3) 14(1964) 154–164 (MR 30, 3167).

1964b *Matrix representations of inverse semigroups.* Proc. London Math. Soc. (3) 14(1964) 165–181 (MR 30, 3168).

Murata, K.

1950 *On the quotient semi-group of a noncommutative semi-group.* Osaka Math. J. 2(1950) 1–5 (MR 12, 155).

Neumann, B. H.

1960 *Embedding theorems for semigroups.* J. London Math. Soc. 35(1960) 184–192 (MR 29, 1268).

Pierce, R. S.

1954 *Homomorphisms of semigroups.* Ann. of Math. (2) 59(1954) 287–291 (MR 15, 930).

Preston, G. B.

1954a *Inverse semigroups.* J. London Math. Soc. 29(1954) 396–403 (MR 16, 215).

1954b *Inverse semigroups with minimal right ideals.* J. London Math. Soc. 29(1954) 404–411 (MR 16, 215).

1954c *Representations of inverse semigroups.* J. London Math. Soc. 29(1954) 411–419 (MR 16, 216).

1956 *The structure of normal inverse semigroups.* Proc. Glasgow Math. Assoc. 3(1956) 1–9 (MR 18, 717).

1958 *Matrix representations of semigroups.* Quart. J. Math. Oxford Ser. (2) 9(1958) 169–176 (MR 20, 5814).

1959 *Embedding any semigroup in a $\mathscr{D}$-simple semigroup.* Trans. Amer. Math. Soc. 93(1959) 351–355 (MR 22, 74).

1961 *Congruences on completely* 0*-simple semigroups.* Proc. London Math. Soc. (3) 11(1961) 557–576 (MR 24, A2628).

1962 *A characterization of inaccessible cardinals.* Proc. Glasgow Math. Assoc. 5(1962) 153–157 (MR 25, 2965).

1965 *Chains of congruences on a completely* 0*-simple semigroup.* J. Austral. Math. Soc. 5(1965) (MR 32, 5759).

1969 *Matrix representations of inverse semigroups.* J. Australian Math. Soc. 9(1969) 29–61.

PTÁK, V.

1949 *Immersibility of semigroups.* Acta Fac. Nat. Univ. Carol. Prague no. 192(1949) 16 pp. (MR 12, 155).

RÉDEI, L.

1952 *Die Verallgemeinerung der Schreierschen Erweiterungstheorie.* Acta Sci. Math. Szeged 14(1952) 252–273 (MR 14, 614).

1963 *Theorie der endlich erzeugbaren kommutativen Halbgruppen.* Hamburger mathematische Einzelschriften, Heft 41, Physica-Verlag, Würzburg, 1963, 228 pp. (MR 28, 5130).

REES, D.

1940 *On semi-groups.* Proc. Cambridge Philos. Soc. 36(1940) 387–400 (MR 2, 127).

REILLY, N. R.

1965 *Contributions to the theory of inverse semigroups.* Doctoral Thesis, University of Glasgow, 1965.

1966 *Bisimple ω-semigroups.* Proc. Glasgow Math. Assoc. 7(1966) 160–167 (MR 32, 7665).

RICH, R. P.

1949 *Completely simple ideals of a semigroup.* Amer. J. Math. 71(1949) 883–885 (MR 11, 327).

ŠAĬN, B. M. (ШАЙН, Б. М. = SCHEIN, B. M.)

1961 *Вмещение полугрупп в обобщенные группы.* (*Embedding semigroups in generalized groups*). Mat. Sbornik (N.S.) 55(97)(1961) 379–400 (MR 25, 3104).

1962 *Представления обобщенных групп.* (*Representations of generalized groups.*) Izv. Vysš. Učebn. Zaved. Matematika no. 3 (28)(1962) 164–176 (MR 25, 3105).

1963 *О транзитивных представлениях полугрупп.* (*On transitive representations of semigroups.*) Uspehi Mat. Nauk 18(1963) no. 3 (111) 215–222 (MR 28, 150).

SAITÔ, T. AND HORI, S.

1958 *On semigroups with minimal left ideals and without minimal right ideals.* J. Math. Soc. Japan 10(1958) 64–70 (MR 20, 1717).

SCHIEFERDECKER, E.

1955 *Zur Einbettung metrischer Halbgruppen in ihre Quotientenhalbgruppen.* Math. Z. 62(1955) 443–468 (MR 17, 384).

SCHREIER, J.

1937 *Über Abbildungen einer abstrakten Menge auf ihre Teilmengen.* Fund. Math. 28(1937) 261–264.

SCHREIER, J. AND ULAM, S.

1933 *Über die Permutationsgruppe der natürlichen Zahlenfolge.* Studia Math. 4(1933) 134–141.

SCHÜTZENBERGER, M. P.

1955/6 *Une théorie algébrique du codage.* (1) Sém. Dubreil-Pisot 1955/56, exposé no. 15, Fac. Sci. Paris. (2) C. R. Acad. Sci. Paris 242(1956) 862–864 (MR 17, 702).

1957 *$\mathscr{D}$ représentation des demi-groupes.* C. R. Acad. Sci. Paris 244(1957) 1994–1996 (MR 19, 249).

SCHWARZ, Š.

1943 *Zur Theorie der Halbgruppen.* (Slovakian, German summary.) Sborník prác Prírodovedekej Fakulty Slovenskej Univerzity v Bratislave No. 6(1953) 64 pp. (MR 10, 12).

1951 *On semigroups having a kernel.* Czechoslovak Math. J. 1(76)(1951) 229–264 (MR 14, 444).

1960 *Dual semigroups.* Czechoslovak Math. J. 10(1960) 201–230 (MR 22, 8075).

1962 *Homomorphisms of a completely simple semigroup onto a group.* Mat.-Fyz. Časopis Sloven. Akad. Vied. 12(1962) 293–300 (MR 28, 2164).

SEIDEL, H.

1965 *Über das Radikal einer Halbgruppe.* Math. Nachr. 29(1965) 255–263 (MR 32, 1276).

1965 *Über das Radikal einer Halbgruppe.* Math. Nachr. 29(1965) 255–263.

STEINFELD, O.

1966 *On semigroups which are unions of completely 0-simple semigroups.* Czechoslovak Math. J. 16(91)(1966) 63–69 (MR 32, 5767).

STOLL, R. R.

1944 *Representations of finite simple semigroups.* Duke Math. J. 11(1944) 251–265 (MR 5, 229).

1951 *Homomorphisms of a semigroup onto a group.* Amer. J. Math. 73(1951) 475–481 (MR 12, 799).

TAMURA, T.

1960 *Decompositions of a completely simple semigroup.* Osaka Math. J. 12(1960) 269–275 (MR 23, A3190).

TAMURA, T. AND GRAHAM, N.

1964 *Certain embedding problems of semigroups.* Proc. Japan Acad. 40(1964) 8–13 (MR 29, 1280).

TEISSIER, MARIANNE

1951 *Sur les équivalences régulières dans les demi-groupes.* C. R. Acad. Sci. Paris 232(1951) 1987–1989 (MR 12, 799).

1953a *Sur les demi-groupes admettant l'existence du quotient d'un côté.* C. R. Acad. Sci Paris 236(1953) 1120–1122 (MR 14, 721).

1953b *Sur les demi-groupes ne contenant pas d'élément idempotent.* C. R. Acad. Sci. Paris 237(1953) 1375–1377 (MR 15, 598).

THIERRIN, G.

1953 *Sur la caractérisation des équivalences régulières dans les demi-groupes.* Acad. Roy. Belg. Bull. Cl. Sci. (5) 39(1953) 942–947 (MR 15, 680).

THURSTON, H. A.

1952 *Equivalences and mappings.* Proc. London Math. Soc. (3) 2(1952) 175–182 (MR 15, 241).

TULLY, E. J., JR.

1960 *Representation of a semigroup by transformations of a set.* Doctoral Dissertation, The Tulane University of Louisiana, 123 pp. (1960).

1961 *Representation of a semigroup by transformations acting transitively on a set.* Amer. J. Math. 83(1961) 533–541 (MR 25, 135).

VAGNER, V. V. (ВАГНЕР, В. В.)

1952 *Обобщенные группы.* (*Generalized groups.*) Doklady Akad. Nauk SSSR (N.S.) 84(1952) 1119–1122 (MR 14, 12).

1953 *Теория обобщенных груд и обобщенных групп.* (*Theory of generalized heaps and generalized groups.*) Mat. Sbornik (N.S.) 32(1953) 545–632 (MR 15, 501).

1956 *Представление упорядоченных полугрупп.* (*Representations of ordered semigroups.*) Mat. Sbornik (N.S.) 38(1956) 203–240 (MR 17, 942). Amer. Math. Soc. Translations (2)36(1964) 295–336.

1957 *Полугруппы частичных преобразований с симметричным отношением транзитивности.* (*Semigroups of partial mappings with a symmetric transitivity relation.*) Izv. Vysš. Učebn. Zaved. Matematika (1957) no. 1, 81–88 (MR 28, 4049).

Venkatesan, P. S.

1962 *On a class of inverse semigroups.* Amer. J. Math. 84(1962) 578–582 (MR 26, 5082).

1963 *The algebraic theory of semigroups.* Doctoral Thesis, University of Madras, 1963.

1966 *On decomposition of semigroups with zero.* Math. Zeitschr. 92(1966) 164–174 (MR 33, 216).

Warne, R. J.

1964 *Homomorphisms of d-simple inverse semigroups with identity.* Pacific J. Math. 14(1964) 1111–1122 (MR 29, 4826).

1965 *A characterization of certain regular d-classes in semigroups.* Illinois J. Math. 9(1956) 304–306 (MR 30, 4848).

1966a *Regular D-classes whose idempotents obey certain conditions.* Duke Math. J. 33(1966) 187–195 (MR 32, 5770).

1966b *A class of bisimple inverse semigroups.* Pacific J. Math. 18(1966) 563–577 (MR 36, 1562).

1966 *Regular D-classes whose idempotents obey certain conditions.* Duke Math. J. 33(1966) 187–195.

1967 *A class of bisimple inverse semigroups.* Pacific J. Math. (To appear.)

Zassenhaus, H.

1949 *The theory of groups.* Translation by S. Kravetz from the German, Chelsea, New York (1949), vii + 159 pp. (MR 11, 77).

ERRATA TO VOLUME I

The authors are indebted for most of the following corrections to R. Croisot, J. M. Howie, S. Lajos, D. D. Miller, W. D. Munn, R. J. Warne, and, above all, to Tôru Saitô. These corrections have been made in the text of the Second Edition of Vol. I (1964).

A negative line number means from the bottom of the page.

p. xv, *l.* 8. For "$y \in A$" read "$y \notin A$".

p. 12, *l.* –1. Insert "if not empty" after "$J\phi^{-1}$".

p. 26, *l.* –7. For "[1941]" read "[1949]".

p. 32, *l.* 14. For "$\mathscr{I}_S$" read "$\mathscr{I}_{S^1}$".

p. 33, *l.* 16. For "S" read "S^1".

p. 33, *l.* 26. Insert "such that $|S| > 1$" at end of first sentence.

p. 40, *l.* –13. Insert "universal" before "interior".

p. 41, *l.* –9. For "by ρ" read "by ρ_0".

p. 43, *l.* 14. Same as above.

p. 51, *l.* 1. For "integers" read "rationals".

p. 51, *ll.* 13-14. For "if and only if" read "implies" in both places.

p. 51, *l.* 14. Insert: "The converse of the first [second] implication holds if $b \neq d[ba^{-1} \neq dc^{-1}]$."

p. 55, *l.* –10. For "(4331)" read "(4113)".

p. 61, *l.* 23. For "$y = H_f$" read "$y \in H_f$".

p. 66, *l.* 18. For "S" read "S^1".

p. 72, *l.* 6. Insert "onto" after "homomorphism".

p. 72, *ll.* 9-11. Replace the last sentence of this paragraph by: "Finally, if A and B are ideals of S containing J, and $A\theta = B\theta$, then $A \backslash J = B \backslash J$, and hence $A = B$."

p. 81, *l.* 19. Move the words "of the" from the end of the line to the beginning.

p. 84, *l.* 9. Insert at end of line: "(Schwarz [1951].)"

p. 85, *ll.* 7-8. Replace these two lines by:

"(d) A semigroup S contains a minimal quasi-ideal if and only if it contains a completely simple kernel K. If this is so, then the minimal quasi-ideals of S are just the maximal subgroups of K."

p. 85, *l.* 17. Delete the words "and only if".

p. 85, *l.* 18. Insert at end: "(Lajos [1961].)"

p. 85. Insert footnote: S. Lajos, *Generalized ideals in semigroups*, Acta Sci. Math. Szeged 22 (1961), 217–222.

p. 86, *l.* 7. For "row-by-product" read "row-by-column".

p. 102, *l.* 23. For "$\mathscr{R}$" read "$\mathscr{L}$".

p. 103, *l.* 11. After "semigroup" insert "or a group".

p. 104, *ll.* –9 and –5. For "$p_{\lambda^* i^*}^{-1}$" read "$p_{\lambda^* i^*}^{*-1}$".

p. 108, *l.* –3. For "$v_{\lambda\phi}$" read "$v_{\lambda\psi}$".

p. 111, *ll.* 10, 13, 14, 19, and 22. For "S" read "S^1".

p. 112, *ll.* 9 and 12. For "$M(s)$" read "$M_D(s)$".

p. 113, *l.* 3. For "$H_{1\lambda}st \cap R_1 = \square$" read "$H_{1\lambda}st \neq H_{1\nu}$".

p. 113, *ll.* 6-8. Replace the last two sentences of this paragraph by: "Since $m_{\lambda\mu}(s) = 0$ for every $\mu \neq \kappa$, it suffices to show that $m_{\kappa\nu}(t) = 0$. But this is so because $H_{1\kappa}t = H_{1\lambda}st \neq H_{1\nu}$."

p. 115, *l.* 16. For "q'_λ of S" read "q'_λ of S^1".

p. 132, *ll.* 23-24. For "$\leqq$" read "$\geqq$" (twice).

p. 146, *l.* 20. For "$v\gamma = (y_{i\lambda}\omega')v'_{\lambda\tau}b^{-1}$"
read "$v_\lambda\gamma = (y_\lambda\omega')v'_{\lambda\tau}b^{-1}$".

p. 149, *l.* 18. For "1957b" read "1957a".

p. 150, *l.* 2. For "F. Blum" read "T. J. Benac".
For "1949" read "1950".

p. 153, *l.* -9. At end of line, for "W_n" read "W_m".

p. 168, *l.* -12. For italic A read German $\mathfrak{A}$. For "$\bar{\Phi}_0$" read "Φ_0".

p. 169, *l.* -14. For "1955b" read "1957a".

p. 173, *l.* -1. For "$\Phi[S]$" read "$\Phi[J]$".

p. 186, *l.* 11. For "Γ" read "Γ^*".

p. 192, *l.* 6. Before "Then" insert: "Let Φ be a field of characteristic $\neq 3$."

p. 193, *ll.* 16-17. Replace these two lines by:
"Then, for all i, j in $N\backslash 1$,

$$\Omega_{ii} = \begin{pmatrix} 0 & 0 \\ 1 & 0 \end{pmatrix}, \qquad \Omega_{ij} = \begin{pmatrix} 0 & 0 \\ 0 & 0 \end{pmatrix} \quad \text{if} \quad i \neq j."$$

p. 195, *l.* 20. For "$H_{P \cup P'}$" read "$H^*_{P \cup P'}$".

p. 195, *l.* 21. For "H_P" read "H^*_P".
For "$H_{P'}$" read "$H^*_{P'}$".

p. 205, *l.* -13. After "for Y" insert: "and that S has an identity element."

p. 205, *l.* -11. Insert: "For present purposes, include $\square$ but exclude S as semiprime ideals of S."

p. 206, *l.* 3. Insert "proper" after "Every".

p. 206, *l.* 11. Replace "$S^*/A\pi$" by "$S^*\backslash A\pi$ if A has an identity element, and with $S^*/A\pi$ otherwise."

p. 208, *l.* 1. For "S" read "K".

p. 219, *l.* 6. For "xi" read "xiii".

In addition to the foregoing, it has been pointed out by R. O. Fulp [1967] that Theorem 5.65 is false. One must assume that, for each archimedean component T_α of T, $S \cap T_\alpha$ is either empty or an archimedean semigroup. See also D. B. McAlister [1967].

It has also been pointed out, by Mr. J. W. Nienhuys, that the "third law of exponents" (p. 3, *l.* 20) does not necessarily hold in a commutative groupoid; the authors were assuming associativity.

FURTHER ERRATA TO VOLUME 1 (MAY, 1970).

(Thanks to R. W. Ball, S. Lajos, S. Tóyama, and others.)

p. 27, *l.* 6. For "296–300" read "707–713".

p. 71, *l.* 24. At end of line insert "such that $J \cap T \neq \square$".

p. 103, *l.* -1. For "$R\theta \neq 0$" read $R\theta \neq 0'$".

p. 117, *l.* 5. For "$S \to N(s)$" read "$s \to N(s)$".

p. 131, *l.* -5 and p. 132, *l.* 11. For "S" read "S^1".

AUTHOR INDEX

Page numbers which include a reference in the exercises are printed in italics.

INDEX

Terms are listed primarily under the broad concept involved, such as *congruence, group, ideal,* and *semigroup*. One-sided concepts are listed under the stem word.

Page numbers which include a reference in the exercises are printed in italics. The dots and dashes stand for previous italicised terms (possibly of several words), the dashes being used for the earlier, and the dots for the later, terms.

For symbols, see the list of notation on page xi.

APPENDIX FOR 1971 REPRINT

Only small alterations were possible for this reprint. We include here some comments and changes too long to incorporate in the text. Included are several references, kindly supplied by B. M. Šaĭn, to Russian literature, of which we were unaware at the time of writing the original text. We also thank J. C. Meakin, D. D. Miller, D. G. FitzGerald, T. E. Hall, R. Šulka, and especially T. Saitô for their comments.

p. 37. Exercise 3 for §6.6. The version of this exercise that appeared in the first printing of this volume also asserted (incorrectly) that S satisfied M_J $[M_L^*, M_R^*]$ if and only if S/A and A both do. That the statement for M_J is false, both ways, is shown in the paper of Saitô cited. For M_L^* $[M_R^*]$ the "only if" statement is true but, as pointed out by T. E. Hall, the "if" statement is false. Hall gives as a counterexample the semigroup $S = A \cup B \cup \{0\}$, where 0 is a zero for S, $B = \{b^j: j = 0, 1, 2, \dots\}$, $A = \{a_i: i = 0, \pm 1, \pm 2, \dots\}$, and where $A^2 = 0$, $b^j b^k = b^{j+k}$, $b^j a_i = a_{i-j}$, and $a_i b^j = a_{i+j}$. Then S/A and A both satisfy M_L^* but A is a $\mathscr{J}$-class of S for which the hypothesis M_L^* (on S) fails.

p. 41. Theorem 7.5 follows from the result, due to B. M. Šaĭn, that an inverse semigroup is a union of groups if its set of idempotents is (naturally) well-ordered. (A minor modification of the proof in the text gives this latter result.) See B. M. Šaĭn (Boris M. Schein), *Generalized groups with the well-ordered set of idempotents*, Mat.-Fyz. Časopis Sloven. Akad. Vied 14(1964) 259–262 (MR 31, 5914). The problem of characterizing those semilattices E for which every inverse semigroup with semilattice of idempotents isomorphic to E must be a union of groups, has been solved by J. M. Howie and B. M. Schein (Šaĭn), *Anti-uniform semilattices*, Bull. Austral. Math. Soc. 1(1969) 263–268.

pp. 47 and 56. See remarks about p. 70 below.

p. 58. Corollary 7.37 coincides with Theorem 4.18 of Chapter 7 of Lyapin [1960a].

p. 69. Theorem 7.58 is a special case of the result of W. D. Munn (*A certain sublattice of the lattice of congruences on a semigroup*, Proc. Cambridge Philos. Soc. 60(1964) 385–391 (MR 29, 176): for regular semigroups the lattice of congruences contained in $\mathscr{H}$ is modular. B. M. Šaĭn writes that this result was independently proved by G. I. Žitomirskiĭ (*On the lattice of congruence relations in a generalized heap*, Izv. Vysš. Učebn. Zaved. Matematika 1965, no. 1(44), 55–61 (MR 31, 2190)). G. Lallement (*Congruences et équivalences de Green sur un demi-groupe régulier*, C.R. Acad. Sci. Paris Sér. A-B 262(1966) A613–A616 (MR 34, 7686)) proved that, for regular semigroups, a congruence is contained in $\mathscr{H}$ if and only if it is idempotent-separating. Thus Theorem 7.58 holds if we replace "inverse" by "regular".

p. 70. For another treatment of representations of inverse semigroups see I. S. Ponizovskiĭ, *Representations of inverse semigroups by partial one-to-one transformations*, Izv. Akad. Nauk SSSR Ser. Mat. 28(1964) 989–1002 (MR 30, 179).

p. 86. It has been pointed out by T. E. Hall that the example in Exercise 10 on this page answers in the negative (see also Clark [1965]) the question raised in the remark on p. 70 of Volume 1.

p. 111. È. G. Šutov (*Embeddings of semigroups into simple and complete semigroups*, Mat. Sb. 62(104)(1963) 496–511 (MR 30, 2100)) has shown that any inverse semigroup can be embedded in a congruence simple (i.e. having no congruences other than the identity and universal congruences) inverse semigroup.

p. 129. The paper Dickson [1913] referred to in this paragraph is:
L. E. Dickson, *Finiteness of the odd perfect and primitive abundant numbers with n distinct prime factors,* Amer. J. Math. 35(1913) 413–422.

p. 136. For a brief and elegant deduction of Theorem 9.28 from the Hilbert Basis Theorem applied to integral semigrouprings, see Peter Freyd, *Rédei's finiteness theorem for commutative semigroups,* Proc. Amer. Math. Soc. 19(1968) 1003.

p. 201. Theorem 10.36 is a special case of a result in B. M. Šaĭn (B. M. Schein), *Homomorphisms and subdirect decompositions of semigroups,* Pacific J. Math. 17(1966) 529–547 (MR 33, 5768).

p. 249. The authors are indebted to B. M. Šaĭn for the following references to the pioneering work of A. K. Suschkewitsch:
Theory of action as the general theory of groups (Russian), Dissertation, Voronez, 1922. (Here "groups" means "semigroups.")
Über die Darstellung der eindeutig nicht umkehrbaren Gruppen mittels der verallgemeinerten Substitutionen, Rec. Math. Moscow 33(1926) 371–374.

p. 275. The question here—whether a maximal subgroup H_e of the completely simple kernel K of a semigroup S is a maximal group homomorphic image of S (in the new sense) if it is a homomorphic image of S—has been answered in the negative by R. J. Plemmons, *On a conjecture concerning semigroup homomorphisms,* Canad. J. Math. (to appear).

p. 278. The result of Exercise 1 for §11.6 also follows from the results of B. M. Šaĭn, *A class of commutative semigroups,* Publ. Math. Debrecen 12(1965) 87–88 (MR 32, 5764). Šaĭn showed that every commutative, cancellative semigroup belongs to $\mathscr{G}'$.